LIPIDOMICS

WILEY SERIES ON MASS SPECTROMETRY

A complete list of the titles in this series appears at the end of this volume.

LIPIDOMICS

Comprehensive Mass Spectrometry of Lipids

XIANLIN HAN

Sanford Burnham Prebys Medical Discovery Institute
Orlando, Florida

Published by John Wiley & Sons, Inc., Hoboken, New Jersey
Published simultaneously in Canada

For general information on our other products and services or for technical support, please contact our Customer Care Department within the United States at (800) 762-2974, outside the United States at (317) 572-3993 or fax (317) 572-4002.

Wiley also publishes its books in a variety of electronic formats. Some content that appears in print may not be available in electronic formats. For more information about Wiley products, visit our web site at www.wiley.com.

Library of Congress Cataloging-in-Publication Data:

Names: Han, Xianlin.
Title: Lipidomics : comprehensive mass spectrometry of lipids / Xianlin Han.
Other titles: Mass spectrometry of lipids
Description: Hoboken, New Jersey : John Wiley & Sons, Inc., [2016] | Includes index.
Identifiers: LCCN 2015040958 | ISBN 9781118893128 (cloth)
Subjects: LCSH: Lipids–Analysis. | Lipids–Spectra. | Mass spectrometry.
Classification: LCC QP751 .H295 2016 | DDC 572/.57–dc23 LC record available at http://lccn.loc.gov/2015040958

Typeset in 10/12pt TimesLTStd by SPi Global, Chennai, India

Printed in the United States of America

10 9 8 7 6 5 4 3 2 1

To my family

Hongping Ye
Rowland Hua
Rachel Jing

CONTENTS

FOREWORD

We live in exciting times of biomedical science. In the past 100 years, the basic complexity in the chemistry of the living organism has now been elucidated and the interaction of diverse biomolecules such as DNA, RNA, proteins, carbohydrates, and lipids are beginning to be understood. Each of these diverse molecular classes has experienced their own biochemical renaissance within the past two decades as the result of the advances in molecular biology and even mass spectrometry. It is now possible to obtain a detailed picture, at the intimate structure level, of both proteins and DNA and their assembly into machines that operate within the living cell. However, the emerging stories of biochemistry now require an understanding of the essential nature of lipids and the role they play in such biochemical complexes.

The paradigm shift in lipid biochemistry that has driven the emergence of the field of lipidomics has been the remarkable ability to ionize nonvolatile molecules and permit their entry into the powerful analyzers that are used in state-of-the-art mass spectrometry. The driving ionizing process was electrospray ionization, yet significant contributions have been obtained using matrix-assisted laser desorption/ionization (MALDI). These two ionization techniques were recognized by the Nobel Prize winner in Chemistry in 2002. All biologically derived lipid molecules, without exception, can now be analyzed by mass spectrometry. This means being able to precisely determine molecular weight, elemental composition, abundance, and even structure of a vast array of lipids that could not previously be investigated.

This monograph by Dr Xianlin Han steps boldly into the area of lipidomics by providing important insights, information, and directions into how one can analyze lipids by mass spectrometry. This is a very complex field that requires understanding of not only basic biochemistry but also physical chemistry and gas phase ion chemistry. Instrument details need to be understood in terms of what type of information can be gleaned and how mass spectrometric experiments can be initiated and controlled to generate the enormous amount of information possible with this physicochemical tool. Lipidomics embraces the important topics of structure elucidation, quantitation, and qualitative analysis of lipids as they present themselves in very complex mixtures, which is the nature of a biological matrix. Upon entering this field, one needs to know how to extract lipids of interest, separate them from other lipids (advanced chromatography), obtain mass spectral data, and interpret mass spectral data. This book very nicely engages these and more topics that are absolutely essential if one has to use this approach to further unravel and marvel at the mysteries of the living system.

ROBERT C. MURPHY, PH.D.

University Distinguished Professor
University of Colorado
August 2015

PREFACE

Lipid analysis has undergone a long history passing from gas chromatography, thin-layer chromatography, and nuclear magnetic resonance spectroscopy to mass spectrometry. Analysis of intact lipid species, particularly in a large scale, has always been fascinating, but challenging. After many years of struggles and efforts from our pioneers in the area of lipid analysis, the term "lipidomics" explosively emerged in the early 2000s, which is defined as the large-scale analysis-based research discipline for studying the underlying mechanism(s) leading to alterations in cellular lipid metabolism, trafficking, and homeostasis in a biological system after a stimulus or growth. Specifically, lipidomics involves identification and quantitation of thousands of cellular lipid molecular species in their intact forms, as well as their interactions with other lipids, proteins, and additional moieties *in vivo*. Investigators in the discipline determine the structures, functions, interactions, dynamics of cellular lipids, and the progressive changes that occur during pathophysiologic perturbations. The emergence of this science was largely due to the recognition of its importance, readiness of its required technologies, and facilitation with other omics.

After more than 10 years of development, the fundamentals and methodologies of lipidomics strategies have greatly advanced. The advancements and discoveries made in the field have been well recognized in a great number of publications, special issues in a variety of prestigious journals, and several books edited by experts in the field. However, systematic and detailed description of these fundamentals, technologies, advancements, and applications is still missing until now. Such materials are highly in demand by novice lipidomics analysts and current investigators to aid in understanding the fundamentals, the principles of rapidly developing methods, and existing tools in order to develop novel approaches in the field. I was fundamentally motivated with these demands and hopes to fill this gap by providing these details in a systematic manner.

Although several technologies have been used in lipidomics to identify, quantify, and understand the structure and function of lipids in biological systems, it is clear that the progress of lipidomics has been accelerated by the development of modern mass spectrometry (e.g., electrospray ionization (ESI) and matrix-assisted laser desorption/ionization). Mass spectrometric analysis of lipids plays a key role in the discipline. Therefore, this book is focused on the mass spectrometry of lipids that has occurred in these years. Other technologies for analysis of lipids, particularly those with chromatography, can be found in the book entitled *Lipid Analysis: Isolation, Separation, Identification and Lipidomic Analysis* written by Drs William W. Christie and Xianlin Han. Readers who are interested in classical techniques and applications of mass spectrometry for analysis of lipids should refer to Dr Robert C. Murphy's book entitled *Mass Spectrometry of Lipids*.

The most prominent analytical strategies in lipidomics science are those developed based on ESI mass spectrometry (MS). These strategies are classified into two major categories: (1) direct infusion-based approaches, which are generally termed "shotgun lipidomics", and (2) HPLC-coupled approaches, which are usually called "LC-MS-based lipidomics." The unique feature of shotgun lipidomics is its constant concentration of lipid solution during prolonged analysis. This feature makes all individual species of a class to experience an essentially identical environment in the ion source, makes the analysis of these species with different MS modes possible, and makes the "ion suppression" effects of abundant coexisting lipid species on low-abundance species resolvable in many cases. On the other hand, LC-MS-based lipidomics greatly exploits the separation science to solve these difficulties. I was glad that I have extensive experiences in both chromatographic separation and mass spectrometry, and I did my best to balance both categories of lipidomics approaches and discuss the principles, advantages, and drawbacks of individual approaches in great detail in this monograph.

The content of this book is classified into four sections: introduction, characterization, quantification, and application. The first part provides the fundamentals of lipids, lipidomics, and mass spectrometry. This section extensively covers the approaches of different lipidomics strategies, the important variables of mass spectrometry to be considered for successful lipid analysis, and the bioinformatic tools for processing lipidomics data. In the second section, I emphasized the concept of "pattern recognition" for characterization of lipids instead of the detailed description of fragmentation mechanisms, which has been the focus of a recently published book entitled *Tandem Mass Spectrometry of Lipids: Molecular Analysis of Complex Lipids* by Dr Robert C. Murphy. I believe that recognizing the fragmentation pattern of a class is easier than understanding the fragmentation mechanisms of these species for the majority of the investigators in lipidomics. Recognizing the fragmentation patterns is helpful and useful to develop strategies and design experiments for identification and quantitative analysis of individual molecular species of a lipid class of interest. Appropriate sampling, good practice of lipid extraction, addition of internal standards, practical methods for accurate quantification, data quality control, and others are the topics of the third section. I comprehensively discussed potential factors that should be recognized and carefully addressed to increase accurate quantification. In the third section,

I also discussed how to interpret the obtained lipidomics data from the angles of metabolic pathways, lipid functions, and other omics supports. The application of lipidomics strategies for biological and biomedical research is the last section of the book. I summarized some examples of a variety of diseases including metabolic syndrome, neurological and neurodegenerative diseases, and cancer. This section also covers lipidomics in plants and yeast strains. Lastly, lipidomics in subcellular organelles and membrane fractions is also discussed to a great degree in this section.

Selection of the topics and references discussed in the book only reflects my interests, experiences, and training, and not necessarily the current status of the lipidomics field. Obviously missing from the book are the topics of steroids, vitamins, and other complex classes of lipids, such as prenols, saccharolipids, polyketides, and lipidomics science in bacteria and algae. To keep the references cited within a reasonable number, I selected the updated publications and review articles on each topic. My apologies if your work was not discussed and/or cited, as we all recognize that we build on the substantial foundations provided by others. It is my sincere hope that the content presented in the monograph will be able to provide useful insights and foundations for investigators to advance the discipline.

I would like to thank all of the very many individuals who have given so freely of their time and expertise during the experiments and made this book possible. Thank you to my former laboratory colleagues in the Division of Bioorganic Chemistry and Molecular Pharmacology in Washington University School of Medicine (WUMS) and my current laboratory colleagues at Sanford Burnham Prebys Medical Discovery Institute (SBP). I sincerely thank my former colleagues, Drs Richard W. Gross and Kui Yang (WUMS), for their support, stimulating discussions, and constructive comments. Thank you to Drs Miao Wang and Chunyan Wang (SBP) for providing many of the mass spectral data support; Drs Jessica Frisch-Daiello and Juan Pablo Palavicini (SBP) for carefully proofreading the book chapters; and Ms Imee Tiu for her administrative assistance during the preparation of the manuscript. I am very grateful to Dr Robert Murphy who provided insightful foreword for the book.

Some figures presented were adapted from the published articles. The permission for reprinting these materials from the authors and publishers is gratefully acknowledged. Financial support from the National Institutes of Health (AG31675 and GM105724) and Sanford Burnham Prebys Medical Discovery Institute has been critical for my continued work in lipidomics. Their interest and support for my research efforts are gratefully acknowledged. Without their assistance, this book could never have been written.

XIANLIN HAN

Orlando, FL

ABBREVIATIONS

AD: Alzheimer's disease
AMPP: *N*-(4-aminomethylphenyl)pyridinium
AP: atmospheric pressure
aPC: alkyl-acyl PC (i.e., plasmanylcholine)
aPE: alkyl-acyl PE (i.e., plasmanylethanolamine)
ASG: acyl steryl glycosides
BMP: bis(monoacylglycero)phosphate
CAS: Chemical Abstract Service
CDP: cytidine diphosphate
Cer: ceramide
CHCA: α-cyano-4-hydroxycinnamic acid
ChEBI: Chemical Entities of Biological Interest
CID: collision-induced dissociation
CL: cardiolipin
DAG: diacylglycerol or diglyceride
DESI: desorption electrospray ionization
DGD1: digalactosyl DAG synthase 1
DGDG: digalactosyldiacylglycerol
DHCer: dihydroceramide
DHB: 2,5-dihydroxybenzoic acid
D*Me*PE: *N*,*N*-dimethylphosphatidylethanolamine

DMG: dimethylglycine
DMS: differential mobility spectrometry
dPC: diacyl PC
dPE: diacyl PE
DRM: detergent-resistant membrane
ER: endoplasmic reticulum
ESI: electrospray ionization
FA: fatty acyl or fatty acid
Fmoc: fluorenylmethoxylcarbonyl
FT ICR: Fourier transform ion cyclone resonance
G-3-P: glycerol-3-phosphate
GalCer: galactosylceramide
GC: gas chromatography
GIPC: glycosylinositol phosphorylceramide
GluCer: glucosylceramide
GPL: glycerophospholipid(s)
HDL: high-density lipoprotein
HETE: hydroxyeicosatetraenoic acid
HexCer: hexosylceramide
HexDAG: hexosyl diacylglycerol (see also MGDG)
HILIC: hydrophilic interaction chromatography
HMDB: Human Metabolome Database
HPLC: high-performance liquid chromatography
HPTLC: high-performance thin layer chromatography
IM-MS: ion-mobility MS
IMS: imaging mass spectrometry
IP_3: inositol triphosphate
IPC: inositol phosphorylceramide
IUPAC: International Union of Pure and Applied Chemistry
KEGG: Kyoto Encyclopedia of Genes and Genomes
LacCer: lactosylceramide
LCB: long chain bases
LDL: low-density lipoproteins
LIT: linear trap
LMSD: Lipid MAPS Structure Database
LOD: limit of detection
lysoGPL: lysoglycerophospholipid(s)
lysoPA: lysophosphatidic acid
lysoPC: choline lysoglycerophospholipid(s)

lysoPE: ethanolamine lysoglycerophospholipid(s)

lysoSM: lysosphingomyelin

MAG: monoacylglcyerol or monoglyceride

MALDI: matrix-assisted laser desorption/ionization

MANOVA: multivariate analysis of variance

m:n: a fatty acyl chain containing *m* carbon atoms and *n* double bonds

MDMS: multidimensional mass spectrometry

MDMS-SL: multidimensional mass spectrometry-based shotgun lipidomics

MGDG: monogalactosyldiacylglycerol

MIPC: mannosyl-inositolphosphoceramide

$M(IP)_2C$: mannosyl-diinositolphosphoceramide

M*Me*PE: *N*-monomethyl phosphatidylethanolamine

MMSE: mini-mental state examination

MS: mass spectrometric or mass spectrometry

MS/MS: tandem mass spectrometry

MTBE: methyl-*tert*-butyl ether

NEFA: nonesterified fatty acid(s)

NLS: neutral loss scan or scanning

NMR: nuclear magnetic resonance

OPDA: oxo-phytodienoic acid

PA: phosphatidic acid

PC: choline glycerophospholipid(s)

PCA: principal component analysis

PE: ethanolamine glycerophospholipid(s)

PG: phosphatidylglycerol

PI: phosphatidylinositol

PIP: phosphatidylinositol phosphate

PIP_2: phosphatidylinositol diphosphate (or bisphosphate)

PIS: precursor-ion scan or scanning

PLA_2: phospholipase A_2

PLC: phospholipase C

PLD: phospholipase D

PLS-DA: partial least square-based discriminant analysis

pPC: alkenyl-acyl PC (i.e., plasmenylcholine)

pPE: alkenyl-acyl PE (i.e., plasmenylethanolamine)

PS: serine glycerophospholipid(s)

ROS: reactive oxygen species

Q: quadrupole

QqQ: triple quadrupoles

SAR: systemic acquired resistance

S1P: sphingoid-1-phosphate

S/N: signal/noise

SG: steryl glycosides

SIM: selected ion monitoring

SIMS: secondary ion mass spectrometry

SM: sphingomyelin

sn: stereospecific numbering

SPE: solid phase extraction

SRM/MRM: selected/multiple reaction monitoring

ST: sulfatide

TAG: triacylglycerol or triglyceride

THAP: 2,4,6-trihydroxyacetophenone

TLC: thin layer chromatography

TOF: time of flight

U(H)PLC: ultra (high)-performance liquid chromatography

UV: ultraviolet

VLDL: very low-density lipoproteins

PART I

INTRODUCTION

1

LIPIDS AND LIPIDOMICS

1.1 LIPIDS

1.1.1 Definition

It is well known that lipids play many essential roles in life [1]. They possess functions to

- Constitute cellular membranes in biological organisms that provide hydrophobic barriers to separate cellular compartments.
- Serve as an optimal matrix to facilitate transmembrane protein function.
- Facilitate as a source of precursors for lipid second messengers during signal transduction.
- Provide the storage and/or supplement of fuel for biological processes.

More and more lines of evidence support a rationale that lipids are associated with many human diseases (e.g., diabetes and obesity, atherosclerosis and stroke, cancer, psychiatric disorders, neurodegenerative diseases and neurological disorders, and infectious diseases) (see Chapter 17). Therefore, the research on lipids has become a unique new discipline called "lipidomics" nowadays.

The majority of lipids are composed of two components. One part is largely hydrophobic ("water-fearing"), meaning that it is not suitably soluble in polar solvents (e.g., water), while the other part is often polar or hydrophilic ("water-loving")

Lipidomics: Comprehensive Mass Spectrometry of Lipids, First Edition. Xianlin Han.
© 2016 John Wiley & Sons, Inc. Published 2016 by John Wiley & Sons, Inc.

and is readily soluble in polar solvents. Therefore, lipids are amphiphilic molecules (having both hydrophobic and hydrophilic portions). However, prominent exceptions are also present, including waxes, triacylglycerol (TAG), cholesterol, cholesteryl esters, all of which are predominantly hydrophobic except for their hydroxyl or carbonyl groups.

In general, lipids are defined as a group of organic compounds in living organisms, most of which are insoluble in water but soluble in nonpolar solvents. Based on this definition, any petroleum products obtained from fossil materials or synthetic organic compounds are excluded in the category of lipids. Indeed, lipids are one of the main constituents of biological cells and the major components of lipoproteins in serum. Lipids are often conjugated with carbohydrates, which are known as lipopolysaccharides.

The historical origins of the term "lipid" and its early definitions can be found elsewhere if the readers are interested [2]. The precise definition of lipids is difficult to give, as no satisfactory or widely accepted definition exists. Thus, many varying definitions about lipids can be found. For example, Merriam-Webster dictionary defines lipids as "any of various substances that are soluble in nonpolar organic solvents (such as hexane, chloroform, and ether), that with proteins and carbohydrates constitute the principal structural components of living cells, and that include fats, waxes, phospholipids, cerebrosides, and related and derived compounds." Wikipedia (http://en.wikipedia.org/wiki/Lipid) describes it as "Lipids may be broadly defined as hydrophobic or amphiphilic small molecules; the amphiphilic nature of some lipids allows them to form structures such as vesicles, liposomes, or membranes in an aqueous environment." General textbooks describe lipids as a group of naturally occurring compounds, which have in common a ready solubility in organic solvents such as chloroform, benzene, ethers, and alcohols. Unfortunately, such a definition is misleading because there are many compounds that are now widely accepted as lipids, which may be more soluble in water than in organic solvents (e.g., lysoglycerophospholipids, acyl CoA, gangliosides).

The most recent definition of lipids was provided by a group of lipid chemists who formed the consortium of lipid metabolites and pathways strategy (Lipid MAPS). They defined lipids based on the origin of the lipid structures as hydrophobic or amphipathic small molecules that may originate entirely or, in part, by carbanion-based condensations of thioesters (fatty acids, polyketides, etc.) and/or by carbocation-based condensations of isoprene units (prenols, sterols, etc.). In this book, this definition, its classification (see the following), and its recommended nomenclature are largely accepted.

1.1.2 Classification

With the different definitions, different kinds of lipid classification are frequently used in the field. For example, many lipid chemists simply classify lipids into polar and nonpolar lipids based on the overall hydrophobicity of the lipids. The nonpolar lipids include fatty acids and their derivatives (e.g., long-chain alcohols and waxes), glycerol-derived lipids (e.g., monoacylglycerols (MAG), diacylglycerols (DAG),

TAG (i.e., fats or oils)), and steroids. These nonpolar lipids are generally soluble in very nonpolar solvents such as hexane, ether, and ester. The polar lipids usually contain a polar head group, such as phosphocholine in choline glycerophospholipids (PC) (see the following), and are usually soluble in relatively polar solvents, such as alcohol, and even water.

Based on the features of chromatographic separation, lipids are classified into simple and complex molecules [2]. "Simple lipids" are those that yield mostly two types of primary products per molecule upon hydrolysis (e.g., fatty acids and their derivatives, MAG); "complex lipids" yield three or more primary hydrolysis products per molecule (e.g., PC, TAG, DAG). These hydrolysis products include fatty acids, phosphoric acid, organic bases, carbohydrates, glycerol, and many more components.

According to the functions of cellular lipids, many biochemists also refer lipids to

- Membrane lipids, which largely constitute the cellular membrane and are usually present in relatively high contents.
- Energy lipids, which are usually involved in energy storage and metabolism.
- Bioactive lipids, which serve as lipid second messengers and are generally present in low or very low abundance.

A more detailed classification is achieved by grouping lipids based on their chemical properties. **Individual lipid molecular species** (each of which has a unique molecular structure) are commonly categorized into small groups, that is, **lipid classes**, based on their chemical structural similarities. For example, individual lipid molecular species that possess an identical polar head group (e.g., phosphocholine, phosphoethanolamine, or phosphoserine) linked to a common glycerol backbone are categorized into a specific lipid class (e.g., PC, ethanolamine glycerophospholipid (PE), serine glycerophospholipid (PS), respectively) (Figure 1.1).

Among each individual lipid class, due to the presence of a unique linkage or another unique feature, these species are further classified into smaller groups, that is, the **subclasses** of the lipid class (Figure 1.2). For example, the oxygen atom of glycerol at *sn*-1 position (here *sn* means stereospecific numbering) is connected to a fatty acyl chain through an ester, ether, or vinyl ether bond in both glycerophospholipids (GPL) and glycerolipids. These different linkages define the subclasses of a GPL class (Figure 1.2a), which are called phosphatidyl-, plasmanyl-, and plasmenyl-according to the recommended nomenclature by International Union of Pure and Applied Chemistry (IUPAC), corresponding to the ester, alkyl ether, and vinyl ether linkage, respectively [3]. These subclasses are abbreviated as prefix "d," "a," and "p," respectively, throughout this book. To date, the plasmanyl and plasmenyl subclasses have only been identified in mammalian lipidomes for the classes of choline, ethanolamine, and serine glycerophospholipids (PC, PE, and PS, respectively) and may be present in the class of phosphatidic acid (PA) and cardiolipin (CL). However, these subclasses have been found in other lipid classes in other species [4]. These different linkages have also been found in DAG and TAG [5, 6]. The presence or absence of a double bond between C4 and C5 of sphingoid base (see the following)

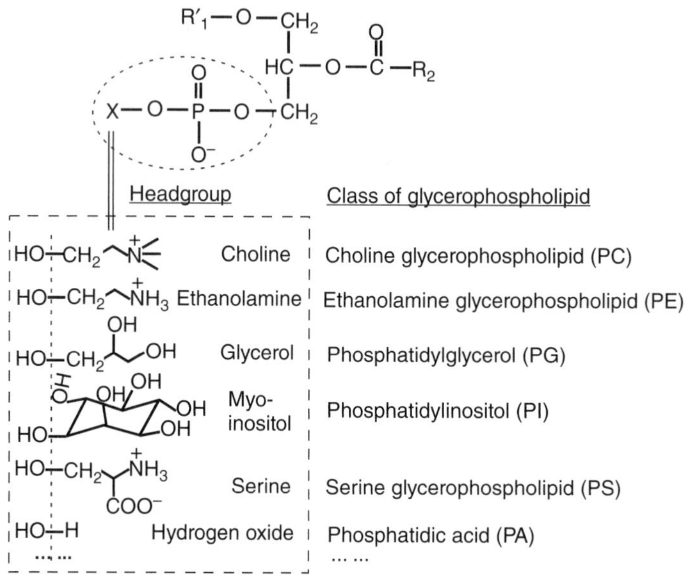

Figure 1.1 Examples of glycerophospholipid classes. Different structures of the moiety X, which are connected to the phosphate and exemplified in the box, determine the individual classes of GPL as indicated with abbreviations that are commonly used in the literature and adapted by the Lipid MAPS consortium.

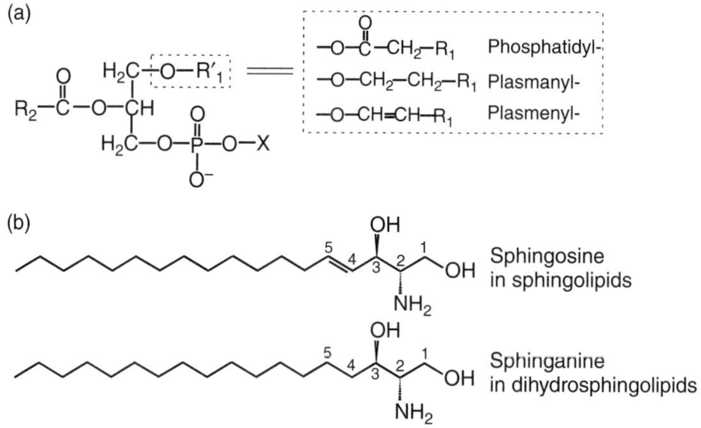

Figure 1.2 Example of lipid subclasses, which are classified based on the different linkages at a certain position or a unique structural feature of a lipid class. (a) The subclasses of phosphatidyl-, plasmanyl-, and plasmenyl- are present in GPL as a result of the different linkages (i.e., ester, ether, and vinyl ether) of a fatty acyl chain to the hydroxyl group at *sn*-1 position of glycerol. (b) The different core structures of sphingoid bases in the presence or absence of a double bond between C4 and C5 carbon atoms lead to the common subclasses of sphingolipids and dihydrosphingolipids. Other less common subclasses of sphingolipids are also present due to other structures of the sphingoid bases (see Figure 1.6).

leads to the classification of the individual sphingolipid class into sphingolipid and dihydrosphingolipid subclasses (Figure 1.2b).

Following are the two major classification systems defined based on chemical properties of lipids that are largely used in the book.

1.1.2.1 Lipid MAPS Approach   Based on their mission, the Lipid MAPS consortium has classified lipids into eight categories, including fatty acyls, glycerolipids, GPL, sphingolipids, sterol lipids, prenol lipids, saccharolipids, and polyketides [7]. Importantly, individual lipid molecular species in this comprehensive classification bears a unique 12-digit identifier, which facilitates the systematization of lipid biology and enables the cataloging of lipids and their properties in a way that is compatible with other macromolecular databases.

The **fatty acyls** are a diverse group of molecules synthesized by chain elongation of an acetyl coenzyme A (acetyl-CoA) primer with malonyl-CoA (or methylmalonyl-CoA) groups that may contain a cyclic functionality and/or are substituted with heteroatoms. Fatty acyls are characterized by a repeating series of methylene groups and are structurally the simplest lipids. This category includes various classes of fatty acids, eicosanoids, docosanoids, fatty alcohols, fatty aldehydes, fatty esters, fatty amides, fatty nitriles, fatty ethers, and hydrocarbons. Fatty acyls, in general, and fatty acids, in particular, are the basic building blocks of more complex lipids such as GPL, (glyco)sphingolipids, glycerolipids, and glycolipids. The presence of modified fatty acyls in complex lipids has been well documented [8–10].

The **glycerolipids** are the lipid species that can only be hydrolyzed into glycerol, a sugar group, fatty acid(s), and/or alkyl variants. Glycerolipids include the species of MAG, DAG, TAG, and glycolipids. The MAG, DAG, and TAG species typically have a glycerol backbone with fatty acid chains linked to the hydroxyl groups of glycerol. However, fatty alcohols linked by an ether bond are also found in these neutral lipids in low abundance [5, 6]. Glycolipid is defined by the IUPAC as a lipid in which the fatty acyl portion of the molecule contains a glycosidic linkage [3].

The **glycerophospholipids** are defined by the presence of at least one phosphate (or phosphonate) group esterified to one of the glycerol hydroxyl groups. GPL species are ubiquitous in nature, are key components of cellular membranes, and are also involved in metabolism and signaling. The complexity of GPL species is illustrated with the presence of different classes, subclasses, and individual molecular species (different fatty acyl chain structures) (Figures 1.1 and 1.2). As illustrated by its name, individual molecular species in this category of lipids contain three components: "glycero-" (i.e., at least one glycerol molecule is centered in each individual species); "phospho-" (i.e., at least one phosphate or phosphodiester is linked to a hydroxyl group of glycerol at the *sn*-3 position); and one or two aliphatic chains that are connected to the *sn*-1, *sn*-2, or both hydroxyl groups of glycerol. There are over 10 varieties of the moieties esterified with the phosphate (i.e., over 10 different classes) (Figure 1.1) and over 30 kinds of possible fatty acyl chains containing different numbers of carbon atoms (i.e., chain length), different degree of unsaturation, and different locations of these double bonds. In addition, there exist three different linkages of the fatty acyl chain with the hydroxyl group of glycerol at the *sn*-1 position (i.e., three different subclasses). Accordingly, we can easily estimate that the possible number of individual molecular species in the category of GPL should be approximately 30,000 ($10 \times 30 \times 30 \times 3$). In practice, mass spectrometric (MS) analysis has detected the presence of a large number of individual lipid species (e.g., plasmalogen, CL, TAG) [11–13].

Sphingolipids are another category of complex cellular lipids. The sphingolipid species contain common long-chain sphingoid bases (Figure 1.2b) as their core structures. These sphingoid bases are first synthesized *de novo* from serine and a long-chain fatty acyl CoA to yield sphinganine and then dihydroceramides, which convert into ceramides, phosphosphingolipids, glycosphingolipids, and other species (Figure 1.3). The polar moieties (which also appear in GPL classes and glycolipids (see above)) that are linked to the hydroxyl group of sphingoid base at position C1 represent the individual sphingolipid classes.

The **sterol lipids** are a group of compounds that carry a core signature of four fused rings (Figure 1.4a) and are subdivided into cholesterol and derivatives, steroids, secosteroids, bile acids and their derivatives, and others (Figure 1.4) [7]. Cholesterol and its derivatives that are the most widely studied sterol lipids in mammalian systems constitute an important component of membrane lipids, along with the GPL and SM [14]. Unique sterols are present in plant, fungal, and marine sources [7]. The steroids, which also contain the same fused four-ring core structure as

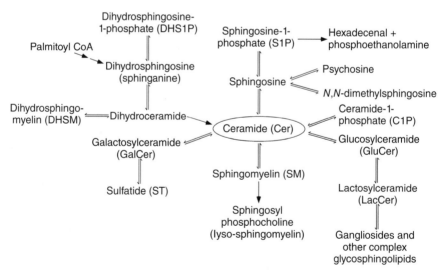

Figure 1.3 Simplified pathways and network of the common sphingolipid classes and other related lipids. The network and the pathways are derived from the Kyoto Encyclopedia of Genes and Genomes (KEGG) pathway databases. The pathways indicate the origins of sphingoid base core structures and their derivatives. Other sphingoid bases can be biosynthesized from other fatty acyl CoA by the replacement of palmitoyl CoA or some other amino acids replacing serine. The structures of individual lipid classes and their abbreviations used in the book are indicated in Figure 1.6.

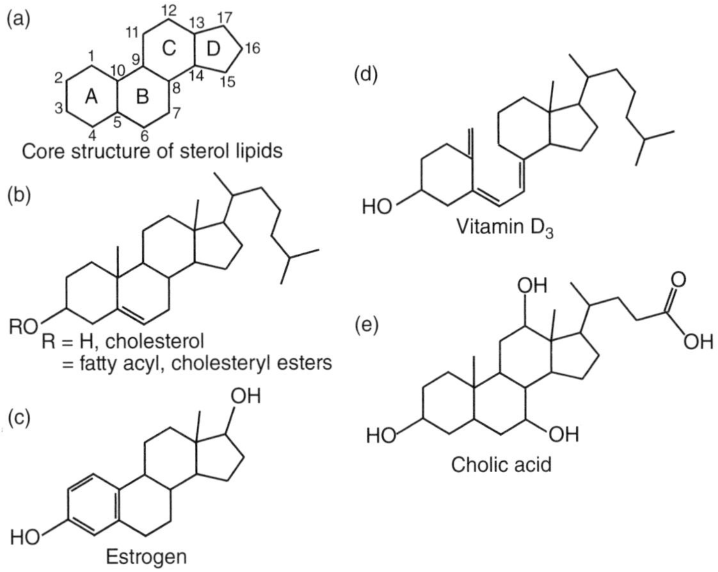

Figure 1.4   The core structure and representatives of sterol lipids. (a) The core structure of the majority of the sterol species or from which the sterol species are resulted. (b) The representative structures of cholesterol (R = H) and cholesteryl esters (R = a fatty acyl). (c) A representative structure of the steroid subgroup species (i.e., estrogen). (d) A representative structure of the secosteroid subgroup (i.e., vitamin D_3). (e) A representative structure of the species among the bile acid subgroup of sterol (i.e., cholic acid).

cholesterol, have different biological roles and function as hormones and signaling molecules [15]. The secosteroids, comprising various forms of vitamin D, are a group of molecules similar to steroids but with a "broken" B ring, hence the "seco" prefix [16]. Bile acids are primarily derivatives of cholan-24-oic acid synthesized from cholesterol in the liver and their conjugates (sulfuric acid, taurine, glycine, glucuronic acid, and others) [17].

In addition to the above-mentioned five categories of lipids, there are three more categories, including prenols, saccharolipids, and polyketides, that are relatively less studied at the current stage of lipidomics. The **prenol lipids** are synthesized from the five carbon precursors isopentenyl diphosphate and dimethylallyl diphosphate that are produced mainly via the mevalonic acid pathway [18]. Prenols are subdivided into isoprenoids, quinones and hydroquinones (e.g., unibiquinones, vitamins E, K), polprenoils, and others [7]. The category of **saccharolipids** accounts for lipids in which fatty acids are linked directly to a sugar backbone. In the saccharolipids, a sugar substitutes for the glycerol backbone that is present in glycerolipids and GPL. Saccharolipids can occur as glycan or as phosphorylated derivatives. The most familiar saccharolipids are the acylated glucosamine precursors of the lipid A component of the lipopolysaccharides in Gram-negative bacteria [19]. Typical lipid A molecules are disaccharides of glucosamine, which are derivatized with as many as seven fatty acyl chains [19]. The **polyketides** are a diverse group of metabolites from plant and microbial sources and contain a much greater diversity of natural product structures, many of which have the character of lipids [7].

Some key features of this lipid classification, which are adapted in this book, include the following:

- The use of stereospecific numbering (*sn*) method for the glycerol-based lipids (e.g., glycerolipids and glycerophospholipids) [3]. Acyl or alkyl chains are typically linked to the *sn*-1 and/or *sn*-2 positions of glycerol, with the exception of some lipids that contain three acyl or alkyl chains or contain more than one glycerol group and archaebacterial lipids in which *sn*-2 and/or *sn*-3 modification occurs.

- The use of sphinganine and sphing-4-enine (i.e., sphingosine) as core structures for the category of sphingolipid species, where the d-*erythro* or 2*S*, 3*R* configuration and 4*E* geometry (in the case of sphing-4-enine) are implied.

- The use of "*d*" and "*t*" designations as the shorthand notation of sphingolipids, which refer to 1,3-dihydroxy and 1,3,4-trihydroxy long-chain bases, respectively.

- The use of *E*/*Z* designations to define double-bond geometry.

- The use of *R*/*S* designations (as opposed to α/β or D/L) to define stereochemistries. The exceptions are those describing substituents on glycerol (*sn*) and sterol core structures and anomeric carbons on sugar residues. In these latter special cases, the α/β format is firmly established.

- The use of the common term "lyso" denoting the position lacking a radyl group in glycerolipids and GPL.

1.1.2.2 *Building Block Approach*

1.1.2.2.1 Building Block Concept and Classification In this classification method, the majority of biologically occurring lipids are the combinations of some building blocks, which represent some kinds of hydrolysis products or their analogs. The commonly recognized building blocks include fatty acyls as categorized by Lipid MAPS classification (see above), a variety of polar head groups (e.g., phosphoesters (including phosphate, phosphocholine, phosphoethanolamine, phosphoglycerol, phosphoserine, and phosphoinositol) and sugar molecules (e.g., glucose, galactose, lactose)), as well as a few backbones (such as glycerol, sphingoid base, and cholesterol) as core structures. With this concept, the molecular species of an entire lipid class or a category of lipid classes could be represented by a common chemical structure.

For example, molecular species of all glycerol-centered lipid classes (e.g., GPL and glycerolipids (see Lipid MAPS classification)) are the combination of three different building blocks connected to three hydroxyl groups of glycerol backbone (Figure 1.5). In this general structure, the building blocks I and II can be a hydrogen or a fatty acyl connected to *sn*-1 and 2 positions of glycerol with an ester, ether, or vinyl ether linkage. Building block III at the *sn*-3 position of glycerol can be a hydrogen atom, a fatty acyl, or one of the various sugar ring(s) and their derivatives

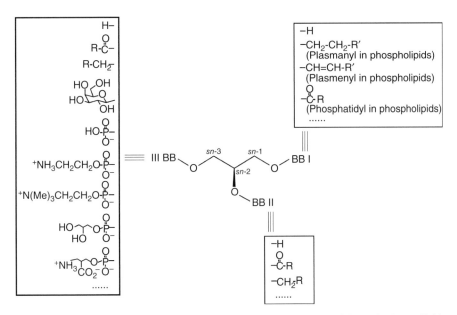

Figure 1.5 General structure of glycerolipids and glycerophospholipids. Both glycerolipids and GPL classes are centered with a glycerol molecule. Three building blocks (BB), which are separately exemplified in the boxes, are connected to the hydroxyl groups of glycerol. Building block (BB) I represents a hydrogen or a fatty acyl moiety connected to *sn*-1 position of glycerol with an ester, ether, or vinyl ether linkage, which defines the subclass as phosphatidyl-, plasmanyl-, or plasmenyl-, respectively, in glycerophospholipids. Building block (BB) II represents a hydrogen or a fatty acyl moiety connected to *sn*-2 positions of glycerol with an ester, ether, or vinyl ether linkage. Building block (BB) III represents a hydrogen atom, a fatty acyl, or one of the various sugar ring(s) and their derivatives in glycerolipids, or phosphoesters in glycerophospholipids and lysoglycerophospholipids. R' and R are usually unbranched saturated or unsaturated aliphatic chain containing 12–20 and 13–21 carbon atoms, respectively.

in glycerolipids, or phosphoesters in GPL and lysoGPL. Here, the fatty acyl chain typically contains 12–24 carbon atoms with variable degrees of unsaturation or modifications.

Similar to glycerol-centered lipids, the majority of the sphingolipid species can be represented with a general structure composed of three building blocks (Figure 1.6). Building block I represents a different polar moiety that links to the oxygen at the C1 position of a sphingoid base. These polar moieties include hydrogen, phosphoethanolamine, phosphocholine, galactose, glucose, lactose, sulfated galactose/lactose, and other complex sugar groups, which correspond to ceramide, ceramide phosphoethanolamine, sphingomyelin (SM), galactosylceramide (GalCer), glucosylceramide (GluCer), lactosylceramide (LacCer), sulfatide (ST), and other glycosphingolipids such as gangliosides, respectively (Figure 1.6). These polar moieties can readily make over 20 sphingolipid classes. Building block II represents a fatty acyl moiety, which is acylated to the primary amine at the C2 position of

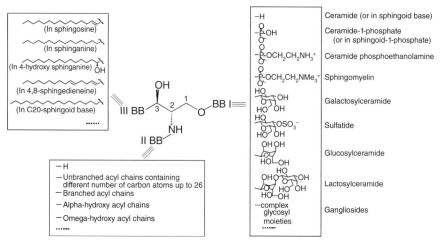

Figure 1.6 General structure of sphingoid-based lipids with three building blocks. Building block (BB) I represents a different polar moiety (linked to the oxygen at the C1 position of sphingoid backbone). These moieties determine the poplar head groups of sphingolipid classes as indicated. Building block II represents fatty acyl chains (acylated to the primary amine at the C2 position of sphingoid backbone) with or without the presence of a hydroxyl group, which is usually located at the alpha or omega position. Building block III represents the fatty acyl chains in all of possible sphingoid backbones, which are carbon–carbon linked to the C3 position of sphingoid backbones and vary with the aliphatic chain length, degree of unsaturation, location of double bonds, presence of branching, and presence of an additional hydroxyl group.

the sphingoid backbone. A variety of fatty acyl chains, including those that contain a hydroxyl group (usually located at the alpha or omega position) (Figure 1.6), are linked to this position. Building block III represents the aliphatic chain present in all sphingoid bases. This building block is connected through a carbon–carbon bond to the C3 position. This aliphatic chain varies in alkyl chain length and branching, the number and positions of double bonds, the presence of additional hydroxyl groups, and other features (Figure 1.6). Over 100 types of this aliphatic chain, after considering the hydroxyl-containing varieties, can also be readily counted. Therefore, a combination of these three factors would yield at least 200,000 ($20 \times 100 \times 100$) sphingolipid molecular species, whereas thousands of possible sphingolipid species can be theoretically constructed from the combination of these three building blocks by using only common aliphatic chains [20]. At the current stage, tens to hundreds of sphingolipid molecular species are readily analyzed by using different lipidomics approaches [21–23].

Sterols are a class of lipids containing a common steroid core of a fused four-ring structure with a hydrocarbon side chain and an alcohol group. Cholesterol is the primary sterol lipid in mammals and is an important constituent of cellular membranes. Oxidization and/or metabolism of cholesterol yield numerous oxysterols, steroids, bile acids, etc., many of which are important signaling molecules in biological systems. Cholesteryl esters esterified with a variety of fatty acyls are enriched in lipoprotein particles, such as low-density lipoproteins (LDL) and very low-density lipoproteins (VLDL).

1.1.2.2.2   The Significance of Building Block Classification   The significance of this classification method is twofold: (1) ready to construct theoretical lipid databases that are expandable; and (2) effectively identify a large number of individual lipid species through identification of the relatively smaller number of building blocks. Luckily, these building blocks can be identified with their corresponding characteristic fragments by using two powerful tandem MS techniques (i.e., neutral loss scan (NLS) and precursor-ion scan (PIS)) [24]. These techniques and their applications for identification of lipid species are described in Chapters 2 and 6 in details.

1.2   LIPIDOMICS

1.2.1   Definition

The entire collection of chemically distinct lipid species in a cell, an organ, or a biological system has been referred to as a lipidome [25]. By analogy to other "omics" disciplines, lipidomics is an analytical chemistry-based research field studying lipidomes in a large scale and at the levels of intact molecular species. The research in lipidomics involves the following:

- Precisely identifying the structures of cellular lipid species including the number of atoms, the number and location of double bonds, the core structures and head groups, individual fatty acyl chains, and the regiospecificity of each isomer, etc.
- Accurately quantifying individual identified lipid species for pathway analysis, comparably profiling the lipid samples for biomarker discovery.
- Determining the interactions of individual lipid species with other lipids, proteins, and metabolites *in vivo*.
- Disclosing the nutritional or therapeutic status for prevention or therapeutic intervention of diseases.

Owing to the different utilities of lipidomics in its research, some subcategories of lipidomics are also frequently named in the literature as molecular/structural lipidomics [26–28], functional lipidomics [29, 30], nutritional lipidomics [31], dynamic lipidomics [32], oxidized lipidomics [33, 34], mediator lipidomics [35], neurolipidomics [36], sphingolipidomics [23, 37, 38], fatty acidomics [39], etc., to reflect their particular focus on the lipidomic studies. The analysis of lipid structures, mass levels, cell functions, and interactions in a spatial and temporal manner provides the dynamic changes of lipids during physiological (e.g., nutritional) or pathological perturbations or cell growth. Accordingly, lipidomics plays an essential role in defining the biochemical mechanisms underlying lipid-related disease processes through identification of alterations in cellular lipid signaling, metabolism, trafficking, and homeostasis.

Overall, lipids are considered as biological metabolites. Hence, lipidomics is covered under the umbrella of the general field of "metabolomics." However, lipidomics is a distinct discipline because of the uniqueness and functional specificity of lipids relative to other metabolites. For example, most components in the cellular lipidome are extractable with organic solvents, so they are readily recovered and separated from other water-soluble metabolites. Lipids form aggregates (i.e., dimers, oligomers, micelles, bilayers, or other aggregated states) in all solvents essentially as their con-centrations increase [1]. This unique property results in substantial difficulties for the quantitative analysis of individual lipid species in their intact forms by mass spec-trometry (MS). This topic is addressed in detail in Chapters 15 and 16.

Cellular lipidomes are variable and highly complex. Tens of thousands of possible lipid molecular species are predictably present in the cellular lipidome at the level of attomole to nanomole of lipids per milligram of protein [20, 38, 40] (see above). These individual molecular species belong to a variety of different lipid classes and subclasses and comprise different lengths, degrees of unsaturation, different locations of double bonds, and potential branching in aliphatic chains. Moreover, additional factors make the study of this already complex and diverse system even more diffi-cult. These include the following facts: (1) cellular lipid molecular species and com-position are quite different among different species, cell types, cellular organelles, membranes, and membrane microdomains (e.g., caveola and/or rafts); and (2) the cel-lular lipidome is dynamic, depending on nutritional status, hormonal concentrations, health conditions, and many others [41].

Recent studies in lipidomics have largely focused on the following areas [42]:

- Identification of novel lipid classes and molecular species.
- Development of quantitative methods for the analysis of attomole to femtomole levels of lipids in cells, tissues, or biological fluids.
- Network analysis that clarifies metabolic adaptation in health and disease and biomarker analysis that facilitates diagnosis of disease states and determination of treatment efficacy.
- Tissue mapping of altered lipid distribution present in complex organs.
- Bioinformatics approaches for the automated high-throughput processing and molecular modeling with lipidomics data.

1.2.2   History of Lipidomics

Although the terms lipidome and lipidomics did not appear in the literature until the early 2000, researchers have initiated the study of cellular lipids on a large scale and at the intact molecular levels at much earlier times [43–52]. These pioneering studies truly demonstrated the possibilities of lipidomic analysis by using a variety of tools. Most importantly, these studies also provided initial insight into the utility of identifying alterations in membrane structure and function that mediate biological responses to cellular adaptation in health and maladaptive alterations during disease, thereby providing the foundation for development of the new discipline, lipidomics. The role of MS in characterization and analysis of lipids can be found in the classical book written by Dr Robert Murphy in 1993 [53].

Most early studies focused on one species, one lipid class, or one enzyme-catalyzed pathway. During these studies, investigators have clearly recognized that the metabolism of individual lipid molecular species or individual lipid classes is interwoven. To conduct research on lipid metabolism only from an isolated system, or only being focused on one molecular species, or one lipid class, has substantial limitations. The metabolism of the entire lipidome of the organelle, the cell type, the organ, the system, or the species should be investigated in a systems biology approach. Therefore, the need for such a comprehensive approach for studies of lipid metabolism greatly catalyzes the emerging of lipidomics and accelerates its development.

Investigators in lipidomics examine the structures, functions, interactions, and dynamics of a vast majority of cellular lipids and identify their cellular organization (i.e., subcellular membrane compartments and domains). The number of lipids in a cellular lipidome is estimated to be in the tens of thousands to millions [20, 38, 40]. Thus, in lipidomic research, a vast amount of information describing the spatial and temporal alterations in the content and composition of different lipid species in a selected system is accrued after perturbation of a cell through changes in its physiological (e.g., nutritional status, hormonal influences, health condition, metabolic levels) or pathological (diabetes, ischemia, neurodegeneration, etc.) state. The information obtained is processed by bioinformatics, which provides mechanistic insights into changes in cellular function. Therefore, lipidomic studies play an essential role in defining the biochemical mechanisms of lipid-related physiological/pathological processes through identifying alterations in cellular lipid metabolism, trafficking, and homeostasis in the selected system.

The term "lipidome" first appeared in the literature in 2001 [25]. In 2002, Rilfors and Lindblom [54] coined the term "functional lipidomics" as "the study of the role played by membrane lipids." In 2003, the field bloomed with different definitions [41, 55], demonstrations of technologies [41, 56], and biological applications [41, 57, 58]. Han and Gross first defined the field of lipidomics through integrating the specific chemical properties inherent in lipid species with a comprehensive mass spectrometric approach [41]. Since then, all areas of the field have been greatly accelerated.

Many modern technologies (including mass spectrometry (MS), nuclear magnetic resonance (NMR), fluorescence spectroscopy, high-performance liquid chromatography (HPLC), and microfluidic devices) have been used in lipidomic research. An edited book using these technologies for lipidomics is available [30]. MS, in part due to the development of new types of instruments and techniques (see Chapter 2), has greatly accelerated the progress of lipidomics. The website http://lipidlibrary.aocs .org/ constantly updates the publications including review papers that utilize modern MS methods for lipidomics. Several special issues on lipidomics have been published including the following:

- *Frontiers in Bioscience*, Volume 12, January 2007.
- *Methods in Enzymology*, Volumes 432 and 434, November 2007.

- *European Journal of Lipid Science and Technology*, Volume 111(1), January 2009.
- *Journal of Chromatography B*, Volume 877(26), September 2009.
- *Methods in Molecular Biology* (Springer Protocols), Volume 579–580, September 2011.
- *Biochimica et Biophysica Acta*, Volume 1811(11), November 2011.
- *Analytical Chemistry, Virtual Issue: Lipidomics*, http://pubs.acs.org/page/vi/2014/Lipidomics.html.
- *Analytical and Bioanalytical Chemistry*, Volume 407 (17), July 2015.

A few edited books on the areas of lipid analysis and lipidomics written by the experts and/or pioneers in the field have also been published [30, 59–61]. The current book provides a comprehensive description of the lipidomics discipline by using MS, from the fundamental, theory, and methods for identification and quantification, to applications.

REFERENCES

1. Vance, D.E. and Vance, J.E. (2008) Biochemistry of Lipids, Lipoproteins and Membranes. Elsevier Science B.V., Amsterdam. pp 631.
2. Christie, W.W. and Han, X. (2010) Lipid Analysis: Isolation, Separation, Identification and Lipidomic Analysis. The Oily Press, Bridgwater, England. pp 448.
3. (a) IUPAC-IUB (1978) Nomenclature of Lipids. Biochem. J. 171, 21–35.
   (b) IUPAC-IUB (1978) Nomenclature of Lipids. Chem. Phys. Lipids 21, 159–173.
   (c) IUPAC-IUB (1977) Nomenclature of Lipids. Eur. J. Biochem. 79, 11–21.
   (d) IUPAC-IUB (1977) Nomenclature of Lipids. Hoppe-Seyler's Z. Physiol. Chem. 358, 617–631.
   (e) IUPAC-IUB (1978) Nomenclature of Lipids. J. Lipid Res. 19, 114–128.
   (f) IUPAC-IUB (1977) Nomenclature of Lipids. Lipids 12, 455–468.
   (g) IUPAC-IUB (1977) Nomenclature of Lipids. Mol. Cell. Biochem. 17, 157–171.
4. Rezanka, T., Matoulkova, D., Kyselova, L. and Sigler, K. (2013) Identification of plasmalogen cardiolipins from Pectinatus by liquid chromatography-high resolution electrospray ionization tandem mass spectrometry. Lipids 48, 1237–1251.
5. Bartz, R., Li, W.H., Venables, B., Zehmer, J.K., Roth, M.R., Welti, R., Anderson, R.G., Liu, P. and Chapman, K.D. (2007) Lipidomics reveals that adiposomes store ether lipids and mediate phospholipid traffic. J. Lipid Res. 48, 837–847.
6. Yang, K., Jenkins, C.M., Dilthey, B. and Gross, R.W. (2015) Multidimensional mass spectrometry-based shotgun lipidomics analysis of vinyl ether diglycerides. Anal. Bioanal. Chem. 407, 5199–5210.
7. Fahy, E., Subramaniam, S., Brown, H.A., Glass, C.K., Merrill, A.H., Jr., Murphy, R.C., Raetz, C.R., Russell, D.W., Seyama, Y., Shaw, W., Shimizu, T., Spener, F., van Meer, G., VanNieuwenhze, M.S., White, S.H., Witztum, J.L. and Dennis, E.A. (2005) A comprehensive classification system for lipids. J. Lipid Res. 46, 839–861.
8. Thomas, C.P. and O'Donnell, V.B. (2012) Oxidized phospholipid signaling in immune cells. Curr. Opin. Pharmacol. 12, 471–477.

9. O'Donnell, V.B. and Murphy, R.C. (2012) New families of bioactive oxidized phospholipids generated by immune cells: Identification and signaling actions. Blood 120, 1985–1992.

10. Aldrovandi, M. and O'Donnell, V.B. (2013) Oxidized PLs and vascular inflammation. Curr. Atheroscler. Rep. 15, 323.

11. Yang, K., Zhao, Z., Gross, R.W. and Han, X. (2007) Shotgun lipidomics identifies a paired rule for the presence of isomeric ether phospholipid molecular species. PLoS ONE 2, e1368.

12. Kiebish, M.A., Bell, R., Yang, K., Phan, T., Zhao, Z., Ames, W., Seyfried, T.N., Gross, R.W., Chuang, J.H. and Han, X. (2010) Dynamic simulation of cardiolipin remodeling: Greasing the wheels for an interpretative approach to lipidomics. J. Lipid Res. 51, 2153–2170.

13. Han, R.H., Wang, M., Fang, X. and Han, X. (2013) Simulation of triacylglycerol ion profiles: Bioinformatics for interpretation of triacylglycerol biosynthesis. J. Lipid Res. 54, 1023–1032.

14. Bach, D. and Wachtel, E. (2003) Phospholipid/cholesterol model membranes: Formation of cholesterol crystallites. Biochim. Biophys. Acta 1610, 187–197.

15. Tsai, M.J. and O'Malley, B.W. (1994) Molecular mechanisms of action of steroid/thyroid receptor superfamily members. Annu. Rev. Biochem. 63, 451–486.

16. Jones, G., Strugnell, S.A. and DeLuca, H.F. (1998) Current understanding of the molecular actions of vitamin D. Physiol. Rev. 78, 1193–1231.

17. Russell, D.W. (2003) The enzymes, regulation, and genetics of bile acid synthesis. Annu. Rev. Biochem. 72, 137–174.

18. Kuzuyama, T. and Seto, H. (2003) Diversity of the biosynthesis of the isoprene units. Nat. Prod. Rep. 20, 171–183.

19. Raetz, C.R. and Whitfield, C. (2002) Lipopolysaccharide endotoxins. Annu. Rev. Biochem. 71, 635–700.

20. Yang, K., Cheng, H., Gross, R.W. and Han, X. (2009) Automated lipid identification and quantification by multi-dimensional mass spectrometry-based shotgun lipidomics. Anal. Chem. 81, 4356–4368.

21. Cheng, H., Jiang, X. and Han, X. (2007) Alterations in lipid homeostasis of mouse dorsal root ganglia induced by apolipoprotein E deficiency: A shotgun lipidomics study. J. Neurochem. 101, 57–76.

22. Jiang, X., Cheng, H., Yang, K., Gross, R.W. and Han, X. (2007) Alkaline methanolysis of lipid extracts extends shotgun lipidomics analyses to the low abundance regime of cellular sphingolipids. Anal. Biochem. 371, 135–145.

23. Merrill, A.H., Jr., Sullards, M.C., Allegood, J.C., Kelly, S. and Wang, E. (2005) Sphingolipidomics: High-throughput, structure-specific, and quantitative analysis of sphingolipids by liquid chromatography tandem mass spectrometry. Methods 36, 207–224.

24. Han, X. and Gross, R.W. (2005) Shotgun lipidomics: Electrospray ionization mass spectrometric analysis and quantitation of the cellular lipidomes directly from crude extracts of biological samples. Mass Spectrom. Rev. 24, 367–412.

25. Kishimoto, K., Urade, R., Ogawa, T. and Moriyama, T. (2001) Nondestructive quantification of neutral lipids by thin-layer chromatography and laser-fluorescent scanning: Suitable methods for "lipidome" analysis. Biochem. Biophys. Res. Commun. 281, 657–662.

26. Jung, H.R., Sylvanne, T., Koistinen, K.M., Tarasov, K., Kauhanen, D. and Ekroos, K. (2011) High throughput quantitative molecular lipidomics. Biochim. Biophys. Acta 1811, 925–934.

27. Llorente, A., Skotland, T., Sylvanne, T., Kauhanen, D., Rog, T., Orlowski, A., Vattulainen, I., Ekroos, K. and Sandvig, K. (2013) Molecular lipidomics of exosomes released by PC-3 prostate cancer cells. Biochim. Biophys. Acta 1831, 1302–1309.

28. Mitchell, T.W., Brown, S.H.J. and Blanksby, S.J. (2012) Structural lipidomics. In Lipidomics, Technologies and Applications. (Ekroos, K., ed.) pp. 99–128, Wiley-VCH, Weinheim

29. Gross, R.W., Jenkins, C.M., Yang, J., Mancuso, D.J. and Han, X. (2005) Functional lipidomics: The roles of specialized lipids and lipid-protein interactions in modulating neuronal function. Prostaglandins Other Lipid Mediat. 77, 52–64.

30. Feng, L. and Prestwich, G.D., eds. (2006) Functional Lipidomics. CRC Press, Taylor & Francis Group, Boca Raton, FL

31. Smilowitz, J.T., Zivkovic, A.M., Wan, Y.J., Watkins, S.M., Nording, M.L., Hammock, B.D. and German, J.B. (2013) Nutritional lipidomics: Molecular metabolism, analytics, and diagnostics. Mol. Nutr. Food Res. 57, 1319–1335.

32. Postle, A.D. and Hunt, A.N. (2009) Dynamic lipidomics with stable isotope labelling. J. Chromatogr. B 877, 2716–2721.

33. Kagan, V.E. and Quinn, P.J. (2004) Toward oxidative lipidomics of cell signaling. Antioxid. Redox. Signal. 6, 199–202.

34. Kagan, V.E., Borisenko, G.G., Tyurina, Y.Y., Tyurin, V.A., Jiang, J., Potapovich, A.I., Kini, V., Amoscato, A.A. and Fujii, Y. (2004) Oxidative lipidomics of apoptosis: Redox catalytic interactions of cytochrome c with cardiolipin and phosphatidylserine. Free Radic. Biol. Med. 37, 1963–1985.

35. Serhan, C.N. (2005) Mediator lipidomics. Prostaglandins Other Lipid Mediat. 77, 4–14.

36. Han, X. (2007) Neurolipidomics: Challenges and developments. Front. Biosci. 12, 2601–2615.

37. Merrill, A.H., Jr., Stokes, T.H., Momin, A., Park, H., Portz, B.J., Kelly, S., Wang, E., Sullards, M.C. and Wang, M.D. (2009) Sphingolipidomics: A valuable tool for understanding the roles of sphingolipids in biology and disease. J. Lipid Res. 50, S97–S102.

38. Han, X. and Jiang, X. (2009) A review of lipidomic technologies applicable to sphingolipidomics and their relevant applications. Eur. J. Lipid Sci. Technol. 111, 39–52.

39. Wang, M., Han, R.H. and Han, X. (2013) Fatty acidomics: Global analysis of lipid species containing a carboxyl group with a charge-remote fragmentation-assisted approach. Anal. Chem. 85, 9312–9320.

40. Yetukuri, L., Katajamaa, M., Medina-Gomez, G., Seppanen-Laakso, T., Vidal-Puig, A. and Oresic, M. (2007) Bioinformatics strategies for lipidomics analysis: Characterization of obesity related hepatic steatosis. BMC Syst. Biol. 1, 12.

41. Han, X. and Gross, R.W. (2003) Global analyses of cellular lipidomes directly from crude extracts of biological samples by ESI mass spectrometry: A bridge to lipidomics. J. Lipid Res. 44, 1071–1079.

42. Han, X., Yang, K. and Gross, R.W. (2012) Multi-dimensional mass spectrometry-based shotgun lipidomics and novel strategies for lipidomic analyses. Mass Spectrom. Rev. 31, 134–178.

43. Wood, R. and Harlow, R.D. (1969) Structural studies of neutral glycerides and phosphoglycerides of rat liver. Arch. Biochem. Biophys. 131, 495–501.

44. Wood, R. and Harlow, R.D. (1969) Structural analyses of rat liver phosphoglycerides. Arch. Biochem. Biophys. 135, 272–281.

45. Gross, R.W. (1984) High plasmalogen and arachidonic acid content of canine myocardial sarcolemma: A fast atom bombardment mass spectroscopic and gas chromatography-mass spectroscopic characterization. Biochemistry 23, 158–165.

46. Gross, R.W. (1985) Identification of plasmalogen as the major phospholipid constituent of cardiac sarcoplasmic reticulum. Biochemistry 24, 1662–1668.

47. Han, X., Gubitosi-Klug, R.A., Collins, B.J. and Gross, R.W. (1996) Alterations in individual molecular species of human platelet phospholipids during thrombin stimulation: Electrospray ionization mass spectrometry-facilitated identification of the boundary conditions for the magnitude and selectivity of thrombin-induced platelet phospholipid hydrolysis. Biochemistry 35, 5822–5832.

48. Han, X., Abendschein, D.R., Kelley, J.G. and Gross, R.W. (2000) Diabetes-induced changes in specific lipid molecular species in rat myocardium. Biochem. J. 352, 79–89.

49. Maffei Facino, R., Carini, M., Aldini, G. and Colombo, L. (1996) Characterization of the intermediate products of lipid peroxidation in phosphatidylcholine liposomes by fast-atom bombardment mass spectrometry and tandem mass spectrometry techniques. Rapid Commun. Mass Spectrom. 10, 1148–1152.

50. Fenwick, G.R., Eagles, J. and Self, R. (1983) Fast atom bombardment mass spectrometry of intact phospholipids and related compounds. Biomed. Mass Spectrom. 10, 382–386.

51. Robins, S.J. and Patton, G.M. (1986) Separation of phospholipid molecular species by high performance liquid chromatography: Potentials for use in metabolic studies. J. Lipid Res. 27, 131–139.

52. McCluer, R.H., Ullman, M.D. and Jungalwala, F.B. (1986) HPLC of glycosphingolipids and phospholipids. Adv. Chromatogr. 25, 309–353.

53. Murphy, R.C. (1993) Mass Spectrometry of Lipids. Plenum Press, New York. pp 290.

54. Lindblom, G., Oradd, G., Rilfors, L. and Morein, S. (2002) Regulation of lipid composition in *Acholeplasma laidlawii* and *Escherichia coli* membranes: NMR studies of lipid lateral diffusion at different growth temperatures. Biochemistry 41, 11512–11515.

55. Lagarde, M., Geloen, A., Record, M., Vance, D. and Spener, F. (2003) Lipidomics is emerging. Biochim. Biophys. Acta 1634, 61.

56. Lee, S.H., Williams, M.V., DuBois, R.N. and Blair, I.A. (2003) Targeted lipidomics using electron capture atmospheric pressure chemical ionization mass spectrometry. Rapid Commun. Mass Spectrom. 17, 2168–2176.

57. Esch, S.W., Williams, T.D., Biswas, S., Chakrabarty, A. and Levine, S.M. (2003) Sphingolipid profile in the CNS of the twitcher (globoid cell leukodystrophy) mouse: A lipidomics approach. Cell. Mol. Biol. 49, 779–787.

58. Cheng, H., Xu, J., McKeel, D.W., Jr. and Han, X. (2003) Specificity and potential mechanism of sulfatide deficiency in Alzheimer's disease: An electrospray ionization mass spectrometric study. Cell. Mol. Biol. 49, 809–818.

59. Byrdwell, W.C., ed. (2005) Modern Methods for Lipid Analysis by Liquid Chromatography/Mass Spectrometry and Related Techniques. AOCS Press, Champaign, IL.

60. Mossoba, M.M., Kramer, J.K.G., Brenna, J.T. and McDonald, R.E., eds. (2006) Lipid Analysis and Lipidomics: New Techniques and Applications. AOCS Press, Champaign, IL.

61. Ekroos, K., ed. (2013) Lipidomics: Technologies and Applications. John Wiley & Sons, Weiheim, Germany

2

MASS SPECTROMETRY FOR LIPIDOMICS

Mass spectrometry (MS) is an analytical discipline that studies the mass-to-charge (m/z) ratio of individual analytes for structural elucidation and quantification by means of mass spectrometer(s). A mass spectrometer generally consists of an ion source, a mass analyzer system, a detector, and a data processing system (Figure 2.1). Samples are introduced to the ion source through an inlet system, which is further described in Chapter 3. The analytes introduced are ionized/"vaporized" in the ion source. The ions in the gas phase are then separated according to their mass to charge (m/z) ratios by the mass analyzer system and detected. The detected signals are displayed in a mass spectrum that is a plot of ion intensity *vs.* m/z by the data processing system.

Accordingly, the principle of MS is the generation of molecular ions and the related fragments, separation of these ions according to their m/z, and measurement of the intensities of individual ions. In this chapter, the basic components of a mass spectrometer most often used in lipidomics, from ion source to detector, are introduced. A variety of MS/MS techniques and their interrelationships are discussed. Finally, a few recent advances of MS, which have impacted lipidomics, are summarized.

2.1 IONIZATION TECHNIQUES

An ion source is the part of a mass spectrometer where analytes are ionized. The resulting ions are then transmitted to the mass analyzer. Many technologies for ion generation have been developed and nearly every ionization technique has been

Lipidomics: Comprehensive Mass Spectrometry of Lipids, First Edition. Xianlin Han.
© 2016 John Wiley & Sons, Inc. Published 2016 by John Wiley & Sons, Inc.

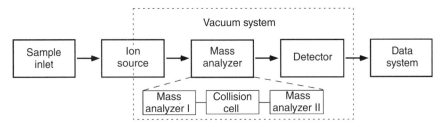

Figure 2.1 Schematic diagram of a mass spectrometer. The ion source may be under vacuum or at atmospheric pressure. The inset of the mass analyzer illustrates a common setting of the majority of mass spectrometers, which affords the tandem MS capability.

applied for lipid analysis [1]. However, the most prominently used techniques in lipidomics are the electrospray ionization (ESI) and matrix-assisted laser desorption/ionization (MALDI). Accordingly, only these two techniques are discussed to a certain extent in this chapter. The books written by Murphy [1] and Christie and Han [2] may be consulted if anyone would like to learn more about other ionization techniques.

2.1.1 Electrospray Ionization

2.1.1.1 Principle of Electrospray Ionization   In ESI, a solution containing the analytes of interest is introduced into the ion source through an inlet. The narrowed orifice at the end of the inlet and the mechanical forces imparted as the solution passes through the narrow orifice facilitate the formation of the Taylor cone and subsequently sprayed small droplets in the ionization chamber (Figure 2.2). The term "electrospray" is used for an apparatus that employs a high voltage to disperse a liquid or for the fine aerosol resulting from this process. Due to the high voltage, the sprayed aerosol can carry net charges due to oxidation/reduction processes. If a positive electric potential is applied to the end of the inlet and a negative electric potential is present at the entrance of the mass analyzer, which is the setting in the positive-ion mode, the droplets carry net positive charges (Figure 2.2). The opposite occurs in the negative-ion mode. In this process and later on, extensive solvent evaporation (i.e., desolvation) is involved due to the high temperature and/or vacuum applied. The solvent evaporates from a charged droplet so that its size decreases and the droplet becomes unstable upon reaching its Rayleigh limit. At this point, the droplet deforms as the electrostatic repulsion of like charges, in an ever-decreasing droplet size, becomes more powerful than the surface tension holding the droplet together [3]. The droplet then undergoes Coulomb fission, whereby the original droplet "explodes" creating many smaller, more stable droplets. The new droplets undergo desolvation and subsequently further Coulomb fissions. Although many physicochemical features of the ionization and fragmentation process are still unclear, droplet surface tension and the spatial proximity of surface charges on sprayed droplets are critical determinants of the ionization process. It should also

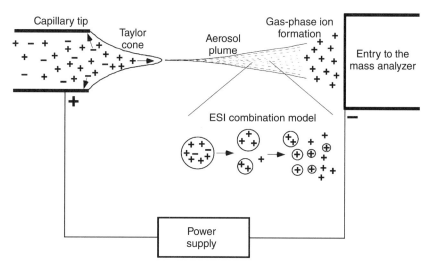

Figure 2.2   Schematic diagram of the principle of electrospray ionization in the positive-ion mode.

be mentioned that a novel ambient ESI technique, in which introduction of an analyte solution into the ion source is avoided, has been developed for desorption of electrospray ionization (DESI, see the following).

There are three major theories that explain the final formation of gas-phase ions:

- The ion evaporation model [4, 5] proposes that as the droplet reaches a certain radius, the field strength at the surface of the droplet becomes large enough to assist the field desorption of solvated ions.

- The charge residue model [6] suggests that electrospray droplets undergo evaporation and fission cycles, eventually leading to progeny droplets that contain on average one analyte ion or less. The gas-phase ions form after the remaining solvent molecules evaporate, leaving the analyte with the charges that the droplet carried.

- A third model invoking combined charged residue-field emission has been proposed [7]. In this model, the ions observed by MS may be (quasi)molecular ions created by the addition of a hydrogen cation (i.e., $[M+H]^+$), or of another cation (e.g., sodium ion to form $[M+Na]^+$), or the removal of a hydrogen nucleus to form $[M-H]^-$.

It is generally accepted that low molecular weight ions are liberated into the gas phase through the ion evaporation mechanism [5], while larger ions form by charged residue mechanism [8]. The third model appears well demonstrated in the ionization of lipid species.

Since the presence of volatile organic solvent(s) (e.g., methanol, chloroform, isopropanol) is favorable for desolvation, it enhances ionization based on the above-described mechanisms. This feature of ESI is indeed favorable for lipid analysis by ESI-MS since organic solvents increase lipid solubility. Moreover, to decrease the initial droplet size and facilitate the ionization process, compounds that can increase the conductivity (e.g., formic acid, acetic acid, ammonium acetate) are customarily added to the solution, which is extensively described below for lipid analysis.

In light of lipid analysis, the third model of ESI could be revised to include the compounds that contain an ionic bond (Figure 2.3). Under high electrical potentials such as in an ESI ion source (typically ~4 kV), charge separation can occur in those compounds containing an ionic bond in the ion source [9, 10]. Specifically, in the positive-ion mode, the component of the compound carrying the cation is selectively dispersed in the fine aerosol in the electrospray ion source, whereas the anionic component of the compound is selectively retained at the end of an emitter as a charge-neutral molecule after oxidation/reduction process and is eventually disposed of as waste (Figure 2.3) [11]. Similarly, in the negative-ion mode, the anionic parts of the compounds are selectively dispersed in the fine aerosol in the ion source in the gas phase and the cationic moieties are removed to waste after redox chemistry occurs. Therefore, conducting with a continuously equilibrating mobile phase, an electrospray ion source is functionally analogous to an electrophoretic device to a certain degree [9–11].

For the charge-neutral, but polar, compounds, the ion source can generate gas-phase ions based on the availability of counterions present in the mobile phase according to the third model. Specifically, in the high electric field, the compounds, which are present in the sprayed solution and carry no separable charge(s) but

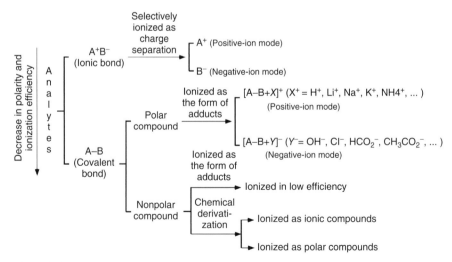

Figure 2.3 The relationship of ion formation and ionization efficiency with the electrical propensity of an analyte. The formation of an adduct ion of a covalent-linked polar compound depends on the availability of the small cation (X^+) or anion (Y^-) in the solution and the affinity of X^+ (or Y^-) with the analyte.

possess intrinsic dipoles, can be induced to interact with small cation(s) (e.g., H^+, Li^+, Na^+, NH_4^+, K^+) or anion(s) (e.g., OH^-, Cl^-, formate, acetate) (whatever is available in the matrix) to yield adduct ions in the positive- or negative-ion mode, respectively (Figure 2.3). Obviously, the ionization efficiencies of these charge-neutral compounds depend on their inherent dipoles, the electrochemical properties of the resultant adducts, the concentration of the small matrix ions, and the affinity of the small ions to the analytes [12].

Based on this model, different lipid classes can be grouped into three different categories based on their charge propensities [12, 13]. The first category of lipid classes includes those carrying at least one net negative charge under weakly acidic pH conditions (i.e., near pH 5). These classes of lipids are generally called anionic lipids (e.g., CL, PA, phosphatidylglycerol (PG), phosphatidylinositol (PI) and its polyphosphate derivatives, PS, ST, acyl CoA, anionic lysoGPL). The second category of lipid classes lacks a net charge under weakly acidic pH conditions, but become negatively charged under alkaline pH conditions. Therefore, they are referred to as weakly anionic lipids. Lipid classes in this category include phosphoethanolamine-carried lipid classes, nonesterified fatty acids and their derivatives, ceramide, and bile acids. It is noted that there are no ionic lipid compounds that carry one net positive charge (i.e., the type A^+ in Figure 2.3) since the majority of ionic lipids contain a phosphate or carboxyl group that is usually negatively charged under proper pH conditions. The remaining lipid classes belong to the third category, which are referred to as charge-neutral, but polar or polarizable, lipids. Lipid classes in this category include PC, lysoPC, SM, hexosylceramide, acylcarnitine, DAG, TAG, and cholesterol and its esters. Lipid classes that possess less polar intrinsic dipoles (e.g., DAG and cholesterol) in the third category, as well as those of nonpolar compounds, can be chemically derivatized to introduce a more polar moiety for enhanced ionization efficiency (Figure 2.3).

Based on this classification of lipid classes possessing differential intrinsic charge properties, a practical strategy for separation of these categories of lipids during mass spectrometric analysis has been practiced [12–15], utilizing the fact that different charge properties result predominantly from the different polar head groups of different lipid classes (Figure 2.4). Specifically, anionic lipids, due to their structural nature of possessing an ionic bond and carrying one or more net negative charge(s) at physiological pH, are directly and selectively ionized from the diluted lipid solution by negative-ion ESI-MS. Weakly anionic lipids, after being rendered to mildly basic extracts by the addition of a small amount of LiOH (or other suitable base), are analyzed from the solution by negative-ion ESI-MS. The rest of the lipid classes, mainly the charge-neutral, but polar, lipids, are analyzed directly from the mildly alkalinized, diluted lipid solution used in the above step by positive-ion ESI-MS, while lipids in the first and second categories are now anionic under the conditions and cannot be ionized. Some charge-neutral, but polar, lipids may also be analyzed directly from the diluted lipid solution without alkalinization through analyzing their protonated ions or sodium adducts. For the lipid classes that are nonpolar or less polar or lack of sensitive and specific fragments (i.e., building blocks), chemical derivatization can

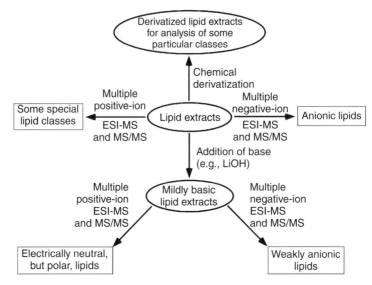

Figure 2.4   Schematic of the experimental strategy used for global analyses of cellular lipidomes directly from crude extracts of biological samples.

be exploited to either enhance the polarity and thereby the ionization efficiency or introduce sensitive and specific fragments for effective MS/MS analyses [16–19].

Such a separation of different lipid classes in the ion source (which is termed as intrasource separation [20]) is similar to the use of an ion-exchange column [21] or an electrophoretic device to separate individual lipid classes. However, compared to ion-exchange chromatography and electrophoresis, the intrasource separation has many advantages such as rapid, direct, *in situ*, reproducible, avoiding artifacts inherent in chromatography-based separations [22].

Such a strategy can be demonstrated with a model mixture of glycerophospholipids (e.g., PG, PE, and PC) that represent three categories of lipids aforementioned. The mixture comprises the three lipid classes at a molar ratio of 1:15:10, with each class containing two different molecular species at equal molar concentration (Figure 2.5). ESI mass spectra were acquired from the analysis of the mixture in the presence or absence of LiOH in the infusion solution. The analysis demonstrated an approximately 30-fold selective ionization of PG over PE species in the negative-ion mode without addition of base, considering that PG is 15-fold less in concentration but over twofold higher in ion intensity than PE (Figure 2.5a). The analysis also clearly demonstrated a selective ionization of PC over the other two classes in the positive-ion mode (Figure 2.5b and d).

It should be specifically pointed out that, although the molecular ion profiles of different lipid classes are different under different experimental conditions, the ion intensity ratio of individual molecular species in each class is virtually consistent. For example, the response factors of PE species are very different under different experimental conditions (Figure 2.5a compared to Figure 2.5c). However, the ratios

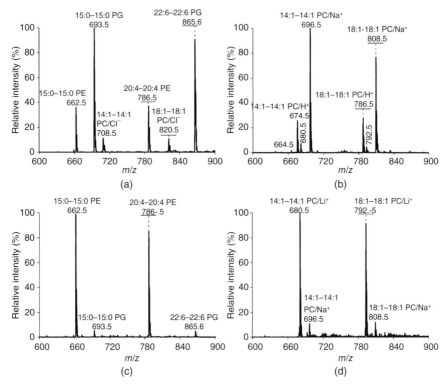

Figure 2.5 Intrasource separation of a model mixture of glycerophospholipids. The mixture comprises di15:0 and di22:6 PG (1 pmol/μL each), di14:1 and di18:1 PC (10 pmol/μL each), and di15:0 and di20:4 PE (15 pmol/μL each) species in 1:1 CHCl₃/MeOH. (a and c) show the mass spectra acquired in the negative-ion mode, and (b and d) show the mass spectra acquired in the positive-ion mode in the absence (a and b) or presence (c and d) of LiOH in the infusion solution. The horizontal bars indicate the ion peak intensities after ^{13}C de-isotoping and normalization of molecular species in each class to the one with lower molecular weight. (Han et al. [20]. Adapted with permission from Springer.)

of individual PE species (e.g., di15:0 PE *vs.* di20:4 PE at *m/z* 662.5 and 786.5 in Figure 2.5a and c) are essentially constant under these conditions after correction for differential distribution of ^{13}C isotopologues [12], even though these two species have very different chain lengths and unsaturation. Similarly, these constant ratios can be readily observed between the PC species such as the pair of di14:1 and di18:1 PC as chlorinated at *m/z* 708.5 and 820.5 (Figure 2.5a), as protonated at *m/z* 674.5 and 786.5 and as sodiated at *m/z* 696.5 and 808.5 (Figure 2.5d), and as lithiated at *m/z* 680.5 and 792.5 (Figure 2.5d), respectively. These observations strongly support the notion that response factors of individual molecular species of a class minimally depends on its aliphatic chain composition during a full-mass scan of the molecular ions under certain conditions, which are extensively described in Chapter 15. Thus, individual molecular species within a class can be quantified through ratiometric comparisons (i.e., ion intensity comparisons) with a pre-selected internal standard of the class under the experimental conditions.

Such a strategy utilizing intrasource separation has been broadly exploited for the analysis of biological lipid extracts and greatly facilitates the global analysis of individual lipid species directly from extracts of biological samples [12, 13, 15]. In summary, a comprehensive series of mass spectra corresponding to each category of lipids with respect to each of the aforementioned conditions for intrasource separation can be obtained from any biological samples by applying this approach.

2.1.1.2 Features of Electrospray Ionization for Lipid Analysis In addition to intrasource separation for lipid analysis, there exist numerous other features of ESI for lipid analysis as follows:

First, the ESI ion source is among the least destructive techniques in which (quasi)molecular ions of analytes become ionized with minimal in-source fragmentation. This feature has many impacts on MS analysis of lipids in lipidomics as follows:

- The minimal in-source fragmentation contributes to a higher ion intensity of ESI-MS, in general, in comparison to those methods possessing substantial amounts of in-source decay.
- Diminished in-source fragmentation contributes to the quantitative analysis of lipid species to a certain degree since fragmentation after collision-induced dissociation (CID) is a process that depends on the thermodynamics of the analytes, which in turn varies with the molecular structure of these analytes.
- ESI is so delicate that solvent adducts, dimers, and other complexes of lipids that usually only form as weakly bound complexes in solution can be detected as noncovalent aggregates or complexes. Therefore, ESI-MS can be employed to investigate lipid interactions and aggregation. Although there have been many applications of this feature in proteomics (see Ref. [23] for a review), the application to studying lipid interactions is at an early stage [24, 25]. Future studies in this area will likely appear. However, it should also be recognized that an aggregated state might cause many difficulties in quantitative analysis of lipids (Chapter 15).
- It should be noted that although the ESI source can be tuned to make in-source fragmentation (or decay) negligible in most cases, in-source fragmentation could be induced through applying a harsh condition in the ion source if needed [26, 27]. Moreover, due to the minimal in-source fragmentation, CID in a collision cell becomes an essential tool for qualitative and quantitative analyses of lipid species.

Second, ESI possesses relatively high adaptability in a broad range of flow rates of the mobile phase, a wide variety of solvents, and a large variety of modifiers including acids, bases, and/or buffers such as ammonium or other salts. Indeed, addition of a specific modifier aids both chromatographic separation and generation of particular adducts of lipid species in either positive- or negative-ion mode to enhance ionization efficiency or analytical sensitivity in tandem mass analysis or both.

Third, due to the easiness for neutral lipids to form adducts with small cations and anions in the ESI source, almost all nonvolatile lipids can be ionized. This feature can be further enhanced in combination with chemical derivatization [16–19]. Thus, ESI-MS perfectly complements GC-MS, a mature technique for the analysis of volatile compounds, for lipid analysis. It is well recognized that cellular lipids generally carry either no net charge (i.e., charge-neutral) or negative charge(s). Accordingly, anionic and weakly anionic lipids selectively yield deprotonated species, $[M-H]^-$, in the negative-ion mode (Figure 2.3), although (quasi)molecular ions with two adducts, $[M-H+2X]^+$ (where $X = H$, NH_4, Na, K, etc.) can be generated in the positive-ion mode under certain experimental conditions (see Part II). For the charge-neutral lipid species, ESI tends to yield molecular adducts of lipids, in the form of $[M+X]^+$ (where $X = H$, NH_4, Na, K, etc.) in the positive-ion mode and $[M+Y]^-$ (where $Y = Cl$, formate, acetate, etc.) in the negative-ion mode (Figure 2.3). The affinity of a small ion with lipid species contributes to the generation of the adduct. Its availability in the analytical solution and the inlet system is another major factor.

Finally, individual species of a polar lipid class (e.g., PC), which essentially have identical intrinsic charge properties as the head group of the class, possess very similar ionization efficiencies (or response factors) when ESI-MS analysis is performed under certain experimental conditions. This feature makes the ESI-MS a suitable tool for the quantitative analysis of lipid species when comparisons of ion intensities are made between molecules with similar functionalities and dipole moments as demonstrated earlier (also see Part III).

Taken together, with these features of ESI, the advantages of ESI-MS for lipid analysis are numerous [12, 13, 28]. The following are a few of these advantages:

- Its ion source could serve as a separation device to selectively ionize a class or a certain category of lipid species based on the charge propensity of lipid classes, thereby leading to the analysis of different lipid classes and individual molecular species with high efficiency and low ion suppression without prior LC separation.

- The sensitivity of ESI-MS for lipid analysis is remarkably high in comparison to other traditional MS approaches with a limit of detection (LOD) at a concentration of amol/µL (i.e., pM) to low fmol/µL (i.e., nM), which continues to be improved as the instruments become more sensitive.

- Instrumentation response factors of individual molecular species in a polar lipid class are essentially identical within experimental error after ^{13}C de-isotoping if the experiment is performed in a low lipid concentration region, which avoids lipid aggregation. Hence, it is feasible to quantitate individual molecular species of a polar lipid class through direct comparison of ion peak intensities with that of a selected internal standard in the same class or through the peak area measurement from the reconstructed total ion current chromatograph in comparison to a minimal set of external calibration curves.

- Linear dynamic relationship between an ion peak intensity (or ion counts) of a polar lipid species and the concentration of the compound is very broad, depending on the instrument sensitivity at the low end and the aggregation concentration at the high end (see Chapter 15). It should be emphasized that for the lipids without large dipoles, correction factors or calibration curves for each individual molecular species have to be pre-determined. Alternatively, derivatization can be employed to modify the charge properties of these less ionizable lipid classes to enhance ionization (Figure 2.4).
- A reproducibility of >95% from a prepared lipid extract in the presence of internal standards can be readily achieved after direct infusion (generally referred to as shotgun lipidomics), which should be independent of storage time, laboratory, analyst, and instrument. This high reproducibility guarantees the accuracy of analysis and reduces the required number of samples for replication.

Accordingly, it is evident that ESI-MS-based lipid analysis has become an essential tool for measuring cellular lipidomes during cellular perturbations, growth, and disease states (see [12, 29–33] for recent reviews).

2.1.1.3   Advent of ESI for Lipid Analysis: Nano-ESI and Off-Axis Ion Inlets

The advantage of nano-electrospray ionization (nano-ESI) over ESI at the flow rate of microliter per minute has been well recognized, including enhanced ionization efficiency, stabling ion current, reduced ion suppression, and utilizing a small amount of samples [34–37]. Therefore, a nano-ESI source has been largely available from the majority of commercially available mass spectrometers. A nano-ESI apparatus for automated sample delivery and infusion is described in Section 3.2.1 in detail. Overall, the term ESI in this book covers all types of electrospray techniques including nano-ESI.

Another advent related to the ionization techniques is the general adaptation of an off-axis electric field-assisted spray device in the commercially available ion source for atmospheric pressure (AP) ionization [38]. This improvement increases ionization efficiency dramatically through separations of ions from neutrals and establishment of complete desolvation. In particular, this improvement has a large impact on global analysis of individual lipid species from biological extracts without chromatographic pre-separation [12].

2.1.2   Matrix-Assisted Laser Desorption/Ionization

MALDI is another soft ionization technique used in MS. MALDI involves a two-step process. First, the firing of an ultraviolet (UV) laser beam induces desorption. In this process, matrix material absorbs the UV laser energy in firing, leading to the ablation of the upper layer (~1 μm) of the matrix material. The hot plume produced during ablation contains many species: neutral or ionized matrix molecules, protonated or deprotonated matrix molecules, and matrix clusters (even nanodroplets). In the second process, the analyte molecules are ionized (e.g., protonated or deprotonated in the hot plume). Ablated species may participate in the ionization of analyte through mechanisms of MALDI that are currently unclear. A detailed review about the desorption process given by Dreisewerd [39] could be consulted.

Generally, the matrix is thought to be ionized by the addition of a proton or other small cations, or by the depletion of a proton after absorbing the laser energy. The matrix then transfers the charge to the analytes (e.g., lipid molecules) [40]. Ions observed after this process are quasimolecular ions, consisting of a neutral molecule [M] and an added or removed ion, for example, $[M+H]^+$ in the case of an added proton, $[M+Na]^+$ in the case of an added sodium ion, or $[M-H]^-$ in the case of a removed proton. To interpret the ionization process by using a neutral matrix such as 9-aminoacridine and 1,5-diaminonapthalene, which has been demonstrated to be useful in the analysis of lipids [41–50], this ionization model needs to be modified since the neutral matrix molecules are not first deprotonated and ionized prior to transferring the charge to the analytes. Instead, the neutral matrix likely aids in the formation of quasimolecular ions by the depletion of a proton from the analyte due to the basicity that the matrix (e.g., 9-aminoacridine) possesses with the help of the energy absorbed from laser firing.

In early times, the MALDI ion source was under a high vacuum. Lately, atmospheric pressure (AP) MALDI has been developed, which in contrast to vacuum MALDI operates at a normal atmospheric environment [51]. The mechanism of AP-MALDI ion production is similar to that of conventional MALDI. The main difference between vacuum MALDI and AP-MALDI is that AP-MALDI produces ions under atmospheric pressure conditions outside of the instrument vacuum housing. In vacuum MALDI, ions are typically produced at 10 mTorr or less, whereas in AP-MALDI, ions are formed at the atmosphere pressure.

The consequences of this development are numerous:

- The AP-MALDI source is an external ionization source and can be easily interchangeable with an ion-trap mass spectrometer [52] or any other MS system equipped with ESI or nano-ESI sources [51, 52].

- The replacement of target (sample) plates is a simple and quick process due to the nature of atmospheric pressure.

- AP-MALDI inherits all the power of particular mass spectrometers (e.g., quadrupole–quadrupole-time of flight (QqTOF)), including high sensitivity, broad mass detection range, MS/MS capability, etc.

- The main disadvantage of AP-MALDI relative to vacuum MALDI was limited sensitivity; however, significant changes regarding ion transferring have been made, leading to profound improvement of detection limits [53].

MALDI-MS has been widely used for lipid analysis since it was introduced in the late 1980s. Its applications, advantages and disadvantages, and new improvements for lipid analysis are described to a certain degree in Chapter 3.

2.2 MASS ANALYZERS

After the ions are generated in the ion source (see above), they are transmitted to
the mass analyzer that operates by separating ions according to their m/z. Although
the analyzers based on magnetic sector fields play an important role in the history of
mass spectrometry, the applications of this type of analyzer in lipidomics are very
limited due to its large size, cost ineffectiveness, and relative hardness of operation. Fourier transform ion cyclotron resonance (FTICR) analyzers also use magnetic
fields. Although they can provide high mass accuracy and resolution, which is important for lipid analysis [54, 55], their applications in lipidomics are now replaced by
the recently developed Orbitrap-based instruments, mostly due to high cost and its
arduous tasks in operation and maintenance. Accordingly, only those mass analyzers such as quadrupole (Q), TOF, and ion trap, which separate ions based on their
behaviors in electrical fields, are introduced in the following section.

2.2.1 Quadrupole

Quadrupole mass analyzers are relatively small and inexpensive. They use oscillating electrical fields to selectively stabilize or destabilize the paths of the ions passing
through a radio frequency (RF) quadrupole field created between four parallel rods
(Figure 2.6a). Only the ions in a certain range of m/z ratios are passed through the
system at a time, but changes to the potentials on the rods allow a wide range of
m/z values to be swept rapidly, either continuously or in a succession of discrete
hops. A quadrupole mass analyzer acts as a mass-selective filter, and for that reason

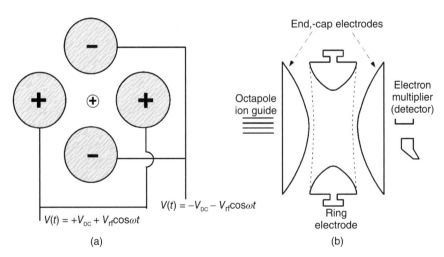

Figure 2.6 Schematic diagrams of quadrupole (a) and ion-trap (b) mass analyzers. $V(t)$ is
the voltage at time t; V_{DC} is the direct current voltage component; and $V_{rf}\cos\omega t$ is the radio
frequency voltage component.

it is referred to as a transmission quadrupole. If a quadrupole is made to rapidly and repetitively cycle through a range of mass filter settings, all the ions in the mass range of interest can be detected and a full-mass spectrum can be displayed.

A triple quadrupole (QqQ) mass spectrometer is a common setting of quadrupoles for ion transmission, in which three consecutive quadrupole stages are used. The first one acts as a mass filter to transmit a particular incoming ion to the second quadrupole that is used as a collision cell, wherein the transmitted ion can be broken into fragments after colliding with a type of inert gas such as helium, nitrogen, or argon present in the collision cell. The third quadrupole also acts as a mass filter that transmits a particular fragment ion at a time to the detector. Accordingly, a triple quadrupole mass spectrometer can be employed to perform various types of MS/MS (see the following section). The figures of merit of quadrupole mass analyzers can be summarized as follows:

- Mass resolution: unit or higher.
- Mass accuracy: 100 ppm at m/z 1000.
- m/z range: up to 4000.
- Linear dynamic range: 10^5.
- Scan speed: ~ second.
- Efficiency (transmission × duty cycle): <1% (scanning) to 95%.
- Compatible with ionization techniques: AP/vacuum (continuous).
- MS/MS: CID at eV.
- Cost: low to moderate.
- Size/weight: bench top.

A QqQ-type mass spectrometer is the most popular instrument utilized for lipid analysis in lipidomics. Their versatility offers the great power for both LC-MS and shotgun lipidomics platforms (Chapter 3) to analyze a variety of lipids [14, 56–58]. The QqQ configuration allows for sensitive identification and quantitation through selected/multiple reaction monitoring (SRM/MRM), neutral-loss scan (NLS), precursor-ion scan (PIS), and product-ion analysis with high selectivity and broad linear dynamic range. The disadvantages of this type of instrument include limited mass accuracy, low resolving power, and only allowing one to perform tandem MS analysis to the second stage [59], although three stages of tandem MS could possibly be conducted with efforts such as induced in-source fragmentation [60–63].

2.2.2 Time of Flight

The TOF analyzer uses an electrical field to accelerate the ions flying through the drift tube with the same potential and then determines the time taken for them to get to the detector. The principle of TOF analyzer is that if an analyte with a mass of m carries the number of charges (z), the kinetic energy ($mv^2/2$) of this compound should be equal to the potential energy of this ion present in the electrical field (V), which follows the law of conservation of energy as

$$mv^2/2 = zeV \qquad (2.1)$$

where e is the charge of an electron and v is the speed of the ion flying through the drift tube. Since the speed defines as

$$v = d/t \qquad (2.2)$$

where d is the length of the drift tube and t is the time that it takes the ion to travel down the drift tube to the detector, integrating these formulas leads to

$$m/z = 2eV(t/d)^2 \qquad (2.3)$$

Therefore, the mass of an individual analyte can be derived from its determined flight time. For example, the heavier ions reach the detector later than the lighter ions [64]. The figures of merit of TOF analyzers can be summarized as follows:

- Mass resolution: 10^4.
- Mass accuracy: 2–50 ppm.
- m/z range: unlimited.
- Linear dynamic range: 10^4.
- Scan speed: ms.
- Efficiency (transmission × duty cycle): 1–95%.
- Compatible with ionization techniques: AP/vacuum (continuous/pulsed).
- MS/MS: CID at eV or keV.
- Cost: low to high.
- Size/weight: bench top or floor standing.

TOF mass analyzers have been broadly used for lipid analysis in lipidomics [33, 65, 66]. Although TOF techniques are improved and can measure masses with high mass accuracy/resolution, high sensitivity, and high efficiency, instruments constructed by TOF alone have difficulty in performing MS/MS experiments for lipid analysis. Therefore, hybrid instruments with quadrupoles (i.e., QqTOF) or liner trap (i.e., LIT-TOF) are required to overcome this difficulty.

The development of hybrid instruments has improved product ion analysis to a great extent in comparison to both QqQ and ion-trap mass analyzers. For example, QqTOF instruments have good mass accuracy and resolving power for determining product ions, whereas QqLIT instruments allow for MS^n analysis in addition to NLS and PIS analyses [59]. It should be noted that QqTOF mass spectrometers are incapable of virtual NLS and PIS analyses, but can extract NLS- and PIS-like dataset from the array of product ion analysis data.

2.2.3   Ion Trap

A three-dimensional ion trap (e.g., Q ion trap) bears the same physical principles as the quadrupole mass analyzer, but the ions are trapped and sequentially ejected mainly by using an RF field, within a space defined by a ring electrode between two end-cap electrodes (Figure 2.6b). A linear quadrupole ion trap is similar to a three-dimensional ion trap, but it traps ions in a two-dimensional, instead of a three-dimensional, quadrupole field. Some of the figures of merit of the ion-trap instrument are as follows:

- Mass resolution: unit.
- Mass accuracy: 100 ppm.
- m/z range: 2000.
- Linear dynamic range: 10^2–10^5.
- Scan speed: ~ second.
- Efficiency (transmission × duty cycle): <1% (scanning) to 95%.
- Compatible with ionization techniques: AP/vacuum (continuous/pulsed).
- MS/MS: CID at eV, capable of MS^n, one-third cutoff rule.
- Cost: low.
- Size/weight: bench top.

Both 3D and linear ion-trap analyzers are widely used in lipidomics [67–72]. Generally, these analyzers have good sensitivity, possess high-throughput capability, and allow multistage tandem MS analyses. However, these ion-trap analyzers suffer from poor mass resolution, low dynamic range, and space–charge effects that do not allow very accurate mass determination or quantitation.

Ions in Orbitrap are electrostatically trapped in an orbit around a central, spindle shaped electrode [73]. The electrode confines the ions so that they orbit around the central electrode and also oscillate back and forth along the central electrode's long axis. This oscillation generates image currents, the frequencies of which depend on the mass-to-charge ratios of the ions. Mass spectra are obtained by Fourier transformation of the recorded image currents. Some of the figures of merit of the Orbitrap in the hybrid instruments such as LIT-Orbitrap or Qq-Orbitrap are as follows:

- Mass resolution: 10^4–10^5.
- Mass accuracy: 1–5 ppm.
- m/z range: 4000–6000.
- Linear dynamic range: 10^2–10^5.
- Scan speed: ~0.1 s.
- Efficiency (transmission × duty cycle): 1–95%.
- Compatible with ionization techniques: API.
- MS/MS: eV, capable of MS^n.
- Cost: moderate to high.
- Size/weight: bench top to floor standing.

Although the Orbitrap analyzer is the newest addition to the family of high-resolution mass spectrometers, it has already made a great impact on lipidomics [65, 74–77]. Advantages include high sensitivity, high mass accuracy/resolution, bench-top size/weight, and being considerably less expensive than FTICR-MS.

2.3   DETECTOR

The detector is the final component of a mass spectrometer, which records either the induced charge or the produced current when an ion passes by or hits a surface. Typically, some type of electron multiplier (e.g., discrete electron multiplier, continuous dynode electron multiplier, and microchannel plate) is commonly used to amplify the signals, although other types of detectors including Faraday cups, scintillation counter, and postacceleration (Daly Knob) detectors are also used. Because the number of ions guided to the detector at a particular instant is typically quite small, considerable amplification is often necessary to get a meaningful signal. To this end, microchannel plate detectors are commonly used in modern commercial instruments [78]. This type of detector is a planar component used for detection of ions and impinging radiation. Because a microchannel plate detector has many separate channels, it can additionally provide spatial resolution [79]. In FTICR-MS and Orbitrap mass spectrometers, the detector usually consists of a pair of metal surfaces within the mass analyzer/ion-trap region and measure the electrical signal of ions which pass near them as the ions oscillate. No direct current but only a weak AC image current is produced in a circuit between the electrodes [73, 80]. Table 2.1 illustrates a general comparison among the commonly used detectors.

TABLE 2.1   A General Comparison of the Commonly Used Detectors

Detector Type	Advantages	Trade-Offs
Faraday cup	Robustness, stable sensitivity, and good for measuring ion transmission	Low amplification ($\sim$10)
Scintillation counter	Extremely robust, long lifetime (>5 yrs), good sensitivity ($\sim$$10^6$)	Sensitive to light
Electron multiplier (EM)	Fast response, good sensitivity ($\sim$$10^6$)	Short lifetime (1–2 yrs)
High-energy dynodes w/EM	Increased sensitivity for measuring high mass	May shorten lifetime of EM
Array	Fast response, good sensitivity, simultaneous detection	Low resolution ($\sim$0.2 amu), expensive, short lifetime (<1 yr)
FT-MS (Orbitrap)	Mass analyzer serves as the detector of high resolution	Used only for the specific instruments

2.4 TANDEM MASS SPECTROMETRY TECHNIQUES

Unlike conventional ionization techniques (e.g., electron or chemical ionization), soft ionization techniques (e.g., ESI and MALDI) (see above) yield minimal in-source fragmentation under appropriate experimental conditions. In-source fragmentation can provide structural information when used properly, but generally leads to complication of lipid analysis since frequently multiple lipid species enter the ion source simultaneously even when coupled with chromatographic separation (i.e., an LC-MS lipidomics approach) (see Chapter 3). Therefore, the absence of in-source fragmentation becomes a big advantage of these ionization techniques for lipid analysis in lipidomics.

The trade-off of this advantage is that identification and characterization of lipids using a mass spectrometer with such a soft-ion source depend heavily on tandem MS analysis. Conduction of tandem MS analysis requires the mass spectrometer possessing multiple mass analyzers or an ion trap. In fact, a variety of hybrid combinations of mass analyzers have been developed and have greatly facilitated lipidomic analyses [33, 59].

There are four main MS/MS modes (including product ion analysis, NLS, PIS, and SRM) that are particularly useful in lipidomics (Figure 2.7). The general principles of these MS/MS techniques can be easily explained by using a QqQ-type mass spectrometer, as briefly described below. Although different parameters are used for mass spectrometers with different hybrid mass analyzers, the underlying chemical principles are quite similar. Comparisons among these techniques are summarized in Table 2.2.

2.4.1 Product-Ion Analysis

In the product ion analysis mode, the first mass analyzer is used to select a particular precursor ion of interest (m_x) by setting the mass analyzer to transmit only this ion. The selected ion is typically accelerated to higher kinetic energy with an electrical potential to induce collisional heating and subsequently fragments after colliding with inert gases (often helium, nitrogen, or argon) in a collision cell. Some of the

TABLE 2.2 Comparisons among the Scan Modes in Tandem Mass Spectrometry

Mode	Mass Analyzer 1	Mass Analyzer 2	Application
Product ion	Selecting	Scanning	To obtain structural information about the precursor ions
PIS	Scanning	Selecting	To detect the analytes yielding an identical fragment ion after CID
NLS	Scanning	Scanning	To detect the analytes losing a common neutral fragment after CID
SRM	Selecting	Selecting	To monitor a particular CID reaction

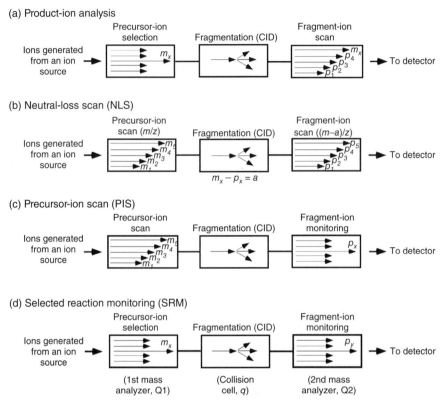

Figure 2.7 Schematic diagram of tandem mass spectrometric techniques. CID denotes collision-induced dissociation. The m_x and p_x $(x = 1, 2, 3, \dots)$ stand for molecular (or precursor) and product ions, respectively. The letter "a" in the neutral-loss scan mode denotes the mass of the neutral-loss fragment.

smaller fragments from fragmentation of the precursor ion after CID will carry an intrinsic charge (i.e., product ions), whereas the others are neutral fragments carrying no charge but representing a mass difference between the precursor ion and a particular product ion. The m/z values of the resultant product ions (p_1, p_2, p_3, etc.) are then detected with the second mass analyzer (Figure 2.7a). The structure of the selected precursor ion thus can be elucidated from the reconstruction of the product ions and/or the fragmentation patterns in conjunction with the mass of the precursor ion. This MS/MS mode is useful for the characterization of available fragmentation pathways and for the analysis of the fragmentation kinetics of discrete molecule ions. This MS/MS technique can be iteratively performed with sequential selection of the resultant product ions for fragmentation in multistage tandem MS (MS^n) experiments. The MS^n technique can provide critical structural information of individual molecular species through chemically defined fragmentation mechanisms and enable researchers to confirm the structural identity of individual lipid species and identify novel lipids. There are numerous applications of this mode in lipidomics (see Part II).

2.4.2 Neutral-Loss Scan

In the NLS mode, both the first and second mass analyzers are scanned simultaneously, but set with a constant mass offset of "a" between the two analyzers. Specifically, the first mass analyzer continuously transmits a particular ion (e.g., at m_x) at a time. The transmitted ion is fragmented in the collision cell with CID. The second mass analyzer is set to only monitor the fragment ion at $m_x - a$ (Figure 2.7b). The mass spectrometer only records the precursor ion (m_x) when this transmitted precursor ion yields a product ion (p_x) where $p_x = m_x - a$ (i.e., through the loss of a neutral fragment that has a mass of "a" from the precursor ion) after CID. This mode is primarily accomplished with QqQ-type mass spectrometers. This MS/MS mode has been extensively used in shotgun lipidomics to effectively detect a class or a group of lipids that possess an identical neutral-loss fragment, typically derived from the unique head group of the lipid class or group [14, 81–83].

2.4.3 Precursor-Ion Scan

In the PIS mode, the first mass analyzer is scanned whereas the second analyzer focuses on monitoring a particular product ion of interest (p_x) after CID. Specifically, the first mass analyzer continuously transmits a particular ion at each time; the transmitted ion is fragmented in the collision cell with CID; and the second mass analyzer is set to only monitor a particular fragment ion (p_x) (Figure 2.7c). All of the precursor ions (m_1, m_2, m_3, etc.) that produce the selected product ion (p_x) after CID are thus recorded in this MS/MS mode (Figure 2.7c). This mode can be achieved with QqQ type as well as with many other hybrid mass spectrometers by using a quadrupole as the first mass analyzer. This MS/MS mode has also been extensively used in shotgun lipidomics to effectively detect a class or a group of lipids that yield a given product ion after CID [14, 81–83].

2.4.4 Selected Reaction Monitoring

In the SRM mode, both the first and the second mass analyzers monitor the selected precursor and product ions of m_x and p_x, respectively (Figure 2.7d). Specifically, the first mass analyzer only selectively transmits a particular ion (m_x); the transmitted ion is fragmented in the collision cell with CID; and the second mass analyzer is set to only monitor a particular fragment ion (p_x) (Figure 2.7d). This mode is primarily accomplished with QqQ-type mass spectrometers where mass-resolving Q_1 isolates the precursor, q acts as a collision cell, and mass-resolving Q_2 monitors a particular product ion. A precursor/product pair is often referred to as a *transition*. This mode yields high specificity and sensitivity via a high duty cycle to detect a pair of ion transition of interest. When either the first or the second mass analyzer or both are set to monitor multiple ions to achieve the detection of multiple pairs of ion transitions, the term "multiple reaction monitoring (MRM)" has been widely used. It should be noted that it is not accurate to use this term to indicate that more than one generation of product ions are being monitored [84].

It should be emphasized that the transitions must be pre-determined, and much work goes into ensuring that the transitions selected have maximum specificity. Moreover, SRM could be considered a special case of PIS in which the first analyzer is fixed at a certain m/z, or a special case of product ion analysis in which the second analyzer is focused at a specific product ion, or a special case of NLS in which both analyzers are separately fixed at the ions of a transition. Accordingly, from an integrated chemical perspective, it should be recognized that the SRM mode only represents a special case of the other three MS/MS techniques with particular advantages associated with and essential to LC-MS analysis where only a limited amount of time is available for data acquisition, and therefore effective duty cycles are critical. The SRM/MRM techniques have been widely used for quantitative analysis of individual lipid species in lipidomics when a mass spectrometer is coupled with LC [57, 85–87].

2.4.5 Interweaving Tandem Mass Spectrometry Techniques

The connection of SRM/MRM mode to other MS/MS modes is briefly mentioned earlier. In fact, the other MS/MS techniques (i.e., product-ion analysis, NLS, and PIS) are also interrelated. This interrelationship provides a foundation to multidimensional mass spectrometry-based shotgun lipidomics (MDMS-SL) and can be schematically illustrated with a simplified model system that comprises three molecular ions (m_1, m_2, and m_3) of a lipid class (Figure 2.8).

In this model, each of the three molecular ions has a different m/z, and therefore each yields a different mass spectrum in the product ion analysis mode after CID. Because these molecular ions belong to the same lipid class, these ions possess virtually identical fragmentation patterns. We assume that the fragmentation pattern of these molecule ions shows three features of product ions as highlighted in the broken-line box (Figure 2.8). First, these molecule ions yield product ions that correspond to the loss of a common neutral fragment with a mass of a. This loss gives rise to product ions p_{1a}, p_{2a}, and p_{3a} from the molecular ions m_1, m_2, and m_3, respectively, where $a = m_1 - p_{1a} = m_2 - p_{2a} = m_3 - p_{3a}$. Second, these molecular ions also yield a common product ion p_c (i.e., $p_{1c} = p_{2c} = p_{3c} = p_c$). These two features commonly result from the head groups of the GPL classes in their product ion analysis after CID (see Part II). Finally, each individual molecular ion yields a specific fragment ion from a common constituent (e.g., a fatty acyl chain). This specific fragment ion leads to an array of product ions p_{1b}, p_{2b}, and p_{3b} that result from the individual molecular ions m_1, m_2, and m_3, respectively. The structure of each individual species, including its backbone, can be derived from these fragments in combination with the m/z of each molecular ion by product-ion analysis after CID (Figure 2.8).

Conceivably, the set of product ions from a molecular ion can also be detected in the NLS mode, because each product ion represents the loss of neutral fragment(s) from its corresponding molecular ion. Therefore, all of the product ions of the model

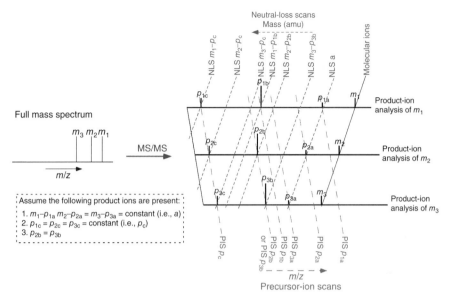

Figure 2.8 Schematic illustration of the interrelationship among the MS/MS techniques for the analysis of individual molecular species of a class of interest. The analysis of only three species (m_1, m_2, and m_3 in the full-mass spectrum) of a class is illustrated for simplicity, whereas there exist up to hundreds of individual molecular species within a class. In the model, the class of lipid species, similar to a class of GPL species, possesses a common neutral-loss fragment with mass of a (i.e., $m_1 - p_{1a} = m_2 - p_{2a} = m_3 - p_{3a} = a$ (a constant)), a common fragment ion at m/z p_c (i.e., $p_{1c} = p_{2c} = p_{3c} = p_c$), and a specific ion to individual species at m/z p_{1b}, p_{2b}, and p_{3b}, respectively, which might not be identical to each other. Both the common neutrally lost fragment and the common fragment ion result from the head group of the class, whereas the individual species-specific ions represent the fatty acyl moieties of the species. The identity of each individual species can be derived from these fragments in combination with the m/z of each molecule ion. All of these patterns are illustrated in the product ion analyses of these molecule ions as indicated. The scans of the individual neutral-loss fragment between a specific molecule ion and its individual fragment ion are shown with fine broken lines and the scans of each individual fragment ion are shown with wide broken lines. It should be recognized that, although the analyses of fragments with either neutral-loss scan (NLS) or precursor-ion scan (PIS) in this model are more complicated than those in product ion analysis, the analyses by NLS or PIS are much simpler than that by product ion analysis of biological samples.

system detected in the product ion analysis can be determined with NLS of masses in an ascending order of p_a, $m_3 - p_{3b}$, $m_2 - p_{2b}$, $m_1 - p_{1b}$, $m_3 - p_c$, $m_2 - p_c$, and $m_1 - p_c$ (fine broken lines (Figure 2.8)). If this series of NLS mass spectra are mapped along the masses, a pseudo-mass spectrum along the black line crossed with each of the molecular ions (i.e., m_1, m_2, or m_3) mimics the product-ion mass spectrum of the corresponding molecular ion.

Similarly, all product ions can also be determined in the PIS mode by scanning all m/z of p_c, p_{2b}, p_{3b}, p_{1b}, p_{3a}, p_{2a}, and p_{1a} in an ascending order (wide broken lines

(Figure 2.8)). Thus, if this series of PIS mass spectra is mapped along the m/z of the scanned fragment ions, a pseudo-mass spectrum along a black line crossed with one of the molecule ions (i.e., m_1, m_2, or m_3) again mimics the product-ion mass spectrum of the corresponding molecular ion.

This simplified model for analysis of a class of lipids illustrates the interrelationship among the modes of product ion scan, PIS, and NLS. Thus, a detailed map of all product ions of interest can be obtained within a mass range of interest either with multiple individual product-ion analyses of individual molecular ions unit by unit, or with NLS through scanning the neutral fragments unit by unit, or with PIS through scanning the fragment ions unit by unit. Each of these analyses alone can map the complete fragment ions of individual lipid species within the mass range of interest as demonstrated previously [29].

QqTOF-type mass spectrometers have been used in mapping the product ions in the product ion analysis mode [88, 89]. The execution of this type of mapping by using QqQ-type instruments is very labor intensive, but the interrelationship among the modes of the MS/MS techniques allows one to perform the complete mapping of product ions with combined use of NLS and PIS. This approach has been well developed by Han, Gross, and their colleagues [14, 15, 83, 90]. In the approach, the building blocks of biologically occurring lipids of a lipid class of interest (Chapter 1) are analyzed with either NLS or PIS of the specific building blocks or their combination [14, 83, 90–93]. It should be recognized that an array of pseudo-product-ion mass spectra generated with NLS or PIS of the specific building blocks or their combination is different from a series of product ion analyses (see Chapter 3). Specifically, a fewer number of NLS and/or PIS are required for the analysis of analytes in biological samples to generate pseudo-product ion mass spectra for all the molecular ions of a class (i.e., higher efficiency) [83].

2.5 OTHER RECENT ADVANCES IN MASS SPECTROMETRY FOR LIPID ANALYSIS

Developments on different parts of a mass spectrometer (e.g., sample introduction, ion source, analyzer, and detector) or new types of instruments occur almost on a yearly basis. As the literature and history has shown, any new development in mass spectrometry has been applied to lipid analysis. It can be predicted that this trend will continue. For example, the Orbitrap Fusion (Lumos) Tribrid mass spectrometer from Thermo Fisher Scientific should give the researchers in lipidomics some new momentum due to its extra high mass accuracy/resolution and friendly operation as recently demonstrated to a certain degree [94]. In the future, research in lipidomics will rely on technologies with automation and high throughput and be more focused on the spatial and temporal changes at a cellular level. Developments in these areas will have great impacts on lipid analysis in general, and lipidomics in particular. In this section, two of such advances in mass spectrometry (i.e., ion-mobility MS (IM-MS) and DESI) are briefly described.

2.5.1 Ion-Mobility Mass Spectrometry

IM-MS has emerged as an important analytical method in the last decade [95]. IM-MS is a postionization separation technique used to separate ionized molecules in the gas phase based on their mobility (i.e., a function of an ion's mass, charge, size, and shape) in a carrier buffer gas under differential voltages or other factors [96]. For example, in the traditional drift-time IM-MS, the migration time through the drift tube is characteristic of different ions, leading to the ability to distinguish distinct analyte species. The area of an ion that gas molecules strike is an ion's collision cross section, which is directly related to the size and shape of the ion and is an indicator of the mass and structure of the ion. The greater this collision cross section is, the more area available for buffer gas to collide and subsequently impede the ion's drift. The ion then requires a longer time to drift through the tube. Therefore, IM-MS provides not only a new dimension of separation but also shape (or structural) information because separation is based on the conformation of a molecule in addition to its mass [97, 98].

IM-MS can allow rapid profiling of complex lipid mixtures including structural isomers [99] and provide rapid 2D analysis of lipids in which each lipid class falls along a trend line in a plot of IM drift time *vs. m/z* [100, 101]. In a recent study [66], Jackson and colleagues have used MALDI-IM/TOF-MS for the analysis of complex mixtures of GPL species in which 2D separation of molecular species based on drift time and *m/z* values was achieved rapidly. They found that the changes in drift time of GPL species are associated with the fatty acyl chain length, the degree of unsaturation, the head group, and the cationization of individual species. Furthermore, the coupling of MALDI with IM-MS allows researchers to directly probe tissue, map the distribution of lipids, and elucidate molecular structure with minimal preparation and within an interval of a few hundred microseconds between the applications of each focused laser desorption pulse to the sample [101]. For example, it was demonstrated that GPL species could be profiled directly from rat brain tissue sections, and 22 GPL species in the classes of PC, PE, PS, PI, and SM were identified. This topic is further discussed in Chapter 12.

In addition to this enabling advance for direct tissue lipid analysis, the significance of IM-MS in lipidomics is numerous. First, separation of isomers, isobars, and conformers is rendered possible with the addition of IM cells to mass spectrometers [97], which allows identification of novel lipid classes and species in a high-throughput manner. Moreover, analysis of chiral isomers could be achieved by the introduction of chiral reagents into an ion-mobility cell as demonstrated in other studies [98]. Finally, the duty cycle of IM-MS is short relative to LC separations and can thus be coupled to such techniques to form 3D modalities such as LC-IM-MS [102]. A recent review paper [103], which extensively discusses the principles of IM-MS technology and its applications for lipidomics, should be consulted for the readers who are interested in this area of research.

2.5.2 Desorption Electrospray Ionization

DESI is an ambient ionization technique that can be used in MS for analysis of compounds with little sample preparation. This technique was developed in 2004 by Professor Graham Cooks' group [104]. DESI is a combination of two MS ionization methods: ESI and desorption ionization. Instead of a laser beam or a primary ion beam, DESI uses energetic, charged electrosprayed solvent droplets to desorb the molecules from the sample surface. The ion source, made by Prosolia Inc., is now commercially available. Ionization occurs by directing a charged mist to the sample surface (Figure 2.9) [105]. The electrospray mist is attracted to the surface by applying a voltage on the sample holder. After ionization, the ions travel through air into the atmospheric pressure interface, which is connected to the mass spectrometer.

There are two kinds of ionization mechanisms proposed to interpret the ion formation of low and high molecular weight analytes in DESI [105]. For example, for the low molecular weight compounds such as lipids, it is believed that ionization occurs by charge transfer through either an electron or a proton exchange. There are three possibilities for the charge transfer to occur: between (1) a solvent ion and an analyte on the surface; (2) a gas-phase ion and an analyte on the surface where the solvent is evaporated before reaching the sample surface; and (3) a gas-phase ion and a gas-phase analyte molecule. The latter charge transfer occurs when a sample has a high vapor pressure. The ionization efficiency of DESI is complex and depends on several factors such as surface properties, electrospray parameters, sprayed solvent composition, and geometric settings (e.g., α, β, and d (Figure 2.9)) [105].

DESI-MS has been used to systematically evaluate the characterization of GPLs and sphingolipids [106]. The effects of surface and solvents on DESI-MS analysis of these lipids were extensively investigated [106]. In this study, a total lipid extract from porcine brain was subjected to the analyses in the positive- and negative-ion modes. The ions such as deprotonated PS, PI, and ST species dominate the spectrum in the negative-ion mode. Similar to ESI-MS, PC species are predominant in the positive-ion mode. The identities of these detected ions in the negative- and positive-ion modes were performed after CID [106].

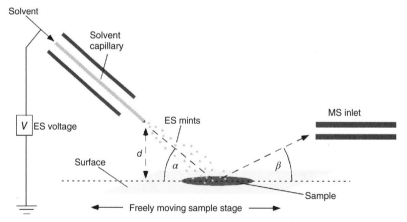

Figure 2.9   Schematic diagram of desorption electrospray ionization source. The letters "α," "β," and "d" represent the spray impact angle, desorbed ions collection angle, and the distance between spray tip and the surface, respectively.

DESI offers numerous advantages for lipid analysis including little or no sample preparation, not requiring the addition of a matrix, atmospheric pressure ionization outside the mass spectrometer, and readily ionizing many lipid species. Thus, this technology can be conveniently used for the direct analysis and imaging of those ionizable lipid species present in biological samples such as tissue [105, 107–109]. This topic is extensively described in Chapter 12.

REFERENCES

1. Murphy, R.C. (1993) Mass Spectrometry of Lipids. Plenum Press, New York. pp 290.
2. Christie, W.W. and Han, X. (2010) Lipid Analysis: Isolation, Separation, Identification and Lipidomic Analysis. The Oily Press, Bridgwater, England. pp 448.
3. Cole, R.B., ed. (2010) Electrospray and MALDI Mass Spectrometry: Fundamentals, Instrumentation, Practicalities, and Biological Applications. Wiley, Hoboken, NJ
4. Iribarne, J.V. and Thomson, B.A. (1976) On the evaporation of small ions from charged droplets. J. Chem. Phys. 64, 2287–2294.
5. Nguyen, S. and Fenn, J.B. (2007) Gas-phase ions of solute species from charged droplets of solutions. Proc. Natl. Acad. Sci. U.S.A. 104, 1111–1117.
6. Dole, M., Mack, L.L., Hines, R.L., Mobley, R.C., Ferguson, L.D. and Alice, M.B. (1968) Molecular beams of macroions. J. Chem. Phys. 49, 2240–2249.
7. Hogan, C.J., Jr., Carroll, J.A., Rohrs, H.W., Biswas, P. and Gross, M.L. (2009) Combined charged residue-field emission model of macromolecular electrospray ionization. Anal. Chem. 81, 369–377.
8. De la Mora, J.F. (2000) Electrospray ionization of large multiply charged species proceeds via Dole's charged residue mechanism. Anal. Chim. Acta 406, 93–104.
9. Ikonomou, M.G., Blades, A.T. and Kebarle, P. (1991) Electrospray-ion spray: A comparison of mechanisms and performance. Anal. Chem. 63, 1989–1998.
10. Tang, L. and Kebarle, P. (1991) Effect of the conductivity of the electrosprayed solution on the electrospray current: Factors determining analyte sensitivity in electrospray mass spectrometry. Anal. Chem. 63, 2709–2715.
11. Gaskell, S.J. (1997) Electrospray: Principles and practice. J. Mass Spectrom. 32, 677–688.
12. Han, X. and Gross, R.W. (2005) Shotgun lipidomics: Electrospray ionization mass spectrometric analysis and quantitation of the cellular lipidomes directly from crude extracts of biological samples. Mass Spectrom. Rev. 24, 367–412.
13. Han, X. and Gross, R.W. (2003) Global analyses of cellular lipidomes directly from crude extracts of biological samples by ESI mass spectrometry: a bridge to lipidomics. J. Lipid Res. 44, 1071–1079.
14. Han, X., Yang, J., Cheng, H., Ye, H. and Gross, R.W. (2004) Towards fingerprinting cellular lipidomes directly from biological samples by two-dimensional electrospray ionization mass spectrometry. Anal. Biochem. 330, 317–331.
15. Han, X. and Gross, R.W. (2005) Shotgun lipidomics: Multi-dimensional mass spectrometric analysis of cellular lipidomes. Expert Rev. Proteomics 2, 253–264.

16. Leiker, T.J., Barkley, R.M. and Murphy, R.C. (2011) Analysis of diacylglycerol molecular species in cellular lipid extracts by normal-phase LC-electrospray mass spectrometry. Int. J. Mass Spectrom. 305, 103–109.

17. Wang, M., Fang, H. and Han, X. (2012) Shotgun lipidomics analysis of 4-hydroxyalkenal species directly from lipid extracts after one-step in situ derivatization. Anal. Chem. 84, 4580–4586.

18. Wang, M., Han, R.H. and Han, X. (2013) Fatty acidomics: Global analysis of lipid species containing a carboxyl group with a charge-remote fragmentation-assisted approach. Anal. Chem. 85, 9312–9320.

19. Wang, M., Hayakawa, J., Yang, K. and Han, X. (2014) Characterization and quantification of diacylglycerol species in biological extracts after one-step derivatization: A shotgun lipidomics approach. Anal. Chem. 86, 2146–2155.

20. Han, X., Yang, K., Yang, J., Fikes, K.N., Cheng, H. and Gross, R.W. (2006) Factors influencing the electrospray intrasource separation and selective ionization of glycerophospholipids. J. Am. Soc. Mass Spectrom. 17, 264–274.

21. Gross, R.W. and Sobel, B.E. (1980) Isocratic high-performance liquid chromatography separation of phosphoglycerides and lysophosphoglycerides. J. Chromatogr. 197, 79–85.

22. DeLong, C.J., Baker, P.R.S., Samuel, M., Cui, Z. and Thomas, M.J. (2001) Molecular species composition of rat liver phospholipids by ESI-MS/MS: The effect of chromatography. J. Lipid Res. 42, 1959–1968.

23. Loo, J.A. and Robinson, C.V. (2004) Review of the 19th Asilomar conference on mass spectrometry: Bimolecular interactions: Identification and characterization of protein complexes. J. Am. Soc. Mass Spectrom. 15, 759–761.

24. Han, X. and Gross, R.W. (1994) Electrospray ionization mass spectroscopic analysis of human erythrocyte plasma membrane phospholipids. Proc. Natl. Acad. Sci. U.S.A. 91, 10635–10639.

25. Thomas, M.C., Mitchell, T.W. and Blanksby, S.J. (2005) A comparison of the gas phase acidities of phospholipid headgroups: experimental and computational studies. J. Am. Soc. Mass Spectrom. 16, 926–939.

26. Hsu, F.-F. and Turk, J. (1999) Structural characterization of triacylglycerols as lithiated adducts by electrospray ionization mass spectrometry using low-energy collisionally activated dissociation on a triple stage quadrupole instrument. J. Am. Soc. Mass Spectrom. 10, 587–599.

27. Kim, H.Y., Wang, T.C. and Ma, Y.C. (1994) Liquid chromatography/mass spectrometry of phospholipids using electrospray ionization. Anal. Chem. 66, 3977–3982.

28. Zehethofer, N. and Pinto, D.M. (2008) Recent developments in tandem mass spectrometry for lipidomic analysis. Anal. Chim. Acta 627, 62–70.

29. Han, X., Yang, K. and Gross, R.W. (2012) Multi-dimensional mass spectrometry-based shotgun lipidomics and novel strategies for lipidomic analyses. Mass Spectrom. Rev. 31, 134–178.

30. Wenk, M.R. (2010) Lipidomics: New tools and applications. Cell 143, 888–895.

31. Blanksby, S.J. and Mitchell, T.W. (2010) Advances in mass spectrometry for lipidomics. Annu. Rev. Anal. Chem. 3, 433–465.

32. Dennis, E.A. (2009) Lipidomics joins the omics evolution. Proc. Natl. Acad. Sci. U.S.A. 106, 2089–2090.

33. Stahlman, M., Ejsing, C.S., Tarasov, K., Perman, J., Boren, J. and Ekroos, K. (2009) High throughput oriented shotgun lipidomics by quadrupole time-of-flight mass spectrometry. J. Chromatogr. B 877, 2664–2672.

34. Karas, M., Bahr, U. and Dulcks, T. (2000) Nano-electrospray ionization mass spectrometry: Addressing analytical problems beyond routine. Fresenius J. Anal. Chem. 366, 669–676.

35. Gangl, E.T., Annan, M.M., Spooner, N. and Vouros, P. (2001) Reduction of signal suppression effects in ESI-MS using a nanosplitting device. Anal. Chem. 73, 5635–5644.

36. El-Faramawy, A., Siu, K.W. and Thomson, B.A. (2005) Efficiency of nano-electrospray ionization. J. Am. Soc. Mass Spectrom. 16, 1702–1707.

37. Southam, A.D., Payne, T.G., Cooper, H.J., Arvanitis, T.N. and Viant, M.R. (2007) Dynamic range and mass accuracy of wide-scan direct infusion nanoelectrospray fourier transform ion cyclotron resonance mass spectrometry-based metabolomics increased by the spectral stitching method. Anal. Chem. 79, 4595–4602.

38. Sjöberg, P.J.R., Bökman, C.F., Bylund, D. and Markides, K.E. (2001) Factors influencing the determination of analyte ion surface partitioning coefficients in electrosprayed droplets. J. Am. Soc. Mass Spectrom. 12, 1001.

39. Dreisewerd, K. (2003) The desorption process in MALDI. Chem. Rev. 103, 395–426.

40. Knochenmuss, R. (2006) Ion formation mechanisms in UV-MALDI. Analyst 131, 966–986.

41. Sun, G., Yang, K., Zhao, Z., Guan, S., Han, X. and Gross, R.W. (2008) Matrix-assisted laser desorption/ionization time-of-flight mass spectrometric analysis of cellular glycerophospholipids enabled by multiplexed solvent dependent analyte-matrix interactions. Anal. Chem. 80, 7576–7585.

42. Fuchs, B., Bischoff, A., Suss, R., Teuber, K., Schurenberg, M., Suckau, D. and Schiller, J. (2009) Phosphatidylcholines and -ethanolamines can be easily mistaken in phospholipid mixtures: A negative ion MALDI-TOF MS study with 9-aminoacridine as matrix and egg yolk as selected example. Anal. Bioanal. Chem. 395, 2479–2487.

43. Angelini, R., Babudri, F., Lobasso, S. and Corcelli, A. (2010) MALDI-TOF/MS analysis of archaebacterial lipids in lyophilized membranes dry-mixed with 9-aminoacridine. J. Lipid Res. 51, 2818–2825.

44. Cheng, H., Sun, G., Yang, K., Gross, R.W. and Han, X. (2010) Selective desorption/ionization of sulfatides by MALDI-MS facilitated using 9-aminoacridine as matrix. J. Lipid Res. 51, 1599–1609.

45. Lobasso, S., Lopalco, P., Angelini, R., Baronio, M., Fanizzi, F.P., Babudri, F. and Corcelli, A. (2010) Lipidomic analysis of porcine olfactory epithelial membranes and cilia. Lipids 45, 593–602.

46. Eibisch, M. and Schiller, J. (2011) Sphingomyelin is more sensitively detectable as a negative ion than phosphatidylcholine: A matrix-assisted laser desorption/ionization time-of-flight mass spectrometric study using 9-aminoacridine (9-AA) as matrix. Rapid Commun. Mass Spectrom. 25, 1100–1106.

47. Marsching, C., Eckhardt, M., Grone, H.J., Sandhoff, R. and Hopf, C. (2011) Imaging of complex sulfatides SM3 and SB1a in mouse kidney using MALDI-TOF/TOF mass spectrometry. Anal. Bioanal. Chem. 401, 53–64.

48. Angelini, R., Vitale, R., Patil, V.A., Cocco, T., Ludwig, B., Greenberg, M.L. and Corcelli, A. (2012) Lipidomics of intact mitochondria by MALDI-TOF/MS. J. Lipid Res. 53, 1417–1425.

49. Cerruti, C.D., Benabdellah, F., Laprevote, O., Touboul, D. and Brunelle, A. (2012) MALDI imaging and structural analysis of rat brain lipid negative ions with 9-aminoacridine matrix. Anal. Chem. 84, 2164–2171.

50. Thomas, A., Charbonneau, J.L., Fournaise, E. and Chaurand, P. (2012) Sublimation of new matrix candidates for high spatial resolution imaging mass spectrometry of lipids: Enhanced information in both positive and negative polarities after 1,5-diaminonapthalene deposition. Anal. Chem. 84, 2048–2054.

51. Laiko, V.V., Baldwin, M.A. and Burlingame, A.L. (2000) Atmospheric pressure matrix-assisted laser desorption/ionization mass spectrometry. Anal. Chem. 72, 652–657.

52. Laiko, V.V., Moyer, S.C. and Cotter, R.J. (2000) Atmospheric pressure MALDI/ion trap mass spectrometry. Anal. Chem. 72, 5239–5243.

53. Galicia, M.C., Vertes, A. and Callahan, J.H. (2002) Atmospheric pressure matrix-assisted laser desorption/ionization in transmission geometry. Anal. Chem. 74, 1891–1895.

54. He, H., Conrad, C.A., Nilsson, C.L., Ji, Y., Schaub, T.M., Marshall, A.G. and Emmett, M.R. (2007) Method for lipidomic analysis: p53 expression modulation of sulfatide, ganglioside, and phospholipid composition of U87 MG glioblastoma cells. Anal. Chem. 79, 8423–8430.

55. Fauland, A., Kofeler, H., Trotzmuller, M., Knopf, A., Hartler, J., Eberl, A., Chitraju, C., Lankmayr, E. and Spener, F. (2011) A comprehensive method for lipid profiling by liquid chromatography-ion cyclotron resonance mass spectrometry. J. Lipid Res. 52, 2314–2322.

56. Postle, A.D., Wilton, D.C., Hunt, A.N. and Attard, G.S. (2007) Probing phospholipid dynamics by electrospray ionisation mass spectrometry. Prog. Lipid Res. 46, 200–224.

57. Quehenberger, O., Armando, A.M., Brown, A.H., Milne, S.B., Myers, D.S., Merrill, A.H., Bandyopadhyay, S., Jones, K.N., Kelly, S., Shaner, R.L., Sullards, C.M., Wang, E., Murphy, R.C., Barkley, R.M., Leiker, T.J., Raetz, C.R., Guan, Z., Laird, G.M., Six, D.A., Russell, D.W., McDonald, J.G., Subramaniam, S., Fahy, E. and Dennis, E.A. (2010) Lipidomics reveals a remarkable diversity of lipids in human plasma. J. Lipid Res. 51, 3299–3305.

58. Massey, K.A. and Nicolaou, A. (2013) Lipidomics of oxidized polyunsaturated fatty acids. Free Radic. Biol. Med. 59, 45–55.

59. Bou Khalil, M., Hou, W., Zhou, H., Elisma, F., Swayne, L.A., Blanchard, A.P., Yao, Z., Bennett, S.A. and Figeys, D. (2010) Lipidomics era: Accomplishments and challenges. Mass Spectrom. Rev. 29, 877–929.

60. Hsu, F.F. and Turk, J. (2000) Structural determination of sphingomyelin by tandem mass spectrometry with electrospray ionization. J. Am. Soc. Mass Spectrom. 11, 437–449.

61. Hsu, F.F. and Turk, J. (2001) Structural determination of glycosphingolipids as lithiated adducts by electrospray ionization mass spectrometry using low-energy collisional-activated dissociation on a triple stage quadrupole instrument. J. Am. Soc. Mass Spectrom. 12, 61–79.

62. Hsu, F.F. and Turk, J. (2001) Studies on phosphatidylglycerol with triple quadrupole tandem mass spectrometry with electrospray ionization: fragmentation processes and structural characterization. J. Am. Soc. Mass Spectrom. 12, 1036–1043.

63. Hsu, F.F., Turk, J., Rhoades, E.R., Russell, D.G., Shi, Y. and Groisman, E.A. (2005) Structural characterization of cardiolipin by tandem quadrupole and multiple-stage quadrupole ion-trap mass spectrometry with electrospray ionization. J. Am. Soc. Mass Spectrom. 16, 491–504.

64. Wollnik, H. (1993) Time-of-flight mass analyzers. Mass Spectrom. Rev. 12, 89–114.

65. Ejsing, C.S., Moehring, T., Bahr, U., Duchoslav, E., Karas, M., Simons, K. and Shevchenko, A. (2006) Collision-induced dissociation pathways of yeast sphingolipids and their molecular profiling in total lipid extracts: A study by quadrupole TOF and linear ion trap-orbitrap mass spectrometry. J. Mass Spectrom. 41, 372–389.

66. Jackson, S.N., Ugarov, M., Post, J.D., Egan, T., Langlais, D., Schultz, J.A. and Woods, A.S. (2008) A study of phospholipids by ion mobility TOFMS. J. Am. Soc. Mass Spectrom. 19, 1655–1662.

67. Larsen, A., Uran, S., Jacobsen, P.B. and Skotland, T. (2001) Collision-induced dissociation of glycero phospholipids using electrospray ion-trap mass spectrometry. Rapid Commun. Mass Spectrom. 15, 2393–2398.

68. Zarrouk, W., Carrasco-Pancorbo, A., Zarrouk, M., Segura-Carretero, A. and Fernandez-Gutierrez, A. (2009) Multi-component analysis (sterols, tocopherols and triterpenic dialcohols) of the unsaponifiable fraction of vegetable oils by liquid chromatography-atmospheric pressure chemical ionization-ion trap mass spectrometry. Talanta 80, 924–934.

69. Hsu, F.F. and Turk, J. (2010) Electrospray ionization multiple-stage linear ion-trap mass spectrometry for structural elucidation of triacylglycerols: Assignment of fatty acyl groups on the glycerol backbone and location of double bonds. J. Am. Soc. Mass Spectrom. 21, 657–669.

70. Holcapek, M., Dvorakova, H., Lisa, M., Giron, A.J., Sandra, P. and Cvacka, J. (2010) Regioisomeric analysis of triacylglycerols using silver-ion liquid chromatography-atmospheric pressure chemical ionization mass spectrometry: Comparison of five different mass analyzers. J. Chromatogr. A 1217, 8186–8194.

71. Hsu, F.F., Wohlmann, J., Turk, J. and Haas, A. (2011) Structural definition of trehalose 6-monomycolates and trehalose 6,6'-dimycolates from the pathogen Rhodococcus equi by multiple-stage linear ion-trap mass spectrometry with electrospray ionization. J. Am. Soc. Mass Spectrom. 22, 2160–2170.

72. Tatituri, R.V., Brenner, M.B., Turk, J. and Hsu, F.F. (2012) Structural elucidation of diglycosyl diacylglycerol and monoglycosyl diacylglycerol from Streptococcus pneumoniae by multiple-stage linear ion-trap mass spectrometry with electrospray ionization. J. Mass Spectrom. 47, 115–123.

73. Zubarev, R.A. and Makarov, A. (2013) Orbitrap mass spectrometry. Anal. Chem. 85, 5288–5296.

74. Taguchi, R. and Ishikawa, M. (2010) Precise and global identification of phospholipid molecular species by an Orbitrap mass spectrometer and automated search engine Lipid Search. J. Chromatogr. A 1217, 4229–4239.

75. Schuhmann, K., Herzog, R., Schwudke, D., Metelmann-Strupat, W., Bornstein, S.R. and Shevchenko, A. (2011) Bottom-up shotgun lipidomics by higher energy collisional dissociation on LTQ Orbitrap mass spectrometers. Anal. Chem. 83, 5480–5487.

76. Nygren, H., Seppanen-Laakso, T., Castillo, S., Hyotylainen, T. and Oresic, M. (2011) Liquid chromatography-mass spectrometry (LC-MS)-based lipidomics for studies of body fluids and tissues. Methods Mol. Biol. 708, 247–257.

77. Shahidi-Latham, S.K., Dutta, S.M., Prieto Conaway, M.C. and Rudewicz, P.J. (2012) Evaluation of an accurate mass approach for the simultaneous detection of drug and metabolite distributions via whole-body mass spectrometric imaging. Anal. Chem. 84, 7158–7165.

78. Dubois, F., Knochenmuss, R., Zenobi, R., Brunelle, A., Deprun, C. and Beyec, Y.L. (1999) A comparison between ion-to-photon and microchannel plate detectors. Rapid Commun. Mass Spectrom. 13, 786–791.

79. Wiza, J. (1979) Microchannel plate detectors. Nuclear Instr. Methods 162, 587–601.

80. Marshall, A.G., Hendrickson, C.L. and Jackson, G.S. (1998) Fourier transform ion cyclotron resonance mass spectrometry: A primer. Mass Spectrom. Rev. 17, 1–35.

81. Brugger, B., Erben, G., Sandhoff, R., Wieland, F.T. and Lehmann, W.D. (1997) Quantitative analysis of biological membrane lipids at the low picomole level by nano-electrospray ionization tandem mass spectrometry. Proc. Natl. Acad. Sci. U.S.A. 94, 2339–2344.

82. Welti, R., Shah, J., Li, W., Li, M., Chen, J., Burke, J.J., Fauconnier, M.L., Chapman, K., Chye, M.L. and Wang, X. (2007) Plant lipidomics: Discerning biological function by profiling plant complex lipids using mass spectrometry. Front. Biosci. 12, 2494–2506.

83. Yang, K., Cheng, H., Gross, R.W. and Han, X. (2009) Automated lipid identification and quantification by multi-dimensional mass spectrometry-based shotgun lipidomics. Anal. Chem. 81, 4356–4368.

84. Sparkman, O.D. (2000) Mass Spectrometry Desk Reference. Global View Publishing, Pittsburgh, PA. pp 106.

85. Merrill, A.H., Jr., Sullards, M.C., Allegood, J.C., Kelly, S. and Wang, E. (2005) Sphingolipidomics: High-throughput, structure-specific, and quantitative analysis of sphingolipids by liquid chromatography tandem mass spectrometry. Methods 36, 207–224.

86. Bielawski, J., Szulc, Z.M., Hannun, Y.A. and Bielawska, A. (2006) Simultaneous quantitative analysis of bioactive sphingolipids by high-performance liquid chromatography-tandem mass spectrometry. Methods 39, 82–91.

87. Mesaros, C., Lee, S.H. and Blair, I.A. (2009) Targeted quantitative analysis of eicosanoid lipids in biological samples using liquid chromatography-tandem mass spectrometry. J. Chromatogr. B 877, 2736–2745.

88. Ejsing, C.S., Duchoslav, E., Sampaio, J., Simons, K., Bonner, R., Thiele, C., Ekroos, K. and Shevchenko, A. (2006) Automated identification and quantification of glycerophospholipid molecular species by multiple precursor ion scanning. Anal. Chem. 78, 6202–6214.

89. Schwudke, D., Oegema, J., Burton, L., Entchev, E., Hannich, J.T., Ejsing, C.S., Kurzchalia, T. and Shevchenko, A. (2006) Lipid profiling by multiple precursor and neutral loss scanning driven by the data-dependent acquisition. Anal. Chem. 78, 585–595.

90. Han, X. and Gross, R.W. (2001) Quantitative analysis and molecular species fingerprinting of triacylglyceride molecular species directly from lipid extracts of biological samples by electrospray ionization tandem mass spectrometry. Anal. Biochem. 295, 88–100.

91. Yang, K., Zhao, Z., Gross, R.W. and Han, X. (2009) Systematic analysis of choline-containing phospholipids using multi-dimensional mass spectrometry-based shotgun lipidomics. J. Chromatogr. B 877, 2924–2936.

92. Su, X., Han, X., Mancuso, D.J., Abendschein, D.R. and Gross, R.W. (2005) Accumulation of long-chain acylcarnitine and 3-hydroxy acylcarnitine molecular species in diabetic

myocardium: Identification of alterations in mitochondrial fatty acid processing in diabetic myocardium by shotgun lipidomics. Biochemistry 44, 5234–5245.

93. Han, X., Yang, K., Cheng, H., Fikes, K.N. and Gross, R.W. (2005) Shotgun lipidomics of phosphoethanolamine-containing lipids in biological samples after one-step in situ derivatization. J. Lipid Res. 46, 1548–1560.

94. Almeida, R., Pauling, J.K., Sokol, E., Hannibal-Bach, H.K. and Ejsing, C.S. (2015) Comprehensive lipidome analysis by shotgun lipidomics on a hybrid quadrupole-orbitrap-linear ion trap mass spectrometer. J. Am. Soc. Mass Spectrom. 26, 133–148.

95. Stach, J. and Baumbach, J.I. (2002) Ion mobility spectrometry – Basic elements and applications. Int. J. Ion Mobility Spectrom. 5, 1–21.

96. Mclean, J.A., Schultz, J.A. and Woods, A.S. (2010) Ion mobility-mass spectrometry. In Electrospray and MALDI Mass Spectrometry: Fundamentals, Instrumentation, Practicalities, and Biological Applications. (Cole, R. B., ed.) pp. 411–439, John Wiley & Sons, Inc., Hoboken, NJ

97. Kanu, A.B., Dwivedi, P., Tam, M., Matz, L. and Hill, H.H., Jr. (2008) Ion mobility-mass spectrometry. J. Mass Spectrom. 43, 1–22.

98. Howdle, M.D., Eckers, C., Laures, A.M. and Creaser, C.S. (2009) The use of shift reagents in ion mobility-mass spectrometry: Studies on the complexation of an active pharmaceutical ingredient with polyethylene glycol excipients. J. Am. Soc. Mass Spectrom. 20, 1–9.

99. Kliman, M., May, J.C. and McLean, J.A. (2011) Lipid analysis and lipidomics by structurally selective ion mobility-mass spectrometry. Biochim. Biophys. Acta 1811, 935–945.

100. Woods, A.S., Ugarov, M., Egan, T., Koomen, J., Gillig, K.J., Fuhrer, K., Gonin, M. and Schultz, J.A. (2004) Lipid/peptide/nucleotide separation with MALDI-ion mobility-TOF MS. Anal. Chem. 76, 2187–2195.

101. Jackson, S.N. and Woods, A.S. (2009) Direct profiling of tissue lipids by MALDI-TOFMS. J. Chromatogr. B 877, 2822–2829.

102. Sowell, R.A., Koeniger, S.L., Valentine, S.J., Moon, M.H. and Clemmer, D.E. (2004) Nanoflow LC/IMS-MS and LC/IMS-CID/MS of protein mixtures. J. Am. Soc. Mass Spectrom. 15, 1341–1353.

103. Paglia, G., Kliman, M., Claude, E., Geromanos, S. and Astarita, G. (2015) Applications of ion-mobility mass spectrometry for lipid analysis. Anal. Bioanal. Chem. 407, 4995–5007.

104. Takats, Z., Wiseman, J.M., Gologan, B. and Cooks, R.G. (2004) Mass spectrometry sampling under ambient conditions with desorption electrospray ionization. Science 306, 471–473.

105. Takats, Z., Wiseman, J.M. and Cooks, R.G. (2005) Ambient mass spectrometry using desorption electrospray ionization (DESI): Instrumentation, mechanisms and applications in forensics, chemistry, and biology. J. Mass Spectrom. 40, 1261–1275.

106. Manicke, N.E., Wiseman, J.M., Ifa, D.R. and Cooks, R.G. (2008) Desorption electrospray ionization (DESI) mass spectrometry and tandem mass spectrometry (MS/MS) of phospholipids and sphingolipids: Ionization, adduct formation, and fragmentation. J. Am. Soc. Mass Spectrom. 19, 531–543.

107. Dill, A.L., Ifa, D.R., Manicke, N.E., Ouyang, Z. and Cooks, R.G. (2009) Mass spectrometric imaging of lipids using desorption electrospray ionization. J. Chromatogr. B 877, 2883–2889.

108. Lanekoff, I., Heath, B.S., Liyu, A., Thomas, M., Carson, J.P. and Laskin, J. (2012) Automated platform for high-resolution tissue imaging using nanospray desorption electrospray ionization mass spectrometry. Anal. Chem. 84, 8351–8356.

109. Lanekoff, I., Burnum-Johnson, K., Thomas, M., Short, J., Carson, J.P., Cha, J., Dey, S.K., Yang, P., Prieto Conaway, M.C. and Laskin, J. (2013) High-speed tandem mass spectrometric in situ imaging by nanospray desorption electrospray ionization mass spectrometry. Anal. Chem. 85, 9596–9603.

3

MASS SPECTROMETRY-BASED LIPIDOMICS APPROACHES

3.1 INTRODUCTION

In this chapter, the lipidomics approaches developed based on the electrospray ionization (ESI) and matrix-assisted laser desorption/ionization (MALDI) technologies (see Chapter 2) are described. Although matrix-assisted laser desorption/ionization mass spectrometry (MALDI-MS) has played many important roles in lipidomics and is also discussed in the chapter, it is no doubt that the majority of the platforms for lipidomics analysis currently used are based on ESI-MS in conjunction with MS/MS analyses.

Ions are generated in an electrospray ion source after a stream of lipid solution is pushed into the ion source chamber by a mechanical force (see Chapter 2). Depending on whether lipid analysis conducted with the lipid solution delivered to the ion source chamber is under a constant lipid concentration condition, these ESI-MS-based lipidomics approaches can be classified into two major categories: (1) direct infusion in which the concentration of lipid solution is constant and (2) high-performance liquid chromatography (HPLC) based on which the concentration of lipid solution delivered to the ion source is constantly changing. These platforms are termed "shotgun lipidomics" and "LC-MS-based lipidomics," respectively, in the literature. The approaches in these two categories are described in detail.

Lipidomics: Comprehensive Mass Spectrometry of Lipids, First Edition. Xianlin Han.
© 2016 John Wiley & Sons, Inc. Published 2016 by John Wiley & Sons, Inc.

3.2   SHOTGUN LIPIDOMICS: DIRECT INFUSION-BASED APPROACHES

3.2.1   Devices for Direct Infusion

The most basic device to accomplish direct infusion is a syringe pump. This device is relatively low cost and can be constructed to deliver a few microliters of solution per minute. Generally, the higher the flow rate delivered, the more stable flow can be achieved. A tightly sealed, high-quality glass syringe is always recommended for this purpose. Its major weakness is that automation of lipid analysis is difficult to achieve with this delivery system. Moreover, clogging of the delivery capillary line occurs frequently even if the samples are prepared carefully. The consumption of samples is also relatively large for the maintenance of a high flow rate as mentioned earlier.

Development of chip-based devices revolutionizes the lipid analysis by ESI-MS after direct infusion, leading to a high throughput and automated manner. For example, the NanoMate device (a silicon-based integrated nanoelectrospray microchip device) from Advion BioSciences utilizes an ESI chip consisting of an array of 400 nanoelectrospray emitters with consistent nozzles allowing for the delivery of a flow rate of ~100 nL/min. These devices not only automate sample injection and reduce substantially any sample clogging but also dramatically reduce the sample size and cross-contamination [1, 2]. Infusing 5–10 µL of sample solution is sufficient for nearly an hour of analysis with a stable spray, thereby guaranteeing high reproducibility, sequential runs of a series of requested mass spectra, and accurate quantitation [1, 2]. Such a stable ion current can be readily obtained for lipid analysis by using a solvent mixture of chloroform–methanol–isopropanol in an approximate volume ratio of 1:2:4 [1]. A D-chip (nozzle size 4.1 µm diameter) is best used with these solvents [1, 2]. The spray voltage is commonly set between ±1.2 and 1.4 kV, using 0.2–0.4 psi back-pressure when infusing total lipid extracts at an approximate concentration of 0.05 µg total protein per microliter in chloroform–methanol (1:2, v/v) containing 5 mM ammonium acetate [2]. The stated parameters including solvent composition and NanoMate settings can always be varied for different instruments and different applications. However, the aforementioned parameters can be used as a starting point to optimize an operational condition to achieve maximal signals for specific applications. A major drawback of utilizing chip-based devices is the relatively high cost. Solvent evaporation during long automated analysis could also be a concern of handling small volumes of lipid samples in these systems [2]. Inclusion of less volatile solvents such as isopropanol has proved to be helpful in improving solvent preservation, lipid solubility, and ionization efficiency [3]. Sealing of the sample plates with thin aluminum foil has also been found useful to minimize solvent loss. Moreover, storage of the sample plate at a low temperature (e.g., 4–10 °C) through a cooling device carried out by the NanoMate system is another choice of operation. With these measures, sealed 96-well plates can be stored at −20 °C for up to 4 weeks without noticeable solvent evaporation and alterations in the measured lipid composition [2]. It is advised that only the solvent-resistant plastic tips should be used for loading samples to the plate. Glass-coated plates or tips are a safer choice if cost is not an issue.

It should be recognized that loop injection (delivering sample solution with an LC system, but without an LC column) is also used in lipidomics and carries some features of direct infusion [4]. Since this sample delivery method is unable to continuously maintain a constant lipid concentration as solvent is pushed through the sample loop, this method is excluded in the category of shotgun lipidomics. In contrast, direct infusion of individual fractions collected after LC separation (including those from a solid-phase extraction (SPE) column) falls under shotgun lipidomics due to its maintenance of a constant concentration condition.

3.2.2 Features of Shotgun Lipidomics

Direct infusion was originally used to efficiently deliver lipid samples and to avoid difficulties from alterations in concentration, chromatographic anomalies, and ion-pairing alterations in the early 1990 [5–8]. Around 2004, the platforms developed based on direct infusion have been separately named as "shotgun lipidomics" by Han and Gross [9] and Ejsing et al. [10]. Since then, this technology has become one of the widely used approaches in lipidomics, particularly for high-throughput analysis of lipids [2, 9, 11–14]. The principles of shotgun lipidomics are to maximally exploit the unique chemical and physical properties of each lipid class and individual molecular species to facilitate the high-throughput analysis of a cellular lipidome directly from organic extracts of biological samples on a large scale [9].

These principles of shotgun lipidomics can only be achieved in conjunction with the major feature of direct infusion, that is, ESI-MS analysis of lipids is conducted at a constant concentration of the solution. This feature in shotgun lipidomics provides many advantages for lipid analysis, particularly for the quantification of individual lipid species. Some of these advantages are as follows. First, constant interactions between lipid species are maintained under a constant concentration condition; therefore, contribution of individual lipid species to the ion current in an ESI source is constant, thereby leading to a constant ratio of ion peak intensities between lipid species of a class. Such a constant ratio can be achieved under different experimental conditions (see Chapter 4), on different MS instruments, and in different laboratories. Second, also due to the constant interactions between lipid species under the condition, ion suppression between each other within a lipid class or between lipid classes is constant. Third, lipid aggregation, which is a big concern for lipid quantification, can be well controlled and minimized.

The unique feature of shotgun lipidomics (i.e., MS analysis is performed under a constant lipid concentration) allows us to have virtually unlimited time to improve mass spectral signal/noise (S/N) ratio, to perform detailed MS/MS mapping with multiple fragmentation techniques (i.e., product-ion analysis, precursor-ion scan (PIS), and neutral-loss scan (NLS)), and to conduct multistage MS/MS analyses. Ramping different instrumental variables (e.g., fragmentation energies, collision gas pressure, reagent gases, ion-mobility parameters) (see Chapter 4 for details) can all be performed during the same infusion at a constant ratio of solvents to analytes that obviates difficulties typically encountered from the time constraints present in the "on-the-fly" analysis during chromatographic elution.

Another major feature of shotgun lipidomics is that the molecular ions of all the species of a lipid class of interest can be displayed in a single full mass spectrum. This feature of shotgun lipidomics makes these molecular species easily visualized, as well as quantified, by direct comparison with their selected internal standard(s). Owing to the minimal source fragmentation and the selective ionization that is largely determined by the charge property of the polar head group, the response factors for the species of a polar lipid class are essentially identical under certain experimental conditions that are discussed in details in Chapters 4, 14, and 15. Hence, it is feasible to quantitate individual molecular species of a polar lipid class through direct comparison of ion peak intensities with that of a selected internal standard in the identical full mass spectrum. This feature also leads to the analysis of molecular species different from those using the liquid chromatography mass spectrometry (LC-MS) approach. Specifically, in shotgun lipidomics, PIS of the particular fragment ions and/or NLS of the neutrally lost fragments of interest (see Chapter 2) can be acquired directly to analyze the individual species of a lipid class or a category of lipid classes due to the appearance of these lipid species in the same mass spectrum.

3.2.3   Shotgun Lipidomics Approaches

Based on these unique features, at least three different approaches of shotgun lipidomics are developed and well documented in the literature, including tandem MS-based shotgun lipidomics, high mass accuracy-based shotgun lipidomics, and multidimensional MS-based shotgun lipidomics.

3.2.3.1   Tandem Mass Spectrometry-Based Shotgun Lipidomics   A characteristic fragment associated with the head group of a lipid class (i.e., one of the building blocks (see above)) is generally present. NLS or PIS of this fragment specifically detects individual species of the class. Following this line of reasoning, researchers have developed a shotgun lipidomics method to "isolate" the individual species of a class of interest through the specific NLS or PIS to the lipid class [15]. After the double filtering process of MS/MS, the mass spectral *S/N* ratio can be greatly enhanced (typically with over an order of magnitude). A workflow of this approach has been illustrated [16]. The characteristic fragments, which can be used for profiling individual lipid classes, particularly those in plant lipidomics, have been comprehensively tabulated in a protocol provided by Welti and colleagues [17].

3.2.3.2   High Mass Accuracy-Based Shotgun Lipidomics   Currently, commercially available hybrid instruments (e.g., quadrupole-time-of-flight (Q-TOF) or Q-Exactive (i.e., quadrupole-Orbitrap) mass spectrometers) offer an improved duty cycle that increases the detection sensitivity and/or high mass resolution/accuracy [18, 19]. These instruments can thus be employed to quickly perform product-ion

MS analysis of a small mass window (e.g., one or a couple of mass units) step by step to map all the fragments or fragments in the entire mass region of interest [3, 20–22]. The high mass resolution and accuracy inherent in these instruments record the accurate mass of fragment ions (0.1 amu or higher) to minimize any false positive identification. Identification and quantification of individual lipid species could be directly conducted from the product-ion mass spectra, for which a name of data-dependent acquisition shotgun lipidomics has been given [3, 22]. Alternatively, any interesting PIS and/or NLS can be extracted from the acquired data array of the product-ion mass spectra. Terms such as multi-PIS high mass accuracy shotgun lipidomics, "top-down lipidomics," or "bottom-up shotgun lipidomics" have been given to this approach [21, 23, 24]. The analyses can be conducted in the positive-ion and negative-ion modes in the presence of ammonium acetate in the infused solution [25]. In multi-PIS high mass accuracy shotgun lipidomics, identification can be performed from bioinformatic reconstruction of the fragments from PIS or NLS. Software packages (e.g., LipidProfiler [21] and LipidInspector [22]) are reported for data processing. Quantification can be achieved with a comparison of the sum of the intensities of extracted fragments of an ion to that of a pre-selected internal standard. In the case of high mass resolution/accuracy-based shotgun lipidomics, software packages such as LipidXplorer [26] and ALEX [27] have been developed separately to process full mass spectral and product-ion analysis data. A schematic workflow of the approach for lipid analysis has been provided by Ekroos and colleagues [2].

3.2.3.3 *Multidimensional MS-Based Shotgun Lipidomics*   The third well-recognized shotgun lipidomics platform in the current practice of lipidomics is the multidimensional MS (MDMS)-based shotgun lipidomics [11, 28–30]. This technology maximally exploits the unique chemistries inherent in discrete lipid classes or subclasses for analysis of lipids, including low-abundance molecular species. In this approach, differential hydrophobicity, stability, and reactivity of different lipid classes and subclasses are exploited during sample preparation (a multiplexed extraction approach) [31]. For example, the presence of a primary amine in phosphoethanolamine-containing species is unique in the cellular lipidome and has been exploited to tag the phosphoethanolamine-containing lipid species with fluorenylmethoxylcarbonyl (Fmoc) chloride [32]. The facile loss of Fmoc from the tagged lipid species allows one to readily identify and quantify those phosphoethanolamine-containing species with unprecedented sensitivity at the level of amol/μL.

The differential charge properties of different lipid classes (which are predominant with the head groups of polar lipid classes) are exploited to selectively ionize a certain category of lipid classes under multiplexed experimental conditions to separate many lipid classes in the ion source (i.e., intrasource separation) [33]. This separation method is analogous to the electrophoretic separation of different compounds that possess different pI values [33] (see Chapter 2 for details).

The concept of building blocks in lipid structure (see above) is fully employed for the identification of individual lipid species in MDMS [11, 34] because these building blocks can be determined with two powerful tandem MS techniques (i.e., NLS and PIS) in a way ramping the neutrally-lost masses and fragment ions, respectively (see the following) [9, 11]. Accordingly, after collision-induced dissociation (CID), the majority of lipid classes possess a unique fragment pattern that can usually be predicted based on the covalent structures of these lipid classes [35, 36]. The informative fragment ion(s) from either the head group or resulted from the neutral loss of the head group are used to identify the lipid class of interest, and PIS or NLS of fatty acyl (FA) chains is used to identify the individual molecular species present within the class.

It is now well known that identification of all individual molecular ions displayed in a full mass spectrum can be achieved with a 2D mass map of this full mass spectrum through product-ion MS analysis of individual molecular ions unit by unit. Alternatively, NLS or PIS as mass or m/z varied unit by unit within the entire mass range of interest can identically be used to map the complete and informative fragments for each individual molecular species since these tandem MS techniques are interwoven (see Chapter 2). In those 2D mass maps, the cross peaks of a given primary molecule ion in the full mass spectrum (i.e., the first dimension) with the second dimension represent the fragments of this given molecule ion. Analysis of these cross peaks (i.e., the individual fragments) determines the structure of a given molecule, as well as its isomers and isobaric species if present [11]. Han and Gross have referred to these kinds of two-dimensional maps as two-dimensional MS [11, 28, 29] because they are entirely analogous to two-dimensional NMR spectroscopy. The only difference between these mapping approaches is that the former is in the mass domain, whereas the latter is in the frequency domain.

Mapping the entire mass ranges of interest by NLS or PIS is time consuming. An alternative approach to solve this issue is to selectively map the building blocks of a class or a category of lipid classes in the mass range of interest. The building blocks as discussed in Chapter 1 can be represented with the fragments characteristic of individual lipid classes and detected by using NLS and PIS as described earlier. Specifically, the building blocks can be monitored with either the specific loss of a neutral fragment in the NLS mode or the yield of a fragment ion in the PIS mode. Mapping of these building blocks yields a two-dimensional mass spectrum, the vertical dimension of which might have a discrete mass or m/z unit instead of a continuous unit to efficiently monitor only naturally occurring building blocks. Accordingly, after the fragmentation pattern of a lipid class is characterized (see Part II), numerous individual molecular species of a lipid class can be determined with 2D MS analysis.

For example, if the interest is to identify the anionic glycerophospholipid species present in lipid extracts of rat myocardium, we have to definitively identify the underneath lipid species of all ions displayed in the full mass spectrum in the mass region between m/z 550 and 1000 acquired from diluted lipid extract solution in the negative-ion mode. From characterization of anionic glycerophospholipids, we learned that the head group building blocks are the neutral loss of serine (87 Da) from PS, the precursor ion of inositol phosphate (m/z 241) from PI, and the precursor ion of glycerophosphate (m/z 153) from all anionic GPL species in addition to the fatty acyl (FA) carboxylates from all fatty acyl chains. Detection of all of these building blocks constitutes a 2D mass spectrum (Figure 3.1). One dimension (x-axis) is the mass of molecular ions and the other dimension (y-axis) is the building blocks.

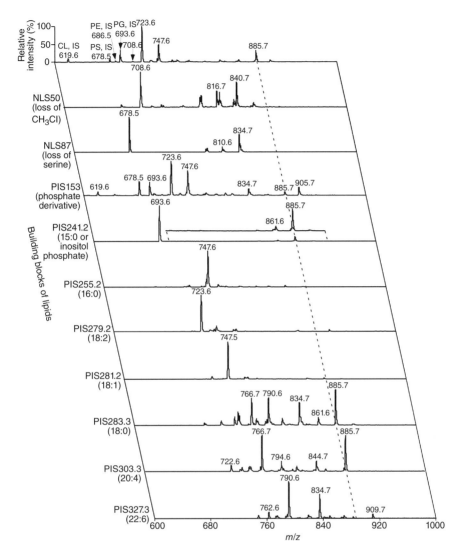

Figure 3.1 An example of a 2D electrospray ionization mass spectrum of lipid extract of rat myocardium acquired in the negative-ion mode. A conventional ESI mass spectrum was acquired in the negative-ion mode directly from a diluted myocardial lipid extract prior to the analysis of lipid building blocks in the second dimension by PIS and NLS as indicated. IS denotes internal standard; $m:n$ indicates a fatty acyl chain containing m carbon atoms and n double bonds. All mass spectral scans were displayed after normalization to the base peak in each individual spectrum.

The crossing peaks of a given primary molecular ion in the first dimension with the second dimension determine the building blocks of this given molecular ion. For example, the broken line in Figure 3.1 indicates that the molecular ion at m/z 885.6 only crosses with the building blocks of PIS241 (inositol phosphate), PIS153 (glycerophosphate), PIS283 (18:0 FA), and PIS303 (20:4 FA) with a >1 ratio of these fatty acyl fragments. Hence, this set of information identifies the molecular ion at m/z 885.7 as 18:0–20:4 PI. It can be readily recognized that the analysis of a handful of building blocks is much more effective than product-ion analysis of individual ion in this mass region.

Table 3.1 lists some examples of the PIS and/or NLS for the analyses of the building blocks of each lipid class with MDMS. Many lipid classes have been characterized with different adducts and/or ionized in the different ion modes under different experimental conditions (see Part II). Therefore, identification of a particular lipid class might be achieved with different characteristic fragment ions or informative neutral losses for the analysis of building blocks resulted from these different adducted molecules and/or ionized in the different ion modes under different experimental conditions. It should be recognized that the use of multiple complementary fragmentation modes could be employed to minimize false discovery rates for abundant species and enhance coverage for extremely low-abundance molecular species.

In theory, to fully investigate the effects of ionization conditions on ionization efficiency and/or the effects of collision conditions on fragmentation processes or other effects, a variety of ionization voltages, ionization temperatures, collision energies, collision gas pressures, etc., should be employed in an experiment (see Chapter 4). These variables can all be logically varied unit by unit within a certain range. Therefore, from a new set of spectra when an individual variable of MS is ramped, a new dimension is added to the basic 2D MS. All of these dimensions form the family of MDMS [11]. Specifically, MDMS is defined as the comprehensive MS analyses conducted under a variety of instrumental variables that collectively comprise an MDMS spectrum.

At the current stage, MDMS is decomposed into multiple 2D MS for ease of use and displayed by varying only one variable at a time while keeping the others fixed under experimental conditions. It is anticipated that advanced computational technology can eventually facilitate the direct use of 3D MS or MDMS and provide a new level of information directly obtainable from the MS analysis in the next generation of computational MDMS-based shotgun lipidomics.

A workflow and the detail protocol of this technology can be found in the literature [30, 48]. At its current stage of development, this platform can identify and quantify thousands of individual lipid species (including many regioisomers) of over 40 lipid classes in cellular lipidomes [30, 49], which represents >95% of the total lipid mass levels of a cellular lipidome, directly from solvent extracts of biological materials from a limited amount of biological source materials (e.g., 10–50 mg of tissue, a million of cells, 100 μL body fluids) in an automated, unbiased, and relatively high-throughput manner [1, 30].

TABLE 3.1 Summary of the Building Blocks in Each Lipid Class Used to Identify Individual Molecular Species[a]

Lipid Class [References]	Ion Format	Scans for Class-Specific Prescreen	Scans for Identification of Acyl Chain and/or Regioisomers	Preliminary Scans for the Second-Step Quantitation
PC [37]	[M+Li]+	NLS189.1, −35 eV	NLS(59.0+FA), −40 eV	NLS183.1, −35 eV for polyunsaturated acyl chain containing species NLS59.0, −24 eV for plasmalogen species NLS189.1, −35 eV for all the other species
lysoPC [37]	[M+Na]+	NLS59.0, −22 eV NLS205.0, −34 eV	PIS104.1, −34 eV PIS147.1, −34 eV	NLS59.0, −22 eV NLS205.0, −34 eV
PE, lysoPE [32]	[M−H]− [M−H+Fmoc]− ([M+C₁₅H₉O₂]−)	PIS196.1, 50 eV for [M−H]− NLS222.2, 30 eV	PIS(FA−H), 30 eV	NLS222.2, 30 eV for [M−H+Fmoc]−
PI, lysoPI [29]	[M−H]−	PIS241.1, 45 eV	PIS(FA−H), 47 eV	PIS241.1, 45 eV
PS, lysoPS [29]	[M−H]−	NLS87.1, 24 eV	PIS(FA−H), 30 eV	NLS87.1, 24 eV
PG, PA, lysoPG, lysoPA [29]	[M−H]−	PIS153.1, 35 eV	PIS(FA−H), 30 eV	PIS153.1, 35 eV
CL, mono-lysoCL [38]	[M−2H]2−	Full MS at high resolution	PIS(FA−H) at high resolution, 25 eV; NLS(FA−H₂O) at high resolution, 22 eV	
TAG [28]	[M+Li]+		NLS(FA), −35 eV	
Sphingomyelin (SM) [37]	[M+Li]+	NLS213.2, −50 eV	NLS(neutral fragments from sphingoid backbone)	NLS213.2, −50 eV

(continued)

TABLE 3.1 (*Continued*)

Lipid Class [References]	Ion Format	Scans for Class-Specific Prescreen	Scans for Identification of Acyl Chain and/or Regioisomers	Preliminary Scans for the Second-Step Quantitation
Ceramide (Cer) [39]	[M−H]⁻	NLS(neutral fragments from sphingoid backbone) (e.g., NLS256.2, 32 eV for *d*18:1 nonhydroxyl species)	NLS(neutral fragments from sphingoid backbone) (e.g., NLS256.2, 32 eV for *d*18:1 nonhydroxyl species)	NLS(neutral fragments from sphingoid backbone) (e.g., NLS256.2, 32 eV for *d*18:1 nonhydroxyl species)
Hexosyl ceramide (HexCer) [40, 41]	[M+Li]⁺	NLS162.2, −50 eV	NLS(neutral fragments from sphingoid backbone)	NLS162.2, −50 eV
Sulfatide (ST) [42]	[M−H]⁻	PIS 97.1, 65 eV	NLS(neutral fragments from sphingoid backbone)	PIS97.1, 65 eV
Sphingoid base-1-phosphate (S1P) [43]	[M−H]⁻	PIS79.1, 24 eV		PIS79.1, 24 eV
Sphingoid base [31]	[M+H]⁺	NLS48.0, −18 eV		NLS48.0, −18 eV
Psychosine [44]	[M+H]⁺	NLS180.0, −24 eV		NLS180.0, −24 eV
Cholesterol [45]	[Cholesteryl methoxyacetate +MeOH+Li]⁺	PIS97.1, −22 eV		PIS97.1, −22 eV
Acyl carnitine [46]	[M+H]⁺	PIS85.1, −30 eV	PIS85.1, −30 eV for all species; PIS145.1, −30 eV for hydroxyl species	PIS85.1, −30 eV
Acyl CoA [47]	[M−H]⁻, [M−2H]²⁻ [M−3H]³⁻	PIS134.0, 30 eV	PIS134.0, 30 eV	PIS134.0, 30 eV

[a]FA and (FA−H) stand for free fatty acid and fatty acyl carboxylate anion, respectively. The abbreviations of glycerophospholipid classes are given in Figure 1.1. Yang et al. [30]. Reproduced with permission of the American Chemical Society.

3.2.4 Advantages and Drawbacks

3.2.4.1 Tandem Mass Spectrometry-Based Shotgun Lipidomics The advantages of this method include simplicity, efficiency, high sensitivity, ease of management, and less expensive instrumental requirements. All individual species in a particular class can be detected in one MS/MS acquisition directly from a total lipid extract with any commercially available triple-quadrupole (i.e., QqQ)-type mass spectrometer. This shotgun lipidomics approach provides global determination of all the species of any targeted class. Because of its great advantages, many laboratories have adopted this platform for lipidomics analysis. The applications by using this approach are provided in the later chapters of this book.

Several concerns are associated with this approach, which are also well recognized. Some of these concerns include the following:

- The fatty acyl substituents of lipid species are not identified.
- The detection with the so-called specific MS/MS scanning might not be entirely specific to the class or the category of classes of interest, whereas this nonspecificity might introduce some artifacts.
- Some altered ionization conditions cannot be easily recognized during and after the experiments.
- Accurate quantification of the detected lipid species might not be as simple as expected because of the differential fragmentation mechanisms manifest in individual lipid species within each lipid class.

Since fragmentation depends on the chemical and physical properties of individual species of a class (including the fatty acyl chain length, the number of double bonds, the location of double bond(s)), at least two internal standards that can cover the variation of the individual species of the class should be employed to achieve relatively accurate measurements of their mass levels as originally demonstrated [15].

3.2.4.2 High Mass Accuracy-Based Shotgun Lipidomics By means of the specified mass spectrometers, this shotgun lipidomics approach provides efficient, broad, and sensitive measurement of lipid species. This approach could be conducted in an untargeted manner to analyze any lipid species present in a cellular lipidome if the dynamic range of the instrument is permitted and the software package is able to cover all those species. This technology has recently been applied to many biological studies [19, 25, 50–53].

The following points should be considered in an experimental design by using this approach:

- Since this approach is essentially a method based on tandem MS techniques, multiple (at least two) internal standards for each lipid class should be included as described in the last section.
- Differential ionization responses of different species among a nonpolar lipid class are well known. Correction for these differential ionization responses in quantification of these species should be considered.

- Linear dynamic range of quantification largely depends on the used instrument for the analysis of fragment ions since detection of the low-abundance fragment ion(s) is affected by the presence of high abundant fragment ion(s) due to the isomeric overlapping, or the presence of multiple species within a selected mass window for fragmentation.

- The effects of isotopologues with the selected mass window for fragmentation should also be considered.

3.2.4.3 Multidimensional Mass Spectrometry-Based Shotgun Lipidomics

MDMS-based shotgun lipidomics overcomes the majority of the limitations of other shotgun lipidomics approaches and possesses many obvious advantages as follows:

- Directly acquire high-density information about the identity and mass levels of individual lipid species directly from biological extracts, in which the mass spectrometer is used as a separation device in addition to being an analyzer, thereby avoiding the need for chromatography.

- In comparison to LC-MS approaches, a large increase in S/N through averaging signals from the unlimited time frame is achieved.

- Do not require any previous knowledge of the species of a lipid class present in the biological extracts, which are identified through *in situ* analysis of their building blocks by using PIS and/or NLS under various MS conditions (i.e., MDMS).

- Use the peak contours in multidimensional space that facilitates refinements in quantitation through two-step quantification approach (see Chapter 15) in combination with corrections for mass offset and isotopologue analysis in a bioinformatics manner.

- Ideally exploit the distinctive chemical characteristics of many lipid classes. Prominent examples include the use of the $[M-2H+1]^{2-}$ isotopologue approach for CL analyses [38], the use of Fmoc derivatization for the analysis of phosphoethanolamine-containing species [32], the use of specifically deuterated amine selective reagents for dynamic lipidomics through PIS analysis of the particular reagents [54], the use of alkaline hydrolysis to greatly enhance penetrance into the sphingolipidome [31], and the use of charge-remote fragmentation for fatty acidomics [49], among others.

Limitations in MDMS-based shotgun lipidomics along with other shotgun lipidomics approaches include the following:

- Although the MDMS-based shotgun lipidomics takes advantage of the four orders of magnitude in the linear dynamic range inherent in mass spectrometry, enrichment approaches are necessary for the examination of extremely low-abundance lipids that shotgun lipidomics is unable to access with the sensitivities of currently available mass spectrometers due to so-called ion suppression. To extend this point, in general, shotgun lipidomics is not ideal for the analysis of poorly ionized lipids in low abundance. However, these classes of lipids can be accessed after derivatization in MDMS-based shotgun lipidomics. It should also be pointed out that MDMS-based shotgun lipidomics could be performed for any fractionated samples. Therefore, chromatographic separation, liquid–liquid partitioning, SPE, and other enrichment methods can be used to greatly extend the limits of detection for extremely low-abundance molecular species.

- It is unable to distinguish isomeric species when their fragmentation patterns are identical.
- Although MDMS-based shotgun lipidomics identifies and quantifies all individual species of a characterized lipid class in an unbiased manner within the limits of instrumentation sensitivities, the approach is not ideal for identification and quantitation of species of an unknown or uncharacterized lipid class since identification of the building blocks of a lipid class has to be pre-determined. Its throughput is relatively lower compared to the other two shotgun lipidomics approaches.

3.3 LC-MS-BASED APPROACHES

3.3.1 General

The LC-MS approach is currently the most popular approach for lipid analysis [55–57], although it may not be the best choice for global lipid analysis (i.e., lipidomics). The rationale behind the LC-MS approaches is to maximally exploit the LC separation technology with the most sensitive detection power of MS currently available. Three major factors are generally considered for successful development of LC-MS methodology in lipidomics, as well as understanding the principles of those developments.

The *first factor* is to select a suitable column and optimize the separation conditions (e.g., mobile phase(s) and their gradient(s), see below) to achieve the best separation of lipid classes, lipid molecular species, or both. All types of columns including normal-phase, reversed-phase, hydrophilic interaction (HILIC), ion-exchange, affinity, etc., or even multidimensional LC have been employed for this purpose [58]. In general, normal-phase HPLC is selected to resolve individual lipid classes and reversed-phase HPLC is selected for the separation of individual species of a particular class. However, with recent advances in HPLC technology (e.g., ultra-performance LC (UPLC)) including the improvement of materials, particle sizes, and packing skills, separation of individual lipid classes and/or molecular species becomes more and more effective, and employing only one-column step may be achievable. For example, many researchers have used UPLC technology for lipidomics once it appears to replace the sequential separation with both normal-phase and reversed-phase HPLC [27, 59, 60]. The description of the types of chromatography with these columns for separation of lipids can be found in a classic book of lipid analysis written by Christie and Han [61]. The principles and representative applications of these chromatographic techniques are described in the next section.

The *second factor* is to consider the appropriate coupling of the LC elution conditions with a mass spectrometer. For example, the ion strength in the mobile phase cannot be too high when HPLC is coupled to a mass spectrometer since the effects of ion suppression (Chapter 15) on ionization efficiency become severe as the concentration of ions, particularly inorganic ions, increases. Under certain experimental conditions, introduction of ion(s) in the mobile phase becomes essential, e.g., introducing ionization modifiers in the case of normal-phase LC to facilitate adduct formation or running ion gradients to enhance separation with a reversed-phase column. Employing a low ion strength and/or volatile acid (e.g., formic or acetic acid), base (e.g., ammonium hydroxide, trimethylamine, piperidine), or salt (e.g., ammonium acetate) is always preferable to a nonvolatile compound. The flow rate of LC complying with MS analysis is also important. Many of these factors associated with LC and MS coupling have been discussed in the book entitled *Liquid Chromatography – Mass Spectrometry* written by Dr Niessen, which could be consulted for those interested [62].

The *third factor* is to set up the MS (or MS/MS) parameters to identify and quantify the eluted individual lipid species as many as possible. The particular features in LC-MS analysis are that the lipid concentrations in eluents are constantly changing, and identification and quantification of lipid species have to be done in a very limited time frame. These features are in contrast to shotgun lipidomics; therefore, totally different settings and methodologies from those in shotgun lipidomics have to be employed.

There are three different, but commonly used, methods in the analysis of lipids by LC-MS related to the MS settings as follows:

3.3.1.1 Selected Ion Monitoring for LC-MS   Global lipid analysis by LC-MS could be conducted by using selected ion monitoring (SIM) in which any ion of interest could be extracted from the total ion chromatograph. This approach determines the entire body of lipid species that have identical molecular weight. Specifically, a mass spectrum is continuously acquired during column elution, and ions of interest are extracted from the acquired data array after a chromatographic separation. The combination of ESI-MS detection with HPLC separation and the sensitivity of SIM compared to other detection modalities makes this approach a judicious choice for lipid profiling and quantitation in many cases where extremely low-abundance lipids are targets for identification. In practice, such a combination has been employed for many applications to identify lipid species.

Limitations of the method include the following:

- Lipid quantitation with this methodology on a large scale is quite limited [63], although targeted analysis of a small number of lipids whose standard curves can be generated is quite common [64].
- It should be recognized that the identity of an individual extracted ion is not definitively identified in the approach.
- The specificity of the extracted ion to the compound of interest is generally a concern with SIM due to multiple interferences that are commonly present.

To reduce any artificial ion extraction, very high mass accuracy/high mass resolution instrumentation is always preferable. Alternatively, analysis of individual lipid species may be limited to one lipid class that is pre-fractionated. A good example of this alternative is the analysis of human skin ceramide species. Masukawa and colleagues quantified ceramide species in human stratum corneum with LC-MS after SPE separation of the ceramide fraction [65]. Over 182 molecule ions corresponding to the diverse ceramide species in the stratum corneum based on *m/z* were measured by using the SIM method.

3.3.1.2 Selected/Multiple Reaction Monitoring for LC-MS   Selected/multiple reaction monitoring (SRM/MRM), which is the special case of either PIS or NLS (see Chapter 2), becomes the preferred method for detection of a particular ion since it can be done in a very short time frame. SRM/MRM could be very specific if the monitored fragment ion is specific to the precursor in combination with an LC separation and no interfering transitions are concomitantly present.

Limitations of the method include the following:

- Pre-determination of individual lipid species at a certain elution time is required. Therefore, this approach does not allow conducting global analysis of individual lipid species for any sample without pre-determination in addition to the time limit for determining the number of pairs of transition ions.
- This approach only determines a body of species containing the pairs of transition ions.

3.3.1.3 Data-Dependent Analysis after LC-MS   In LC-MS analysis, individual lipid species of a class are eluted from a selected column more or less at different times regardless of the column type used. For example, the species containing saturated fatty acyl substituents are usually eluted earlier than those containing unsaturated fatty acyl substituents with normal-phase HPLC, while even the deuterium-labeled isotopologues could be separated sparingly with reversed-phase HPLC from their unlabeled counterparts. Thus, species have to be identified one by one in the LC-MS domain through the product-ion MS analysis of the detected ion that represents the species. In the approach of data-dependent analysis, a fixed number of molecular ions whose *m/z* values were detected in a survey scan at a time of elution are selected using the rules pre-set by an operator and are subject to product-ion MS analyses [66]. This is an ideal approach for identification and quantification of individual lipid species by LC-MS.

However, there exist some limitations:

- It is unfortunate that the short elution time does not allow one to extensively identify all the eluted species. Data-dependent analysis is typically limited to a small number of abundant ions.
- When the elution time becomes narrower, as is observed in UPLC, it is more difficult to attain extensive identification by this method since a normal ion peak shape has to be maintained for the purpose of quantification.

3.3.2   LC-MS-Based Approaches for Lipidomics

3.3.2.1   *Normal-Phase LC-MS-Based Approaches*   Normal-phase HPLC separates compounds based on the polar interactions between the analytes and the stationary phase of a column. Typically, the stationary phase in normal-phase LC is the unmodified silica particles in which free silanol groups are the interacting functional groups. However, silanol groups modified by various organic moieties, such as diol, nitrile, nitro, methylcyano, or phenylcyano, bonded chemically via a short spacer to the surface are also commonly used in LC-MS analysis of lipids. Such bonded phases tend to give much more reproducible separations with less tailing of peaks, and they also equilibrate much more rapidly with the mobile phase in gradient applications.

In the most common application of this separation mode, components are separated according to the number and nature of the polar functional groups (e.g., ester bonds, phosphate, hydroxyl, and amine groups) in lipid molecules. Since the head group of an individual lipid class predominantly determines the polar interactions with stationary phase, normal-phase HPLC separates a lipid extract solution into the lipid classes rather than into molecular species.

A solvent system consisting of chloroform and methanol or hexane and isopropanol with or without addition of a small volume of water is commonly used for this mode to separate lipid classes. These solvent systems can be readily compiled with an ESI ion source. Unfortunately, due to the toxicity of chloroform, inclusion of this solvent in LC-MS is seldom. The use of inorganic salts is incompatible with MS, a small amount of organic salt or acid may be used as a modifier. Isocratic elution by using a mobile phase with a constant composition may be used for a certain lipid class, but gradient elution in which the polarity of the mobile phase is increased at a controlled rate affords greater versatility.

For example, Hermansson et al. [63] employed a diol-modified silica column (250×1.0 mm, 5-μm particles) and used an isocratic elution with a mobile phase comprising hexane–isopropanol–water–formic acid–triethylamine ($628:348:24:2:0.8$; v/v). The mass spectrometric detection was in the negative-ion mode, in which either $[M-H]^-$ or $[M+HCOO]^-$ molecular ions were monitored. The investigators separated most of the common lipid classes and developed a method with MS detection to automatically identify and quantify over 100 lipid species through two-dimensional mapping of elution time and the masses of lipid species (i.e., an SIM approach).

The Lipid MAPS consortium provided an extensive protocol for the analysis of GPL species by employing normal-phase chromatography [67]. In the protocol, a Luna Silica column (250×2.0 mm, 5 μm particle size) was used. Lipid classes were separated using a binary nonlinear gradient consisting of isopropanol–hexane–ammonium bicarbinate (100 mM) from mobile phase A ($58:40:2$, v/v) to mobile phase B ($50:40:10$). The MS spectra were acquired in the negative-ion mode, and an SIM approach was utilized for determining the presence of lipid species. The total number of lipid species that were detected in the study was not given.

In another example of normal-phase LC-MS analysis of human and monkey plasma lipids, a Luna Silica column (150×2.0 mm, 3-μm particle size) was employed with a linear gradient from mobile phase A (chloroform–methanol–ammonium hydroxide, 89.5:10:0.5) to B (chloroform–methanol–ammonium hydroxide–water, 55:39:0.5:5.5) [68]. Mass spectra were recorded in both positive-ion and negative-ion ESI modes with an MRM method. PC, SM, ceramide, and GluCer were monitored in the positive-ion mode; and PE, PI, PS, PG, PA, and ganglioside M3 were measured in the negative-ion mode. A total of 153 lipid species were determined in the study.

Overall, when LC-MS with a normal-phase column is employed for lipid analysis, the SIM approach is the likely choice for MS detection due to a short elution time of an individual lipid class. Although the MRM approach could be successfully applied, great efforts on setting up the ion transitions are needed. It appears that the data-dependent analysis approach is less favored, also likely due to the limited elution time to perform product-ion analysis of all lipid species of a class.

3.3.2.2 Reversed-Phase LC-MS-Based Approaches

In reversed-phase HPLC, the separation is based on the selective interactions of analytes with a relatively non-polar liquid stationary phase and a relatively polar liquid mobile phase. Accordingly, in contrast to normal-phase LC, HPLC in the reversed-phase mode separates lipid species based more on their hydrophobicity of the fatty acyl chains (such as the chain lengths, the number and configuration of the double bonds) than the polarity of their head groups. As a result, the lipid species of different classes, but possessing similar overall mass-to-charge values, may possess similar hydrophobicity, thereby resulting in potential complications of data interpretation and quantitation for the analysis of complex mixtures. Owing to this reason, the majority of lipid analyses employing reversed-phase LC are conducted for a specific class rather than a biological lipid extract.

Although there are many nonpolar stationary phases used in reversed-phase HPLC, the most widely employed and most important for lipid analysis are those in which long-chain hydrocarbons are covalently bonded to the surface of silica particles at a particle size of 3–10 μm. Among these, the most widely used stationary phase is the one consisting of octadecylsilyl (i.e., C18 or ODS) groups. The composition of the mobile phase used for reversed-phase LC is also crucial because it is known that solvent molecules penetrate between the bonded chains, interact with them through dispersive forces, and determine their conformation and structure. Either acetonitrile or methanol is commonly used as a major component of the mobile phase along with a modifier (such as organic salt, organic acid, organic base, or their combination) for coupling with MS. It should be recognized that due to potential hydrolysis of the bonded long alkyl chain under the strong acid or base conditions, a pH value of 2–9 is optimal.

The history and many applications of reversed-phase LC for lipid analysis could be found in a review paper [56]. Here only a few examples with LC-ESI/MS analysis of lipid species are given.

At the earliest stage of LC-ESI/MS analysis of lipids, Kim et al. employed a C18 column (150 × 2.1 mm, 3-μm particle size) with a mobile phase consisting of the 0.5% ammonium hydroxide in a water–methanol–hexane mixture employing a linear gradient from 12:88:0 to 0:88:12 [69]. They detected ions that were shown as the protonated or sodiated form in the positive-ion mode. They determined PI, PS, PC, PE, and SM with an SIM approach.

An elegant example was the identification of oxidized PC molecules after exposure of a plasmalogen PC species to the free radical initiator 2,2′-azobis(2-amidinopropane)hydrochloride by Khaselev and Murphy [70]. The researchers employed a 5-μm C18 (150 × 1.0 mm) column with the mobile phase consisting of methanol–water–acetonitrile (60:20:20, v/v/v) containing 1 mm ammonium acetate as mobile phase A and 1 mm methanolic ammonium acetate solution as solvent B running a linear gradient from 0% B to 100% for 40 min followed by isocratic elution of 100% B for 20 min. They detected the oxidized PC species in both positive-ion and negative-ion modes with precursor-ion monitoring as well as with product-ion analysis. Numerous oxidized species were detected and identified. They concluded that the oxidation of plasmenyl phospholipids esterified with polyunsaturated fatty acid groups at *sn*-2 likely undergo unique and specific free radical oxidation at the 1′-alkenyl position as well as oxidation of the double bond closest to the ester moiety at *sn*-2.

Analysis of individual lipid species of a class in the Lipid MAPS protocols was largely developed based on reversed-phase LC-MS. For example, in the protocol for eicosanoid analysis, a C18 column (250 × 2.1 mm) was employed and a specific gradient from mobile phase A (water–acetonitrile–formic acid, 63:37:0.02, v/v/v) to B (acetonitrile–isopropanol, 50:50, v/v) was applied [71]. Individual eicosanoid species were detected by using the MRM approach and quantified in comparison to the relevant internal standard.

Sommer et al. [72] used reversed-phase LC-MS or LC-MS/MS with a C18 capillary column to fully characterize the individual lipid species including fatty acid compositions after fractionation of different lipid classes utilizing normal-phase LC-MS on an offline setting. Similar approaches using 2D LC-MS online or offline were also used by Byrdwell for a total lipid analysis [73] as well as for others [74–77].

As development of the instrumentation and stationary phases for LC, UPLC becomes a popular tool, replacing conventional reversed-phase LC. UPLC uses reduced particle size to improve chromatographic resolution, overcoming the increased back-pressure by a new generation of HPLC pumps, capable of pumping at 15,000 psi. Many researchers attempted to use this technology to replace the sequential separation with both normal-phase and reversed-phase HPLC [27, 59, 60, 78]. For example, Laaksonen et al. [27] used an Acquity UPLC C18 column (50 × 1.0 mm, 1.7-μm particle size) with a binary solvent system of A (water with 1% 1 M ammonium acetate, 0.1% formic acid) and B (acetonitrile–isopropanol, 5:2 with 1% 1 M ammonium acetate, 0.1% formic acid) to profile lipids of human plasma (10 μL) in the positive-ion mode. An 18-min gradient run was performed. A total of 132 lipid species including PC, PE, SM, PS, TAG, and cholesteryl esters were identified.

Overall, LC-MS analysis with a reversed-phase column is very powerful and common for the analysis of individual species after fractionation of an interest class, whereas a UPLC system is popularly used for global lipid analysis. Regarding the MS detection, the former is more associated with an MRM approach while the SIM approach is the likely choice in the latter case.

3.3.2.3 Hydrophilic Interaction LC-MS-Based Approaches

Hydrophilic interaction LC (HILIC) is a variant of normal-phase LC. HILIC uses hydrophilic stationary phases, but employs reversed-phase type eluents. Any polar chromatographic surface can be used for HILIC separations, even nonpolar bonded silicas. A typical mobile phase for HILIC includes acetonitrile with a small amount of water. However, any aprotic solvent miscible with water (e.g., tetrahydrofunan or dioxane) can be used. Alcohols can also be used with a higher concentration. Ionic additives, such as ammonium acetate and ammonium formate, are usually used to serve as the modifiers for controlling the pH and ion strength of the mobile phase.

Regarding the mechanism of separation in HILIC, it is commonly believed that the mobile phase forms a water-rich layer on the surface of the polar stationary phase *vs.* the water-deficient mobile phase, creating a liquid/liquid extraction system. Therefore, the analytes are distributed between these two layers. The more polar compounds have a stronger interaction with the stationary aqueous layer than the less polar compounds. Accordingly, separation in HILIC depends on both the polarity and the degree of solvation of an analyte, whereas the modifiers can also contribute to the polarity of the analyte, thereby affecting the retention time. This nature of the mode accounts for its popular usage recently, particularly for those attempting global lipidomic analysis [75, 77, 79–82].

For example, a protocol [79] was developed for rapid and simultaneous quantification of sphingolipid species by employing an HILIC silica column (1.8-μm particle size, 50×2.1 mm) with a mobile phase consisted of water containing 0.2% formic acid and 200 mM ammonium formate (A) and acetonitrile containing 0.2% formic acid (B) in a specific gradient run. The study demonstrated the separation of sphingosine, sphinganine, phytosphingosine, di- and trimethyl-sphingosine, SM, HexCer, LacCer, ceramide-1-phosphate, and dihydroceramide-1-phosphate. The authors believed that this protocol is superior to those based on reversed-phase LC-MS in good peak shapes, a short analysis time, and, most importantly, coelution of analytes with their respective internal standards, which could avoid an overestimation of species concentrations due to mutual ion suppression.

A very recent study by using LC-MS/MS with an HILIC column (150×2.1 mm, 3 μm; Waters Corporation, Milford, MA) was reported to determine lysoGPL species including regioisomers [83]. The mobile phase used is identical to that described in the previous paragraph with a specific gradient run. The researchers designed a scheduled MRM approach for the analysis. In the study, 68 lysoGPL species consisting of 110 regioisomers were detected in plasma and 43 lysoGPL species consisting of 67 regioisomers in skin samples were determined. Intriguingly, the study revealed that most of the lysoGPL species present in skin were 2-acyl isomers, whereas 1-acyl isomers were the dominant species in plasma.

3.3.2.4 Other LC-MS-Based Approaches Silver-ion or argentation chromatography is very useful for lipid analysis based on the interaction of silver ions with the *pi* electrons of double bonds to form polar complexes; the greater the number of double bonds in a molecule, the stronger the complex formation and the longer it is retained on the column. This type of chromatography has been coupled with MS for lipid analysis. For example, silver-ion LC coupled to MS was used for regioisomeric analysis of TAG species [84], identification of conjugated linoleic acid isomers [85], and separation of other fatty acid isomers [86].

Chiral chromatography is widely used for the resolution of chiral isomers of eicosanoids and enantiomeric isomers of di- and monoacylglycerols. The technique and its applications were extensively reviewed by Prof. Ian Blair in relation to the targeted analysis of these lipid classes and individual species employing LC-MS [87–89].

Ion-exchange chromatography was successfully used to separate lipid classes [90]. Because of the utilization of high ion strength in the mobile phase, it is relatively difficult to directly couple to a mass spectrometer for lipid analysis. However, ion-exchange SPE column could be used for fractionation of lipid classes followed by reversed-phased LC-MS or shotgun lipidomics of a particular lipid class.

3.3.3 Advantages and Drawbacks

The major advantage of LC-MS analysis is the usage of chromatographic separation to simplify the complex lipid extracts. For example, a normal-phase column could be applied for resolving a lipid mixture to an individual lipid class, or a reversed-phase column may be used to separate lipid species based on their different hydrophobicities. Of course, no single column is able to totally resolve a biological lipid extract into single species. Thus, combination of different types of columns was explored for two-dimensional or multidimensional separations either online or offline [58, 72]. Generally, offline separation approaches (including step fractionations) are broadly used for enrichment and analysis of lipid classes that are not easily ionized or are present in low abundances. Particularly, LC-MS plays an essential role in the discovery of novel lipid classes and molecular species.

The following few concerns associated with LC-MS analysis should be recognized:

- The determined ionization efficiency of analytes is generally measured at different elution times in the LC separation, which introduces variations in ionization efficiency from different mobile phase compositions.
- When a normal-phase HPLC column is employed for the separation of different lipid classes, different lipid species in a class are not uniformly distributed in the eluted peak (i.e., each individual molecular species of a class possesses its

own distinct retention time and peak shape due to differential interactions with the stationary phase and the ion-pairing agents employed).

- A solvent gradient is usually employed to resolve individual molecular species with a reversed-phase HPLC column. Changes in the components of the mobile phase might also cause ionization instability and/or affect ionization efficiency of lipid species eluted at different mobile phase composition.
- Reversed-phase HPLC gradients are generally initiated largely with an aqueous mobile phase, which induces solubility problems in a molecular species-dependent manner because reversed-phase HPLC is typically used to concentrate samples up to the limit of solubility in the aqueous phase, which leads to aggregation and differential ionization efficiencies (see Chapter 15) under such conditions.
- Differential loss of lipids on the column is also not unusual [91].
- Differential response factors of different molecular species may be present in the SRM/MRM approach due to its tandem MS nature (see Chapter 15). Multiple internal standards that represent the diversity of molecular species of a class should be used in this case.

These practical difficulties limit the use of LC-MS for a large-scale analysis of lipids in particular for their absolute quantitation, although there are many examples of LC-MS in applications of discovery and identification of novel lipids, particularly those present in low or very low abundance in a small scale [92–95].

3.3.4 Identification of Lipid Species after LC-MS

Column separation could provide a dimension of elution time for the identification of individual lipid species. However, definitive identification of the species eluted from a column has to be achieved through product-ion MS analysis to match the unique fragmentation pattern(s) of each lipid class. For a less experienced analyst, database searching of a product-ion MS spectrum should be helpful. To this end, generation of libraries and/or databases containing information about the structures, masses, isotope patterns, and MS/MS spectra in different ionization modes of lipids along with possible LC retention time range is critical. This is analogous to the libraries of gas chromatography (GC)-MS spectra, which can be used to search a compound of interest after GC-MS analysis. A few libraries/databases containing these types of information are available for this type of application and are extensively discussed in Chapter 5. Through these tools, therefore, one can potentially identify a lipid species by matching a fragmentation pattern of the species along with other available information manually or automatically. Unfortunately, identification of the eluted species is complicated when there is incomplete resolution of individual molecular species of a class. In this case, either improving the resolution of individual species or training for the analysis of the product-ion MS spectra acquired from standard mixtures consisting of a few species is necessary.

3.4 MALDI-MS FOR LIPIDOMICS

3.4.1 General

MALDI-MS has also been used for lipid analysis since it was introduced in the late 1980s [92, 93]. MALDI-MS has been applied for characterizing almost every lipid class (e.g., nonesterified fatty acids, glycerolipids (e.g., DAG and TAG), cholesterol and its derivatives, GPL, and sphingolipids) and to study oxidized lipids and biological lipid extracts (see Refs [94–98] for comprehensive reviews). In general, classic MALDI mass spectral analysis of these lipid classes almost all yielded reasonable responses in the positive-ion mode, but the majority of individual lipid species showed a cluster of $[M+H]^+$, $[M+Na]^+$, $[M+K]^+$, or other ions. PC species generally yield the best signals in the positive-ion MALDI-MS, which led to the suppressed signals of other coexisting lipid classes. This high sensitivity mainly results from the stable quaternary ammonium moiety [99]. Much simpler mass spectra in the molecular ion region are usually obtained in the negative-ion mode, where $[M-H]^-$ ions are usually predominant. However, postsource decay is more severe in the negative-ion mode than that in the positive-ion mode, and fatty acyl carboxylates resulted from the postsource decay are usually present as the base peak when the species of complex lipid classes are analyzed. This issue makes the analysis of acidic GPL classes (e.g., phosphatidylethanolamine) difficult and relatively less sensitive. Accordingly, prechromatographic separation of different lipid classes by HPLC or thin-layer chromatography (TLC) seems necessary followed by the analysis of these individual lipid classes in the positive-ion or negative-ion mode [100].

3.4.2 Analysis of Lipid Extracts

Positive-ion MALDI-MS analysis of PC and SM species usually shows two (quasi)molecular ions corresponding to proton and sodium adducts, respectively. The peak intensity ratio of these ions depends on the availability of sodium in the matrix. The proton adducts of these GPL species yield an exclusive fragment at m/z 184 corresponding to phosphocholine. However, the sodium adduct gives a more informative fragmentation pattern as previously described [93], which are essentially identical to the fragmentation pattern obtained by ESI-MS/MS. Although phosphocholine-containing lipid molecular species can result in an intense positive-ion signal, desorption/ionization of these molecules in the negative-ion mode is very poor [99].

Positive-ion MALDI mass spectrum of PE species is characterized by the presence of a specific fragment ion corresponding to the loss of the phosphoethanolamine head group [93]. PE species can also be ionized in the negative-ion mode with a lower sensitivity in comparison to that in the positive-ion mode. Negative-ion MALDI mass spectra of PE species are usually dominated by the matrix adducts of PE.

Positive-ion MALDI mass spectrum of PS species usually displays an additional ion peak corresponding to [M−H+2Na]+ in addition to the presence of protonated and sodiated molecular species. Moreover, a fragment ion (similar to that of PE) corresponding to the loss of phosphoserine head group is also present in the positive-ion MALDI mass spectrum of PS species, indicating that the cleavage of the polar head group is the predominant fragmentation pathway of these GPL classes [93]. Negative-ion MALDI mass spectrum of PS species shows a base peak corresponding to [M−H]− and other intense ion peaks corresponding to [M+Na−2H]− and matrix adducts in the molecular ion region.

Many classes of polar lipids have been characterized by MALDI-MS [94]. MALDI-MS has also been employed for characterization of nonpolar lipid classes such as cholesterol and TAG species. Positive-ion MALDI mass spectra of TAG species exclusively display the sodiated species. MALDI mass spectra of TAG species also display the ions corresponding to the loss of sodium fatty acyl carboxylate(s) due to postsource decay [93]. An extensive list of lipid classes detected by MALDI-MS and corresponding matrices has been documented [101], which should be consulted for those interested. Previously, postsource decay techniques have been commonly employed for characterization of polar lipids [93]. However, since both MALDI-TOF/TOF and MALDI Q-TOF mass spectrometers are now commercially available, true tandem MS analysis of lipids by MALDI-MS is feasible [102–104].

MALDI-MS has already been successfully applied for direct analysis of lipids in tissue samples. This topic is extensive described in Chapter 12.

3.4.3   Advantages and Drawbacks

MALDI-MS possesses multiple advantages for lipid analysis [94, 96]. These advantages include the following:

- MALDI-MS analysis is rapid, and each sample can be assessed in less than a minute.
- The experiments with MALDI-MS analysis are convenient since spotting lipid samples to a MALDI plate is relatively easy.
- The sensitivity of MALDI-MS is high in comparison to that of other ionization techniques since picomole amounts of samples may be sufficient for MALDI-MS analysis.
- Lipid samples spotted on a MALDI plate can often be reanalyzed at a later time.
- Automation of MALDI-MS is currently available from plate spotting to data acquisition.

However, this technique also has multiple drawbacks for lipidomics [96], which include the following:

- Matrix compound(s) are usually ionized, which result in complications for lipid analysis in the low m/z region, such as the fragments that result from postsource decay, and are used for characterization of lipid structures.

- The post-source decay present in MALDI-MS is a double-edged sword as it is quite useful for identification of lipid structures, but results in a problem for quantitation due to the differential fragmentational kinetics of different lipid species.

- The presence of numerous adducts and/or ion forms of individual lipid species (see above) complicates the mass spectral analysis and leads to reduction of analysis sensitivities in both detection and quantification.

- The presence of lipid aggregation during crystallization makes analytes distributed heterogeneously in the sample spot.

- Accurate quantitative results of lipid analyses by MALDI-MS are less than ideal and, thus, little progress has been made toward the direct quantitation of lipids by MALDI-MS, particularly from complex lipid mixtures.

Therefore, classic MALDI-MS is largely used to rapidly screen the lipid profile of a sample, and its application for lipidomics is limited.

3.4.4   Recent Advances in MALDI-MS for Lipidomics

3.4.4.1   Utilization of Novel Matrices   Great efforts have been made to eliminate these drawbacks. To improve the spot homogeneity for lipid analysis, one of the efforts is the introduction of ionic-liquid (or ionic-solid) matrices [105–107]. In general, studies by applying ionic-liquid matrices have demonstrated multiple advantages for lipid analysis in comparison to commonly used matrices. The spot homogeneity is a key to improve the reproducibility for lipid analysis, thereby achieving the quantitative analysis of lipid species by MALDI-MS.

Exploiting new matrices as well as developing novel spotting skills is another area of new improvements. By utilizing a type of neutral matrix (i.e., 9-aminoacridine) under different acidic/basic conditions, selective analyses of different lipid classes were achieved [108, 109]. Use of the less polar solvents to crystallize the matrix results in both an increase in solubility and homogeneous distribution of lipids (both reduce aggregation). This neutral matrix substantially facilitates the desorption/ionization of lipids in the negative-ion mode with minimal postsource decay. The larger conjugated system in 9-aminoacridine due to a better dispersion of laser energy in comparison to commonly used matrices (e.g., α-cyano-4-hydroxycinnamic acid (CHCA) and 2,5-dihydroxybenzoic acid (DHB)) is attributed to the reduction of postsource decay. This neutral matrix only yields low matrix backgrounds in the low mass range of interest, which further enhances the *S/N* of mass spectral analysis. Essentially identical spectra of a biological extract acquired by both ESI-MS and MALDI-MS with 9-aminoacridine evidenced the quantitative analysis in nature (Figure 3.2). Similar to the principles that underpin "intrasource separation" [33], many of the lipid classes can be high selectively ionized, including PC, SM, and TAG in the positive-ion mode [108]. Furthermore, in the negative-ion mode, this neutral matrix has proven to be highly sensitive for the analysis of negatively charged lipids, including CL, PI, and ST directly from extracts of mammalian

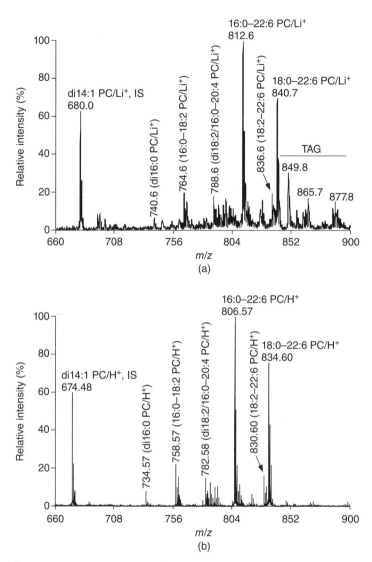

Figure 3.2 Mass spectral comparison of PC molecular species present in mouse heart lipid extracts acquired by either ESI or MALDI. Extracts of mouse myocardium were prepared by a modified Bligh and Dyer procedure and analyzed by ESI-MS in the presence of LiOH (a) or MALDI-MS utilizing 9-aminoacridine as matrix dissolved in isopropanol/acetonitrile (60/40, v/v) (b). "IS" denotes internal standard. Sun et al. [108]. Reproduced with permission of the American Chemical Society.

tissue samples without any prior chromatography [108, 109]. Use of this matrix to analyze lipids from different sources has recently been reported from different laboratories [98, 109–117]. Accordingly, the utilization of this matrix indeed opens the new door for lipidomics by MALDI-MS. Similarly, another neutral matrix (i.e., 1,5-diaminonapthalene, DAN) showed enhanced sensitivity of MALDI-MS for analysis of lipids in both positive-ion and negative-ion modes [118].

3.4.4.2 (HP)TLC-MALDI-MS   To resolve the difficulties in the analysis of acidic GPL species due to ion suppression by PC species [94], investigators have demonstrated that high-performance thin-layer chromatography (HPTLC)-separated lipid classes can be conveniently analyzed with MALDI-MS directly on the TLC plates [119–121]. For example, Rohlfing et al. readily detected CL, PG, PE, PA, PC, and SM after spotting glycerol onto the bands of interest [120]. Briefly, developed HPTLC plates are cut to small pieces to fit into the sample plate. Acquisition of mass spectra is directly performed after the sample plate is placed into the ion source of a mass spectrometer. However, it is important to note that the same effect can lead to rapid sample depletion and also increases the likelihood of silica contamination of the ion source. With the same line of reasoning, an alternative approach that couples TLC blot with MALDI-TOF/MS has been developed to image the bands and to quantify the separated lipid classes [122–124].

3.4.4.3 Matrix-Free Laser Desorption/Ionization Approaches   To eliminate the matrix background as well as other perturbations and labor intense from the introduction of matrix, great interests have been directed toward the development of methods that permit soft laser desorption/ionization from surfaces without a matrix. These techniques use an active nanostructured surface to couple the laser energy to the desorption/ionization of analytes present on the surface [125]. These surfaces are typically composed of carbon or silicon and replace the standard MALDI target plates. A variety of similar approaches have been described including desorption/ionization from porous silicon [126], nanowire-assisted laser desorption/ionization [127], nanostructure-initiator mass spectrometry [128–131], and graphite-assisted laser desorption/ionization [132], with the primary difference between the techniques being the type of substrate.

For example, by using the nanostructure-initiator mass spectrometry, Patti and colleagues [130] detected intact cholesterol and its derivative molecules *in situ* to provide the first images of brain sterol localization in a knockout mouse model of 7-dehydrocholesterol reductase. Other nanomaterials (e.g., silver nanoparticles) have also been used to analyze lipids directly from tissue samples [133]. From a thin film of colloidal graphite on rat brain tissue, direct lipid profiling was performed by graphite-assisted laser desorption/ionization mass spectrometry, which allowed to detect 22 HexCer species, whereas only eight HexCer species are detected with MALDI-TOF/MS [132]. Chemically selective analysis for HexCer and ST species was successfully obtained.

Collectively, the primary advantage of these approaches compared to standard MALDI is the production of significantly fewer matrix-related ions in the resulting spectrum, greatly simplifying the detection of low mass compounds such as nonesterified fatty acids and significantly enhancing the limit of detection [134, 135].

REFERENCES

1. Han, X., Yang, K. and Gross, R.W. (2008) Microfluidics-based electrospray ionization enhances intrasource separation of lipid classes and extends identification of individual molecular species through multi-dimensional mass spectrometry: Development of an automated high throughput platform for shotgun lipidomics. Rapid Commun. Mass Spectrom. 22, 2115–2124.

2. Stahlman, M., Ejsing, C.S., Tarasov, K., Perman, J., Boren, J. and Ekroos, K. (2009) High throughput oriented shotgun lipidomics by quadrupole time-of-flight mass spectrometry. J. Chromatogr. B 877, 2664–2672.

3. Schwudke, D., Liebisch, G., Herzog, R., Schmitz, G. and Shevchenko, A. (2007) Shotgun lipidomics by tandem mass spectrometry under data-dependent acquisition control. Methods Enzymol. 433, 175–191.

4. Bowden, J.A., Bangma, J.T. and Kucklick, J.R. (2014) Development of an automated multi-injection shotgun lipidomics approach using a triple quadrupole mass spectrometer. Lipids 49, 609–619.

5. Han, X. and Gross, R.W. (1994) Electrospray ionization mass spectroscopic analysis of human erythrocyte plasma membrane phospholipids. Proc. Natl. Acad. Sci. U.S.A. 91, 10635–10639.

6. Duffin, K.L., Henion, J.D. and Shieh, J.J. (1991) Electrospray and tandem mass spectrometric characterization of acylglycerol mixtures that are dissolved in nonpolar solvents. Anal. Chem. 63, 1781–1788.

7. Kerwin, J.L., Tuininga, A.R. and Ericsson, L.H. (1994) Identification of molecular species of glycerophospholipids and sphingomyelin using electrospray mass spectrometry. J. Lipid Res. 35, 1102–1114.

8. Weintraub, S.T., Pinckard, R.N. and Hail, M. (1991) Electrospray ionization for analysis of platelet-activating factor. Rapid Commun. Mass Spectrom. 5, 309–311.

9. Han, X. and Gross, R.W. (2005) Shotgun lipidomics: Electrospray ionization mass spectrometric analysis and quantitation of the cellular lipidomes directly from crude extracts of biological samples. Mass Spectrom. Rev. 24, 367–412.

10. Ejsing, C.S., Ekroos, K., Jackson, S., Duchoslav, E., Hao, Z., Pelt, C.K.v., Simons, K. and Shevchenko, A. (2004) Shotgun lipidomics: High throughput profiling of the molecular composition of phospholipids ASMS Abstract Achieves, p 25.

11. Han, X. and Gross, R.W. (2005) Shotgun lipidomics: Multi-dimensional mass spectrometric analysis of cellular lipidomes. Expert Rev. Proteomics 2, 253–264.

12. Welti, R., Shah, J., Li, W., Li, M., Chen, J., Burke, J.J., Fauconnier, M.L., Chapman, K., Chye, M.L. and Wang, X. (2007) Plant lipidomics: Discerning biological function by profiling plant complex lipids using mass spectrometry. Front. Biosci. 12, 2494–2506.

13. Shevchenko, A. and Simons, K. (2010) Lipidomics: Coming to grips with lipid diversity. Nat. Rev. Mol. Cell Biol. 11, 593–598.

14. Han, X., Yang, K. and Gross, R.W. (2012) Multi-dimensional mass spectrometry-based shotgun lipidomics and novel strategies for lipidomic analyses. Mass Spectrom. Rev. 31, 134–178.

15. Brugger, B., Erben, G., Sandhoff, R., Wieland, F.T. and Lehmann, W.D. (1997) Quantitative analysis of biological membrane lipids at the low picomole level by nano-electrospray ionization tandem mass spectrometry. Proc. Natl. Acad. Sci. U.S.A. 94, 2339–2344.

16. Welti, R. and Wang, X. (2004) Lipid species profiling: A high-throughput approach to identify lipid compositional changes and determine the function of genes involved in lipid metabolism and signaling. Curr. Opin. Plant Biol. 7, 337–344.

17. Samarakoon, T., Shiva, S., Lowe, K., Tamura, P., Roth, M.R. and Welti, R. (2012) Arabidopsis thaliana membrane lipid molecular species and their mass spectral analysis. Methods Mol. Biol. 918, 179–268.

18. Chernushevich, I.V., Loboda, A.V. and Thomson, B.A. (2001) An introduction to quadrupole-time-of-flight mass spectrometry. J. Mass Spectrom. 36, 849–865.

19. Ejsing, C.S., Moehring, T., Bahr, U., Duchoslav, E., Karas, M., Simons, K. and Shevchenko, A. (2006) Collision-induced dissociation pathways of yeast sphingolipids and their molecular profiling in total lipid extracts: A study by quadrupole TOF and linear ion trap-orbitrap mass spectrometry. J. Mass Spectrom. 41, 372–389.

20. Ekroos, K., Chernushevich, I.V., Simons, K. and Shevchenko, A. (2002) Quantitative profiling of phospholipids by multiple precursor ion scanning on a hybrid quadrupole time-of-flight mass spectrometer. Anal. Chem. 74, 941–949.

21. Ejsing, C.S., Duchoslav, E., Sampaio, J., Simons, K., Bonner, R., Thiele, C., Ekroos, K. and Shevchenko, A. (2006) Automated identification and quantification of glycerophospholipid molecular species by multiple precursor ion scanning. Anal. Chem. 78, 6202–6214.

22. Schwudke, D., Oegema, J., Burton, L., Entchev, E., Hannich, J.T., Ejsing, C.S., Kurzchalia, T. and Shevchenko, A. (2006) Lipid profiling by multiple precursor and neutral loss scanning driven by the data-dependent acquisition. Anal. Chem. 78, 585–595.

23. Schwudke, D., Hannich, J.T., Surendranath, V., Grimard, V., Moehring, T., Burton, L., Kurzchalia, T. and Shevchenko, A. (2007) Top-down lipidomic screens by multivariate analysis of high-resolution survey mass spectra. Anal. Chem. 79, 4083–4093.

24. Schuhmann, K., Herzog, R., Schwudke, D., Metelmann-Strupat, W., Bornstein, S.R. and Shevchenko, A. (2011) Bottom-up shotgun lipidomics by higher energy collisional dissociation on LTQ Orbitrap mass spectrometers. Anal. Chem. 83, 5480–5487.

25. Ejsing, C.S., Sampaio, J.L., Surendranath, V., Duchoslav, E., Ekroos, K., Klemm, R.W., Simons, K. and Shevchenko, A. (2009) Global analysis of the yeast lipidome by quantitative shotgun mass spectrometry. Proc. Natl. Acad. Sci. U.S.A. 106, 2136–2141.

26. Herzog, R., Schwudke, D., Schuhmann, K., Sampaio, J.L., Bornstein, S.R., Schroeder, M. and Shevchenko, A. (2011) A novel informatics concept for high-throughput shotgun lipidomics based on the molecular fragmentation query language. Genome Biol. 12, R8.

27. Laaksonen, R., Katajamaa, M., Paiva, H., Sysi-Aho, M., Saarinen, L., Junni, P., Lutjohann, D., Smet, J., Van Coster, R., Seppanen-Laakso, T., Lehtimaki, T., Soini, J. and Oresic, M. (2006) A systems biology strategy reveals biological pathways and plasma biomarker candidates for potentially toxic statin-induced changes in muscle. PLoS One 1, e97.

28. Han, X. and Gross, R.W. (2001) Quantitative analysis and molecular species fingerprinting of triacylglyceride molecular species directly from lipid extracts of biological samples by electrospray ionization tandem mass spectrometry. Anal. Biochem. 295, 88–100.

29. Han, X., Yang, J., Cheng, H., Ye, H. and Gross, R.W. (2004) Towards fingerprinting cellular lipidomes directly from biological samples by two-dimensional electrospray ionization mass spectrometry. Anal. Biochem. 330, 317–331.

30. Yang, K., Cheng, H., Gross, R.W. and Han, X. (2009) Automated lipid identification and quantification by multi-dimensional mass spectrometry-based shotgun lipidomics. Anal. Chem. 81, 4356–4368.

31. Jiang, X., Cheng, H., Yang, K., Gross, R.W. and Han, X. (2007) Alkaline methanolysis of lipid extracts extends shotgun lipidomics analyses to the low abundance regime of cellular sphingolipids. Anal. Biochem. 371, 135–145.

32. Han, X., Yang, K., Cheng, H., Fikes, K.N. and Gross, R.W. (2005) Shotgun lipidomics of phosphoethanolamine-containing lipids in biological samples after one-step in situ derivatization. J. Lipid Res. 46, 1548–1560.

33. Han, X., Yang, K., Yang, J., Fikes, K.N., Cheng, H. and Gross, R.W. (2006) Factors influencing the electrospray intrasource separation and selective ionization of glycerophospholipids. J. Am. Soc. Mass Spectrom. 17, 264–274.

34. Han, X. (2007) Neurolipidomics: Challenges and developments. Front. Biosci. 12, 2601–2615.

35. Song, H., Hsu, F.F., Ladenson, J. and Turk, J. (2007) Algorithm for processing raw mass spectrometric data to identify and quantitate complex lipid molecular species in mixtures by data-dependent scanning and fragment ion database searching. J. Am. Soc. Mass Spectrom. 18, 1848–1858.

36. Kind, T., Liu, K.H., Lee do, Y., Defelice, B., Meissen, J.K. and Fiehn, O. (2013) LipidBlast in silico tandem mass spectrometry database for lipid identification. Nat. Methods 10, 755–758.

37. Yang, K., Zhao, Z., Gross, R.W. and Han, X. (2009) Systematic analysis of choline-containing phospholipids using multi-dimensional mass spectrometry-based shotgun lipidomics. J. Chromatogr. B 877, 2924–2936.

38. Han, X., Yang, K., Yang, J., Cheng, H. and Gross, R.W. (2006) Shotgun lipidomics of cardiolipin molecular species in lipid extracts of biological samples. J. Lipid Res. 47, 864–879.

39. Han, X. (2002) Characterization and direct quantitation of ceramide molecular species from lipid extracts of biological samples by electrospray ionization tandem mass spectrometry. Anal. Biochem. 302, 199–212.

40. Han, X. and Cheng, H. (2005) Characterization and direct quantitation of cerebroside molecular species from lipid extracts by shotgun lipidomics. J. Lipid Res. 46, 163–175.

41. Hsu, F.F. and Turk, J. (2001) Structural determination of glycosphingolipids as lithiated adducts by electrospray ionization mass spectrometry using low-energy collisional-activated dissociation on a triple stage quadrupole instrument. J. Am. Soc. Mass Spectrom. 12, 61–79.

42. Hsu, F.-F., Bohrer, A. and Turk, J. (1998) Electrospray ionization tandem mass spectrometric analysis of sulfatide. Determination of fragmentation patterns and characterization of molecular species expressed in brain and in pancreatic islets. Biochim. Biophys. Acta 1392, 202–216.

43. Jiang, X. and Han, X. (2006) Characterization and direct quantitation of sphingoid base-1-phosphates from lipid extracts: A shotgun lipidomics approach. J. Lipid Res. 47, 1865–1873.

44. Jiang, X., Yang, K. and Han, X. (2009) Direct quantitation of psychosine from alkaline-treated lipid extracts with a semi-synthetic internal standard. J. Lipid Res. 50, 162–172.

45. Cheng, H., Jiang, X. and Han, X. (2007) Alterations in lipid homeostasis of mouse dorsal root ganglia induced by apolipoprotein E deficiency: A shotgun lipidomics study. J. Neurochem. 101, 57–76.

46. Su, X., Han, X., Mancuso, D.J., Abendschein, D.R. and Gross, R.W. (2005) Accumulation of long-chain acylcarnitine and 3-hydroxy acylcarnitine molecular species in diabetic myocardium: Identification of alterations in mitochondrial fatty acid processing in diabetic myocardium by shotgun lipidomics. Biochemistry 44, 5234–5245.

47. Kalderon, B., Sheena, V., Shachrur, S., Hertz, R. and Bar-Tana, J. (2002) Modulation by nutrients and drugs of liver acyl-CoAs analyzed by mass spectrometry. J. Lipid Res. 43, 1125–1132.

48. Wang, M. and Han, X. (2014) Multidimensional mass spectrometry-based shotgun lipidomics. Methods Mol. Biol. 1198, 203–220.

49. Wang, M., Han, R.H. and Han, X. (2013) Fatty acidomics: Global analysis of lipid species containing a carboxyl group with a charge-remote fragmentation-assisted approach. Anal. Chem. 85, 9312–9320.

50. Zech, T., Ejsing, C.S., Gaus, K., de Wet, B., Shevchenko, A., Simons, K. and Harder, T. (2009) Accumulation of raft lipids in T-cell plasma membrane domains engaged in TCR signalling. EMBO J. 28, 466–476.

51. Klemm, R.W., Ejsing, C.S., Surma, M.A., Kaiser, H.J., Gerl, M.J., Sampaio, J.L., de Robillard, Q., Ferguson, C., Proszynski, T.J., Shevchenko, A. and Simons, K. (2009) Segregation of sphingolipids and sterols during formation of secretory vesicles at the trans-Golgi network. J. Cell Biol. 185, 601–612.

52. Klose, C., Ejsing, C.S., Garcia-Saez, A.J., Kaiser, H.J., Sampaio, J.L., Surma, M.A., Shevchenko, A., Schwille, P. and Simons, K. (2010) Yeast lipids can phase separate into micrometer-scale membrane domains. J. Biol. Chem. 285, 30224–30232.

53. Sampaio, J.L., Gerl, M.J., Klose, C., Ejsing, C.S., Beug, H., Simons, K. and Shevchenko, A. (2011) Membrane lipidome of an epithelial cell line. Proc. Natl. Acad. Sci. U.S.A. 108, 1903–1907.

54. Postle, A.D. and Hunt, A.N. (2009) Dynamic lipidomics with stable isotope labelling. J. Chromatogr. B 877, 2716–2721.

55. Myers, D.S., Ivanova, P.T., Milne, S.B. and Brown, H.A. (2011) Quantitative analysis of glycerophospholipids by LC-MS: Acquisition, data handling, and interpretation. Biochim. Biophys. Acta 1811, 748–757.

56. Brouwers, J.F. (2011) Liquid chromatographic-mass spectrometric analysis of phospholipids. Chromatography, ionization and quantification. Biochim. Biophys. Acta 1811, 763–775.

57. Zoerner, A.A., Gutzki, F.M., Batkai, S., May, M., Rakers, C., Engeli, S., Jordan, J. and Tsikas, D. (2011) Quantification of endocannabinoids in biological systems by chromatography and mass spectrometry: A comprehensive review from an analytical and biological perspective. Biochim. Biophys. Acta 1811, 706–723.

58. Guo, X. and Lankmayr, E. (2010) Multidimensional approaches in LC and MS for phospholipid bioanalysis. Bioanalysis 2, 1109–1123.

59. Yin, P., Zhao, X., Li, Q., Wang, J., Li, J. and Xu, G. (2006) Metabonomics study of intestinal fistulas based on ultraperformance liquid chromatography coupled with Q-TOF mass spectrometry (UPLC/Q-TOF MS). J. Proteome Res. 5, 2135–2143.

60. Rainville, P.D., Stumpf, C.L., Shockcor, J.P., Plumb, R.S. and Nicholson, J.K. (2007) Novel application of reversed-phase UPLC-oaTOF-MS for lipid analysis in complex biological mixtures: A new tool for lipidomics. J. Proteome Res. 6, 552–558.

61. Christie, W.W. and Han, X. (2010) Lipid Analysis: Isolation, Separation, Identification and Lipidomic Analysis, The Oily Press, Bridgwater, England. pp 448.

62. Niessen, W.M.A. (1999) Liquid Chromatography-Mass Spectrometry. Marcel Dekker, Inc., New York.

63. Hermansson, M., Uphoff, A., Kakela, R. and Somerharju, P. (2005) Automated quantitative analysis of complex lipidomes by liquid chromatography/mass spectrometry. Anal. Chem. 77, 2166–2175.

64. Liebisch, G., Drobnik, W., Reil, M., Trumbach, B., Arnecke, R., Olgemoller, B., Roscher, A. and Schmitz, G. (1999) Quantitative measurement of different ceramide species from crude cellular extracts by electrospray ionization tandem mass spectrometry (ESI-MS/MS). J. Lipid Res. 40, 1539–1546.

65. Masukawa, Y., Narita, H., Sato, H., Naoe, A., Kondo, N., Sugai, Y., Oba, T., Homma, R., Ishikawa, J., Takagi, Y. and Kitahara, T. (2009) Comprehensive quantification of ceramide species in human stratum corneum. J. Lipid Res. 50, 1708–1719.

66. Mann, M., Hendrickson, R.C. and Pandey, A. (2001) Analysis of proteins and proteomes by mass spectrometry. Annu. Rev. Biochem. 70, 437–473.

67. Ivanova, P.T., Milne, S.B., Byrne, M.O., Xiang, Y. and Brown, H.A. (2007) Glycerophospholipid identification and quantitation by electrospray ionization mass spectrometry. Methods Enzymol. 432, 21–57.

68. Shui, G., Stebbins, J.W., Lam, B.D., Cheong, W.F., Lam, S.M., Gregoire, F., Kusonoki, J. and Wenk, M.R. (2011) Comparative plasma lipidome between human and cynomolgus monkey: Are plasma polar lipids good biomarkers for diabetic monkeys? PLoS One 6, e19731.

69. Kim, H.Y., Wang, T.C. and Ma, Y.C. (1994) Liquid chromatography/mass spectrometry of phospholipids using electrospray ionization. Anal. Chem. 66, 3977–3982.

70. Khaselev, N. and Murphy, R.C. (2000) Structural characterization of oxidized phospholipid products derived from arachidonate-containing plasmenyl glycerophosphocholine. J. Lipid Res. 41, 564–572.

71. Deems, R., Buczynski, M.W., Bowers-Gentry, R., Harkewicz, R. and Dennis, E.A. (2007) Detection and quantitation of eicosanoids via high performance liquid chromatography-electrospray ionization-mass spectrometry. Methods Enzymol. 432, 59–82.

72. Sommer, U., Herscovitz, H., Welty, F.K. and Costello, C.E. (2006) LC-MS-based method for the qualitative and quantitative analysis of complex lipid mixtures. J. Lipid Res. 47, 804–814.

73. Byrdwell, W.C. (2008) Dual parallel liquid chromatography with dual mass spectrometry (LC2/MS2) for a total lipid analysis. Front Biosci 13, 100–120.

74. Nie, H., Liu, R., Yang, Y., Bai, Y., Guan, Y., Qian, D., Wang, T. and Liu, H. (2010) Lipid profiling of rat peritoneal surface layers by online normal- and reversed-phase 2D LC QToF-MS. J. Lipid Res. 51, 2833–2844.

75. Lisa, M., Cifkova, E. and Holcapek, M. (2011) Lipidomic profiling of biological tissues using off-line two-dimensional high-performance liquid chromatography-mass spectrometry. J. Chromatogr. A 1218, 5146–5156.

76. Fauland, A., Kofeler, H., Trotzmuller, M., Knopf, A., Hartler, J., Eberl, A., Chitraju, C., Lankmayr, E. and Spener, F. (2011) A comprehensive method for lipid profiling by liquid chromatography-ion cyclotron resonance mass spectrometry. J. Lipid Res. 52, 2314–2322.

77. Wang, S., Li, J., Shi, X., Qiao, L., Lu, X. and Xu, G. (2013) A novel stop-flow two-dimensional liquid chromatography-mass spectrometry method for lipid analysis. J. Chromatogr. A 1321, 65–72.

78. Sandra, K., Pereira Ados, S., Vanhoenacker, G., David, F. and Sandra, P. (2010) Comprehensive blood plasma lipidomics by liquid chromatography/quadrupole time-of-flight mass spectrometry. J. Chromatogr. A 1217, 4087–4099.

79. Scherer, M., Leuthauser-Jaschinski, K., Ecker, J., Schmitz, G. and Liebisch, G. (2010) A rapid and quantitative LC-MS/MS method to profile sphingolipids. J. Lipid Res. 51, 2001–2011.

80. Okazaki, Y., Kamide, Y., Hirai, M.Y. and Saito, K. (2013) Plant lipidomics based on hydrophilic interaction chromatography coupled to ion trap time-of-flight mass spectrometry. Metabolomics 9, 121–131.

81. Cifkova, E., Holcapek, M., Lisa, M., Ovcacikova, M., Lycka, A., Lynen, F. and Sandra, P. (2012) Nontargeted quantitation of lipid classes using hydrophilic interaction liquid chromatography-electrospray ionization mass spectrometry with single internal standard and response factor approach. Anal. Chem. 84, 10064–10070.

82. Cifkova, E., Holcapek, M. and Lisa, M. (2013) Nontargeted lipidomic characterization of porcine organs using hydrophilic interaction liquid chromatography and off-line two-dimensional liquid chromatography-electrospray ionization mass spectrometry. Lipids 48, 915–928.

83. Koistinen, K.M., Suoniemi, M., Simolin, H. and Ekroos, K. (2015) Quantitative lysophospholipidomics in human plasma and skin by LC-MS/MS. Anal. Bioanal. Chem. 407, 5091–5099.

84. Holcapek, M., Dvorakova, H., Lisa, M., Giron, A.J., Sandra, P. and Cvacka, J. (2010) Regioisomeric analysis of triacylglycerols using silver-ion liquid chromatography-atmospheric pressure chemical ionization mass spectrometry: Comparison of five different mass analyzers. J. Chromatogr. A 1217, 8186–8194.

85. Sun, C., Black, B.A., Zhao, Y.Y., Ganzle, M.G. and Curtis, J.M. (2013) Identification of conjugated linoleic acid (CLA) isomers by silver ion-liquid chromatography/in-line ozonolysis/mass spectrometry (Ag±LC/O3-MS). Anal. Chem. 85, 7345–7352.

86. Momchilova, S.M. and Nikolova-Damyanova, B.M. (2010) Separation of isomeric octadecenoic fatty acids in partially hydrogenated vegetable oils as p-methoxyphenacyl esters using a single-column silver ion high-performance liquid chromatography (Ag-HPLC). Nat. Protoc. 5, 473–478.

87. Lee, S.H., Williams, M.V. and Blair, I.A. (2005) Targeted chiral lipidomics analysis. Prostaglandins Other Lipid Mediat. 77, 141–157.

88. Mesaros, C., Lee, S.H. and Blair, I.A. (2009) Targeted quantitative analysis of eicosanoid lipids in biological samples using liquid chromatography-tandem mass spectrometry. J. Chromatogr. B 877, 2736–2745.

89. Lee, S.H. and Blair, I.A. (2009) Targeted chiral lipidomics analysis of bioactive eicosanoid lipids in cellular systems. BMB Rep. 42, 401–410.

90. Gross, R.W. and Sobel, B.E. (1980) Isocratic high-performance liquid chromatography separation of phosphoglycerides and lysophosphoglycerides. J. Chromatogr. 197, 79–85.

91. DeLong, C.J., Baker, P.R.S., Samuel, M., Cui, Z. and Thomas, M.J. (2001) Molecular species composition of rat liver phospholipids by ESI-MS/MS: The effect of chromatography. J. Lipid Res. 42, 1959–1968.

92. Marto, J.A., White, F.M., Seldomridge, S. and Marshall, A.G. (1995) Structural characterization of phospholipids by matrix-assisted laser desorption/ionization Fourier transform ion cyclotron resonance mass spectrometry. Anal. Chem. 67, 3979–3984.

93. Al-Saad, K.A., Zabrouskov, V., Siems, W.F., Knowles, N.R., Hannan, R.M. and Hill, H.H., Jr. (2003) Matrix-assisted laser desorption/ionization time-of-flight mass spectrometry of lipids: Ionization and prompt fragmentation patterns. Rapid Commun. Mass Spectrom. 17, 87–96.

94. Schiller, J., Suss, R., Arnhold, J., Fuchs, B., Lessig, J., Muller, M., Petkovic, M., Spalteholz, H., Zschornig, O. and Arnold, K. (2004) Matrix-assisted laser desorption and ionization time-of-flight (MALDI-TOF) mass spectrometry in lipid and phospholipid research. Prog. Lipid Res. 43, 449–488.

95. Schiller, J., Suss, R., Fuchs, B., Muller, M., Zschornig, O. and Arnold, K. (2007) MALDI-TOF MS in lipidomics. Front. Biosci. 12, 2568–2579.

96. Fuchs, B. and Schiller, J. (2008) MALDI-TOF MS analysis of lipids from cells, tissues and body fluids. Subcell. Biochem. 49, 541–565.

97. Fuchs, B. and Schiller, J. (2009) Application of MALDI-TOF mass spectrometry in lipidomics. Eur. J. Lipid Sci. Technol. 111, 83–89.

98. Fuchs, B., Suss, R. and Schiller, J. (2010) An update of MALDI-TOF mass spectrometry in lipid research. Prog. Lipid Res. 49, 450–475.

99. Petkovic, M., Schiller, J., Muller, M., Benard, S., Reichl, S., Arnold, K. and Arnhold, J. (2001) Detection of individual phospholipids in lipid mixtures by matrix-assisted laser desorption/ionization time-of-flight mass spectrometry: Phosphatidylcholine prevents the detection of further species. Anal. Biochem. 289, 202–216.

100. Estrada, R. and Yappert, M.C. (2004) Regional phospholipid analysis of porcine lens membranes by matrix-assisted laser desorption/ionization time-of-flight mass spectrometry. J. Mass Spectrom. 39, 1531–1540.

101. Ellis, S.R., Brown, S.H., In Het Panhuis, M., Blanksby, S.J. and Mitchell, T.W. (2013) Surface analysis of lipids by mass spectrometry: More than just imaging. Prog. Lipid Res. 52, 329–353.

102. Jackson, S.N., Wang, H.Y. and Woods, A.S. (2005) In situ structural characterization of phosphatidylcholines in brain tissue using MALDI-MS/MS. J. Am. Soc. Mass Spectrom. 16, 2052–2056.

103. Wang, H.Y., Jackson, S.N. and Woods, A.S. (2007) Direct MALDI-MS analysis of cardiolipin from rat organs sections. J. Am. Soc. Mass Spectrom. 18, 567–577.

104. Jackson, S.N., Wang, H.Y. and Woods, A.S. (2007) In situ structural characterization of glycerophospholipids and sulfatides in brain tissue using MALDI-MS/MS. J. Am. Soc. Mass Spectrom. 18, 17–26.

105. Li, Y.L., Gross, M.L. and Hsu, F.-F. (2005) Ionic-liquid matrices for improved analysis of phospholipids by MALDI-TOF mass spectrometry. J. Am. Soc. Mass Spectrom. 16, 679–682.

106. Jones, J.J., Batoy, S.M., Wilkins, C.L., Liyanage, R. and Lay, J.O., Jr. (2005) Ionic liquid matrix-induced metastable decay of peptides and oligonucleotides and stabilization of phospholipids in MALDI FTMS analyses. J. Am. Soc. Mass Spectrom. 16, 2000–2008.

107. Ham, B.M., Jacob, J.T. and Cole, R.B. (2005) MALDI-TOF MS of phosphorylated lipids in biological fluids using immobilized metal affinity chromatography and a solid ionic crystal matrix. Anal. Chem. 77, 4439–4447.

108. Sun, G., Yang, K., Zhao, Z., Guan, S., Han, X. and Gross, R.W. (2008) Matrix-assisted laser desorption/ionization time-of-flight mass spectrometric analysis of cellular glycerophospholipids enabled by multiplexed solvent dependent analyte-matrix interactions. Anal. Chem. 80, 7576–7585.

109. Cheng, H., Sun, G., Yang, K., Gross, R.W. and Han, X. (2010) Selective desorption/ionization of sulfatides by MALDI-MS facilitated using 9-aminoacridine as matrix. J. Lipid Res. 51, 1599–1609.

110. Lobasso, S., Lopalco, P., Angelini, R., Baronio, M., Fanizzi, F.P., Babudri, F. and Corcelli, A. (2010) Lipidomic analysis of porcine olfactory epithelial membranes and cilia. Lipids 45, 593–602.

111. Dannenberger, D., Suss, R., Teuber, K., Fuchs, B., Nuernberg, K. and Schiller, J. (2010) The intact muscle lipid composition of bulls: An investigation by MALDI-TOF MS and 31P NMR. Chem. Phys. Lipids 163, 157–164.

112. Angelini, R., Babudri, F., Lobasso, S. and Corcelli, A. (2010) MALDI-TOF/MS analysis of archaebacterial lipids in lyophilized membranes dry-mixed with 9-aminoacridine. J. Lipid Res. 51, 2818–2825.

113. Angelini, R., Vitale, R., Patil, V.A., Cocco, T., Ludwig, B., Greenberg, M.L. and Corcelli, A. (2012) Lipidomics of intact mitochondria by MALDI-TOF/MS. J. Lipid Res. 53, 1417–1425.

114. Teuber, K., Schiller, J., Fuchs, B., Karas, M. and Jaskolla, T.W. (2010) Significant sensitivity improvements by matrix optimization: A MALDI-TOF mass spectrometric study of lipids from hen egg yolk. Chem. Phys. Lipids 163, 552–560.

115. Urban, P.L., Chang, C.H., Wu, J.T. and Chen, Y.C. (2011) Microscale MALDI imaging of outer-layer lipids in intact egg chambers from Drosophila melanogaster. Anal. Chem. 83, 3918–3925.

116. Marsching, C., Eckhardt, M., Grone, H.J., Sandhoff, R. and Hopf, C. (2011) Imaging of complex sulfatides SM3 and SB1a in mouse kidney using MALDI-TOF/TOF mass spectrometry. Anal. Bioanal. Chem. 401, 53–64.

117. Cerruti, C.D., Benabdellah, F., Laprevote, O., Touboul, D. and Brunelle, A. (2012) MALDI imaging and structural analysis of rat brain lipid negative ions with 9-aminoacridine matrix. Anal. Chem. 84, 2164–2171.

118. Thomas, A., Charbonneau, J.L., Fournaise, E. and Chaurand, P. (2012) Sublimation of new matrix candidates for high spatial resolution imaging mass spectrometry of lipids: Enhanced information in both positive and negative polarities after 1,5-diaminonapthalene deposition. Anal. Chem. 84, 2048–2054.

119. Fuchs, B., Schiller, J., Suss, R., Zscharnack, M., Bader, A., Muller, P., Schurenberg, M., Becker, M. and Suckau, D. (2008) Analysis of stem cell lipids by offline HPTLC-MALDI-TOF MS. Anal. Bioanal. Chem. 392, 849–860.

120. Rohlfing, A., Muthing, J., Pohlentz, G., Distler, U., Peter-Katalinic, J., Berkenkamp, S. and Dreisewerd, K. (2007) IR-MALDI-MS analysis of HPTLC-separated phospholipid mixtures directly from the TLC plate. Anal. Chem. 79, 5793–5808.

121. Stubiger, G., Pittenauer, E., Belgacem, O., Rehulka, P., Widhalm, K. and Allmaier, G. (2009) Analysis of human plasma lipids and soybean lecithin by means of high-performance thin-layer chromatography and matrix-assisted laser desorption/ionization mass spectrometry. Rapid Commun. Mass Spectrom. 23, 2711–2723.

122. Goto-Inoue, N., Hayasaka, T., Taki, T., Gonzalez, T.V. and Setou, M. (2009) A new lipidomics approach by thin-layer chromatography-blot-matrix-assisted laser desorption/ionization imaging mass spectrometry for analyzing detailed patterns of phospholipid molecular species. J. Chromatogr. A 1216, 7096–7101.

123. Taki, T., Gonzalez, T.V., Goto-Inoue, N., Hayasaka, T. and Setou, M. (2009) TLC blot (far-eastern blot) and its applications. Methods Mol. Biol. 536, 545–556.

124. Goto-Inoue, N., Hayasaka, T., Sugiura, Y., Taki, T., Li, Y.T., Matsumoto, M. and Setou, M. (2008) High-sensitivity analysis of glycosphingolipids by matrix-assisted laser desorption/ionization quadrupole ion trap time-of-flight imaging mass spectrometry on transfer membranes. J. Chromatogr. B 870, 74–83.

125. Peterson, D.S. (2007) Matrix-free methods for laser desorption/ionization mass spectrometry. Mass Spectrom. Rev. 26, 19–34.

126. Wei, J., Buriak, J.M. and Siuzdak, G. (1999) Desorption-ionization mass spectrometry on porous silicon. Nature 399, 243–246.

127. Go, E.P., Apon, J.V., Luo, G., Saghatelian, A., Daniels, R.H., Sahi, V., Dubrow, R., Cravatt, B.F., Vertes, A. and Siuzdak, G. (2005) Desorption/ionization on silicon nanowires. Anal. Chem. 77, 1641–1646.

128. Northen, T.R., Yanes, O., Northen, M.T., Marrinucci, D., Uritboonthai, W., Apon, J., Golledge, S.L., Nordstrom, A. and Siuzdak, G. (2007) Clathrate nanostructures for mass spectrometry. Nature 449, 1033–1036.

129. Woo, H.K., Northen, T.R., Yanes, O. and Siuzdak, G. (2008) Nanostructure-initiator mass spectrometry: A protocol for preparing and applying NIMS surfaces for high-sensitivity mass analysis. Nat. Protoc. 3, 1341–1349.

130. Patti, G.J., Shriver, L.P., Wassif, C.A., Woo, H.K., Uritboonthai, W., Apon, J., Manchester, M., Porter, F.D. and Siuzdak, G. (2010) Nanostructure-initiator mass spectrometry (NIMS) imaging of brain cholesterol metabolites in Smith-Lemli-Opitz syndrome. Neuroscience 170, 858–864.

131. Patti, G.J., Woo, H.K., Yanes, O., Shriver, L., Thomas, D., Uritboonthai, W., Apon, J.V., Steenwyk, R., Manchester, M. and Siuzdak, G. (2010) Detection of carbohydrates and steroids by cation-enhanced nanostructure-initiator mass spectrometry (NIMS) for biofluid analysis and tissue imaging. Anal. Chem. 82, 121–128.

132. Cha, S. and Yeung, E.S. (2007) Colloidal graphite-assisted laser desorption/ionization mass spectrometry and MSn of small molecules. 1. Imaging of cerebrosides directly from rat brain tissue. Anal. Chem. 79, 2373–2385.

133. Hayasaka, T., Goto-Inoue, N., Zaima, N., Shrivas, K., Kashiwagi, Y., Yamamoto, M., Nakamoto, M. and Setou, M. (2010) Imaging mass spectrometry with silver nanoparticles reveals the distribution of fatty acids in mouse retinal sections. J. Am. Soc. Mass Spectrom. 21, 1446–1454.

134. Budimir, N., Blais, J.C., Fournier, F. and Tabet, J.C. (2006) The use of desorption/ionization on porous silicon mass spectrometry for the detection of negative ions for fatty acids. Rapid Commun. Mass Spectrom. 20, 680–684.

135. Muck, A., Stelzner, T., Hubner, U., Christiansen, S. and Svatos, A. (2010) Lithographically patterned silicon nanowire arrays for matrix free LDI-TOF/MS analysis of lipids. Lab Chip 10, 320–325.

4

VARIABLES IN MASS SPECTROMETRY FOR LIPIDOMICS

4.1 INTRODUCTION

It is well known that performing a successful MS experiment depends on the hardware of a mass spectrometer, which is schematically illustrated in Figure 2.1 and extensively discussed in Chapter 2. However, recognizing all the factors that may affect an experiment and then possibly optimizing all of the experimental conditions and instrumental settings are the key for a researcher to successfully conduct an MS experiment by utilizing an available instrument. There exist many variables relevant to the experimental conditions and instrumental settings at every stage of an MS experiment from sample preparation, sample introduction, separation, ionization, and identification and quantification. In this chapter, each of those variables for the analysis of lipid species is discussed to a certain extent except for those related to quantification, which are described in Part III.

4.2 VARIABLES IN LIPID EXTRACTION (I.E., MULTIPLEX EXTRACTION CONDITIONS)

4.2.1 The pH Conditions of Lipid Extraction

The pH conditions during sample preparation, in general, and lipid extraction, in particular, are important as they affect lipid analysis to a great extent since the polarity and/or the charge properties of the species of many lipid classes are pH-dependent.

Lipidomics: Comprehensive Mass Spectrometry of Lipids, First Edition. Xianlin Han.
© 2016 John Wiley & Sons, Inc. Published 2016 by John Wiley & Sons, Inc.

For example, PE species are positively charged under acidic conditions and carry a net negative charge under basic conditions. Another example is the acidic lipid species including those carrying a functional group of phosphate, sulfate, or carboxylate (see Chapter 1). They possess at least one net negative charge at pH greater than 5 while all of these species become charge-neutral under more acidic conditions. Accordingly, by changing the pH condition of an extraction solution, both the polarity and charge properties of the species of different lipid classes can be altered. This variation not only can affect the recovery of different lipid classes during extraction but also have impacts on the ionization efficiency and the formation of ion types. In practice, an acidic condition is favorable for the extraction of acidic lipid species such as PA, while a neutral condition is preferred for the extraction of PE species. A weak acidic condition might be employed for "total" lipid extraction to balance these effects, which is further discussed in Chapter 13. Following are the effects of different pH conditions on ionization efficiency and ion formation that are discussed.

4.2.2 Solvent Polarity of Lipid Extraction

The polarity of the solvent used in lipid extraction has a large impact on the recovery of lipid species. Different types of solvents can be exploited to selectively extract different categories of lipids. For example, nonpolar solvents (e.g., hexane, ethyl ether) can be used to extract nonpolar lipids, including TAG, cholesterol and cholesteryl esters, and nonesterified fatty acids (NEFA). The majority of lipid classes can be extracted into chloroform (or dichloromethane) by using the Foch method, a modified procedure of Bligh and Dyer [1], or into methyl-*tert*-butyl ether (MTBE) [2] (see Chapter 13 for details). Very polar lipids (e.g., acyl CoAs, acyl carnitines, lysoGPLs, PI polyphosphates, gangliosides) can be largely recovered from the aqueous phase of a solvent extraction method aforementioned by employing some types of SPE column (e.g., reversed-phase or affinity) or with special solvents [3–5].

It should be emphasized that adding at least one species of each lipid class of interest prior to extraction to serve as an internal standard is very important for the sake of quantitative analysis. Any incomplete recovery could be compensated for with the internal standard(s) to a certain degree since any differential extraction efficiency of individual species in a class from that of the standard is largely a matter of secondary effect. However, for any analytical method conducted based on external standards, a complete recovery of extraction of a lipid class of interest is essential for accurate quantification.

4.2.3 Intrinsic Chemical Properties of Lipids

The intrinsic chemical stabilities of different lipid classes and subclasses can be considered as a variable in sample preparations to achieve global lipidomic analysis by MS. For example, lipid species containing a sphingoid backbone or ether-linked glycerolipids are stable under alkaline treatment. Therefore, this chemical stability has been well exploited to isolate and enrich sphingolipids and ether-containing lysolipids upon base-treatment of the samples, which hydrolyzes all ester-linked glycerolipids [6–8]. In contrast, vinyl ether-linked lipid species (i.e., plasmalogens) (see Chapter 1) are very sensitive to an acidic environment. Researchers have explored this chemical instability to unambiguously distinguish the presence of plasmalogen species from their plasmanyl counter-isomers by comparing the mass spectra acquired before and after acid treatment of a lipid sample [9–11].

Similarly, the differential reactivities of different functional groups possessed by the different lipid classes and subclasses could be explored to specifically analyze the classes and subclasses. For example, by exploring the specific reactivity of the primary amine group, multiple laboratories have developed methods for specific and enhanced analyses of primary amine-containing species, including PE and lysoPE [12–15]. Moreover, targeted to those lipid classes containing a hydroxyl group such as oxysterols, diacylglycerols (DAG), monoacylglycerols (MAG), anandamide and analogs, different derivative methods have been applied to enhance the ionization efficiency and confer a rich-informative and characteristic fragmentation pattern for effective analysis of these hydroxy-containing lipid classes [16–20]. Through coupling N-(4-aminomethylphenyl)pyridinium or other similar reagents to the carboxylic acid group to form an amide linkage, sensitive measurement of eicosanoids, definitive determination of the double-bond locations in NEFA species, and even analysis of the entire fatty acidomics in a high-throughput manner have been developed [21–24]. By exploring the reaction activity of 4-hydroxyalkenal species with carnosine (a diamino acid), analysis of these reactive lipid oxidative intermediates can be accurately and effectively achieved at the first time [25]. The differential reactivity of vinyl ether double bond present in plasmalogen species in comparison to those present in aliphatic chains can be exploited to unambiguous identification of isomeric ions resulted from plasmenyl- and plasmanyl-ether containing lipid species after derivatization with iodine in methanol [15].

Collectively, the variation of the intrinsic chemical stabilities and/or reactivities of different lipid classes and subclasses is a useful dimension for analysis of these classes and subclasses. Exploring methods in this area likely leads us to achieve the selective analysis of an individual lipid class/subclass in lipidomics.

4.3   VARIABLES IN THE INFUSION SOLUTION

The principles, and the bolts and nuts of the tools used to derive lipid sample solutions to the ion source (i.e., sample inlets), are discussed to a certain degree in Chapter 3. This section discusses the variations present in the lipid sample solutions prior to the ion source, as well as the effects of these variations on MS analysis of lipids. Because these variables are present in the cases of sample solutions delivered by using any type of inlet, the word "infusion" used herein generalizes all of those cases, but not for the case of direct infusion.

4.3.1   Polarity, Composition, Ion Pairing, and Other Variations in the Infusion Solution

Solvent(s) used in the infused lipid solution not only affect the solubility of lipids but also have a large impact on ionization efficiency in addition to the separation in

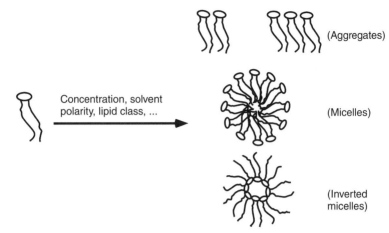

Figure 4.1   Schematic illustration of lipid-forming aggregation. Formation of aggression is a process depending on many factors including lipid concentration, lipid class, lipid molecular species, and solvent polarity and composition.

the case of liquid chromatography-mass spectrometry (LC-MS). It is well known that the solubility of a specific class of lipids is very much dependent on the solvent used. Specifically, the majority of lipids are poorly soluble in polar solvents (Figure 4.1); they form aggregates, micelles, or vesicles as concentration increases, whereas lipids at an aggregated state cannot be effectively and quantitatively ionized (Chapter 15). Accordingly, variation of solvent polarity and/or composition is important to achieve an optimal experimental condition. To this end, familiarity with the chemical and physical properties of the lipid samples is crucial to properly select a solvent or a mixture of solvents. A continuous variation of solvent composition in both shotgun lipidomics and LC-MS could be achieved with a gradient mixer if necessary. In current lipidomic analysis with electrospray ionization mass spectrometry (ESI-MS), the most commonly used solvent system is chloroform (or dichloromethane) and methanol (from 1:2 to 2:1 (v/v)) with or without the addition of modifiers (e.g., organic acid, base, or salt) in the case of shotgun lipidomics, whereas an appropriate solvent system for LC-MS is largely determined by chromatographic mode. Isopropanol is another commonly used solvent, mostly in combination with chloroform and methanol.

The ions present in the infusion solution greatly affect ionization of lipids in addition to separation in the case of LC-MS. Based on the principles of ESI-MS (see Chapter 2), these ions not only affect the ionization efficiency of individual lipid species but also change the formation of lipid adducts. The presence of inorganic salt(s) in the infusion solution may destabilize the ionization current and reduce the ionization efficiency. Therefore, a clean extraction (without aqueous phase contamination) is critical for the majority of the experiments in shotgun lipidomics. It should be recognized that sodium ions are present everywhere (biological samples, glass test tubes, infusion capillary tubing, stationary phases, etc.), and sodium adducts could be the predominant molecular ions of lipid species if no other modifier(s) are added. Even with the addition of modifier(s), low-abundance sodium adducts are still present in many cases. This may complicate the interpretation of results obtained from the LC-MS analysis in certain cases.

4.3.2 Variations of the Levels or Composition of a Modifier in the Infusion Solution

Similar to the residual ions in the infusion solution, the chemical and physical properties of the modifier added to the infusion solution can significantly influence the ionization sensitivity and efficiency, as well as the molecule-ion profiles of different lipid classes. The effects of the modifier on charge-neutral lipid classes are particularly larger in comparison to anionic lipid classes (Chapter 1). For example, PC species are generally ionized as the cation adduct from the modifier in the positive-ion mode and as an anion adduct from the modifier in the negative-ion mode.

With the addition of an acidic modifier (e.g., formic acid, acetic acid), the ionization of PC, SM, and PE in a lipid mixture as the protonated ions is favored in the positive-ion mode (Figure 4.2a). Although the species of other acidic lipid classes (e.g., PS, PI, and PG) can also be ionized in the positive-ion mode under the acidic conditions without the coexistence of charge-neutral lipids such as PC, SM, or PE species in the mixture [27], they are virtually suppressed by the coexisting charge-neutral lipids (Figure 4.2a). In the negative-ion mode, charge-neutral lipids are ionized as the adducts of the anion moiety from the modifier or as de-methyl species from PC and SM, whereas the acidic lipid species are ionized as deprotonated species; but the ionization is not favored under the conditions because an acidic condition prevents the deprotonation of these species. This situation gets worse as the concentration of the acid increases or the acidity gets stronger. Moreover, it should also be recognized that the ionization of the charge-neutral lipids as anion adducts could lead to a very complicated mass spectrum acquired in the negative-ion mode under the condition (Figure 4.2b), and greatly increase the possibility of detecting isobaric or isomeric ions of anionic lipid species. Accordingly, negative-ion ESI-MS analysis of charge-neutral lipids is not favorable for shotgun lipidomics and is rarely employed for LC-MS analysis with addition of an acidic modifier.

With the addition of a neutral modifier (e.g., ammonium acetate, ammonium formate, lithium chloride), phosphocholine-containing lipid species (e.g., PC and SM) are ionized as the protonated ions in the case of a modifier with ammonium or the cationic adducts of the modifier, whereas nonzwitterionic lipids (e.g., HexCer, TAG, DAG) do not form or are only ionized as the cation adducts in the positive-ion mode (Figure 4.2c) [28, 29]. It should be pointed out that protonated TAG species were assigned in a study, but no evidence with the identity, except mass match, was provided [30]. In the positive-ion mode, ionization of anionic lipids are suppressed with the cation adducts of charge-neutral lipid species due to the poor ionization efficiency of the anionic lipids in comparison with charge-neutral lipids (Figure 4.2c). Again, anionic lipids can still be ionized as the protonated species or the cation adducts if there is absence of PC, SM, and PE species (see below), or in the case of LC-MS

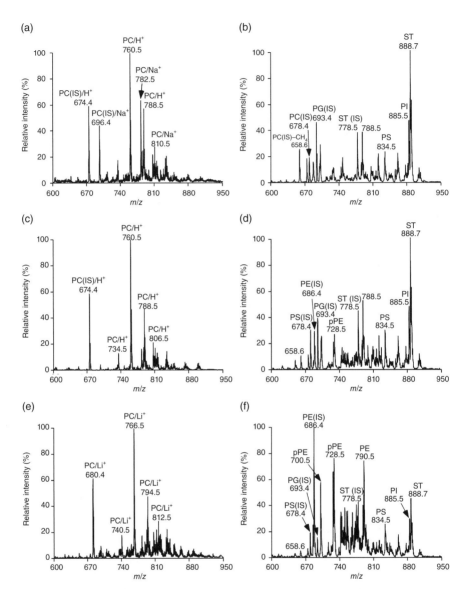

Figure 4.2 Representative positive- and negative-ion ESI mass spectra acquired under weak acidic, neutral, and weak basic conditions. A lipid extract of mouse spinal cord at 48 days was prepared and mass spectrometric analysis was performed [26]. Positive- and negative-ion ESI mass spectra as indicated were acquired after direct infusion in the presence of 0.5% acetic acid (a and b), 5 mM ammonium acetate (c and d), and 10 μM lithium hydroxide (e and f) in the infused solution. IS, pPE, and ST stand for "internal standard," "plasmenylethanolamine," and "sulfatide," respectively.

analysis [27]. The molecular species of phosphoethanolamine-containing lipids possess weak acidic properties (see Chapter 1). Accordingly, ionization of these species falls in between the phosphocholine-containing lipids and anionic lipids in the positive-ion mode under the condition. In the negative-ion mode under the experimental condition, anionic lipids, including phosphoethanolamine-containing lipids, are selectively ionized as the deprotonated molecule ions, whereas phosphocholine-containing lipid species can be ionized as anion adducts to a certain extent (Figure 4.2d). For analysis of an equimolar mixture, the peak intensities of the ions that correspond to anionic lipids are always more abundant than those that correspond to phosphocholine-containing lipid adducts in the negative-ion mode under the condition [31]. As expected, this type of modifier is favorable for detection of phosphocholine-containing lipid species in the positive-ion mode and for detection of anionic lipids in the negative-ion mode in LC-MS analysis of lipid mixtures. Accordingly, these types of modifiers are commonly used in LC-MS analysis with all types of chromatographic modes (see Chapter 3).

With the addition of the basic modifiers (e.g., LiOH, NH_4OH), all the anionic lipids, including phosphoethanolamine-containing lipid species, which are present in the deprotonated form in the solution, are not easy to be ionized in the positive-ion mode; but their ionization is enhanced in the negative-ion mode. Moreover, unlike other anions (e.g., chloride, acetate, formate), ionization of phosphocholine-containing lipid species as OH^- adduct does not seem to be favored. Accordingly, phosphocholine-containing lipid species are selectively ionized as the protonated ions or cation adducts in the positive-ion mode, whereas anionic lipids, including phosphoethanolamine-containing lipid species, are selectively ionized in the negative-ion mode with the basic modifiers (Figure 4.2e and f). Because all the anionic lipids, including phosphoethanolamine-containing lipid species, are present in the deprotonated form in the solution under the condition, their ionization efficiencies in the negative-ion mode are quite similar as previously demonstrated [31]. Hence, the peak intensity ratios well represent the molar ratios of these lipid species (Figure 4.2f). This observation is in contrast to those when acidic or neutral modifiers are added, where anionic lipids such as PG and PS species are selectively ionized in comparison to weakly anionic lipid species (e.g., phosphoethanolamine-containing lipids) (Figure 4.2b and d). These types of selective ionization have been the bases of the intrasource separation technique as previously described [31–33] and are summarized in Chapter 2. Because of the selective ionization of charge-neutral lipids *vs.* anionic lipids in the positive- and negative-ion modes, respectively, basic modifiers are also frequently employed in LC-MS analysis of lipids [34].

In addition to the substantial effects of the types of modifiers on ionization sensitivity and selectivity, the concentration of a selected modifier could also yield substantial influence on ionization, including the formation of ion adducts. For example, ESI-MS analysis of a mixture of GPL species in the negative-ion mode (Figure 4.3a) demonstrated the decrease in the ion peaks corresponding to the PG species when the concentration of lithium hydroxide (LiOH, a modifier) increased. This occurred due to the fact that PE species were rendered anionic after addition of LiOH and the ionization efficiency of anionic PE species was similar to that of anionic GPL species

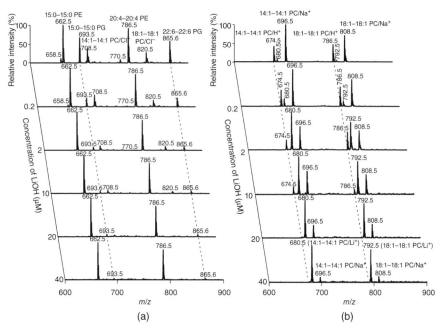

Figure 4.3 The effects of lithium hydroxide concentration on ionization of different lipid classes. A GPL mixture consisting of di15:0 and di22:6 PG, di14:1 and di18:1 PC, and di15:0 and di20:4 PE molecular species in a molar ratio of 1:1:10:10:15:15 in 1:1 (v/v) chloroform/methanol was prepared and analyzed by varying the addition of a different amount of LiOH (as a modifier) as indicated in both negative- (a) and positive-ion (b) modes. ESI-MS analysis of the GPL mixture was conducted utilizing a TSQ ESI mass spectrometer (Thermo Fisher Scientific, San Jose, CA).

[35–37]. Similar results were also obtained by addition of small amounts of other alkaline media including ammonium hydroxide. The analysis from negative-ion ESI mass spectrum also showed the disappearance of ion peaks corresponding to chlorinated PC species as the concentration of LiOH increased in the infusion solution. This observation likely resulted from the combination of selective ionization of PE under the stated conditions and a reduced availability of chloride to PC species in the presence of hydroxyl ion. Intriguingly, no ion peaks corresponding to the hydroxyl adducted PC species were present in the spectrum even at the highest concentration of LiOH examined, indicating that unlike chloride, acetate, or other small anions, the affinity of hydroxyl ion to PC species is very weak.

When the GPL mixture was analyzed in the positive-ion mode, the mass spectrum showed that the substantive ion peaks were comprised of the various adducts of PC species (Figure 4.3b), indicating the selective ionization of PC species from other classes of lipids. These mass spectra also showed that the disappearance of protonated ion peaks is due to the neutralization of protons by LiOH, and the appearance of lithiated adducts is due to the increased availability of lithium ions in the infusion solutions after increasing addition of LiOH. The persistent presence of ion peaks corresponding to sodiated PC species even at the relatively high concentration of LiOH suggests either the ready availability of sodium ions in the system or a high affinity of sodium ions with PC species or both.

Similar to the case of authentic GPL mixtures as shown in Figure 2.5, regardless of the presence of different modifiers that could substantially affect the molecule-ion profiles of different lipid classes displayed in different mass spectra that are acquired under different experimental conditions, the profile of individual molecular species of each lipid class is not influenced. For example, the ionization efficiencies of PE species are very different under different experimental conditions as demonstrated and discussed in Figure 4.2. However, the ratios of individual PE species (e.g., di16:1 PE *vs.* 18:0–22:6 PE at m/z 686.4 and 790.5 in Figure 4.2d and f, respectively) are essentially constant. Similarly, these constant ratios can be readily observed between the PC species such as the pair of di14:0 PC and 16:0–18:1 PC as protonated species at m/z 674.4 and 760.5 (Figure 4.2a and c) and lithium adducts at m/z 680.4 and 766.5 (Figure 4.2e), respectively. Similar results were also demonstrated in the study shown in Figure 4.3. These observations strongly support **the notion** that *the ionization efficiencies (or response factors) of individual molecular species of a polar lipid class as detected by a full mass spectrum are essentially identical, and their fatty acyl chains only minimally contribute to the ionization efficiencies under the experimental conditions.* Thus, individual molecular species of a polar lipid class can be quantified through comparisons of ion peak intensities (i.e., **ratiometric comparison**) with the selected internal standard of the class (Chapter 14). However, in the case of LC-MS analysis, this notion should always be verified prior to the measurements because employing a gradient of mobile phase or constantly changing the concentration of an eluent might result in different response factors of individual lipid species of a class.

Again, if a continuous variation of either the concentration of a modifier or a component of the modifiers in the solution is required, a gradient mixer with a selective switch can be employed in the case of shotgun lipidomics. In this case of LC-MS analysis, such variation can be readily achieved through a gradient of mobile phases containing different modifiers if necessary. Although stepwise addition of different modifiers or different amount of a modifier is frequently used in the practice of lipidomic analysis, a continuous change of modifiers has not been seen in the literature.

4.3.3 Lipid Concentration in the Infusion Solution

Many studies have demonstrated that ESI-MS analysis of lipids is concentration-dependent [35, 38, 39]. At a concentration higher than 0.1 nmol/µL (i.e., 0.1 mM) in chloroform–methanol (1:1, v/v), the effects of acyl chain length and unsaturation on ionization efficiency (then response factor) are apparent [38, 40]. This observation is largely due to a well-known fact that lipids tend to form aggregates (dimers, oligomers, micelles, and even vesicles) as lipid concentration and/or solvent polarity increases (Figure 4.1).

Ionization of these clusters is very complicated from a few aspects as follows:

- The sizes of aggregates (i.e., the number of lipid molecules in each cluster) are unpredictable; they are present in different sizes, which again largely depend on the lipid concentration, lipid composition, solvent polarity, etc. Therefore, these clusters could appear in a broad mass region.

- The composition of an individual aggregate that results from a complex lipid mixture is unpredictable and complex; generally, the more hydrophobic species are much easier to form aggregates than the less hydrophobic species. This could occur in shotgun lipidomics, MALDI-MS analysis, and LC-MS in any modes for global lipidomic analysis.

- The number of small ion adducts or deprotonated lipid species associated with the aggregated clusters are unpredictable and complex. For example, the number of the adducted small ions could vary from zero to a number greater than those of the lipid species in the cluster, leading to a broad range of m/z and a variety of peak shapes, since a peak from the cluster with multi-ions is essentially shown as a bump.

- All these clusters not only lead to higher noise levels due to the aforementioned reasons but also reduce the concentrations of the monomers, resulting in reduced apparent sensitivity. Accordingly, higher concentration than that above which the aggregates form does not increase in ionization efficiency, but indeed decreases the sensitivity of ionization as aforementioned.

It should be emphasized that in the low concentration range (i.e., pmol/μL or lower), a linear correlation of the absolute ion intensity with the concentration of each polar lipid class is obtained [35, 38, 39, 41] (see Chapter 15). Regardless of what the method (i.e., shotgun lipidomics or LC-MS) is employed, this linear correlation should always be determined in the presence of other matrix lipids present in biological lipid extracts along with the internal standard(s). The ionization efficiency (or the response factor of a mass spectrometer) of lipid species depends predominantly on the charge properties of the polar head groups in this concentration range after correction for ^{13}C isotopologue distributions [33, 35, 42, 43].

From the biophysical chemistry point of view, use of the concentration of the infusion solution as a variable could allow one to determine the formation of aggregates from identical or different lipid species, and to study chemical or physical properties of lipid molecules with the competition for proton, small matrix cation, or anion among lipid molecules in a lipid class or molecules in different lipid classes. In practice, a continuous variation of lipid concentration can be achieved with a gradient mixer. In general, titration of lipid concentration in both shotgun lipidomics and LC-MS analysis for determination of the linear dynamic range is always practiced, which is further discussed in Chapter 15.

4.4 VARIABLES IN IONIZATION

4.4.1 Source Temperature

Ion source temperature plays an important role in ion generation. Specifically, desolvation is very important in the ionization process that occurs in an ESI source, whereas source temperature is a critical factor in desolvation and significantly affects the ionization efficiency. Temperature condition for the analysis of any samples with ESI-MS is usually optimized prior to the analysis with the mobile phase solution used by either shotgun lipidomics or LC-MS approaches. However, the required temperature for optimal ionization of each individual lipid class may be different. Moreover, due to the variation in the components, as well as composition of individual samples, the pre-optimized temperature condition might not be the best one for a particular sample, particularly in the case of LC-MS analysis. Accordingly, ramping the temperature should provide the best condition for the analysis of each individual lipid class of a sample or for the analysis of each specific sample. In addition, through a temperature ramping, information about the ionization efficiency of different components present in the samples can be revealed. Furthermore, any interactions among the analytes, solvents, and other matrix components can also be interrogated with data array analysis (i.e., MDMS analysis) acquired at the variety of temperatures. In the case of LC-MS analysis with a gradient of mobile phases, a specific program of temperature variation corresponding to the mobile phase components could be considered to achieve optimal ionization sensitivity.

In addition to the effects of source temperature on ionization, source temperature is also associated with source fragmentation of lipids. The higher the source temperature, the more severe the source fragmentation could occur. Accordingly, optimization of source temperature for ionization efficiency should also consider this degradation fact. Moreover, the influence on source fragmentation is molecular species and lipid class dependent. For example, a loss of serine from PS species resulting in PA species can readily occur if source temperature is not well-controlled, which may complicate the quantification of PA species. Dimethyl PE ions yielding from anionic adducts of PC are also detectable under some experimental conditions.

4.4.2   Spray Voltage

Similar to the variation of source temperature, spray voltage is also a critical factor that affects ion generations and ionization efficiency in all the ionization techniques. This factor is usually pre-determined in one of the steps such as tuning, calibration, or optimization with a standard solution. It should be kept in mind that such a pre-optimized parameter might not be the best condition for both shotgun lipidomics and LC-MS approaches. For the former, the pre-determined parameter(s) may not be the best suited for the analysis of each individual lipid species or individual lipid class in a complex lipid sample that contains different lipid classes possessing different charge properties. For the latter, this parameter may need to be varied based on the variations in the concentration and composition of the analytes and/or matrix components. Accordingly, it is desired to have the spray voltage optimized for the analysis of each individual lipid class in a sample or of each individual sample. In theory, this could be achieved by determining the effects of spray voltage on the ionization efficiency of different components present in the samples with a spray voltage ramping. Unfortunately, it is still currently impractical to perform such a ramping based on the variations in components, samples, and/or eluent conditions. However, this type of ramping should become achievable as instrumentation becomes more and more sensitive and rapidly responsive.

4.4.3   Injection/Eluent Flow Rate

The flow rate of injection in the case of shotgun lipidomics or eluent in the case of LC-MS analysis is also an important variation in ionization. It is well known that nanospray can substantially improve ionization sensitivity, as well as save great amounts of source materials. Our previous studies have demonstrated that flow rates not only affect the ionization efficiency and sensitivity but also influence other chemical or physical consequences of the ionization process, such as intrasource separation [31]. For example, Figure 4.4a shows that although the ionization efficiencies of PG increased as the flow rate increased at any given concentration of the examined GPL mixture in the negative-ion mode, the increases were insignificant at the flow rates of greater than 4 µL/min. The effects of the flow rate on the ionization efficiencies of PE and PC classes at any given concentration in the infusion mixture employed were minimal (Figure 4.4b). These results indicate that the selective ionization of PG from PE or PC was constant within experimental error for the given instrument under a wide range of experimental conditions employed.

It should be recognized that the effects of the lipid concentrations on the selective ionization of anionic lipid classes from PE and PC classes as determined in the study are intriguing. At the low concentration such as 0.1 pmol/µL of each PG species in the examined mixture, the contribution of each GPL class to the spray current (i.e., ionization efficiency) was essentially identical and independent of the infusion flow rates. This result indicates that at the low concentrations employed, the solution behaves as an "ideal" solution and each molecular species that exists in solution has an equal chance to be ionized to maintain the minimal faradic current. As the concentration of analytes in the infusion solution increased, the ionization efficiency of PG exponentially increased, whereas those of PE and PC concordantly decreased, indicating an augmentation of the ionization selectivity of PG molecular species accompanied by the reduction of the ionization efficiency of less readily ionizable coexisting molecular species such as PE and PC to achieve a maximal spray current. This selectivity is likely achieved through a small cation transferring from a neutral anionic GPL species to PC or PE species. This transfer leads to a charge redistribution, which prevents PE and PC species from entering the plate orifice (or other ion inlets). This rationale was experimentally validated in the original study [31].

The variation of injection/eluent flow rate is usually pre-determined at the optimization step using a standard solution with an available instrument. Such a pre-optimized parameter might not be the best condition for the analysis of each individual sample, which always contains the varied concentration and composition of lipids and/or matrix components. Accordingly, it should always be kept in mind that optimization of the flow rate is important for optimal analysis of each individual sample or any eluent stage. Moreover, variation of injection/eluent flow rates may lead us to investigate the interactions among the analytes, solvents, and other matrix components, as well as their chemical/physical consequences.

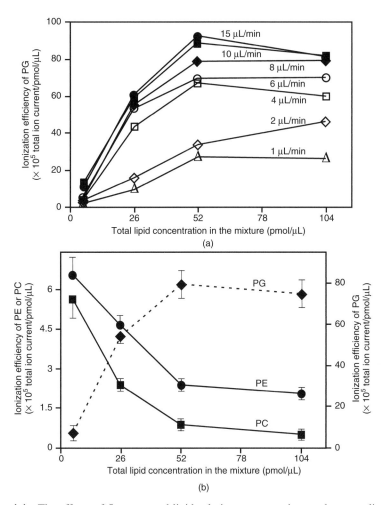

Figure 4.4 The effects of flow rate and lipid solution concentration on the normalized ion count density (i.e., ionization efficiency) of each examined GPL class in the negative-ion mode. The GPL mixture was comprised of di15:0 and di22:6 PG, di14:1 and di18:1 PC, and di15:0 and di20:4 PE molecular species in a molar ratio of 1:1:10:10:15:15 in 1:1 (v/v) chloroform/methanol. Ion intensities in ion current of each GPL species were determined from ESI-MS analyses of three independent injections utilizing a TSQ ESI mass spectrometer (Thermo Fisher Scientific, San Jose, CA). The ionization efficiencies of PG (a) and the broken line in (b), PC (b), and PE (b) were calculated from the normalization of the total ion current of a class to a unit concentration (i.e., pmol/μL) of each class at the indicated lipid concentration and flow rate. The ionization efficiency of PE or PC represents the mean ± SD of the ionization efficiencies of the class at the flow rates of 1, 2, 4, 6, 8, 10, and 15 μL/min. The ionization efficiency of PG in (b) represents the mean ± SD of the ionization efficiencies in (a) at the flow rates of 4, 6, 8, 10, and 15 μL/min. Some of the error bars are within the symbols. Han et al. [31]. Adapted with permission of Springer.

4.5 VARIABLES IN BUILDING-BLOCK MONITORING WITH MS/MS SCANNING

4.5.1 Precursor-Ion Scanning of a Fragment Ion Whose *m/z* Serves as a Variable

The PIS technique has been discussed in details in Chapter 2, which is a powerful and effective tool to detect fragments after CID. By using this technique, a group or a category of lipid species that yield a common fragment ion can be selectively determined through monitoring the common fragment ion. This tool has been widely used in shotgun lipidomics (Chapter 3). Although practical execution of this technique in LC-MS analysis is difficult, extracting a common fragment ion to have a pseudo-PIS analysis from an array of data-dependent analysis of lipid species is possible and should be very useful to detect the presence of the species yielding the common fragment ion.

In theory, this variable in *m/z* should vary from zero to the highest value (M_h) of *m/z* of a mass range of interest. However, the lowest values of an informative fragment ion in lipid analysis are generally the adduct ions (i.e., Li^+, Na^+, Cl^-), whereas the highest value of such a fragment ion should be that corresponding to the neutral loss of a small molecule (e.g., H_2O, CH_4) from the adducted molecule ion. Accordingly, the ramped *m/z* range of this variable can be performed in the range of *m/z* from 50 to M_h-18 if someone would like to map all lipid species as demonstrated previously [44].

In practice, one can learn the resultant product ions of all species of a lipid class of interest from the analysis of a few representative lipid species of the class (i.e., pattern recognition, see Chapter 6). Hence, this variable in PIS analysis can vary only to detect those potential product ions in a pattern, which are largely related to the building blocks of the group of lipids as described in Chapter 1. Indeed, the identity of an individual lipid species can be fully determined with only two or three characteristic product ions that carry the building-block information about the species in addition to the *m/z* of the molecule ion. Accordingly, the task of mapping the fragment ions of a lipid sample could be simplified to only monitor these characteristic building-block fragment ions.

4.5.2 Neutral-Loss Scanning of a Neutral Fragment Whose Mass Serves as a Variable

Similar to PIS, the NLS technique has also been discussed in detail in Chapter 2, and it is a powerful and effective tool to detect fragments after CID. By using this technique, a group or a category of lipid species that carry a common neutrally lost fragment can be selectively determined through monitoring the loss of this common neutral fragment. This tandem MS mode has been widely applied in shotgun lipidomics (Chapter 3). Although it is impractical to perform this technique in LC-MS analysis, extracting a common neutral-loss fragment to obtain a pseudo-NLS analysis from an array of data-dependent analysis of lipid species is possible and should be very useful in detecting the presence of the species yielding the common neutral-loss fragment.

This variable in neutrally lost mass could vary in a mass range from zero to the highest mass value (M_h) of molecule ions of interest. The lowest value of a neutral-loss fragment in lipid analysis could occur with a value of a small molecule lost (e.g., H_2O, CH_4), whereas the highest value of a neutral-loss fragment is the molecule that is lost to yield the smallest fragment ion. Accordingly, this variable varies in a reversed order to that in the PIS mapping. It is obvious that NLS forms a new dimension to map all the fragment ions and interweaves with other tandem MS techniques as aforementioned (see Chapter 2).

As previously discussed (Chapter 2), an apparent neutral-loss fragment is equivalent to the mass difference between a molecular ion and a product ion. Thus, all the neutral-loss fragments can be learned from the product-ion analysis of a few authentic lipid species of a class of interest (i.e., pattern recognition). Therefore, this variable in NLS analysis of a lipid class can only be narrowed to monitor those neutral-loss fragments that correspond to all the potential product ions produced from the lipid species of the class. Again, the identity of a lipid species can be fully determined with only two or three neutral-loss fragments that correspond to the characteristic product ions, which carry the building-block information about the species in addition to the m/z value of molecule ion. Thus, the task of mapping the neutral-loss fragments can be simplified to only monitor those characteristics of building blocks, instead of NLS mapping all the potential neutral-loss fragments in the entire mass range.

4.5.3 Fragments Associated with the Building Blocks are the Variables in Product-Ion MS Analysis

Each individual molecular ion yields different characteristic product ions in a collision condition-dependent manner (see next section). These characteristic product ions are tightly associated with the building blocks of individual lipid molecular species and provide a pattern of individual lipid class from which individual molecular species can be relayed after considering the m/z value. Although it appears that the m/z values in a full scan mass spectrum within the mass range of interest serve as the variables in the product-ion analysis, the new dimension is the resultant product ions that constitute a 2D map from the building-block point of view. Therefore, in product-ion analysis, the yielded product ions could be conceivably recognized as a variable.

To detect the correct product ions, selection of the mass window for a precursor ion is critical in the product-ion analysis mode. To this end, either a small mass window centered on the ion peaks should be selected or a mass spectrometer with high mass accuracy/resolution should be employed to eliminate any complications due to the overflow from the adjacent peaks. With the help of high mass accuracy/resolution possessed by a mass spectrometer, a wider mass window could be employed [45]. For the majority of the unit mass resolution instruments, one can scan only the m/z of the molecule ions present in a full scan mass spectrum.

It should be noted that as the development of mass spectrometers possessing a high duty cycle such as SCIEX TripleTOF 5600 system, which shows a 100-Hz acquisition speed, conducting product-ion analysis of all the ions in a mass range of interest unit by unit or information-dependent acquisition becomes feasible. Alternatively, when an ultra-mass accuracy/resolution instrument such as the Orbitrap Fusion (Lumos) Tribrid mass spectrometer from Thermo Fisher Scientific is available, all-ion-fragmentation can be performed and a neutral-loss pattern of each individual lipid species of interest can be determined for those building blocks as previously described [46, 47] in combination with accurate mass search [48].

4.6 VARIABLES IN COLLISION

4.6.1 Collision Energy

Collision energy provides kinetic energy to the precursor ions in a collision cell. Some of the kinetic energy is converted into internal energy during collision with a small neutral molecule such as helium and nitrogen and results in bond breakage and consequently the fragmentation of the precursor ion into smaller fragments. The effects of collision energy on the tandem MS analysis in any modes aforementioned (Chapter 2) have been well recognized [43, 49, 50] and even applied for the elucidation of chemical and/or physical properties of lipid species [50, 51]. For example, tandem MS analysis of 22:6 fatty acid in the product-ion mode by varying collision energy demonstrated that the intensities of many fragment ions showed a collision energy-dependent pattern (Figure 4.5a). Among these, the fragment ion resulted from the loss of 44 mass unit (corresponding to the loss of CO_2) displayed a Gaussian-like distribution (Figure 4.5b). It was found that the apex and the shape of this Gaussian distribution greatly depend on the chain length, the double-bond number, and the location of double bond(s) of individual fatty acid. Therefore, this collision energy-dependent chemical/physical property was explored to distinguish the isomers of a particular polyunsaturated fatty acid due to differential location of double bonds [51].

Figure 4.6 showed a typical example of how collision energy could affect profiling (or quantification) of individual lipid molecular species in the neutral-loss mode. Notably, positive-ion tandem MS analysis of phosphocholine-containing species present in a lipid extract of mouse liver were performed in the neutral-loss mode through monitoring the neutral loss of 183.1 amu, corresponding to the neutral loss of phosphocholine from the lithium adducts under a variety of collision energy levels. It is shown that the spectra acquired under different collision energy are significantly different. For example, the peak intensity ratio of the ions at m/z 680.5 (a selected internal standard) and m/z 812.6 varied nearly equally at 20 eV to only approximately 25% at 50 eV of collision energy. This example not only clearly illustrates the dependency of tandem MS analysis on collision energy but also strongly evidences the presence of differential fragmentation kinetics of different lipid molecular species, which indicates that quantification of individual lipid species by using tandem MS, in general, including NLS, PIS, and MRM techniques should be very cautious and well justified.

4.6.2 Collision-Gas Pressure

Similar to the collision energy, variation of collision-gas pressure has a large impact on ion fragmentation. Collision-gas pressure mainly affects the collision pathway

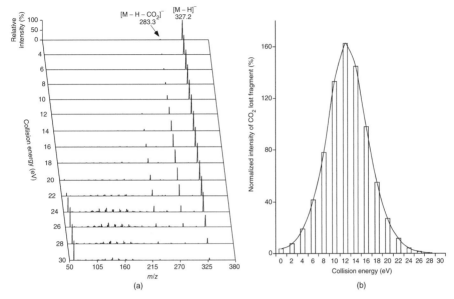

Figure 4.5 Two-dimensional MS analysis of n-3 22:6 fatty acid by varying collision energy in the negative-ion mode. MS analysis was performed with a TSQ Quantum Ultra Plus triple-quadrupole mass spectrometer (Thermo Fisher Scientific, San Jose, CA) equipped with an automated nanospray apparatus (i.e., Nanomate HD, Advion Bioscience Ltd, Ithaca, NY) and Xcalibur system software. A 2D MS analysis of 22:6 fatty acid (a) was performed by varying collision energy as indicated with a fixed collision-gas pressure of 1 mTorr in the product-ion mode. An intensity distribution of the fragment ion corresponding to the neutral loss of CO_2 from the molecular ion *vs.* collision energy (b) was obtained from the 2D MS analysis as displayed in (a) and showed as normalized absolute intensity. Yang et al. [51]. Reproduced with permission of the American Chemical Society.

and collision frequency of an ion in the collision cell. The higher the collision-gas pressure, the shorter the collision path of an ion and the more frequently the ion collides with the collision gas in the cell. This shorter collision path and increased frequency of collisional events can lead to increased fragmentation of the precursor ions, as well as sequential fragmentation of the resultant fragment ions. In other words, as collision-gas pressure increases, product-ion mass spectra display a fragmentation pattern with increased intensities of the fragments possessing lower masses. For example, Figure 4.7 shows an array of product-ion mass spectra of a molecular ion at m/z 305.2 corresponding to 8,11,14-eicosatrienoic acid (n-6) by varying collision-gas pressure. It clearly showed that the tandem MS spectra displayed more fragment ions, as well as the increased intensities of fragment ions having lower masses as increased collision-gas pressure.

Similar to the variation of collision energy, the 2D mass spectrometric analysis by varying collision-gas pressure can be explored to identify the location of the double bonds of the unsaturated fatty acids based on the intensity changes of the fragment ion

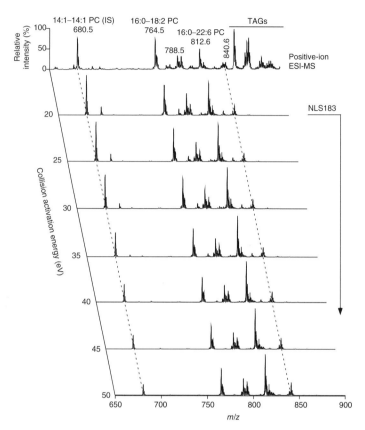

Figure 4.6 An example demonstrating the effects of collision energy on profiling phosphocholine-containing molecular species in the neutral-loss mode. A full ESI mass spectrum in the positive-ion mode was acquired directly from a diluted mouse hepatic lipid extract in the presence of a small amount of lithium hydroxide. Tandem MS profiling of PC and SM molecular species was performed through neutral-loss scanning of 183.1 amu (i.e., neutral loss of phosphocholine from the lithium adducts of PC species) under the aforementioned lipid solution condition with variation of collision energies (as indicated). "IS" denotes internal standard. All mass spectral scans are displayed after normalization to the base peak in each individual spectrum.

corresponding to [M-H-44]⁻, and used to determine the composition of the isomeric mixtures of an unsaturated fatty acid due to the differential double-bond locations [51]. Collectively, although the parameter of collision-gas pressure is determined in optimization of instrument condition and is largely fixed during an experiment, variation of collision-gas pressure could be used to study different fragmentation pathways, fragmentation patterns, and/or other chemical/physical parameters.

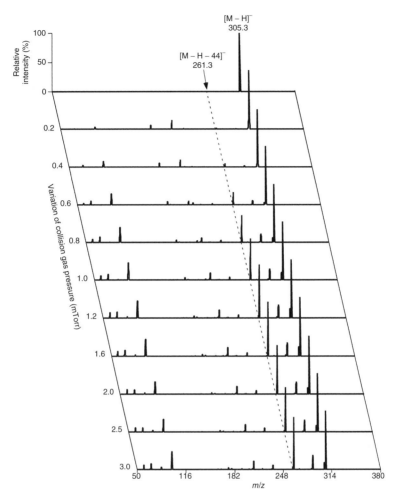

Figure 4.7 Two-dimensional mass spectrometric analysis of polyunsaturated fatty acid fragmentation pattern with variation of collision-gas pressure. MS analysis was performed with a TSQ Quantum Ultra Plus triple-quadrupole mass spectrometer (Thermo Fisher Scientific, San Jose, CA) equipped with an automated nanospray apparatus (i.e., Nanomate HD, Advion Bioscience Ltd., Ithaca, NY) and Xcalibur system software. Product-ion scan of 8,11,14-eicosatrienoic acid (20:3 FA) (5 pmol/µL) was performed after direct infusion in the negative-ion mode at the fixed collision energy of 16 eV and varied collision-gas pressures ranging from 0 to 3 mTorr as indicated. A 2-min period of signal averaging in the profile mode was employed for each scan. All the scans were automatically acquired with a customized sequence subroutine operated under Xcalibur software. All the scans are displayed after being amplified to the 5% of the base peak in each individual scan.

4.6.3   Collision Gas Type

In comparison to collision energy or collision-gas pressure, the impacts of varying collision gas type on fragmentation after CID is relatively small. Therefore, this area of research has not been paid much attention and studies by varying inert collision gas types have never been reported. We believe that this variable could also be used to study fragmentation pathways, fragmentation kinetics, and/or other chemical and physical properties of lipids because different types of collision gases possess inherently different cross-sectional areas, intrinsic energies, and other properties. For example, by using a type of collision gas containing a chiral center, it could be possible to induce strong interactions with one kind of chiral isomer, but not the other, of an analyte to achieve differential fragmentation of these isomers, whereas chiral isomers are commonly seen in eicosanoids.

4.7   VARIABLES IN SEPARATION

4.7.1   Charge Properties in Intrasource Separation

As an inherited physical feature of ESI ion source, which has been extensively discussed in Chapter 2, intrasource separation can be explored to separate lipid classes based on their charge properties. This feature has been used in both shotgun lipidomics and LC-MS analysis (see Chapter 3). In the latter, investigators have used it to selectively detect PC, lysoPC, SM, TAG, etc. in the positive-ion mode and analyze phosphoethanolamine-containing lipids, anionic lipids, etc. in the negative-ion mode [52, 53]. In shotgun lipidomics, a similar selection of lipid classes for determination in both positive- and negative-ion modes was exploited [31–33]. This feature was extensively explored in MDMS-SL as a primary separation tool to simplify the MS analysis and largely avoid the chromatographic separation step to achieve effective, high throughput, and global analysis of an entire cellular lipidome. In this section, therefore, only the principles and applications of intrasource separation in MDMS-SL are described as a representative.

Different charge properties are present in different lipid classes, which largely depend on the nature of their head groups. Based on their charge properties, lipid classes can be classified into three categories: (1) anionic lipids, (2) weakly anionic lipids, and (3) charge-neutral polar lipids (Chapters 1 and 2). The category of anionic lipids includes the lipid classes possessing at least one net negative charge under weakly acidic pH conditions (e.g., ~4). Examples of lipid classes in this category are CL, PG, PI, PS, PA, ST, their lysolipids, acyl CoA, etc. The category of weakly anionic lipids includes the lipid classes that do not possess net charge at weakly acidic pH, but become negatively charged under basic conditions. Examples of lipid classes in this category are PE, lysoPE, NEFA and their derivatives, bile acids, and ceramide. The category of charge-neutral lipids includes the lipid classes that do not carry net charge under any pH conditions. The lipid classes in this category are usually polar, such as PC, lysoPC, SM, HexCer, and acyl carnitine. Some nonpolar lipid classes such as MAG, DAG, TAG, and cholesterol and its esters also belong to this category.

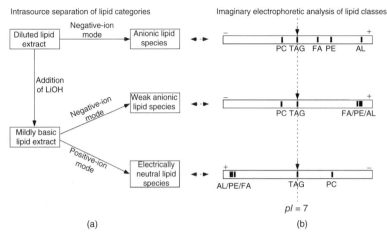

Figure 4.8   Schematic comparison of intrasource separation of lipid categories to the theoretical electrophoretic separation of lipid classes. (a) Schematically shows the selective ionization of different lipid categories under three different experimental conditions with or without adding a small amount of LiOH. (b) Schematically shows the imaginary chromatograms of lipid classes after electrophoretic analyses under corresponding experimental conditions. PC, TAG, FA, PE, and AL stand for phosphatidylcholine, triacylglyceride, nonesterified fatty acid, phosphatidylethanolamine, and anionic lipids, respectively. Christie and Han [1b]. Reproduced with permission of Elsevier.

The separation method in lipidomic analysis by using ESI ion source to selectively ionize these three lipid categories one by one has been referred to as intrasource separation [31]. A strategy to maximally achieve this separation is to vary the pH condition of the lipid solution by addition of a small amount of LiOH as similarly illustrated in Figure 2.4 (Figure 4.8a). This type of separation is comparable to the separation through electrophoresis (Figure 4.8b). A demonstration of this separation with a model system comprised of representative GPL species was given in Figure 2.5. The application of this separation to any lipid extract of a biological sample can also be similarly demonstrated as detailed in the following sections.

After preparing a lipid extract and diluting the extract to a total lipid concentration of approximately 50 pmol/µL [1b, 33] (see Chapter 13 for estimation of the total lipid levels), the category of anionic lipids is ionized in the negative-ion mode. Under the conditions, a nearly 40-fold selective ionization of anionic lipids over weakly anionic lipids was determined by using the model system (see Chapter 2 and Figure 2.5). This can selectively be calculated from the peak intensities as well as the levels of the respective internal standards for anionic lipids and weakly anionic lipids (Figure 4.9a). This selectivity is due to the basicity of the weakly anionic lipids being much weaker in comparison to that of anionic lipids as previously described [31].

Then, a small amount of LiOH methanol solution (at a final concentration of ~30 pmol LiOH/µL) is added to make the aforementioned diluted lipid solution basic. Full scan ESI-MS analysis of this solution is conducted in the negative-ion mode. The

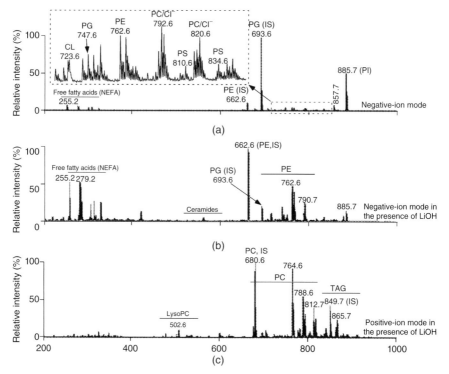

Figure 4.9 Representative ESI-MS analysis of lipid classes resolved by intrasource separation. Lipid extracts from mouse liver samples were prepared by using a modified procedure of Bligh and Dyer [1]. MS analysis was performed with a TSQ Vantage triple-quadrupole mass spectrometer (Thermo Fisher Scientific, San Jose, CA) equipped with an automated nanospray apparatus (i.e., TriVersa, Advion Bioscience Ltd., Ithaca, NY) and Xcalibur system software. Mass spectra were acquired directly from the diluted lipid extract in the negative-ion mode (a), after addition of 50 nmol LiOH/mg of protein in the diluted lipid extract and analyzed in the negative-ion mode (b), or the identical lipid solution to that in (b) in the positive-ion mode (c). "IS" denotes internal standard; PC, PE, PG, PI, PS, TAG, NEFA, and CL stand for phosphatidylcholine, phosphatidylethanolamine, phosphatidylglycerol, phosphatidylinositol, phosphatidylserine, triacylglycerol, nonesterified fatty acid, and doubly charged cardiolipin, respectively.

full mass spectrum acquired displays PE species predominantly and low abundance of anionic lipid species (Figure 4.9b). This is due to the fact that the mass level of PE class is approximately 30–40 mol% of the total lipid mass, whereas the total mass content of all anionic GPL classes only accounts for 5–20 mol%. Moreover, the intensity ratio of the internal standards for respective quantification of anionic and weakly anionic lipid species is essentially identical to the ratio of their mass levels. This suggests that the ionization efficiencies of PE and anionic GPL species are virtually identical under the conditions.

Finally, a full mass scan in the positive-ion mode is acquired from the diluted, LiOH-added lipid solution. The acquired full mass spectrum only displays the ions corresponding to the charge-neutral lipid species such as PC, SM, HexCer, TAG, which are largely ionized as lithium adducts [31] (Figure 4.9c). This selective ionization is specific to those lipid species among the third category of lipid classes because the species of both anionic and weakly anionic lipid classes carry at least one net negative charge under the experimental conditions, and ionization of the negatively charged species is virtually suppressed in the presence of a large amount of lipid species of charge-neutral classes in the positive-ion mode as described earlier in this section.

Collectively, analogous to the electrophoretic separation of charge-carried compounds, intrasource separation allows selective ionization of lipid classes possessing different charge properties *in situ* directly from an injected solution. Therefore, this separation approach leads to the analysis in an effective, high-throughput, and untargeted manner. However, it should be noted that intrasource separation only provides primary separation based on the charge categories, and the lipid classes subsequently detected by full mass spectra are still very complicated. Appropriate methods such as MDMS-SL should be employed for identification and quantification of individual lipid species after separation.

4.7.2 Elution Time in LC Separation

The variation of elution time in lipid analysis is unique to LC-MS and is associated with the column type employed. Elution time from a normal-phase HPLC is proportional to the polarities, dipole moments, and specific interactions of analytes with the stationary phase. Variation of elution time associated with a reversed-phase column is proportional to the different hydrophobicities that result from the differences in a combination of the number of carbon atoms, the number of double bonds, or the location of double bonds present in individual molecular species. The elution time from other modes such as HILIC or ion-exchange columns varies based on particular physical interactions of lipid species with the selected stationary phase and/or a partition between the mobile phase and stationary phase. Generally, the elution time of an individual lipid species depends on the physical property of the species self, the stationary phase, and the mobile phase including the eluent gradient. Although we can predict the order of elution times of individual lipid species after the column and mobile phase are selected, it is still difficult to foresee the exact elution time corresponding to a particular species. Moreover, differences of microenvironments such as minor mobile phase changes, little differences of injected samples, and the column at different ages may also affect the elution time to a certain degree. Therefore, it is common or necessary to correct the elution time to align ion peaks. In the practice of lipidomic analysis, this variation has been well demonstrated in an early study that showed the power of this dimension through identification of hundreds of lipid species in a column run [52]. In any case, the variation of elution time under different column conditions is an important dimension(s) in lipidomics, and it also needs well attention with the small variations from run to run.

4.7.3   Matrix Properties in Selective Ionization by MALDI

Selection of a suitable matrix to analyze a particular lipid class or a category of lipid classes is important for MALDI-MS analysis of lipids and has been well documented [54–57]. It is conventional to use α-cyano-4-hydroxycinnamic acid (CHCA) for analysis of lipids in the positive-ion mode while 2,5-dihydroxybenzoic acid (DHB) in the negative-ion mode. Utilization of these matrices results in some of the obstacles for quantitative and global analyses of cellular lipidomes (see Chapter 2).

Finally, great efforts have been made to overcome these drawbacks largely associated with matrices by employing various matrix compounds, binary and ternary component matrix systems, and multiple sample preparation strategies. Examples of matrix systems that have been examined to overcome these drawbacks include liquid matrixes [58], ionic-liquid matrixes [59–61], solid ionic crystal matrixes [62], 2,4,6-trihydroxyacetophenone (THAP) [63], 2,6-dihydroxyacetophenone [64], p-nitroaniline [65], and nanoparticle surface layers [66–72]. Improvement of S/N and decreases in matrix cluster ionization were achieved by using a binary mixture of CHCA and 9-aminoacridine as a dual-component matrix [73]. Direct comparison of matrices such as CHCA, DHB, THAP, and 9-aminoacridine for analysis of PC species under a constant laser intensity demonstrated the presence of both proton and sodium adducts of the molecular species with a markedly reduced sensitivity and decreased detection limit with CHCA, DHB, or THAP as matrix in comparison to the use of 9-aminoacridine (Figure 4.10). Further studies also demonstrated a nearly 100-fold selective ionization of sulfatide (ST) over PI as well as other anionic lipid species with 9-aminoacridine as matrix [57]. It is intriguing that ST is structurally similar to but configurationally different from PI, suggesting the importance of a suitable matrix in matching with analytes.

Collectively, different matrices possess their different physical properties and could match with different lipid classes in crystallization and ionization process. Appropriate selection of a matrix for ionization of a particular lipid class or a category of lipid classes could lead to ionization in high sensitivity, quantitation, and effectiveness. In addition to the mixed matrix component systems, it appears that the matrix 9-aminoacridine could provide the best ionization for global analysis of cellular lipidomes. In any case, the variation of matrices for ionization of different lipids is an important dimension in lipidomics. Furthermore, understanding the roles of matrices in selective ionization of different lipid classes, enhancing ionization sensitivity, and the resulting reproducible and quantitative data is still warranted.

4.7.4   Drift Time (or Collision Cross Section) in Ion-Mobility Separation

Ion-mobility mass spectrometry (IM-MS) has emerged as an important analytical method in the last decade [74]. In IM-MS, ions are generated by pyrolysis, electrospray, laser desorption, or other ionization techniques prior to their entry into a gas-filled mobility drift cell. In this cell, ions drift at a velocity obtained from an electric field based on their shapes or dipoles in the case of differential mobility spectrometry (DMS). The greater the cross section of an analyte is (i.e., the larger the ion

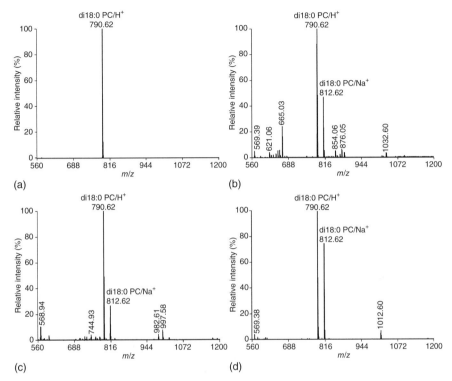

Figure 4.10 Comparison of S/N in MALDI mass spectra of di18:0 PC acquired on a 4800 MALDI-TOF/TOF analyzer in the positive-ion mode using different matrices. (a) 9-aminoacridine (10 mg/mL) dissolved in isopropanol/acetonitrile (60/40, v/v); (b) CHCA (10 mg/mL) dissolved in methanol containing 0.1% TFA; (c) DHB (0.5 M) dissolved in methanol containing 0.1% TFA; and (d) THAP (40 mM) dissolved in methanol.

size), the more area available for buffer gas to collide and impede the ion's drift, the ion then requires a longer time to drift through the tube. The different time of individual ion drift provides the separation of different shaped molecules. In the case of DMS, by combining the variation of a separation voltage and a compensation voltage, different lipid classes can be selectively filtered largely based on the dipole of a lipid class [75]. Therefore, IM-MS provides shape/size [76, 77] or dipole [75] information in addition to its mass. A 2D map of ion-mobility drift time *vs. m/z* can be readily acquired to analyze lipid classes and individual molecular species [78–80]. Recently, researchers use the drift time to calculate the rotationally averaged collision cross section, which represents the effective area for the interaction between an individual ion and the neutral gas through which it travels [81].

IM-MS is particularly useful for analysis of isobaric and isomeric lipid species, which possess different structures or configurations. For example, MALDI-IM MS has been used to analyze gangliosides, a class of complex glycosphingolipids that have different degrees of sialylation [82]. Both ganglioside D1a and D1b are structural isomers, differing in the location of the sialic acids. It was found that the drift times or collision cross sections of these isomers in gas phase are very different due to their different localization of the sialic acids. The effectiveness of ion mobility to separate the different ganglioside species present in a total ganglioside extract from mouse brain was well demonstrated [82].

Collectively, IM-MS clearly provides separation of lipid classes according to their charge properties, individual molecular species of a lipid class based on their molecular size (including chain length and unsaturation), and isobaric/isomeric species possessing different conformational structures [83]. This *in situ* drift time/collision cross section variation could be used as an additional variable to the other separation variables (e.g., intrasource separation, LC-MS elution, and optimal selection of MALDI matrix for ionization) described earlier as an aid to providing 3D analysis of complex lipid mixtures.

4.8   CONCLUSION

From the extensive discussion in this chapter, it should be recognized that there exist numerous variables at all the components of a mass spectrometer system and in all the lipidomics approaches for MS analysis of lipids, particularly of complex biological mixtures. These variables each apparently have great impacts on definitive identification and quantification of individual lipid species including isomers. The effects of many of the variables on lipidomics have not been well investigated or understood due to the limitation of the current development on the instrumentation or computer capacity. However, it should be recognized that the variables present in the different components of an MS system could influence any stage of MS analysis of lipids, such as method development, individual experimental designing, and data processing and interpretation, and recognized that the optimized parameters of individual variables during the method development might not represent the best conditions for analysis of complex biological samples. To this end, it is strongly advised that one should maintain all other variables constant when one interested variable is under investigation, such as ramping the variable in the broadest range possible.

REFERENCES

1. (a) Bligh, E.G. and Dyer, W.J. (1959) A rapid method of total lipid extraction and purification. Can. J. Biochem. Physiol. 37, 911–917.
   (b) Christie, W.W. and Han, X. (2010) Lipid Analysis: Isolation, Separation, Identification and Lipidomic Analysis. The Oily Press, Bridgwater, England. pp 448.
2. Matyash, V., Liebisch, G., Kurzchalia, T.V., Shevchenko, A. and Schwudke, D. (2008) Lipid extraction by methyl-tert-butyl ether for high-throughput lipidomics. J. Lipid Res. 49, 1137–1146.

3. Kalderon, B., Sheena, V., Shachrur, S., Hertz, R. and Bar-Tana, J. (2002) Modulation by nutrients and drugs of liver acyl-CoAs analyzed by mass spectrometry. J. Lipid Res. 43, 1125–1132.

4. Tsui, Z.C., Chen, Q.R., Thomas, M.J., Samuel, M. and Cui, Z. (2005) A method for profiling gangliosides in animal tissues using electrospray ionization-tandem mass spectrometry. Anal. Biochem. 341, 251–258.

5. Wang, C., Wang, M. and Han, X. (2015) Comprehensive and quantitative analysis of lysophospholipid molecular species present in obese mouse liver by shotgun lipidomics. Anal. Chem. 87, 4879–4887.

6. Merrill, A.H., Jr., Sullards, M.C., Allegood, J.C., Kelly, S. and Wang, E. (2005) Sphingolipidomics: High-throughput, structure-specific, and quantitative analysis of sphingolipids by liquid chromatography tandem mass spectrometry. Methods 36, 207–224.

7. Bielawski, J., Szulc, Z.M., Hannun, Y.A. and Bielawska, A. (2006) Simultaneous quantitative analysis of bioactive sphingolipids by high-performance liquid chromatography-tandem mass spectrometry. Methods 39, 82–91.

8. Jiang, X., Cheng, H., Yang, K., Gross, R.W. and Han, X. (2007) Alkaline methanolysis of lipid extracts extends shotgun lipidomics analyses to the low abundance regime of cellular sphingolipids. Anal. Biochem. 371, 135–145.

9. Kayganich, K.A. and Murphy, R.C. (1992) Fast atom bombardment tandem mass spectrometric identification of diacyl, alkylacyl, and alk-1-enylacyl molecular species of glycerophosphoethanolamine in human polymorphonuclear leukocytes. Anal. Chem. 64, 2965–2971.

10. Yang, K., Zhao, Z., Gross, R.W. and Han, X. (2007) Shotgun lipidomics identifies a paired rule for the presence of isomeric ether phospholipid molecular species. PLoS One 2, e1368.

11. Cheng, H., Jiang, X. and Han, X. (2007) Alterations in lipid homeostasis of mouse dorsal root ganglia induced by apolipoprotein E deficiency: A shotgun lipidomics study. J. Neurochem. 101, 57–76.

12. Han, X., Yang, K., Cheng, H., Fikes, K.N. and Gross, R.W. (2005) Shotgun lipidomics of phosphoethanolamine-containing lipids in biological samples after one-step in situ derivatization. J. Lipid Res. 46, 1548–1560.

13. Berry, K.A. and Murphy, R.C. (2005) Analysis of cell membrane aminophospholipids as isotope-tagged derivatives. J. Lipid Res. 46, 1038–1046.

14. Zemski Berry, K.A., Turner, W.W., VanNieuwenhze, M.S. and Murphy, R.C. (2009) Stable isotope labeled 4-(dimethylamino)benzoic acid derivatives of glycerophosphoethanolamine lipids. Anal. Chem. 81, 6633–6640.

15. Fhaner, C.J., Liu, S., Zhou, X. and Reid, G.E. (2013) Functional group selective derivatization and gas-phase fragmentation reactions of plasmalogen glycerophospholipids. Mass Spectrom. (Tokyo) 2, S0015.

16. Johnson, D.W. (2001) Analysis of alcohols, as dimethylglycine esters, by electrospray ionization tandem mass spectrometry. J. Mass Spectrom. 36, 277–283.

17. Jiang, X., Ory, D.S. and Han, X. (2007) Characterization of oxysterols by electrospray ionization tandem mass spectrometry after one-step derivatization with dimethylglycine. Rapid Commun. Mass Spectrom. 21, 141–152.

18. Griffiths, W.J., Liu, S., Alvelius, G. and Sjovall, J. (2003) Derivatisation for the characterisation of neutral oxosteroids by electrospray and matrix-assisted laser desorption/ionisation tandem mass spectrometry: The Girard P derivative. Rapid Commun. Mass Spectrom. 17, 924–935.

19. Griffiths, W.J., Wang, Y., Alvelius, G., Liu, S., Bodin, K. and Sjovall, J. (2006) Analysis of oxysterols by electrospray tandem mass spectrometry. J. Am. Soc. Mass Spectrom. 17, 341–362.

20. Wang, M., Hayakawa, J., Yang, K. and Han, X. (2014) Characterization and quantification of diacylglycerol species in biological extracts after one-step derivatization: A shotgun lipidomics approach. Anal. Chem. 86, 2146–2155.

21. Bollinger, J.G., Thompson, W., Lai, Y., Oslund, R.C., Hallstrand, T.S., Sadilek, M., Turecek, F. and Gelb, M.H. (2010) Improved sensitivity mass spectrometric detection of eicosanoids by charge reversal derivatization. Anal. Chem. 82, 6790–6796.

22. Wang, M., Han, R.H. and Han, X. (2013) Fatty acidomics: Global analysis of lipid species containing a carboxyl group with a charge-remote fragmentation-assisted approach. Anal. Chem. 85, 9312–9320.

23. Yang, K., Dilthey, B.G. and Gross, R.W. (2013) Identification and quantitation of fatty acid double bond positional isomers: A shotgun lipidomics approach using charge-switch derivatization. Anal. Chem. 85, 9742–9750.

24. Bollinger, J.G., Rohan, G., Sadilek, M. and Gelb, M.H. (2013) LC/ESI-MS/MS detection of FAs by charge reversal derivatization with more than four orders of magnitude improvement in sensitivity. J. Lipid Res. 54, 3523–3530.

25. Wang, M., Fang, H. and Han, X. (2012) Shotgun lipidomics analysis of 4-hydroxyalkenal species directly from lipid extracts after one-step in situ derivatization. Anal. Chem. 84, 4580–4586.

26. Wang, C., Wang, M., Zhou, Y., Dupree, J.L. and Han, X. (2014) Alterations in mouse brain lipidome after disruption of CST gene: A lipidomics study. Mol. Neurobiol. 50, 88–96.

27. Hsu, F.-F. and Turk, J. (2009) Electrospray ionization with low-energy collisionally activated dissociation tandem mass spectrometry of glycerophospholipids: Mechanisms of fragmentation and structural characterization. J. Chromatogr. B 877, 2673–2695.

28. Han, X., Abendschein, D.R., Kelley, J.G. and Gross, R.W. (2000) Diabetes-induced changes in specific lipid molecular species in rat myocardium. Biochem. J. 352, 79–89.

29. Ejsing, C.S., Duchoslav, E., Sampaio, J., Simons, K., Bonner, R., Thiele, C., Ekroos, K. and Shevchenko, A. (2006) Automated identification and quantification of glycerophospholipid molecular species by multiple precursor ion scanning. Anal. Chem. 78, 6202–6214.

30. Chansela, P., Goto-Inoue, N., Zaima, N., Hayasaka, T., Sroyraya, M., Kornthong, N., Engsusophon, A., Tamtin, M., Chaisri, C., Sobhon, P. and Setou, M. (2012) Composition and localization of lipids in *Penaeus merguiensis* ovaries during the ovarian maturation cycle as revealed by imaging mass spectrometry. PLoS One 7, e33154.

31. Han, X., Yang, K., Yang, J., Fikes, K.N., Cheng, H. and Gross, R.W. (2006) Factors influencing the electrospray intrasource separation and selective ionization of glycerophospholipids. J. Am. Soc. Mass Spectrom. 17, 264–274.

32. Han, X. and Gross, R.W. (2003) Global analyses of cellular lipidomes directly from crude extracts of biological samples by ESI mass spectrometry: A bridge to lipidomics. J. Lipid Res. 44, 1071–1079.

33. Han, X. and Gross, R.W. (2005) Shotgun lipidomics: Electrospray ionization mass spectrometric analysis and quantitation of the cellular lipidomes directly from crude extracts of biological samples. Mass Spectrom. Rev. 24, 367–412.

34. Shui, G., Bendt, A.K., Pethe, K., Dick, T. and Wenk, M.R. (2007) Sensitive profiling of chemically diverse bioactive lipids. J. Lipid Res. 48, 1976–1984.

35. Han, X. and Gross, R.W. (1994) Electrospray ionization mass spectroscopic analysis of human erythrocyte plasma membrane phospholipids. Proc. Natl. Acad. Sci. U.S.A. 91, 10635–10639.

36. Han, X., Yang, J., Cheng, H., Ye, H. and Gross, R.W. (2004) Towards fingerprinting cellular lipidomes directly from biological samples by two-dimensional electrospray ionization mass spectrometry. Anal. Biochem. 330, 317–331.

37. Han, X., Cheng, H., Mancuso, D.J. and Gross, R.W. (2004) Caloric restriction results in phospholipid depletion, membrane remodeling and triacylglycerol accumulation in murine myocardium. Biochemistry 43, 15584–15594.

38. Koivusalo, M., Haimi, P., Heikinheimo, L., Kostiainen, R. and Somerharju, P. (2001) Quantitative determination of phospholipid compositions by ESI-MS: Effects of acyl chain length, unsaturation, and lipid concentration on instrument response. J. Lipid Res. 42, 663–672.

39. DeLong, C.J., Baker, P.R.S., Samuel, M., Cui, Z. and Thomas, M.J. (2001) Molecular species composition of rat liver phospholipids by ESI-MS/MS: The effect of chromatography. J. Lipid Res. 42, 1959–1968.

40. Zacarias, A., Bolanowski, D. and Bhatnagar, A. (2002) Comparative measurements of multicomponent phospholipid mixtures by electrospray mass spectroscopy: Relating ion intensity to concentration. Anal. Biochem. 308, 152–159.

41. Kim, H.Y., Wang, T.C. and Ma, Y.C. (1994) Liquid chromatography/mass spectrometry of phospholipids using electrospray ionization. Anal. Chem. 66, 3977–3982.

42. Han, X., Gubitosi-Klug, R.A., Collins, B.J. and Gross, R.W. (1996) Alterations in individual molecular species of human platelet phospholipids during thrombin stimulation: Electrospray ionization mass spectrometry-facilitated identification of the boundary conditions for the magnitude and selectivity of thrombin-induced platelet phospholipid hydrolysis. Biochemistry 35, 5822–5832.

43. Han, X. and Gross, R.W. (2005) Shotgun lipidomics: Multi-dimensional mass spectrometric analysis of cellular lipidomes. Expert Rev. Proteomics 2, 253–264.

44. Han, X., Yang, K. and Gross, R.W. (2012) Multi-dimensional mass spectrometry-based shotgun lipidomics and novel strategies for lipidomic analyses. Mass Spectrom. Rev. 31, 134–178.

45. Schwudke, D., Oegema, J., Burton, L., Entchev, E., Hannich, J.T., Ejsing, C.S., Kurzchalia, T. and Shevchenko, A. (2006) Lipid profiling by multiple precursor and neutral loss scanning driven by the data-dependent acquisition. Anal. Chem. 78, 585–595.

46. Zhang, B., Wang, W. and Han, X. (2012) Accurate neutral loss-assisted shotgun lipidomics (ANLA-SL) for ultra-high-throughput analysis of cellular lipidomes. In 60th ASMS Conference on Mass Spectrometry and Allied Topics, May 20–24, 2012, Vancouver, Canada. p. TP20 Poster 400.

47. Han, X. and Wang, M. (2013) High-throughput lipidomics. Patent No. WO/2013/173642.

48. Wang, M., Huang, Y. and Han, X. (2014) Accurate mass searching of individual lipid species candidates from high-resolution mass spectra for shotgun lipidomics. Rapid Commun. Mass Spectrom. 28, 2201–2210.

49. Domingues, M.R., Nemirovskiy, O.V., Marques, M.G., Neves, M.G., Cavaleiro, J.A., Ferrer-Correia, A.J. and Gross, M.L. (1998) High- and low-energy collisionally activated decompositions of octaethylporphyrin and its metal complexes. J. Am. Soc. Mass Spectrom. 9, 767–774.

50. Dayon, L., Pasquarello, C., Hoogland, C., Sanchez, J.C. and Scherl, A. (2010) Combining low- and high-energy tandem mass spectra for optimized peptide quantification with isobaric tags. J. Proteomics 73, 769–777.

51. Yang, K., Zhao, Z., Gross, R.W. and Han, X. (2011) Identification and quantitation of unsaturated fatty acid isomers by electrospray ionization tandem mass spectrometry: A shotgun lipidomics approach. Anal. Chem. 83, 4243–4250.

52. Taguchi, R., Hayakawa, J., Takeuchi, Y. and Ishida, M. (2000) Two-dimensional analysis of phospholipids by capillary liquid chromatography/electrospray ionization mass spectrometry. J. Mass Spectrom. 35, 953–966.

53. Sommer, U., Herscovitz, H., Welty, F.K. and Costello, C.E. (2006) LC-MS-based method for the qualitative and quantitative analysis of complex lipid mixtures. J. Lipid Res. 47, 804–814.

54. Schiller, J., Suss, R., Arnhold, J., Fuchs, B., Lessig, J., Muller, M., Petkovic, M., Spalteholz, H., Zschornig, O. and Arnold, K. (2004) Matrix-assisted laser desorption and ionization time-of-flight (MALDI-TOF) mass spectrometry in lipid and phospholipid research. Prog. Lipid Res. 43, 449–488.

55. Fuchs, B., Suss, R. and Schiller, J. (2010) An update of MALDI-TOF mass spectrometry in lipid research. Prog. Lipid Res. 49, 450–475.

56. Sun, G., Yang, K., Zhao, Z., Guan, S., Han, X. and Gross, R.W. (2008) Matrix-assisted laser desorption/ionization time-of-flight mass spectrometric analysis of cellular glycerophospholipids enabled by multiplexed solvent dependent analyte–matrix interactions. Anal. Chem. 80, 7576–7585.

57. Cheng, H., Sun, G., Yang, K., Gross, R.W. and Han, X. (2010) Selective desorption/ionization of sulfatides by MALDI-MS facilitated using 9-aminoacridine as matrix. J. Lipid Res. 51, 1599–1609.

58. Stubiger, G., Pittenauer, E. and Allmaier, G. (2003) Characterisation of castor oil by on-line and off-line non-aqueous reverse-phase high-performance liquid chromatography-mass spectrometry (APCI and UV/MALDI). Phytochem. Anal. 14, 337–346.

59. Mank, M., Stahl, B. and Boehm, G. (2004) 2,5-Dihydroxybenzoic acid butylamine and other ionic liquid matrixes for enhanced MALDI-MS analysis of biomolecules. Anal. Chem. 76, 2938–2950.

60. Li, Y.L., Gross, M.L. and Hsu, F.-F. (2005) Ionic-liquid matrices for improved analysis of phospholipids by MALDI-TOF mass spectrometry. J. Am. Soc. Mass Spectrom. 16, 679–682.

61. Darsow, K.H., Lange, H.A., Resch, M., Walter, C. and Buchholz, R. (2007) Analysis of a chlorosulfolipid from *Ochromonas danica* by matrix-assisted laser desorption/ionization quadrupole ion trap time-of-flight mass spectrometry. Rapid Commun. Mass Spectrom. 21, 2188–2194.

62. Ham, B.M., Jacob, J.T. and Cole, R.B. (2005) MALDI-TOF MS of phosphorylated lipids in biological fluids using immobilized metal affinity chromatography and a solid ionic crystal matrix. Anal. Chem. 77, 4439–4447.

63. Stubiger, G. and Belgacem, O. (2007) Analysis of lipids using 2,4,6-trihydroxyacetophenone as a matrix for MALDI mass spectrometry. Anal. Chem. 79, 3206–3213.

64. Wang, H.Y., Jackson, S.N. and Woods, A.S. (2007) Direct MALDI-MS analysis of cardiolipin from rat organs sections. J. Am. Soc. Mass Spectrom. 18, 567–577.

65. Estrada, R. and Yappert, M.C. (2004) Alternative approaches for the detection of various phospholipid classes by matrix-assisted laser desorption/ionization time-of-flight mass spectrometry. J. Mass Spectrom. 39, 412–422.

66. Wei, J., Buriak, J.M. and Siuzdak, G. (1999) Desorption-ionization mass spectrometry on porous silicon. Nature 399, 243–246.

67. Go, E.P., Apon, J.V., Luo, G., Saghatelian, A., Daniels, R.H., Sahi, V., Dubrow, R., Cravatt, B.F., Vertes, A. and Siuzdak, G. (2005) Desorption/ionization on silicon nanowires. Anal. Chem. 77, 1641–1646.

68. Northen, T.R., Yanes, O., Northen, M.T., Marrinucci, D., Uritboonthai, W., Apon, J., Golledge, S.L., Nordstrom, A. and Siuzdak, G. (2007) Clathrate nanostructures for mass spectrometry. Nature 449, 1033–1036.

69. Woo, H.K., Northen, T.R., Yanes, O. and Siuzdak, G. (2008) Nanostructure-initiator mass spectrometry: A protocol for preparing and applying NIMS surfaces for high-sensitivity mass analysis. Nat. Protoc. 3, 1341–1349.

70. Patti, G.J., Shriver, L.P., Wassif, C.A., Woo, H.K., Uritboonthai, W., Apon, J., Manchester, M., Porter, F.D. and Siuzdak, G. (2010) Nanostructure-initiator mass spectrometry (NIMS) imaging of brain cholesterol metabolites in Smith–Lemli–Opitz syndrome. Neuroscience 170, 858–864.

71. Patti, G.J., Woo, H.K., Yanes, O., Shriver, L., Thomas, D., Uritboonthai, W., Apon, J.V., Steenwyk, R., Manchester, M. and Siuzdak, G. (2010) Detection of carbohydrates and steroids by cation-enhanced nanostructure-initiator mass spectrometry (NIMS) for biofluid analysis and tissue imaging. Anal. Chem. 82, 121–128.

72. Cha, S. and Yeung, E.S. (2007) Colloidal graphite-assisted laser desorption/ionization mass spectrometry and MSn of small molecules. 1. Imaging of cerebrosides directly from rat brain tissue. Anal. Chem. 79, 2373–2385.

73. Guo, Z. and He, L. (2007) A binary matrix for background suppression in MALDI-MS of small molecules. Anal. Bioanal. Chem. 387, 1939–1944.

74. Stach, J. and Baumbach, J.I. (2002) Ion mobility spectrometry – Basic elements and applications. Int. J. Ion Mobility Spectrom. 5, 1–21.

75. Krylov, E.V., Nazarov, E.G. and Miller, R.A. (2007) Differential mobility spectrometer: Model of operation. Int. J. Mass Spec. 266, 76–85.

76. Kanu, A.B., Dwivedi, P., Tam, M., Matz, L. and Hill, H.H., Jr. (2008) Ion mobility-mass spectrometry. J. Mass Spectrom. 43, 1–22.

77. Howdle, M.D., Eckers, C., Laures, A.M. and Creaser, C.S. (2009) The use of shift reagents in ion mobility-mass spectrometry: Studies on the complexation of an active pharmaceutical ingredient with polyethylene glycol excipients. J. Am. Soc. Mass Spectrom. 20, 1–9.

78. Woods, A.S., Ugarov, M., Egan, T., Koomen, J., Gillig, K.J., Fuhrer, K., Gonin, M. and Schultz, J.A. (2004) Lipid/peptide/nucleotide separation with MALDI-ion mobility-TOF MS. Anal. Chem. 76, 2187–2195.

79. Jackson, S.N. and Woods, A.S. (2009) Direct profiling of tissue lipids by MALDI-TOFMS. J. Chromatogr. B 877, 2822–2829.

80. Jackson, S.N., Ugarov, M., Post, J.D., Egan, T., Langlais, D., Schultz, J.A. and Woods, A.S. (2008) A study of phospholipids by ion mobility TOFMS. J. Am. Soc. Mass Spectrom. 19, 1655–1662.

81. May, J.C., Goodwin, C.R., Lareau, N.M., Leaptrot, K.L., Morris, C.B., Kurulugama, R.T., Mordehai, A., Klein, C., Barry, W., Darland, E., Overney, G., Imatani, K., Stafford, G.C.,

Fjeldsted, J.C. and McLean, J.A. (2014) Conformational ordering of biomolecules in the gas phase: Nitrogen collision cross sections measured on a prototype high resolution drift tube ion mobility-mass spectrometer. Anal. Chem. 86, 2107–2116.

82. Jackson, S.N., Colsch, B., Egan, T., Lewis, E.K., Schultz, J.A. and Woods, A.S. (2011) Gangliosides' analysis by MALDI-ion mobility MS. Analyst 136, 463–466.

83. Paglia, G., Kliman, M., Claude, E., Geromanos, S. and Astarita, G. (2015) Applications of ion-mobility mass spectrometry for lipid analysis. Anal. Bioanal. Chem. 407, 4995–5007.

5

BIOINFORMATICS IN LIPIDOMICS

5.1   INTRODUCTION

Bioinformatics involves the creation and advancement of databases, algorithms, statistics, and theory to solve problems arising from the management and analysis of huge amounts of biological and/or biomedical data. Therefore, it becomes an integral part of research and development in the biomedical sciences, as well as in the development of lipidomics. Large amounts of data are generated in MS analysis of lipids for both identification and quantification. It is difficult to process these data without proper tools and to understand the biological meanings of the data set without systematic analysis and modeling of these obtained data. Bioinformatics in lipidomics roughly include automated data processing, statistical analysis of data sets, pathway and network analysis, and lipid modeling in systems and biophysical context [1]. A recent tutorial article [2] has well described some of these aspects, particularly in the chemometric comparison, which should be consulted for interested readers.

In general, identification of lipids using any LC-MS-based method is largely based on the tandem analysis of individual lipid species in the product-ion mode. To this end, it is critical to generate libraries and/or databases containing information about the structures, masses, isotope patterns, and MS/MS spectra in different ionization modes and different adducts or ion forms of lipids, along with possible LC retention time range. This is analogous to the libraries of GC-MS spectra, which can be used to search a compound of interest after GC-MS analysis. On the other hand, MDMS-SL identifies lipid species *in situ* through analysis of building blocks using

Lipidomics: Comprehensive Mass Spectrometry of Lipids, First Edition. Xianlin Han.
© 2016 John Wiley & Sons, Inc. Published 2016 by John Wiley & Sons, Inc.

PIS and/or NLS. The requirement of the database/libraries for shotgun lipidomics is less restricted in comparison to those serving for the LC-MS methods, and a theoretical database constructed from the known building blocks is enough to fit the needs [3]. In the chapter, these currently available libraries and/or databases are briefly introduced. Through these tools, therefore, one can potentially identify a lipid species by matching a fragmentation pattern of the species along with other available information manually or automatically.

Next, the principles and a few programs and/or software packages are discussed. These tools allow for performing multiple data processing steps such as spectral filtering, peak detection, alignment, normalization, as well as exploratory data analysis and visualization for the needs of both shotgun lipidomics and LC-MS methods. Finally, the tools for and examples of pathway and network analysis and modeling of the obtained lipidomics data are given. The studies on this area are still in an early stage. We sincerely hope researchers can make efforts to create additional ideas and tools to fulfill the need for bioinformatics in lipidomics.

5.2   LIPID LIBRARIES AND DATABASES

Development of databases and the related bioinformatics tools has become an essential part of the omics studies. Over the recent years, empowered by high-throughput technologies for omics fields, the creation of databases devoted to lipids has been undertaken. Consequently, lipid-centered databases have been developed that enable researchers to comfortably analyze lipid species. Following are a few representative examples of lipid-related libraries and databases.

5.2.1   Lipid MAPS Structure Database

The Lipid MAPS consortium has created a number of useful resources for the bioinformatics analysis of lipids. Specifically, the new naming system that assigns a unique 12-character signature for biologically relevant lipid species affords automated processing of lipidomics data [4]. The Lipid MAPS online suite of tools enables the drawing of lipid structures and prediction of possible structures from mass spectrometry data [5]. These tools are useful for providing conformity in the field regarding the generation of lipid structures and nomenclature.

The Lipid MAPS Structure Database (LMSD) is a relational database containing structures and annotations of biologically relevant lipids. As of the mid-2015, LMSD contains over 40,360 unique lipid structures and is the largest public database in the world specifically for lipids. The structures of lipids in the database are from the following sources [6]:

- Lipid MAPS consortium's core laboratories and partners.
- Lipids identified by Lipid MAPS experiments.
- Computationally generated structures for appropriate lipid classes.

- Biologically relevant lipids manually curated from LipidBank, Lipid Library, Cyberlipids, Chemical Entities of Biological Interest (ChEBI), and other public sources.
- Novel lipids submitted to peer-reviewed journals are also added to the database.

The LMSD is publicly available at www.lipidmaps.org/data/structure/.

LMSD has the following distinct features in comparison to other existing lipid databases such as LipidBank:

- Usage of hierarchical classification and consistent nomenclature based on a comprehensive classification scheme proposed by Lipid MAPS.
- A unique Lipid MAPS ID number assigned to individual lipid structure, which reflects its position in the classification hierarchy.
- No duplicate structures.
- The ability to search the database using structure. All the lipid structures in LMSD follow the structure drawing rules proposed by the Lipid MAPS consortium [4].

LMSD offers several structure viewing options including gif image (default), ChemDraw (requires ChemDraw ActiveX/Plugin), MarvinView (Java applet), and JMol (Java applet). Recently, the list of lipid-specific keywords has been expanded. In addition to Gene Ontology (GO) and Kyoto Encyclopedia of Genes and Genomes (KEGG) term description, the description field of the UniProt records and the EntrezGene names has also been scanned for humans and mice.

In addition to a classification-based retrieval of lipids, users can search LMSD using either text-based or structure-based search options. The text-based search implementation supports data retrieval by any combination of these data fields: Lipid MAPS ID, systematic or common name, mass, formula, category, main class, and subclass data fields. The structure-based search, in conjunction with optional data fields, provides the capability to perform a substructure search or exact match for the structure drawn by the user. Search results, in addition to structure and annotations, also include relevant links to external databases.

Taken together, LMSD developed by the Lipid MAPS consortium contains a large body of lipid information and a suite of tools in providing vital conformity in lipid nomenclature and structural analysis. LMSD should play an important role in advancement of lipidomics research.

5.2.2 Building-Block Concept-Based Theoretical Databases

MDMS-SL analyzes lipid molecular species in a combined targeted and nontargeted approach and extensively employs the building-block concept for identification of lipid molecular structures (see Chapter 2). The database for MDMS-SL should be as broad and flexible as possible and can be readily constructed based on the building blocks of individual lipid classes (see Chapter 1) [3]. Such a virtual database consists of every possible combination in each class/category of lipids, contains all of the species currently recognized, and can be readily expanded with additional building blocks (such as modified fatty acyls). Therefore, a database of individual molecular species for each lipid class or a category of lipid classes can be constructed with the variables of the number of carbon atoms (m) and the number of double bonds (n) in individual fatty acyl building block denoted as $m:n$. The database includes the following:

- Total number of carbon atoms
- Total number of double bonds
- Chemical formulas
- Accurate monoisotopic mass
- Building blocks (i.e., fragments).

For example, the glycerol in which three protons are replaced with building blocks has a formula of $C_3H_5O_3$ and is the characteristic backbone for all lipid classes in the categories of both GPL and glycerolipids. The phosphodiester-containing head group (Column 4 of Table 5.1) linked to the oxygen atom of glycerol at its sn-3 position is a class-specific building block (i.e., building block III, Figure 1.5) in the GPL classes (Column 1 of Table 5.1). The difference between glycerolipids and GPL classes is that the building block of glycerolipids at this position is not a phosphodiester, but either an aliphatic chain in TAG, or a proton or an aliphatic chain in DAG and MAG in glycolipids [3]. The oxygen atom of glycerol at sn-1 position is connected to a fatty acyl chain (i.e., building block I, Figure 1.5) through an ester, ether, or vinyl ether bond in both GPL and glycerolipids. These different linkages define the subclasses (Column 2 of Table 5.1) of a GPL class, which are called phosphatidyl-, plasmanyl-, and plasmenyl-, respectively, and prefixed with "d," "a," and "p." This aliphatic building block varies with the number of carbon atoms and the number of double bonds, as well as the location of double bonds in the aliphatic chains. The building block replacing the proton at the sn-2 position of glycerol (i.e., building block II, Figure 1.5) is a fatty acyl chain in all lipid classes of both GPL and glycerolipids except the proton may still exist in the classes of DAG, MAG, and in the case of lysoGPL. Again, this building block of fatty acyl chain varies with the number of carbon atoms and the number of double bonds, as well as the location of double bonds in the fatty acyl chains. Building blocks I and II for each subclass of an individual GPL class are combined with variables of m and n (representing m carbon atoms and n double bonds, respectively, in these two building blocks) (Column 5 of Tables 5.1). Accordingly, the entire subclass of molecular species in each individual aforementioned lipid class can be represented by these two variables of m and n (Column 6 of Table 5.1). The lysoGPL classes are the special cases of the GPL classes. The database of each lysoGPL class, therefore, is constructed with its corresponding parent GPL class (Table 5.1). Table 5.2 lists the theoretical database of sphingolipid species similarly constructed to GPL species as described earlier.

By only using the commonly recognized fatty acyls, the structures of approximately 6500 GPL species, 3200 glycerolipid species, 26,000 sphingolipids,

TABLE 5.1 A Schematic Representation of the Database Used for Glycerophospholipids[a]

Lipid Class	Lipid Subclasses	Backbone	Head Group (Building Block III)	Side Chains (Building Blocks I and II)	Sum Formula	Negative-Ion Mode	Positive-Ion Mode	Number of Possible Species[b]
PC	Diacyl PC	$C_3H_5O_3$	$C_5H_{13}O_3PN$	$C_mH_{2m-2n-2}O_2$	$C_{m+8}H_{2m-2n+16}O_8PN$	$[M+Cl]^-$	$[M+Li]^+$, $[M+Na]^+$	314
	Alkenyl-acyl PC			$C_mH_{2m-2n-2}O$	$C_{m+8}H_{2m-2n+16}O_7PN$			314
	Alkyl-acyl PC			$C_mH_{2m-2n}O$	$C_{m+8}H_{2m-2n+18}O_7PN$			314
PE	Diacyl PE		$C_2H_7O_3PN$	$C_mH_{2m-2n-2}O_2$	$C_{m+5}H_{2m-2n+10}O_8PN$	$[M-H]^-$,		314
	Alkenyl-acyl PE			$C_mH_{2m-2n-2}O$	$C_{m+5}H_{2m-2n+10}O_7PN$	$[M-H+Fmoc]^-$ (i.e.,		314
	Alkyl-acyl PE			$C_mH_{2m-2n}O$	$C_{m+5}H_{2m-2n+12}O_7PN$	$[M+C_{15}H_9O_2]^-$)		314
PS	Diacyl PS		$C_3H_7O_5PN$	$C_mH_{2m-2n-2}O_2$	$C_{m+6}H_{2m-2n+10}O_{10}PN$	$[M-H]^-$		314
	Alkenyl-acyl PS			$C_mH_{2m-2n-2}O$	$C_{m+6}H_{2m-2n+10}O_9PN$			314
	Alkyl-acyl PS			$C_mH_{2m-2n}O$	$C_{m+6}H_{2m-2n+12}O_9PN$			314
PG			$C_3H_8O_5P$	$C_mH_{2m-2n-2}O_2$	$C_{m+6}H_{2m-2n+11}O_{10}P$	$[M-H]^-$		314
PI			$C_6H_{12}O_8P$	$C_mH_{2m-2n-2}O_2$	$C_{m+9}H_{2m-2n+15}O_{13}P$	$[M-H]^-$		314
PA			H_2O_3P	$C_mH_{2m-2n-2}O_2$	$C_{m+3}H_{2m-2n+5}O_8P$	$[M-H]^-$		314

(continued)

TABLE 5.1 (Continued)

Lipid Class	Lipid Subclasses	Backbone	Head Group (Building Block III)	Side Chains (Building Blocks I and II)	Sum Formula	Negative-Ion Mode	Positive-Ion Mode	Number of Possible Species[b]
lysoPC	Acyl-LPC		$C_5H_{13}O_3PN$	C_mH_{2m-2n} O	$C_{m+8}H_{2m-2n+18}O_7PN$	$[M+Cl]^-$	$[M+Li]^+$, $[M+Na]^+$	82
	Alkenyl-LPC			C_mH_{2m-2n}	$C_{m+8}H_{2m-2n+18}O_6PN$			82
	Alkyl-LPC			$C_mH_{2m-2n+2}$	$C_{m+8}H_{2m-2n+20}O_6PN$			82
lysoPE	Acyl-LPE		$C_2H_7O_3PN$	C_mH_{2m-2n} O	$C_{m+5}H_{2m-2n+12}O_7PN$	$[M-H]^-$,		82
	Alkenyl-LPE			C_mH_{2m-2n}	$C_{m+5}H_{2m-2n+12}O_6PN$	$[M-H+Fmoc]^-$ (i.e., $[M+C_{15}H_9O_2]^-$)		82
	Alkyl-LPE			$C_mH_{2m-2n+2}$	$C_{m+5}H_{2m-2n+14}O_6PN$			82
lysoPS	Acyl-LPS		$C_3H_7O_5PN$	C_mH_{2m-2n} O	$C_{m+6}H_{2m-2n+12}O_9PN$	$[M-H]^-$		82
	Alkenyl-LPS			C_mH_{2m-2n}	$C_{m+6}H_{2m-2n+12}O_8PN$			82
	Alkyl-LPS			$C_mH_{2m-2n+2}$	$C_{m+6}H_{2m-2n+14}O_8PN$			82
lysoPG			$C_3H_8O_5P$	C_mH_{2m-2n} O	$C_{m+6}H_{2m-2n+13}O_9P$	$[M-H]^-$		82
lysoPI			$C_6H_{12}O_8P$	C_mH_{2m-2n} O	$C_{m+9}H_{2m-2n+17}O_{12}P$	$[M-H]^-$		82
lysoPA			H_2O_3P	C_mH_{2m-2n} O	$C_{m+3}H_{2m-2n+7}O_7P$	$[M-H]^-$		82
CL		$(C_3H_5O_3)_2$	$C_3H_8O_7P_2$	$C_mH_{2m-2n-4}O_4$	$C_{m+9}H_{2m-2n+14}O_{17}P_2$	$[M-2H]^{2-}$		1081
	MonolysoCL	$(C_3H_5O_3)_2$	$C_3H_8O_7P_2$	$C_mH_{2m-2r-2}O_3$	$C_{m+9}H_{2m-2n+16}O_{16}P_2$	$[M-2H]^{2-}$		622
							Total	**6455**

[a] *Source:* Yang et al. [3]. Adapted with permission of the American Chemical Society. The database is constructed with the building blocks I, II, and III in glycerophospholipids as shown in Figure 1.5 by using the variables m and n. The variable m represents the number of total carbon atoms of acyl chains (m = 12–26, 24–52, 36–78, and 48–104 for species having one, two, three, and four fatty acyl chains, respectively) while the variable n represents the number of total double bonds of the acyl chains (n = 0–7, 0–14, 0–21, and 0–28 for species having one, two, three, and four fatty acyl chains, respectively). The ion modes indicate the ionization mode(s) used to analyze the indicated lipid class in MDMS shotgun lipidomics.

[b] The regioisomers and the isomers resulting from the different locations of double bond(s) are not considered. The number of molecular species are calculated based on the naturally occurring fatty acids containing the highest degree of unsaturation for acyl chains of 12–26 carbon atoms are 12:1, 13:1, 14:3, 15:3, 16:5, 17:3, 18:5, 19:3, 20:6, 21:5, 22:7, 23:5, 24:7, 25:6, and 26:7, respectively, that were previously identified.

TABLE 5.2 A Schematic Representation of the Database Used for Sphingolipids[a]

Lipid Class	Lipid Subclasses	BackBone	Sphingoid Base (Building Block III)	Head Group (Building Block I)	Side Chain (Building Block II)	Sum Formula	Negative-Ion Mode	Positive-Ion Mode	Number of Possible Species[b]
Ceramide (Cer)	Nonhydroxyl Cer	$C_3H_6O_2N$	$C_xH_{2x-2y+1}$	H	$C_mH_{2m-2n-1}O$	$C_{m+x+3}H_{2m-2n+2x-2y+7}O_3N$	$[M-H]^-$		2214
	Hydroxyl Cer				$C_mH_{2m-2n-1}O_2$	$C_{m+x+3}H_{2m-2n+2x-2y+7}O_4N$			2214
Sphingomyelin (SM)	Nonhydroxyl SM			$C_5H_{13}O_3PN$	$C_mH_{2m-2n-1}O$	$C_{m+x+8}H_{2m-2n+2x-2y+19}O_6PN_2$	$[M+Cl]^-$	$[M+Li]^+$, $[M+Na]^+$	2214
	Hydroxyl SM				$C_mH_{2m-2n-1}O_2$	$C_{m+x+8}H_{2m-2n+2x-2y+19}O_7PN_2$			2214
Ceramide phospho-ethanolamine (CerPE)	Nonhydroxyl CerPE			$C_2H_6O_3PN$	$C_mH_{2m-2n-1}O$	$C_{m+x+5}H_{2m-2n+2x-2y+12}O_6PN_2$	$[M-H]^-$, $[M-H+Fmoc]^-$	$[M+H]^+$	2214
	Hydroxyl CerPE				$C_mH_{2m-2n-1}O_2$	$C_{m+x+5}H_{2m-2n+2x-2y+12}O_7PN_2$			2214
Hexosyl-ceramide (HexCer)	Nonhydroxyl HexCer			$C_6H_{11}O_5$	$C_mH_{2m-2n-1}O$	$C_{m+x+9}H_{2m-2n+2x-2y+17}O_9N$	$[M+Cl]^-$	$[M+Li]^+$, $[M+Na]^+$	2214
	Hydroxyl HexCer				$C_mH_{2m-2n-1}O_2$	$C_{m+x+9}H_{2m-2n+2x-2y+17}O_9N$			2214

(continued)

TABLE 5.2 (*Continued*)

Lipid Class	Lipid Subclasses	BackBone	Sphingoid Base (Building Block III)	Head Group (Building Block I)	Side Chain (Building Block II)	Sum Formula	Negative-Ion Mode	Positive-Ion Mode	Number of Possible Species[b]
Sulfatide (ST)	Nonhydroxyl ST			$C_6H_{11}SO_8$	$C_mH_{2m-2n-1}O$	$C_{m+x+9}H_{2m-2n+2x-2y+17}O_{11}NS$	$[M-H]^-$		2214
	Hydroxyl ST				$C_mH_{2m-2n-1}O_2$	$C_{m+x+9}H_{2m-2n+2x-2y+17}O_{12}NS$			2214
Lactosyl-ceramide (LacCer)	Nonhydroxyl LacCer			$C_{12}H_{21}O_{10}$	$C_mH_{2m-2n-1}O$	$C_{m+x+15}H_{2m-2n+2x-2y+27}O_{13}N$	$[M+Cl]^-$	$[M+Li]^+$, $[M+Na]^+$	2214
	Hydroxyl LacCer				$C_mH_{2m-2n-1}O_2$	$C_{m+x+15}H_{2m-2n+2x-2y+27}O_{14}N$			2214
Lyso-SM				$C_5H_{13}O_3PN$	H	$C_{x+8}H_{2x-2y+21}O_5PN_2$	$[M+Cl]^-$	$[M+Li]^+$, $[M+Na]^+$	27
Sphingoid base				H	H	$C_{x+3}H_{2x-2y+9}O_2N$		$[M+H]^+$	27
Sphingoid base 1-phosphate				H_2O_3P	H	$C_{x+3}H_{2x-2y+10}O_5PN$	$[M-H]^-$		27
Psychosine				$C_6H_{11}O_5$	H	$C_{x+9}H_{2x-2y+19}O_7N$		$[M+H]^+$	27
								Total	26676

[a] *Source*: Yang et al. [3]. Adapted with permission of the American Chemical Society. The database is constructed based on the building blocks I, II, and III in sphingolipids as indicated in Figure 1.6 by using the variables of x, y, m, and n. The variable m represents the number of total carbon atoms of fatty amide chain ($m = 12$–26), the variable n represents the number of total double bonds of the fatty amide chain ($n = 0$–7), the variable x represents the number of total carbon atoms of a partial sphingoid base ($x = 11$–19), and the variable y represents the number of total double bonds of the partial sphingoid base ($y = 0$–2). The ion modes indicate the ionization mode(s) used to analyze the indicated lipid class in MDMS shotgun lipidomics. Gangliosides are not included.

[b] The isomers resulting from the different locations of double bond(s) are not considered. The number of molecular species are calculated based on the naturally occurring fatty acids containing the highest degree of unsaturation for acyl chains of 12–26 carbon atoms are 12:1, 13:1, 14:3, 15:3, 16:5, 17:3, 18:5, 19:3, 20:6, 21:5, 22:7, 23:5, 24:7, 25:6, and 26:7, respectively, that were previously identified.

100 sterols, and 410 other lipids that are predominantly involved in energy metabolism are easily constructed. Therefore, a total of over 36,000 molecular species, not counting regioisomers, oxidized lipids, or other covalently modified entities, are included in the initial construction of the database for MDMS-SL [3]. Moreover, by modifying the general chemical formulas, the constructed databases can be easily extended to cover any new species and/or subclasses in each lipid class when the sensitivity of mass spectrometers is further improved, and/or any unusual lipid profiles are analyzed from a biological sample.

5.2.3   LipidBlast – *in silico* Tandem Mass Spectral Library

LipidBlast is a library containing tandem mass spectra in the product-ion mode created *in silico*, validated to a great extent, and maintained by the Fiehn laboratory at University of California-Davis. LipidBlast contains a total of 212,516 tandem mass spectra for 119,200 different lipids in 26 lipid classes [7]. This library is freely available for commercial and noncommercial use at http://fiehnlab.ucdavis.edu/projects/LipidBlast/.

The *in silico* MS/MS library was generated based on the following steps [7].

- Defined the structures to be included and subsequently exhaustively *in silico* generated all possible structures. To this end, they imported approximately half of all the LipidBlast compound structures from Lipid MAPS database or generated using Lipid MAPS tools [6]. This part includes 13 lipid classes of the most common GPL classes and glycerolipids [6]. Because Lipid MAPS database does not cover many bacterial and plant lipids, an additional 54,805 compounds from 13 additional lipid classes were generated in LipidBlast using the combinatorial chemistry algorithms provided by ChemAxon Reactor11 (JChem v.5.5, 2011; http://www.chemaxon.com/) and SmiLib12 to yield a total of 119,200 compounds.

- Experimentally acquired MS/MS spectra on different platforms and theoretically interpreted structural class-specific fragmentations and rearrangements. The group performed MS/MS measurements in the product-ion mode of over 500 highly diverse GPL and glycerolipid standard compounds containing different numbers of carbon atoms and double bonds from individual lipid class. In addition, they selected MS/MS spectra from approximately 300 publications for those lipid classes of which the pure standards were unavailable. They analyzed the fragmentations and rearrangements for individual lipid class, including the precursor ions of $[M+H]^+$, $[M+Na]^+$, $[M+NH_4]^+$, $[M-H]^-$, $[M-2H]^{2-}$, $[M]^+$, and $[M+Li]^+$, and product ions, as well as their relative ion abundances. They found that the examined lipids showed predictable MS/MS spectra, with the dominant fragmentations being the loss of the polar head groups, the product ions resulting from the losses of acyl or alkyl chain from precursor ions, and the product ions corresponding to the fatty acid fragments (best observed in the negative-ion mode as $[FA-H]^-$). They observed many other specific fragments and rearrangements that were subsequently added to the rule-based generation of MS/MS mass spectra in LipidBlast.

- Generated characteristic fragmentations and heuristic modeling of ion abundances for possibly detectable adduct ions of lipid species of individual lipid class based on the rules observed above. Specifically, the LipidBlast MS/MS library was created by extending the obtained knowledge about fragmentations and ion abundances from the lipid standards to the thousands of *in silico* generated lipid structures. Heuristic methods to model precursor and product ions including their relative ion abundances for individual lipid class were used. For each individual precursor ion, the characteristic losses and specific fragment ions together with their accurate masses and molecular formulas were calculated. The library was created according to the observed ion intensities from standards by the corresponding instruments, considering the association of different relative ion intensities with specific types of mass spectrometers. Finally, all MS/MS spectra with lipid species name, adduct name, lipid class, accurate precursor mass, accurate mass fragment, heuristic modeled abundance, and fragment annotation were generated as electronic files.
- Rigorously validated the *in silico* generated MS/MS analysis. The group performed evaluations to detect false positives and false negatives, using decoy database searches and MS/MS analysis of authentic lipid standards measured in-house and from the literature. The search parameters and detailed statistics are available at the website given above.
- Demonstrated the applications of the library for high-throughput lipid identification [7]. The group analyzed lipid extracts of the human plasma using a low-resolution mass spectrometer. Using LipidBlast, they structurally annotated a total of 264 lipids. The data set was cross-checked with manual peak annotations and data available from Lipid MAPS. Using accurate mass LC-MS/MS, they annotated a total of 523 lipid molecular species. A similar number of plasma lipids were obtained in comparison to those previously published [8, 9].

The developers concluded that LipidBlast could be successfully applied to analyze MS/MS data from over 40 different mass spectrometer types and used with other available search engines and scoring algorithms, which represents a paradigm shift in lipidomics because it is not feasible to chemically synthesize all metabolites or natural products as authentic standards for library generation or quantification purposes. Moreover, the current array of MS/MS mass spectra for plant, animal, viral, and bacterial lipids in LipidBlast could be readily extended to many other important lipid classes.

5.2.4   METLIN Database

METLIN [10, 11] is a metabolomics database, which is a repository of metabolite information as well as tandem MS data. The METLIN database was developed and is maintained by Dr Siuzdak laboratory at The Scripps Research Institute. The METLIN metabolite database is available to the public online (http://metlin.scripps.edu) for metabolite searches.

METLIN represents one of the mostly comprehensive metabolite databases in the world today. It includes masses, chemical formulas, and structures for over 15,000 endogenous and exogenous metabolites including numerous lipid species and contains over 64,000 structures. It also contains tandem MS data on more than 10,000 distinct metabolites and over 50,000 ESI-Q-TOF MS/MS spectra in the product-ion mode acquired at four collision energies (0, 10, 20, and 40 eV) in both positive- and negative-ion modes. Moreover, over 160,000 predicted unique fragment structures are also added to the METLIN database using *in silico* fragmentation [12]. Available MS/MS data are expanding continuously as more metabolite information is being deposited and discovered.

The METLIN metabolite database is implemented using the open-source software tool, MySQL. Most compounds are annotated with both a chemical formula and structure. Individual metabolite is linked to outside resources such as the KEGG, Human Metabolome Database (HMDB), and the respective PubChem database entries through the included numbers of KEGG, HMDB, and Chemical Abstract Service (CAS), respectively. This kind of linkage makes it easy for the researchers to find further references and inquiries about the metabolite.

Overall, the METLIN database allows researchers to readily search and characterize metabolites through their chemical and/or physical features such as accurate mass, and single and multiple fragments including neutral-loss fragments. These capabilities of the database greatly facilitate the value of their metabolomics MS and MS/MS data and expedite the identification process. METLIN is one of the most used databases in metabolomics including lipidomics.

5.2.5 Human Metabolome Database

The HMDB [13–15] is a comprehensive online database of small molecule metabolites found in the human body, created by the Human Metabolome Project funded by Genome Canada. HMDB is a freely available electronic database through the website http://www.hmdb.ca/. The database contains information on more than 6500 metabolites including lipids with the data of chemistry, clinical information, and molecular biology/biochemistry. Additionally, approximately 1500 protein (and DNA) sequences are linked to these metabolite entries. Metabolite entries are presented in the HMDB as MetaboCards. Each MetaboCard entry contains over 100 data fields with two-thirds of the information being devoted to chemical/clinical data and the other one-third devoted to enzymatic or biochemical data. Many data fields are hyperlinked to other databases including KEGG, PubChem, MetaCyc, ChEBI, PDB, Swiss-Prot, and GenBank. A variety of pathway and structure viewing applets adds an additional layer of information to this database. The HMDB provides features including text search, sequence search, chemical structure search, and relational query search.

5.2.6 LipidBank Database

LipidBank is the official database of the Japanese Conference on the Biochemistry of Lipids (JCBL), which is publicly available at http://lipidbank.jp/. LipidBank provides information on identified natural lipids such as fatty acids, glycerophospholipids, glycerolipids, sphingolipids, steroids, and various vitamins. Currently, LipidBank contains over 6000 molecules that are classified into 26 groups [16]. The database contains molecular structures in both ChemDraw and MDL molfile (MOL) formats; lipid names in both common and IUPAC nomenclatures; spectral information including molecular mass, UV, IR, NMR, and others if available; and the information about the literature that reports lipid identification. All molecular information has been manually curated and approved by experts in lipid research.

5.3 BIOINFORMATICS TOOLS IN AUTOMATED LIPID DATA PROCESSING

In this section, the fundamentals for automated lipid data processing that is largely associated with LC-MS analysis including spectral filtering, peak detection, alignment, baseline correction are first discussed. Followed are the visualization and biostatical analysis. Finally, a few commonly used software programs including both commercially available and customized programs are overviewed. The principles of automated identification of lipid species are postponed in the following chapter. The principles and methods for lipid quantification are extensively described in Chapter 14.

5.3.1 LC-MS Spectral Processing

Mass spectral data after direction infusion can be averaged in the profile mode from an entire acquisition frame and only a small number of data points can be exported. In contrast to direct infusion, it would be the best to export every single data point from LC-MS analysis since the neighboring data points could represent very different mass spectral information due to the nature of constant concentration changes. This huge array of raw data sets makes it difficult to process and store. Therefore, bioinformatician is the first to compile the data set based on some criteria (e.g., elimination of some redundant information including isotopologues, adducts, in-source fragmentation, or others).

For example, Brown's group handles these raw data as follows [17]: Raw MS files acquired are first converted to readable format by using open-source software from Institute for Systems Biology. Then, a custom Fortran program is used to extract the relevant information and construct averages of the spectra in 10-s binned intervals over 1 amu mass-to-charge (m/z) bins in a trough-to-trough manner by finding the most common trough point in a given file. They found that quantitation is only moderately sensitive to the choice of a temporal averaging interval, and the optimal window is somewhat dependent on the scan rate of the instrument. Other steps must be undertaken before area-under-curve calculations on a species can be carried out. These include retention time alignment, background correction, and deisotoping.

Most of the methods [18–20] commonly used for retention time alignment are based in some manner on the correlation between spectra according to Brown and colleagues [17]. Since the retention times in lipidomic analyses are well constrained within individual classes by the observed retention times of the spiked-in standards, species of interest are bracketed by the standards in time. This information is usually used to effectively time-shift spectra within the time-m/z domain of each class without the need for pairwise spectral computations. The required alignment shifts can then be chosen to maximize the correlation of time-lag-shifted spectrum against one arbitrarily chosen sample from that session of MS analysis. However, the expense of computing the correlations can often be avoided altogether by using the temporal maxima of the internal standards' peaks.

The correction of the background contribution to the peak intensities of a mass spectrum is very important for peak detection and accurate quantification of each analyte content with MS, particularly when the species is in low abundance. Accurate baseline correction could reduce the complications faced by uncertainty about the intercept of the standard curve in LC-MS analysis, given the impact it has on the applicability of standard curves developed in a solvent background for subsequent use in analyzing cellular- and noncellular-based samples. For this reason, it is useful to require that the intercept terms of standard curve linear regressions have no statistical difference from zero. Thus, intercepts are essentially forced through the origin after background correction. This removes reliance on any latent subtraction of noise through the intercept term.

Many methods commonly used for general metabolomic LC-MS analysis (e.g., the CODA algorithm [21], the MEND algorithm [22], and various Gaussian-second derivative-based methods of peak picking [23, 24]) are useful for simultaneous peak identification and noise reduction and, typically, employ filtering or smoothing functions. However, applying such kind of filtering tools, which implies a model of the noise and/or the peak shapes present in the original data, may lead to distortion of peak identification. For example, considering the large variation in how different GPL classes interact with the stationary phase of a silica-based LC column to produce retention time tails, modeling the peak shapes may not be universally reliable at all. Instead, the problem can be reduced to simply finding the appropriate retention time window boundaries for a given m/z.

To this end, Brown and colleagues proposed [17] that a Williams–Kloot test [25] could be used to define the boundaries of the retention time windows. A series of tests are conducted by expanding the putative integration window in each direction of time separately from the central maximum. With each expansion, the peak is compared to the background, and a Williams–Kloot test is conducted to evaluate the differences between the results obtained from successively expanded windows. Once the usefulness of adding one more point to the peak becomes indistinguishable from the addition of a background point as evaluated through the Williams–Kloot test, the window location is defined. The integral under the polygon including these points and bounded by a diagonal line at the bottom of the peak is used as the ion count corresponding to that feature.

5.3.2 Biostatistical Analyses and Visualization

The next step in bioinformatics is to perform a statistical comparison between data sets. Many statistical hypothesis-driving tests can be performed to determine the significance. Two-sample Student's t-test is commonly employed to investigate whether the means of two groups of samples are significantly different from each other. Wilcoxon test investigates the hypothesis on median and can be applied on a single sample or two samples (paired or unpaired samples). In the former, the test determines whether the median of the sample is different from the hypothesized median of the population. In the latter case, it tests if the median of one sample is different from the second sample. Two common nonparametric tests are Wilcoxon signed-rank test for paired data, and the Mann–Whitney U test (also known as Mann–Whitney–Wilcoxon test, the Wilcoxon t-test, the Wilcoxon two-sample test, or the Wilcoxon W test) for unpaired data. These tests are based on ranking of the data and looking at the ranks rather than the actual values of the observations. Analysis variance (ANOVA) can be used to compare the means of two or more groups assuming that sampled population are normally distributed. Correlation can be performed to describe the degree of relationship between two variables and is measured using a correlation coefficient.

A few approaches are commonly used for analysis of multivariate data:

- Principal component analysis (PCA) uncovers simpler patterns from the complex intercorrelated variables.
- Partial least square-based discriminant analysis (PLS-DA) is a widely used, supervised classification algorithm [26] when dimensionality reduction is needed, and discrimination is sought in multivariate analysis.
- Multivariate analysis of variance (MANOVA) is a statistical test procedure for comparing multivariate (population) means of several groups, which uses the variance–covariance between variables in testing the statistical significance of the mean differences.

To this end, any commercially available software (e.g., SAS, NCSS, IBM SPSS Statistics, SIMCA-P) can be used to perform the analysis. It has been well recognized that use of a proper statistical analysis method is essential to improve visualization, accurate classification, and outlier estimation [27].

Graphic display is usually the most effective way to compare and communicate data if done cautiously [28]. When illustrating changes of mass levels between lipid classes, a bar chart is usually used. A heat map is commonly used for showing the differences between individual molecular species. In the heat map format, lipidomic data are often displayed in the order of acyl chain length or summed carbon number, m/z, and sometimes divided into lipid subcategories [29] and/or hierarchial clustering [30]. A pie chart is frequently used to show the changes of composition. Multidimensional or multicolor-coded illustrations are also frequently used to demonstrate the changes of mass levels or composition of lipid classes, and individual molecular species [31–33]. Many drawing software packages such as Excel, Prism, Tableau could be employed for the purpose.

5.3.3 Annotation for Structure of Lipid Species

Although many systematic indices (e.g., Lipid MAPS, Chemical Entries of Biological Interest (ChEBI), IUPAC International Chemical Identifiers (InChI), simplified molecular-input line entry system (SMILES)) were developed to list the chemical compounds, these indices (identifiers) can only be meaningful if the compound is totally identified. However, in practice, lipidomics analysis in many cases can only provide partial identification of lipid molecular structures at the current development of technology. Moreover, different lipidomics approaches provide different levels of structural identification of lipid species. Therefore, how to clearly express and report the information about the levels of identification for the structures of lipid species (which can be derived from MS analysis) is not only helpful for the readers but also important for bioinformatics and data communication. To this end, the analysis by shotgun lipidomics could be used as a typical example to explain these levels. Similar phenomena also exist in the analysis of lipid species employing LC-MS-based approaches.

In the tandem MS-based shotgun lipidomics approach, a head group-specific tandem MS spectrum is usually used for the analysis of a class of lipid species. In this case, only the information of the lipid class is provided without distinction of subclasses if existing. For example, PIS of m/z 184 is commonly exploited to profile phosphocholine-containing lipid species [34]. Such an analysis indeed does not differentiate PC species containing an ether bond from those containing an ester bond. Moreover, such an analysis also does not clearly distinguish SM species from PC species due to the potential overlap of SM with M+1 ^{13}C isotopologue of PC species.

In the high mass accuracy/resolution mass spectrometry-based shotgun lipidomics approach, the overlaps between SM and M+1 ^{13}C isotopologue of PC species, and between PC subclasses can be resolved [35]. Moreover, the linkages of alkyl *vs.* alkenyl at the *sn*-1 position of glycerol and the fatty acyl chains of diacyl species are also identified by product-ion analysis [35, 36]. Clearly, this approach provides much more structural information than that of the tandem MS approach as aforementioned.

In the MDMS-SL approach, in addition to identification of the overlaps and the identities of fatty acyl chains, the location of those FA chains (i.e., regioisomers) in diacyl PC species can also be accessed [3, 37]. Furthermore, even the location of double bonds in the fatty acyl chains can be identified with efforts [38]. Therefore, these types of additional structural information should be reflected from the data report.

To reflect these different levels of structural information, a system of shorthand notation for lipid structures derived from MS analysis was proposed [39] and becomes widely accepted. For example, for analysis of PC species, it was proposed to notate with PC(nominal mass) or PC(m:n) to represent a detected species by the tandem MS approach (where m and n are the total number of carbon atoms and double bonds of aliphatic chains in the species, respectively); with PC(m_1:n_1_m_2:n_2) or PC(o-m_1:n_1_m_2:n_2) to reflect the identification of individual fatty acyl chains and the ether bond by the high mass resolution mass spectrometry-based shotgun lipidomics; and with PC(m_1:n_1/m_2:n_2) or PC(o-m_1:n_1/m_2:n_2) to denote the total identification of PC species by MDMS-SL. Similarly, MRM-based methods after

LC-MS should be only denoted with PC(*m*:*n*) and data-dependent acquisition methods after LC-MS could be expressed the latter cases as described earlier.

5.3.4   Software Packages for Common Data Processing

There exist a few programs and/or software packages that perform multiple data-processing steps such as spectral filtering, peak detection, alignment, normalization, and exploratory data analysis and visualization for the requirements of LC-MS methods [40–46]. These programs were developed and largely depend on the elution times and the determined masses of individual ions. Below are given a few toolboxes used for the development of these programs. Moreover, a few commercially available software packages are also listed for readers' interest.

5.3.4.1   XCMS   XCMS is an open-source, cloud-based metabolomic data-processing platform that provides high-quality metabolomic analysis in a user-friendly, web-based format [18]. Since it is written in R and graphical support is substantial [18], XCMS allows users to easily upload LC-MS metabolomic data and process. Predefined parameter settings for different instruments (e.g., Q-TOF, Orbitrap) are available, as well as options for customization. Results can be viewed online in an interactive, customizable table showing statistics, chromatograms, and putative METLIN identities, which links to MS/MS libraries [18]. XCMS, which incorporates novel nonlinear retention time alignment, matches filtration, peak detection, and peak matching. By correcting retention time, the relative metabolite ion intensities are directly compared to identify changes in specific endogenous metabolites. However, XCMS is not intended for quantification, which is very critical and is a key challenge in lipidomics [47]. For example, the platform has no support for applying standard curves; background correction is not very flexible or straightforward and relies on a smoothing function that bears assumptions about the shape of the peaks; isotopic correction is also not built-in, which is very important in lipidomics (see Chapter 15); among others. XCMS is freely available under an open-source license at http://metlin.scripps.edu/download/.

5.3.4.2   MZmine 2   MZmine 2 is a new generation of a popular open-source data-processing toolbox [19]. A key concept of the MZmine 2 software design is the strict separation of core functionality and data-processing modules. Its main goal is to provide a user-friendly, flexible, and easily extendable framework. It mainly focuses on LC-MS data, covering the entire LC-MS data analysis workflow, and also supports for high-resolution spectral processing. Data-processing modules take advantage of embedded visualization tools, allowing for immediate previews of parameter settings. The functionality includes the identification of peaks using online databases, MS^n data support, improved isotope pattern support, scatter plot visualization, and a new method for peak list alignment based on the random sample consensus algorithm. It has the advantage of being parallelizable for multiple computer processors and has cross-platform support as it is written in Java. However, while data analysis is automated, it is not scriptable. Background correction heavily relies on either polynomial smoothing or a user-defined, straight-line cutoff. MZmine 2 is freely available under a GNU general public license and can be obtained from the project website: http://mzmine.sourceforge.net/.

5.3.4.3 *A Practical Approach for Determination of Mass Spectral Baselines*

Precise determination of the baseline levels of mass spectra is critical for identification and quantification of analytes. In contrast to other previous approaches based on smoothing or random cutoff [48–50], a practical approach based on the fact that an accelerated intensity change exists from noise to signal was developed for determination of the baselines of mass spectra acquired under different conditions [51]. The accelerated intensity change was derived from an accumulative layer thickness curve that was derived from the thicknesses of individually deducted layers one by one, each of which was calculated from the thickness of averaged lowest ion intensities from existing spectral data after deduction of a previous layer. The layer where the accelerated intensity change occurred was defined as a transition layer, which was determined from the polynomial regression in the sixth order of the accumulative layer thickness curve followed by resolving the roots of its fourth derivative. This approach has been widely validated through comparison to the manually determined baselines of all available mass spectra in the author's laboratory. The software program is available upon request.

5.3.4.4 *LipidView*

LipidView™ is the commercial software sold by Sciex. The software is a data-processing tool for characterization and quantification of lipid species from ESI-MS analysis. It enables lipid profiling by searching precursor- and fragment-ion masses against a lipid fragment database containing over 25,000 lipid species covering over 50 lipid classes, and reports a numerical and graphical output for various lipid molecular species, lipid classes, fatty acids, and long-chain bases [35]. LipidView™ software allows users to perform automated data processing from template methods, method editing and selection, lipid species identification, isotopic correction, multiple internal standards-based quantification, visualization, result reporting, etc. It is capable of processing the data from all SCIEX Triple Quad™, QTRAP®, and TripleTOF® Systems. However, since this software is largely generated based on the shotgun lipidomics approach [35], methods for LC-MS analysis of lipids and subsequent data processing may need to be further developed, applications for processing data from other manufacturers' instruments may not be straightforward, and quantitation methods developed based on one internal standard for each lipid class may lead to a large systematic error. A trial version of LipidView™ software can be obtained through https://licensing.absciex.com/download/index.

5.3.4.5 *LipidSearch*

LipidSearch is the commercial software (Thermo Fisher Scientific) developed jointly by Prof. Ryo Taguchi and MKI (Tokyo, Japan). It is a powerful new tool for automatic identification and relative quantification of cellular lipid species from a large amount of mass spectrometric data obtained from both LC-MS and shotgun lipidomics approaches. A lipid database containing over 1.5 million lipid ions and their predicted fragment ions is associated with the software. It supports a variety of instruments and a number of acquisition modes, including PIS, NLS, and product-ion analysis.

The software provides two different identification algorithms:

- A group-specific algorithm identifies lipids based on the polar head groups or fatty acids using a combination of PIS and NLS from lipid mixtures.
- The comprehensive identification algorithm for product-ion scans discriminates each lipid by matching the predicted fragmentation pattern stored in the database.

Identified lipids are quantified by detecting their precursor ions from the full MS scans and integrating extracted ion chromatograms. Accurate peak areas are calculated by denoising and smoothing the peak profiles prior to separating any partially overlapped peaks. Quantified results are compared using t-test statistics. Since the software is newly developed and is still in premature stage, broad validation is still needed to demonstrate its power for identification and quantification of lipid species. The software package can be purchased from Thermo Fisher Scientific Co.

5.3.4.6 SimLipid SimLipid®, developed by PREMIER Biosoft, is the commercial software that allows for high-throughput lipid identification and quantification. It analyzes lipid MS, MS/MS, and MS^n data for structural elucidation, isotopic correction, and quantification of elucidated lipids by comparing them to internal standards. The program accepts experimental MS and MS/MS (m/z and intensity values) obtained by mass spectrometry in any format and from any type of instruments. SimLipid® supports $[M+H]^+$, $[M+NH_4]^+$, $[M+Na]^+$, $[M+C_5H_{12}N]^+$, and $[M+Li]^+$ ions in the positive-ion mode and $[M-H]^-$, $[M+CH_3COO]^-$, $[M+Cl]^-$, $[M-CH_3]^-$, and $[M+HCOO]^-$ in the negative-ion mode. SimLipid® supports LC-MS and LC-MS/MS high-throughput data-processing methods such as peak detection, smoothing, chromatogram deconvolution, peak alignment, peak deisotoping, and adduct identification corresponding to the peaks detected. The software enables lipid identification and profiling by searching precursors against the known lipid structures available in SimLipid® database.

SimLipid® database is a large relational database containing eight lipid categories as classified by Lipid MAPS having 36,299 lipid species. The database links to KEGG, HMDB, ChEBI, PubChem, and LipidBank. The database is continuously being updated. Theoretical fragments of lipids from eight categories of lipids are available along with their theoretical masses and the corresponding fragment structures. Additional information such as lipid ID, lipid abbreviation, systematic name, composition, and other database links are also made available for easy reference. Although it seems the software is very powerful, further validation is still needed to demonstrate the accuracy of identification and quantification. The software package can be purchased through the website of the company: http://www.premierbiosoft .com/index.html.

5.3.4.7 MultiQuant MultiQuant, the commercial software from Applied Biosystems, provides a comprehensive package for quantitation of lipid species using MRM assays and allows for the building of highly customizable standard curve scenarios. However, the program is only applicable to MRM data, not full scan LC-MS output. Although MultiQuant allows for an audit trail, it is not truly scriptable. The baseline correction algorithm provided in the software is not very flexible.

5.3.4.8 Software Packages for Shotgun Lipidomics Multiple programs and/or software packages are developed based on the principles of shotgun lipidomics, including LIMSA [52], LipidProfiler [35], LipidInspector [53], AMDMS-SL [3], LipidXplorer [54], and ALEX [55]. These tools are developed based on the different platforms of shotgun lipidomics. LIMSA, which is available through the website (www.helsinki.fi/science/lipids/software.html), serves as an interface to process data from individual full-MS and tandem MS spectra. The software package LipidXplorer deals with the multiple PIS and NLS data and the software ALEX processes full mass spectral and product-ion analysis data acquired with those instruments with high mass accuracy/high mass resolution (e.g., Q-TOF and Orbitrap). The AMDMS-SL program is developed to identify and quantify individual lipid species from the data obtained from multidimensional MS-based shotgun lipidomics.

5.4 BIOINFORMATICS FOR LIPID NETWORK/PATHWAY ANALYSIS AND MODELING

5.4.1 Reconstruction of Lipid Network/Pathway

To classical lipidology, the majority of existing pathway tools (e.g., KEGG, Ingenuity, and MetaCore) are well suitable to illustrate the network of lipid metabolism. For example, KEGG uses individual lipid classes as the connection nodes and makes the gene(s) involved to connect the nodes. It is unfortunate that lipidomics not only studies lipid classes but also individual lipid species. These pathway tools are unable to illustrate individual lipid species from different lipid classes. For example, KEGG contains a pathway for "Sphingolipid metabolism"; however, this pathway is mainly focused on the biosynthesis of the different sphingolipid classes (e.g., Cer, sphingosine, and SM). The pathway cannot account for the different fatty acyl species of individual classes for which a letter "*R*" is always labeled. As predicted, over 25,000 sphingolipid species (Table 5.2) possibly exist in a cellular lipidome, demonstrating an obvious limitation in the current pathway reconstruction efforts.

To solve this limitation for analyzing these large and increasingly complex data sets, many efforts have been made. Lipid MAPS consortium seeks to develop a systematic and universal classification and nomenclature system for individual lipid species. Special lipid databases (e.g., Lipid MAPS, LipidBank (see above)) are made available. Pathway mapping strategies such as VANTED and KEGG pathway provide the basis for comprehensive pathway reconstruction. Obviously, further advances are required. Although some pathway reconstruction efforts have been made [56, 57], this research area is clearly still at the early stage for lipidomics.

It should be further recognized that the complexity of pathway reconstruction is that alterations in the biochemical pathway levels do not reflect well the cause of lipid concentration changes. In fact, the measured lipid concentrations represent the regulation at multiple spatial and dynamic scales, for example, systemic lipid metabolism, global changes in cell membrane composition, or lipid oxidation. Again, although some strategic efforts in this area have been made through dynamic simulation of lipidomics data (see next section), the inherent difficulty of accounting for such complexity in the analysis of lipidomics data will remain a formidable challenge, as well as a research opportunity for some time to come. Taken together, further developments in reconstructing context-specific lipid network/pathway at the different levels are clearly needed to reflect the altered lipids between different states due to any physiological, pathophysiological, or pathological change(s). To this end, multidimensional or multicolor-coded illustrations of lipidomics data with lipid metabolism pathways to demonstrate the changes of mass levels or composition of lipid classes and individual molecular species would be powerful and useful [31–33].

5.4.2 Simulation of Lipidomics Data for Interpretation of Biosynthesis Pathways

As discussed so far, the majority of the advances in bioinformatics in lipidomics have largely focused on lipid identification and maybe quantification, whereas interpretation of alterations in biological function resulting in adaptive or pathological changes in lipid metabolism is still lagged [40]. Thus, development of bioinformatic and systems biology approaches to link the changes of cellular lipidome to alterations in the biological functions, including the enzymatic activities, which involve the biosynthesis of the altered lipid classes and molecular species. Such development should significantly advance the understanding of the roles of lipids in biological systems and of the biochemical mechanisms underpinning lipid changes [1].

One of the attempts for the purpose has been the recent development in dynamic or steady-state simulation of the obtained lipidomics data [58–61]. The researchers clustered the lipid classes and individual molecular species involved in the biosynthesis of a particular lipid class and utilized the known biosynthesis and/or remodeling pathways to simulate the ion profiles of the lipid class of interest in order to achieve the best match between the simulated and determined ion spectra. Owing to the large set of lipidomics data, numerous parameters involving the biosynthesis pathways can be obtained from the simulation. These parameters are largely associated with biological functions through the model used for simulation.

For example, it is well known that cardiolipin (CL) species are synthesized from the condensation of PG with cytidine diphosphate-diacylglycerol (CDP-DAG) species, and these newly synthesized immature species are remodeled into mature CL species from available donor acyl chains (Figure 5.1). Thus, this synthesis/remodeling process can be simulated by using the determined profiles of PG, PC, PE, and acyl CoA in comparison to the determined ion profile of CL species [58, 59]. The biological parameters in these simulations can be modified to compare CL profiles obtained from lipidomics analysis. Through the simulation, the

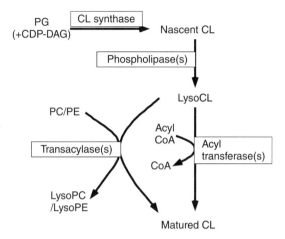

Figure 5.1 Schematic illustration of cardiolipin biosynthesis and remodeling pathways. Newly synthesized cardiolipin (CL) (immature or nascent CL) is formed by the condensation of phosphatidylglycerol (PG) and cytidine diphosphate-diacylglycerol (CDP-DAG) catalyzed by CL synthase. Immature CL is then deacylated to form monolysoCL and then reacylated using acyl chains from acyl CoA or transacylated from the *sn*-2 acyl chain of PC and PE species, leading to the formation of matured CL.

coordinated activities of phospholipase(s), acyltransferase(s), and/or transacylase(s) involved in CL remodeling can be assessed from the simulation utilizing specific distributions of acyl chains. Moreover, all the CL molecular species, including isomeric species, can be readily recapitulated, and the existence of a variety of very low-abundance CL molecular species, many of which cannot be accurately determined using currently available technologies, can be predicted. When applying this dynamic simulation approach to interrogate alterations in CL species under patholo(physio)logical conditions, the mechanisms regulating the complex tissue-specific CL molecular species distributions and underlying alterations in acyl chain selectivity in health and disease can be accrued.

In another case, in order to determine the contributions of individual TAG biosynthesis pathways to TAG pools, thereby recapitulating the enzymatic activities involved in TAG biosynthesis, a steady-state simulation of the ion profiles of TAG species determined by lipidomics was performed [61]. The simulation was based on the known TAG biosynthesis pathways of DAG reacylation with acyl CoA [62] (Figure 5.2), which are as follows:

- The dephosphorylation of phosphatidic acid (PA) (DAG_{PAT}).
- The reacylation of MAG species (DAG_{MAG}), which could result from multiple sources including dephosphorylation of lysoPA, TAG/DAG hydrolysis with lipase activities, and reacylation from glycerol [62, 63].
- To a minimal extent, the hydrolysis of PI (PC) and PI polyphosphate species through phospholipase C (PLC) activities (DAG_{PI}).

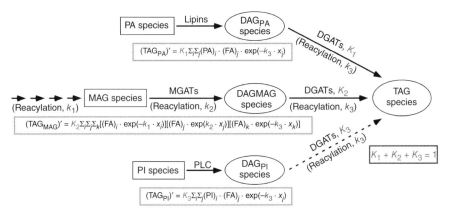

Figure 5.2 Schematic illustration of triacylglycerol biosynthesis model for simulation of triacylglycerol ion profiles. Triacylglycerol (TAG) species are *de novo* synthesized with reacylation of diacylglycerol (DAG) species of different pools produced mainly through dephosphorylation of phosphatidic acid (PA) (DAG_{PA}) and reacylation of monoacylglycerol (MAG) (DAG_{MAG}) as well as, to a less degree (as indicated with a broken line arrow), through hydrolysis of phosphatidylinositol (PI) with phospholipase C (PLC) activities (DAG_{PI}). The contributions of these pathways to the TAG pools were determined through simulation of individual TAG ion profile with parameters of K_1, K_2, and K_3, respectively, which are the probabilities of individual DAG pools being reacylated to TAG. In addition, the parameters of k_1, k_2, and k_3 were used in the *sn*-1, 2, and 3 reacylation steps of TAG species in the forms of $\exp(-k_1 \cdot x_j)$, $\exp(k_2 \cdot x_j)$, and $\exp(-k_3 \cdot x_j)$, respectively, where k_1 and k_3 represented a simulated decay constant, whereas k_2 represented a simulated enhancing constant and x_j is the number of double bonds present in the corresponding FA chain. MGAT and DGAT denote MAG and DAG acyltransferases, respectively. The multiple arrows at the k_1 step indicate that MAG species could yield from a variety of sources.

This approach was extensively validated by the comparison of individual simulated TAG species with those obtained from lipidomics analysis [61]. The simulated K parameters represent the relative contributions of the different DAG pools resulting from the PA, MAG, and PI (PC) pathways to TAG synthesis. Accordingly, the bioinformatic simulation provides a powerful vehicle to determine altered TAG biosynthesis pathways under pathophysiological conditions.

This type of bioinformatic simulation was also applied to the PC and PE biosyntheses [60]. From these studies, it is clearly demonstrated that through simulation, the following information can be assessed:

- The enzymatic activities involving the lipid biosynthesis/remodeling.
- The insights into the biochemical mechanisms underpinning altered lipids under patho(physio)logical states.
- Broader identification of individual lipid species compared to those determined by MS.

Although the clustered simulation is very useful and powerful and provides the foundation for analysis of the entire lipidome, it is clear that simulation approaches or models for analysis of a more comprehensive network for lipidomics are still warranted.

5.4.3   Modeling of Spatial Distributions and Biophysical Context

Spatial modeling of determined lipids in a systems and biophysical context is another challenge to bioinformatics. Although attempts on this topic have appeared in the literature, this area of research is still in a very immature stage in comparison to the other aforementioned components of bioinformatics for lipidomics. For example, Yetukuri and colleagues [64] reconstituted high-density lipoprotein (HDL) particles *in silico* with large-scale molecular dynamics simulations using information about the lipid composition of HDL particles from a lipidomic analysis. Through the simulation, the changes in particle size as lipid composition changes were determined and compared to that directly measured. Moreover, specific spatial distributions of lipids within the HDL particles, including a higher amount of TAG species at the surface of HDL particles in subjects with low HDL cholesterol levels induced with the changes in lipid composition, were revealed from the simulation. Accordingly, it is implicated that modeling of spatial distributions by using lipidomics data could lead to a better understanding of HDL metabolism and function.

5.5   INTEGRATION OF "OMICS"

5.5.1   Integration of Lipidomics with Other Omics

Integration of lipidomics data with genomics and transcriptomics could provide broader information and better connections in studying lipid metabolism and lipid signaling. Although applying this approach to process GPL and glycerolipid is rare, this approach for sphingolipidomics has been well documented and recently been extensively reviewed by Merrill [32]. The reason that such research is particularly powerful for exploring sphingolipid metabolism is likely due to the minimal remodeling of sphingolipid species after *de novo* synthesis while constant remodeling of GPL and glycerolipid species occurs.

The following are a few typical examples of integrating "omics" for sphingolipidomics. Integration of genomic, transcriptomic, and lipidomic data revealed a signaling role for phytosphingosine-1-phosphate in regulating genes required for mitochondrial respiration [65]. Mathematical modeling was applied to *Cryptococcus neoformans* to explore sphingolipid metabolism in the organism under acidic conditions in order to better understand fungal pathogenesis [66]. A mathematical approach that was used model-reference adaptive control was conducted to investigate the dynamics of *de novo* sphingolipid synthesis by HEK cells stably transfected with serine palmitoyl transferase [67]. It was found that the simulated results were comparable to those obtained from mass action kinetics and suggested that an adaptive feedback from increased metabolite levels might exist.

In another study, researchers developed a model for the C16-branch of sphingolipid metabolism by integrating lipidomics and transcriptomics data from RAW264.7 cells, and using a two-step matrix-based approach to estimate the rate constants from experimental data [68]. They selected palmitate as the *N*-acyl-linked fatty acid because it is a major subspecies for all categories of complex sphingolipids in RAW264.7 cells. The rate constants obtained from the first step were further refined using generalized constrained nonlinear optimization (the second step). The model fit the lipidomics data for all species. The study not only provided a better understanding about how these sphingolipid species are made and function, but also aid to interpret (and ultimately predict) the outcomes of changes in precursors, effects of inhibitors, genetic mutations, etc. It is obvious that this type of bioinformatics-based work should provide potentially interesting directions for our future investigation in lipidomics.

5.5.2 Lipidomics Guides Genomics Analysis

In population studies of human plasma by lipidomics, the detected lipid changes do not definitively indicate that the relevant pathways or the network of the changed lipids, thus the genes involved in the pathways, are changed due to the complication of the populations. However, it may be suggested that the detected lipid changes could be due to the genetic variations present in the populations in combination with other factors. Therefore, these relevant genes could be targeted to examine their association with a patho(physio)logical condition, whereas a genome-wide association study directly from the population may miss the target(s).

For example, shotgun lipidomics analysis was recently employed to explore the potential biomarkers of human plasma lipids for early detection of Alzheimer's disease (AD) [69]. In the study, the levels of over 800 molecular species of plasma lipids from 26 AD patients (average mini-mental state examination (MMSE) score of 21) and 26 cognitively normal controls were determined in a nontargeted approach by using MDMS-SL [3, 70, 71]. The data were then correlated with clinic diagnosis, apolipoprotein E4 genotype, and cognitive performance. It was found that an essentially uniform, but opposite, pattern of disruption in the mass levels of SM and Cer species in AD plasma was present. From the findings, a set of genes related to SM and Cer biosynthesis and degradation was enlisted and informed researchers in genetics to study disease pathophysiology (e.g., global cortical amyloid burden, brain regional volume, glucose metabolism) in a systems biology approach in a population of over 1000 individuals who were extensively characterized. It was found that six genes (i.e., PA phosphatase, Cer synthase S4 and S6, sphingosine kinase, choline/ethanolamine phosphotransferase, and serine palmitoyl transferase) were significantly associated with AD pathophysiology based on SNP- and/or gene-level analysis [72]. This example clearly indicates that lipidomics not only provides the assessment and/or phenotypes of genetic variation but can also guide us to conduct targeted genomic analysis.

REFERENCES

1. Niemela, P.S., Castillo, S., Sysi-Aho, M., Oresic, M. (2009) Bioinformatics and computational methods for lipidomics. J. Chromatogr. B 877, 2855–2862.

2. Checa, A., Bedia, C., Jaumot, J. (2015) Lipidomic data analysis: Tutorial, practical guidelines and applications. Anal. Chim. Acta 885, 1–16.

3. Yang, K., Cheng, H., Gross, R.W., Han, X. (2009) Automated lipid identification and quantification by multi-dimensional mass spectrometry-based shotgun lipidomics. Anal. Chem. 81, 4356–4368.

4. Fahy, E., Subramaniam, S., Brown, H.A., Glass, C.K., Merrill, A.H., Jr., Murphy, R.C., Raetz, C.R., Russell, D.W., Seyama, Y., Shaw, W., Shimizu, T., Spener, F., van Meer, G., VanNieuwenhze, M.S., White, S.H., Witztum, J.L., Dennis, E.A. (2005) A comprehensive classification system for lipids. J. Lipid Res. 46, 839–861.

5. Fahy, E., Sud, M., Cotter, D., Subramaniam, S. (2007) LIPID MAPS online tools for lipid research. Nucleic Acids Res. 35, W606-612.

6. Sud, M., Fahy, E., Cotter, D., Brown, A., Dennis, E.A., Glass, C.K., Merrill, A.H., Jr., Murphy, R.C., Raetz, C.R., Russell, D.W., Subramaniam, S. (2007) LMSD: LIPID MAPS structure database. Nucleic Acids Res. 35, D527-532.

7. Kind, T., Liu, K.H., Lee do, Y., Defelice, B., Meissen, J.K., Fiehn, O. (2013) LipidBlast *in silico* tandem mass spectrometry database for lipid identification. Nat. Methods 10, 755–758.

8. Quehenberger, O., Armando, A.M., Brown, A.H., Milne, S.B., Myers, D.S., Merrill, A.H., Bandyopadhyay, S., Jones, K.N., Kelly, S., Shaner, R.L., Sullards, C.M., Wang, E., Murphy, R.C., Barkley, R.M., Leiker, T.J., Raetz, C.R., Guan, Z., Laird, G.M., Six, D.A., Russell, D.W., McDonald, J.G., Subramaniam, S., Fahy, E., Dennis, E.A. (2010) Lipidomics reveals a remarkable diversity of lipids in human plasma. J. Lipid Res. 51, 3299–3305.

9. Gao, X., Zhang, Q., Meng, D., Isaac, G., Zhao, R., Fillmore, T.L., Chu, R.K., Zhou, J., Tang, K., Hu, Z., Moore, R.J., Smith, R.D., Katze, M.G., Metz, T.O. (2012) A reversed-phase capillary ultra-performance liquid chromatography-mass spectrometry (UPLC-MS) method for comprehensive top-down/bottom-up lipid profiling. Anal. Bioanal. Chem. 402, 2923–2933.

10. Smith, C.A., O'Maille, G., Want, E.J., Qin, C., Trauger, S.A., Brandon, T.R., Custodio, D.E., Abagyan, R., Siuzdak, G. (2005) METLIN: A metabolite mass spectral database. Ther. Drug Monit. 27, 747–751.

11. Tautenhahn, R., Cho, K., Uritboonthai, W., Zhu, Z., Patti, G.J., Siuzdak, G. (2012) An accelerated workflow for untargeted metabolomics using the METLIN database. Nat. Biotechnol. 30, 826–828.

12. Wolf, S., Schmidt, S., Muller-Hannemann, M., Neumann, S. (2010) *In silico* fragmentation for computer assisted identification of metabolite mass spectra. BMC Bioinf. 11, 148.

13. Wishart, D.S., Tzur, D., Knox, C., Eisner, R., Guo, A.C., Young, N., Cheng, D., Jewell, K., Arndt, D., Sawhney, S., Fung, C., Nikolai, L., Lewis, M., Coutouly, M.A., Forsythe, I., Tang, P., Shrivastava, S., Jeroncic, K., Stothard, P., Amegbey, G., Block, D., Hau, D.D., Wagner, J., Miniaci, J., Clements, M., Gebremedhin, M., Guo, N., Zhang, Y., Duggan, G.E., Macinnis, G.D., Weljie, A.M., Dowlatabadi, R., Bamforth, F., Clive, D., Greiner, R., Li, L., Marrie, T., Sykes, B.D., Vogel, H.J., Querengesser, L. (2007) HMDB: The Human Metabolome Database. Nucleic Acids Res. 35, D521-526.

14. Wishart, D.S., Knox, C., Guo, A.C., Eisner, R., Young, N., Gautam, B., Hau, D.D., Psychogios, N., Dong, E., Bouatra, S., Mandal, R., Sinelnikov, I., Xia, J., Jia, L., Cruz, J.A., Lim, E., Sobsey, C.A., Shrivastava, S., Huang, P., Liu, P., Fang, L., Peng, J., Fradette, R., Cheng, D., Tzur, D., Clements, M., Lewis, A., De Souza, A., Zuniga, A., Dawe, M., Xiong, Y., Clive, D., Greiner, R., Nazyrova, A., Shaykhutdinov, R., Li, L., Vogel, H.J., Forsythe, I. (2009) HMDB: A knowledgebase for the human metabolome. Nucleic Acids Res. 37, D603-610.

15. Wishart, D.S., Jewison, T., Guo, A.C., Wilson, M., Knox, C., Liu, Y., Djoumbou, Y., Mandal, R., Aziat, F., Dong, E., Bouatra, S., Sinelnikov, I., Arndt, D., Xia, J., Liu, P., Yallou, F., Bjorndahl, T., Perez-Pineiro, R., Eisner, R., Allen, F., Neveu, V., Greiner, R., Scalbert, A. (2013) HMDB 3.0 – The human metabolome database in 2013. Nucleic Acids Res. 41, D801-807.

16. Watanabe, K., Yasugi, E., Oshima, M. (2000) How to search the glycolipid data in LIPID-BANK for Web: The newly developed lipid database. Japan Trend Glycosci. Glycotechnol. 12, 175–184.

17. Myers, D.S., Ivanova, P.T., Milne, S.B., Brown, H.A. (2011) Quantitative analysis of glycerophospholipids by LC-MS: Acquisition, data handling, and interpretation. Biochim. Biophys. Acta 1811, 748–757.

18. Smith, C.A., Want, E.J., O'Maille, G., Abagyan, R., Siuzdak, G. (2006) XCMS: processing mass spectrometry data for metabolite profiling using nonlinear peak alignment, matching, and identification. Anal. Chem. 78, 779–787.

19. Pluskal, T., Castillo, S., Villar-Briones, A., Oresic, M. (2010) MZmine 2: Modular framework for processing, visualizing, and analyzing mass spectrometry-based molecular profile data. BMC Bioinf. 11, 395.

20. Katajamaa, M., Oresic, M. (2007) Data processing for mass spectrometry-based metabolomics. J. Chromatogr. A 1158, 318–328.

21. Windig, W., Phalp, J.M., Payne, A. (1996) A noise and background reduction method for component detection in liquid chromatography/mass spectrometry. Anal. Chem. 68, 3602–3606.

22. Andreev, V.P., Rejtar, T., Chen, H.S., Moskovets, E.V., Ivanov, A.R., Karger, B.L. (2003) A universal denoising and peak picking algorithm for LC-MS based on matched filtration in the chromatographic time domain. Anal. Chem. 75, 6314–6326.

23. Milne, S.B., Tallman, K.A., Serwa, R., Rouzer, C.A., Armstrong, M.D., Marnett, L.J., Lukehart, C.M., Porter, N.A., Brown, H.A. (2010) Capture and release of alkyne-derivatized glycerophospholipids using cobalt chemistry. Nat. Chem. Biol. 6, 205–207.

24. Fredriksson, M.J., Petersson, P., Axelsson, B.O., Bylund, D. (2009) An automatic peak finding method for LC-MS data using Gaussian second derivative filtering. J. Sep. Sci. 32, 3906–3918.

25. Williams, E., Kloot, N. (1953) Interpolation in a series of correlated observations. Aust. J. Appl. Sci. 4, 1–17.

26. Barker, M., Rayens, W. (2003) Partial least squares for discrimination. J. Chemometr. 17, 166–173.

27. Li, X., Lu, X., Tian, J., Gao, P., Kong, H., Xu, G. (2009) Application of fuzzy c-means clustering in data analysis of metabolomics. Anal. Chem. 81, 4468–4475.

28. Tufte, E.R. (2001) The Visual Display of Quantitative Information. Graphics Press, pp 197.

29. Andreyev, A.Y., Fahy, E., Guan, Z., Kelly, S., Li, X., McDonald, J.G., Milne, S., Myers, D., Park, H., Ryan, A., Thompson, B.M., Wang, E., Zhao, Y., Brown, H.A., Merrill, A.H., Raetz, C.R., Russell, D.W., Subramaniam, S., Dennis, E.A. (2010) Subcellular organelle lipidomics in TLR-4-activated macrophages. J. Lipid Res. 51, 2785–2797.

30. Dennis, E.A., Deems, R.A., Harkewicz, R., Quehenberger, O., Brown, H.A., Milne, S.B., Myers, D.S., Glass, C.K., Hardiman, G., Reichart, D., Merrill, A.H., Jr., Sullards, M.C., Wang, E., Murphy, R.C., Raetz, C.R., Garrett, T.A., Guan, Z., Ryan, A.C., Russell, D.W., McDonald, J.G., Thompson, B.M., Shaw, W.A., Sud, M., Zhao, Y., Gupta, S., Maurya, M.R., Fahy, E., Subramaniam, S. (2010) A mouse macrophage lipidome. J. Biol. Chem. 285, 39976–39985.

31. Kapoor, S., Quo, C.F., Merrill, A.H., Jr., Wang, M.D. (2008) An interactive visualization tool and data model for experimental design in systems biology. Conf. Proc. IEEE Eng. Med. Biol. Soc. 2008, 2423–2426.

32. Merrill, A.H., Jr. (2011) Sphingolipid and glycosphingolipid metabolic pathways in the era of sphingolipidomics. Chem. Rev. 111, 6387–6422.

33. Momin, A.A., Park, H., Portz, B.J., Haynes, C.A., Shaner, R.L., Kelly, S.L., Jordan, I.K., Merrill, A.H., Jr. (2011) A method for visualization of "omic" datasets for sphingolipid metabolism to predict potentially interesting differences. J. Lipid Res. 52, 1073–1083.

34. Brugger, B., Erben, G., Sandhoff, R., Wieland, F.T., Lehmann, W.D. (1997) Quantitative analysis of biological membrane lipids at the low picomole level by nano-electrospray ionization tandem mass spectrometry. Proc. Natl. Acad. Sci. U.S.A. 94, 2339–2344.

35. Ejsing, C.S., Duchoslav, E., Sampaio, J., Simons, K., Bonner, R., Thiele, C., Ekroos, K., Shevchenko, A. (2006) Automated identification and quantification of glycerophospholipid molecular species by multiple precursor ion scanning. Anal. Chem. 78, 6202–6214.

36. Ekroos, K., Ejsing, C.S., Bahr, U., Karas, M., Simons, K., Shevchenko, A. (2003) Charting molecular composition of phosphatidylcholines by fatty acid scanning and ion trap MS3 fragmentation. J. Lipid Res. 44, 2181–2192.

37. Yang, K., Zhao, Z., Gross, R.W., Han, X. (2009) Systematic analysis of choline-containing phospholipids using multi-dimensional mass spectrometry-based shotgun lipidomics. J. Chromatogr. B 877, 2924–2936.

38. Yang, K., Zhao, Z., Gross, R.W., Han, X. (2011) Identification and quantitation of unsaturated fatty acid isomers by electrospray ionization tandem mass spectrometry: A shotgun lipidomics approach. Anal. Chem. 83, 4243–4250.

39. Liebisch, G., Vizcaino, J.A., Kofeler, H., Trotzmuller, M., Griffiths, W.J., Schmitz, G., Spener, F., Wakelam, M.J. (2013) Shorthand notation for lipid structures derived from mass spectrometry. J. Lipid Res. 54, 1523–1530.

40. Forrester, J.S., Milne, S.B., Ivanova, P.T., Brown, H.A. (2004) Computational lipidomics: A multiplexed analysis of dynamic changes in membrane lipid composition during signal transduction. Mol. pharmacol. 65, 813–821.

41. Hermansson, M., Uphoff, A., Kakela, R., Somerharju, P. (2005) Automated quantitative analysis of complex lipidomes by liquid chromatography/mass spectrometry. Anal. Chem. 77, 2166–2175.

42. Laaksonen, R., Katajamaa, M., Paiva, H., Sysi-Aho, M., Saarinen, L., Junni, P., Lutjohann, D., Smet, J., Van Coster, R., Seppanen-Laakso, T., Lehtimaki, T., Soini, J., Oresic, M. (2006) A systems biology strategy reveals biological pathways and plasma biomarker candidates for potentially toxic statin-induced changes in muscle. PLoS One 1, e97.

43. Fahy, E., Cotter, D., Byrnes, R., Sud, M., Maer, A., Li, J., Nadeau, D., Zhau, Y., Subramaniam, S. (2007) Bioinformatics for lipidomics. Methods Enzymol. 432, 247–273.

44. Sysi-Aho, M., Katajamaa, M., Yetukuri, L., Oresic, M. (2007) Normalization method for metabolomics data using optimal selection of multiple internal standards. BMC Bioinf. 8, e93.

45. Hubner, G., Crone, C., Lindner, B. (2009) lipID--a software tool for automated assignment of lipids in mass spectra. J. Mass Spectrom. 44, 1676–1683.

46. Hartler, J., Trotzmuller, M., Chitraju, C., Spener, F., Kofeler, H.C., Thallinger, G.G. (2011) Lipid Data Analyzer: Unattended identification and quantitation of lipids in LC-MS data. Bioinformatics 27, 572–577.

47. Yang, K., Han, X. (2011) Accurate quantification of lipid species by electrospray ionization mass spectrometry – Meets a key challenge in lipidomics. Metabolites 1, 21–40.

48. Satten, G.A., Datta, S., Moura, H., Woolfitt, A.R., Carvalho Mda, G., Carlone, G.M., De, B.K., Pavlopoulos, A., Barr, J.R. (2004) Standardization and denoising algorithms for mass spectra to classify whole-organism bacterial specimens. Bioinformatics 20, 3128–3136.

49. Ivanova, P.T., Milne, S.B., Byrne, M.O., Xiang, Y., Brown, H.A. (2007) Glycerophospholipid identification and quantitation by electrospray ionization mass spectrometry. Methods Enzymol. 432, 21–57.

50. Norris, J.L., Cornett, D.S., Mobley, J.A., Andersson, M., Seeley, E.H., Chaurand, P., Caprioli, R.M. (2007) Processing MALDI mass spectra to improve mass spectral direct tissue analysis. Int. J. Mass Spectrom. 260, 212–221.

51. Yang, K., Fang, X., Gross, R.W., Han, X. (2011) A practical approach for determination of mass spectral baselines. J. Am. Soc. Mass Spectrom. 22, 2090–2099.

52. Haimi, P., Uphoff, A., Hermansson, M., Somerharju, P. (2006) Software tools for analysis of mass spectrometric lipidome data. Anal. Chem. 78, 8324–8331.

53. Schwudke, D., Oegema, J., Burton, L., Entchev, E., Hannich, J.T., Ejsing, C.S., Kurzchalia, T., Shevchenko, A. (2006) Lipid profiling by multiple precursor and neutral loss scanning driven by the data-dependent acquisition. Anal. Chem. 78, 585–595.

54. Herzog, R., Schwudke, D., Schuhmann, K., Sampaio, J.L., Bornstein, S.R., Schroeder, M., Shevchenko, A. (2011) A novel informatics concept for high-throughput shotgun lipidomics based on the molecular fragmentation query language. Genome Biol. 12, R8.

55. Husen, P., Tarasov, K., Katafiasz, M., Sokol, E., Vogt, J., Baumgart, J., Nitsch, R., Ekroos, K., Ejsing, C.S. (2013) Analysis of lipid experiments (ALEX): A software framework for analysis of high-resolution shotgun lipidomics data. PLoS One 8, e79736.

56. Wheelock, C.E., Goto, S., Yetukuri, L., D'Alexandri, F.L., Klukas, C., Schreiber, F., Oresic, M. (2009) Bioinformatics strategies for the analysis of lipids. Methods Mol. Biol. 580, 339–368.

57. van Iersel, M.P., Kelder, T., Pico, A.R., Hanspers, K., Coort, S., Conklin, B.R., Evelo, C. (2008) Presenting and exploring biological pathways with PathVisio. BMC Bioinformatics 9, 399.

58. Kiebish, M.A., Bell, R., Yang, K., Phan, T., Zhao, Z., Ames, W., Seyfried, T.N., Gross, R.W., Chuang, J.H., Han, X. (2010) Dynamic simulation of cardiolipin remodeling: Greasing the wheels for an interpretative approach to lipidomics. J. Lipid Res. 51, 2153–2170.

59. Zhang, L., Bell, R.J., Kiebish, M.A., Seyfried, T.N., Han, X., Gross, R.W., Chuang, J.H. (2011) A mathematical model for the determination of steady-state cardiolipin remodeling mechanisms using lipidomic data. PLoS One 6, e21170.

60. Zarringhalam, K., Zhang, L., Kiebish, M.A., Yang, K., Han, X., Gross, R.W., Chuang, J. (2012) Statistical analysis of the processes controlling choline and ethanolamine glycerophospholipid molecular species composition. PLoS One 7, e37293.

61. Han, R.H., Wang, M., Fang, X., Han, X. (2013) Simulation of triacylglycerol ion profiles: bioinformatics for interpretation of triacylglycerol biosynthesis. J. Lipid Res. 54, 1023–1032.

62. Coleman, R.A., Lee, D.P. (2004) Enzymes of triacylglycerol synthesis and their regulation. Prog. Lipid Res. 43, 134–176.

63. Rushdi, A.I., Simoneit, B.R. (2006) Abiotic condensation synthesis of glyceride lipids and wax esters under simulated hydrothermal conditions. Orig. Life Evol. Biosph. 36, 93–108.

64. Yetukuri, L., Soderlund, S., Koivuniemi, A., Seppanen-Laakso, T., Niemela, P.S., Hyvonen, M., Taskinen, M.R., Vattulainen, I., Jauhiainen, M., Oresic, M. (2010) Composition and lipid spatial distribution of HDL particles in subjects with low and high HDL-cholesterol. J. Lipid Res. 51, 2341–2351.

65. Cowart, L.A., Shotwell, M., Worley, M.L., Richards, A.J., Montefusco, D.J., Hannun, Y.A., Lu, X. (2010) Revealing a signaling role of phytosphingosine-1-phosphate in yeast. Mol. Syst. Biol. 6, 349.

66. Garcia, J., Shea, J., Alvarez-Vasquez, F., Qureshi, A., Luberto, C., Voit, E.O., Del Poeta, M. (2008) Mathematical modeling of pathogenicity of Cryptococcus neoformans. Mol. Syst. Biol. 4, 183.

67. Quo, C.F., Moffitt, R.A., Merrill, A.H., Wang, M.D. (2011) Adaptive control model reveals systematic feedback and key molecules in metabolic pathway regulation. J Comput. Biol. 18, 169–182.

68. Gupta, S., Maurya, M.R., Merrill, A.H., Jr., Glass, C.K., Subramaniam, S. (2011) Integration of lipidomics and transcriptomics data towards a systems biology model of sphingolipid metabolism. BMC Syst. Biol. 5, 26.

69. Han, X., Rozen, S., Boyle, S., Hellegers, C., Cheng, H., Burke, J.R., Welsh-Bohmer, K.A., Doraiswamy, P.M., Kaddurah-Daouk, R. (2011) Metabolomics in early Alzheimer's disease: Identification of altered plasma sphingolipidome using shotgun lipidomics. PLoS One 6, e21643.

70. Han, X., Gross, R.W. (2005) Shotgun lipidomics: Electrospray ionization mass spectrometric analysis and quantitation of the cellular lipidomes directly from crude extracts of biological samples. Mass Spectrom. Rev. 24, 367–412.

71. Han, X., Yang, K., Gross, R.W. (2012) Multi-dimensional mass spectrometry-based shotgun lipidomics and novel strategies for lipidomic analyses. Mass Spectrom. Rev. 31, 134–178.

72. Kim, S., Nho, K., Shen, L., Kling, M., Han, X., Zhu, H., Sullivan, P., Arnold, S., Risacher, S., Ramanan, V., Doraiswamy, P.M., Trojanowski, J., Kaddurah-Daouk, R., Saykin, A. (2013) Targeted lipidomic pathway-guided genetic association with Alzheimer's disease–relevant endophenotypes. Alz. Dement. 9(4S), 680–681.

PART II

CHARACTERIZATION OF LIPIDS

6

INTRODUCTION

6.1 STRUCTURAL CHARACTERIZATION FOR LIPID IDENTIFICATION

Characterization of lipid structures is a major and essential part of lipidomic research. This is largely due to the identification and even quantification of individual lipid species present in biological samples that ultimately depend on the information learned from the extensive characterization of representative authentic lipid species of individual class. Accordingly, characterization of common ions of lipid species from the majority of the cellular lipid classes is summarized in this part.

For clarification, a few points related to characterization of lipid species should be made clear.

- Although in-source fragmentation was used for characterization of individual lipid species [1], such studies are largely conducted through tandem MS in the product-ion mode after CID. Extensive characterization of the newly discovered lipid classes including branched fatty acid esters of hydroxy fatty acids [2] and sulfono-containing ether lipids [3] represents the typical examples.

- An individual lipid species may appear in a variety of ion forms under different experimental conditions. Many of these ions are meaningful for ion chemistry to elucidate the fragmentation pathways in the gas phase. However, these ions may be rarely present in routine experiments and not practically useful for biological applications. Therefore, only the common ion forms widely used for analysis of biological samples are discussed in this part.

Lipidomics: Comprehensive Mass Spectrometry of Lipids, First Edition. Xianlin Han.
© 2016 John Wiley & Sons, Inc. Published 2016 by John Wiley & Sons, Inc.

- It is always good to characterize a wide variety of representative molecular species of a lipid class based on the degree of unsaturation, fatty acyl chain length, different regioisomers, etc. However, due to the cost for extensive characterization, only a minimal set of species is usually employed for the purpose and a fragmentation pattern is derived from the limited characterization for general utilization.

- Since the ESI ion source is compatible with nearly all of mass analyzers in which tandem MS can be performed, a variety of types of mass spectrometers are employed for characterization of lipids. Some examples of the instruments that are used to perform tandem MS include tandem sectors, QqQ, ion-trap, ion-cyclotron resonance, TOF/TOF, and hybrid instruments such as Q-TOF. The majority of tandem MS analyses presented in this part are conducted by using the QqQ-type instrument since majority of the early studies on characterization of lipid species were conducted with this type of instrument, and the characterized fragmentation pattern from this type of instrument can well represent those obtained from other types of instruments even including MALDI-MS [4, 5].

In this section, the fundamentals of tandem MS characterization of lipid species are discussed. The fragmentation pattern of a particular compound depends on the kinetics of the decomposing reaction as well as the thermodynamics of the resultant product ions, which can be assessed by quantum mechanics. In other words, the intrinsic chemical properties of the compound determine its fragmentation pattern. This principle of fragmentation indicates that the fragmentation pattern of a lipid class is very similar or even essentially identical due to their structural similarity. The effects of different fatty acyls on the fragmentation pattern of a lipid class are minimal except the further loss of CO_2 from the resultant fatty acyl carboxylate, which depends on the number and location of double bonds [6]. This minimal effect is largely due to the high stability of fatty acyls under low energy CID.

Different from the fragmentation pattern of a lipid class, the appearance of an acquired product-ion mass spectrum of a compound after CID from a selected ESI mass spectrometer could be varied dramatically. The variation of the spectrum largely depends on the kinetic energy of the precursor ion and the number of collisions of both the precursor ion and the resultant product ions as discussed below.

The kinetic energy of the precursor ion largely depends on the collision energy once an instrument is selected since different mass spectrometers could make the precursor ions to carry a different degree of intrinsic energy. It should be recognized that collision energy is controlled with the collision-cell voltage in any mass spectrometer that possesses a collision cell, and it is controlled with the CID voltage in an ion-trap type instrument. In general, the higher the collision energy is, the more abundant the product ions in the lower molecular mass region are yielded after the number of collisions is fixed (see below). To maximize the appearance of product ions in a tandem MS spectrum, a technique by ramping the collision cell (or CID) voltage could be employed [7].

Figure 6.1 shows the product ion ESI mass spectra of lithiated 16:0–18:1 dPC after CID with different levels of collision energy as indicated. The spectra clearly indicate that the intensity ratios of the fragment ions are varied with the collision energy. For example, the fragment ions at the low mass region significantly increase as collision energy increases from 10 to 40 eV (Figure 6.1a–d, respectively). However, all the spectra demonstrate an essentially identical fragmentation pattern as indicated: a loss of 59 amu (i.e., trimethylamine), a loss of 183 amu (i.e., phosphocholine), a loss of 189 amu (i.e., lithium cholinephosphate), two losses of 315 and 341 amu (i.e., 16:0 fatty acid (FA, *sn*-1 FA) + 59 and 18:1 FA (*sn*-2 FA) + 59, respectively), and a few other minor fragment ions. Such an identical fragmentation pattern was

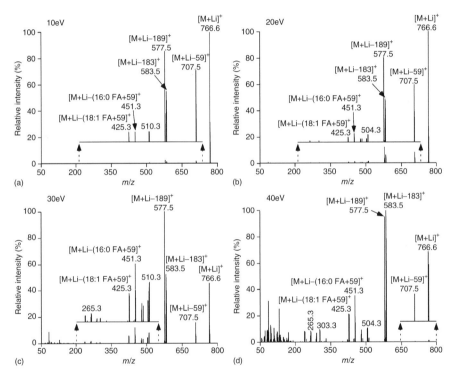

Figure 6.1 Product-ion ESI mass spectra of lithiated 16:0–18:1 phosphatidylcholine after CID at different collision energy. Product-ion ESI-MS analysis of lithiated 16:0–18:1 dPC at *m/z* 766.5 in the presence of lithium hydroxide in the infusion solution was performed on a Thermo Fisher TSQ Vantage mass spectrometer. Collision activation was carried out with collision energy of 10 (a), 20 (b), 30 (c), and 40 (d) eV, and gas pressure of 1 mTorr. These results demonstrate two key points. First, while product-ion mass spectra acquired at different collision energy are very different, the fragmentation pattern of dPC species is identical under different experimental conditions. The pattern includes the neutral losses of 59, 183, 189, *sn*-1 FA plus 59, and *sn*-2 FA plus 59. Second, the low mass fragment ions increase as collision energy.

also obtained from other lithiated molecular species of phosphatidylcholine subclass under different experimental conditions as previously demonstrated [8, 9].

The number of collisions that both the precursor ion and the resultant product ions experience is associated with the instruments employed. It depends on the collision-gas pressure when a quadrupole serves as the collision cell or the collision duration, in that case an ion trap is used as an analyzer as well as a collision cell. In the case of a collision cell, regardless of a very low collision gas pressure is employed, consecutive fragmentation processes always exist. In contrast, the collision duration in most ion-trap instruments tends to be shorter, allowing little or no time for consecutive collision of the resultant product ions. Thus, fewer product ions, particularly those in the low mass region, and less structural information are obtained. This may be a disadvantage of an ion-trap instrument for structural characterization, but it is an advantage for elucidation of fragmentation pathways. Moreover, its ability to perform multistage product-ion analysis through continuous selection and fragmentation of a resultant product ion of interest can compensate for the generation of fewer product ions in an ion-trap instrument. Collectively, ion-trap instrumentation is very powerful for elucidation of fragmentation pathways, since it allows structure to be determined in a very controlled manner. It should be noted that researchers could achieve elucidation of fragmentation pathways to a great degree through ramping CID voltage (i.e., collision energy) as discussed earlier.

It can be conceivably recognized that relatively higher collision energy (e.g., keV translational energy) used in the instruments with tandem sectors or TOF/TOF results in significantly more consecutive fragmentations than other types of instruments with relatively lower collision energy (e.g., less than 100 eV in QqQ type). Therefore, this type of instrument may not be very favorable for elucidation of fragmentation pathways as well as structural characterization of lipid species. But it may be useful for identification of structural cleavage that needs high collision energy (e.g., the location(s) of double bonds in fatty acyl chains as previously demonstrated [10, 11]).

Collectively, while the fragmentation pattern of a lipid class largely relies on its chemical structure and charge properties, the appearance of a product-ion mass spectrum after CID depends on the CID conditions (collision energy and collision-gas pressure) as well as the type of instrument employed. QqQ mass spectrometers generally yield well-balanced, informative fragment ions for both structural characterization and studies of fragmentation processes of complex lipids with the aid of collision energy ramping. Accordingly, due to its early development in coupling with the ESI source, high efficiency in collection of fragment ions in the radio frequency-only collision cell, easiness of operation, and relatively low cost, QqQ-type mass spectrometers are widely used for characterization of lipid structures and elucidation of fragmentation pathways in lipidomics. Moreover, product-ion patterns obtained from the QqQ-type mass spectrometers are similar to those obtained from the Q-TOF type instruments as well as postsource decay MALDI-MS, both of which are also widely used for structural characterization of lipids.

From the aforementioned discussion, it is clear that the intensities of product ions of an individual lipid species after CID depend on both the CID conditions and the employed mass spectrometer. This implies that direct comparison of the product-ion intensities between spectra acquired under different experimental conditions may not be meaningful. This also indicates that constructing a library or database comprised of product-ion mass spectra of individual lipid species is not very practical for ESI-MS analysis of lipids or for identification of individual lipid species present in biological samples by matching the intensities of fragment ions. However, as aforementioned, the gas-phase ion chemistry and the mechanisms underlying the fragmentation processes of a particular lipid species are essentially identical among the various conditions employed. Therefore, recognizing the fragmentation patterns instead of the intensity profiles of the product ions is highly recommended. This point is further discussed in the next section. A recent developed *in silico* library of product-ion mass spectra (i.e., LipidBlast, see Chapter 5) was also constructed based on such a principle, that is, matching the fragmentation pattern [12].

Finally, a few practical points should be noted.

- Any approach for sample introduction (i.e., direct infusion, LC-MS, or loop injection) can be employed for characterization of lipid species, but direct infusion is the primary choice and has been widely applied due to its easiness of operation.
- Synthetic compounds if available are always preferable for structural characterization and fragmentation elucidation since individual lipid ions selected from the analysis of biological samples may contain isomeric species. The presence of isomers at a selected ion definitely leads to a complicated mass spectrum and an inaccurate fragmentation pattern.
- Although solvent(s) could significantly influence the efficiency and stability of lipid ionization (see Chapter 4), the effects of solvent(s) on either elucidation of lipid structures or determination of their fragmentation patterns are minimal. Therefore, there is no particular preference for selection of solvents for this purpose.

6.2 PATTERN RECOGNITION FOR LIPID IDENTIFICATION

6.2.1 Principles of Pattern Recognition

As discussed in the previous section, one of the major purposes in characterization of lipid species is to elucidate the fragmentation patterns for identification of lipid species present in biological samples (i.e., cellular lipidomes). Identification of individual lipid species should generally cover all the essential elements of the lipid structure, including information about class, subclass, fatty acyls, and regiospecificity if present. The lipid class is determined with the structure of the polar head group. The linkages of *sn*-1 fatty acyl chain to the glycerol hydroxy group defines the subclasses of GPL classes; whether a *trans* double bond is present between C4 and C5 of the sphingoid backbone defines the subclasses of sphingolipids as sphingosine-based sphingolipids or sphinganine-based dihydrosphingolipids; the presence of a hydroxyl group at the α-position of the fatty acyl amide chain classifies the sphingolipids into two subclasses: hydroxy sphingolipids and regular sphingolipids. Identification of the structure of individual fatty acyl should include the chain length, unsaturation, and the location of each double bond. The regiospecificity defines the connected position of individual fatty acyl chain. An elucidated fragmentation pattern should include all of these types of information to definitively identify the structure of a lipid species or the structures of an entire class of lipid species. Therefore, this pattern matches well with the building-block concept as described in Chapter 1.

In the current stage of lipidomics practice, the majority of the elucidated fragmentation patterns should be able to define the head groups, the linkages for subclasses, the mass corresponding to a fatty acyl chain, and usually the intensity ratio of the ions carrying the fatty acyl chain information. This suggests that the classes, subclasses, and regioisomers (from the intensity ratios of the ions related to fatty acyl chains) can be well determined. From the detected fatty acyl mass, particularly those obtained from the instruments with high mass accuracy, the information about the total number of carbon atoms and double bonds of an individual fatty acyl chain can be derived. Unfortunately, determining the distribution of double bonds in a fatty acyl chain still remains to be a challenge in most of the cases although great efforts have been made to solve this issue [13, 14].

There are numerous publications in the literature focusing on structural characterization of cellular lipids since the early stages of ESI-MS development. Therefore, fragmentation patterns of the majority of the lipid classes from a variety of ion forms can be summarized from the previous studies. These elucidated fragmentation patterns are presented separately in the chapters of the rest of this part. It is advised that advanced readers should always look for the details from the original studies on the topic and/or study several invaluable review articles [15–20]. It is worth noting that Hsu and Turk have performed a series of studies on structural characterization and elucidation of fragmentation patterns of different lipid classes in an unprecedented detail. Their work has been reviewed [18, 19, 21], and these articles should be consulted for a detailed understanding of structural characterization of specific lipid classes.

Recognizing and remembering these fragmentation patterns are very useful and important. Simply, remembering one pattern from one ion form of a lipid class is much easier than to memorize all the tandem MS spectra yielding from the species of the class with this type of ion form. Indeed, all these tandem MS spectra can be readily derived from the fragmentation pattern if one is familiar with it. Moreover, as aforementioned, the appearance of a product-ion mass spectrum of a lipid species may vary with experimental conditions, whereas the fragmentation pattern of a species is minimally changed with these conditions (Figure 6.1).

Although elution time from a LC separation provides valuable information for a particular species, definitive identification of the lipid species has to be performed based on its product-ion mass spectrum. For those researchers who are using LC-MS for identification and quantification of lipid species, familiar with the fragmentation patterns of lipid classes, therefore deriving the product-ion mass spectra of individual lipid species, is very important although a few databases and/or libraries (see Chapter 5) can be used to aid the identification. Resolving complex biological lipids into a single component is idea, but not practical. In the majority of the cases, a few lipid species are eluted at the same time. It is difficult to search the databases/libraries to definitively identify these species. Manual identification of these species is inevitable. Familiar with the fragmentation patterns of lipid classes would greatly aid the identification of those species in the case. Finally, familiar with the fragmentation patterns of lipid classes would also be essential for establishment of MRM methods for quantification of individual lipid classes after LC-MS.

For those who would like to employ shotgun lipidomics (particularly MDMS-SL) for identification and quantification of individual species of a class, familiar with the fragmentation patterns of individual lipid classes, could enable them to design MS/MS scans for building-block analysis in the PIS, NLS, or both modes to selectively identify individual molecular species of a lipid class of interest.

6.2.2 Examples

The following are a few examples to demonstrate how an elucidated fragmentation pattern can be used for identification of individual species of a lipid class present in biological samples by MDMS-SL. These examples well represent different lipid classes (e.g., glycerophospholipids, sphingolipid, and glycerolipids), different mass levels (from low abundance to abundance), different polarities (from polar to non-polar), etc.

6.2.2.1 Choline Lysoglycerophospholipid   As discussed in Chapter 4, in the absence of any modifiers in an infused solution, sodium adducts are the prominent ion form for the majority of lipid classes including lysoPC when the lipid solution is analyzed by ESI-MS in the positive-ion mode. Product-ion ESI-MS analysis of sodium adducts of either *sn*-1 or *sn*-2 acyl lysoPC species after CID [22] demonstrates a unique fragmentation pattern (Figure 6.2). This fragmentation pattern is present for all examined acyl lysoPC species under a variety of experimental conditions as previously demonstrated [9, 22, 23]. The fragments with this pattern largely correspond to the building blocks of acyl lysoPC species (Figure 6.2), which provides the essential information necessary for identification of acyl lysoPC species present in biological samples.

Specifically, the fragmentation pattern contains fragment ions corresponding to the neutral losses of 59 amu (i.e., trimethylamine) and 205 amu (i.e., sodium cholinephosphate), respectively. These two fragment ions definitively determine the head group of these acyl lysoPC species. Thus, the fatty acyl chain of individual acyl lysoPC species can be deduced from the *m/z* value after conforming its ester linkage to a glycerol backbone (see below). The connection of the acyl chain to the hydroxyl group of glycerol (i.e., regiospecificity) can be determined with the other two fragment ions at *m/z* 104 and 147 present in the fragmentation pattern, which correspond to choline and sodiated five-membered cyclophosphane, respectively (Figure 6.2). This is due to that the intensity ratio of the ions at *m/z* 104 and 147 is as 3.5 for the *sn*-1 acyl lysoPC species and 0.125 for the *sn*-2 acyl lysoPC species as previously determined [9, 22]. These differential intensity ratios of fragment

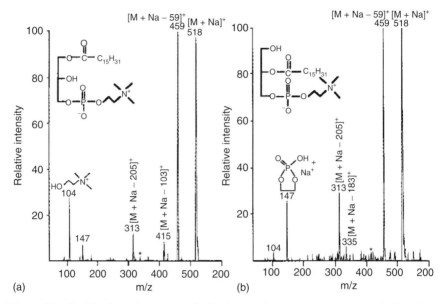

Figure 6.2 Positive-ion electrospray ionization tandem mass spectra of choline lyso-glycerophospholipid regioisomers. Positive-ion ESI tandem MS analyses of sodiated 1-hexadecanoyl-2-hydroxy-*sn*-glycero-3-phosphocholine (a) and sodiated 1-hydroxy-2-hexadecanoyl-*sn*-glycero-3-phosphocholine (b) was performed on a TSQ mass spectrometer (Finnigan TSQ MAT 700). Collision activation was carried out with collision energy of 20 eV and gas pressure of 1 mTorr. The asterisks indicate the product ions at *m/z* 335 in (a) and *m/z* 415 in (b) present in very low abundance.   Han and Gross [22]. Reproduced with permission of the American Chemical Society.

ions present in different lysoPC regioisomers has been well elucidated through charge-remote fragmentation pathways [23, 24].

A similar fragmentation pattern can be obtained from ether-linked lysoPC (i.e., alkenyl and alkyl lysoPC) species. However, the fragment ion corresponding to the neutral loss of 205 amu (i.e., sodium cholinephosphate) is minimal as previously demonstrated [22]. This is due to the fact that less active protons at the C2 position are present in aliphatic chain of ether lipids relative to those present in fatty acyl chains where the C2 protons are activated by the neighboring carbonyl group [23, 24]. Therefore, the peak intensity ratio present in NLS59 and NLS203 readily determines the ether-containing lysoPC molecular species (see below).

Based on such a fragmentation pattern, all lysoPC species including regioisomers present in any biological samples can be effectively, selectively, and sensitively detected and identified through NLS59, NLS205, PIS104, and PIS147. This is due to the presence of a common, but differential, fragmentation pattern for all the lysoPC species and due to the nature of tandem MS. Importantly, these tandem MS scans, each of which represents a building block of lysoPC species, plus a full MS scan, constitute a 2D MS mapping of the building blocks of all lysoPC species present in a biological sample (e.g., mouse plasma in Figure 6.3). In the 2D MS spectrum, the fragment ions detected by the tandem MS scans crossing with a precursor ion displayed in the full MS scan represent the product ions yielded from the precursor ion after CID.

Specifically, a full MS scan is acquired in the positive-ion mode in the mass range between m/z 450 and 610 after direct infusion of a diluted lipid extract since all naturally occurring lysoPC species should fall into this mass range. The NLS59 sensitively and selectively "filters" the low-abundance lysoPC species in this mass range. The NLS205 further confirms the detected lysoPC species with NLS59 due to its specificity to sodiated lysoPC species in this mass region. Moreover, ether-linked lysoPC species can also be differentiated from acyl lysoPC species since the intensity ratio of an acyl lysoPC species relative to the selected internal standard in NLS59 is significantly different from that of the ether-linked counterpart as determined [9, 23]. The aliphatic components in lysoPC species can then be derived from the m/z value. The regioisomeric identity of an acyl lysoPC species or the ratio of regioisomers if present as a mixture of the isomers can be determined from the peak intensity ratio of an ion present in both scans of PIS104 and PIS147.

As in the case showed in Figure 6.3, the intensity of the ion at m/z 502 is markedly reduced in NLS205 relative to that in NLS59, indicating that this ion is an ether-linked lysoPC species. The broken line highlights the building blocks with which the molecular ion at m/z 542 crosses and which identify the ion as acyl lysoPC species since an intense signal was detected with NLS205. Moreover, the comparable intensities of this ion present in both PIS104 and PIS147 indicate that this ion contains both sn-1 and sn-2 acyl lysoPC isomers at a ratio of 0.22, that is, over 80% of 18:2 lysoPC as sn-2 isomer.

It should be noted that although both NLS59 and NLS205 are specific to lysoPC species and sodiated lysoPC species in the mass region, any single tandem MS scan is not enough to definitively determine a detected ion as a lysoPC species. However, combination of these scans should substantially increase the specificity of identification. In other words, all the lysoPC species present in a biological sample can be detected by either NLS59 or NLS205, but a detected ion peak by either NLS59 or NLS205 may not represent a lysoPC species. However, the probability that an artificial ion peak can be detected by both tandem MS scans is negligible.

6.2.2.2 Sphingomyelin Characterization of authentic SM species as lithium adducts by positive-ion ESI-MS in the product-ion mode after CID reveals an informative and unique fragmentation pattern (Figure 6.4). This pattern represents the fragmentation of all examined SM species under different experimental conditions [9, 25, 26]. Importantly, the building blocks necessary for determination of SM structures can be extracted from this pattern. Therefore, the structures of individual SM species as lithium adducts can be readily determined from LC-MS analysis after remembering this fragmentation pattern. Alternatively, MDMS-SL analysis of SM species present in any biological samples can be designed and performed (see below).

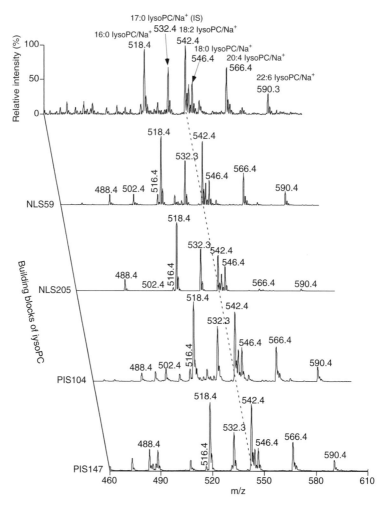

Figure 6.3 Representative two-dimensional mass spectrometric analyses of lysoPC molecular species in lipid extracts of mouse plasma in the positive-ion mode. A full MS scan in the mass range from *m/z* 460 to 610 was acquired first from a diluted lipid extract of mouse plasma in the positive-ion mode, which displayed the abundant sodium adducts of lysoPC species. Neutral loss scan (NLS) of 59.0 amu (i.e., trimethylamine) with a collision energy (CE) of 22 eV or 205 amu (i.e., sodium cholinephosphate, CE 34 eV), precursor-ion scan (PIS) of *m/z* 104.1 (i.e., choline, CE 34 eV), and PIS of *m/z* 147.0 (i.e., sodiated five-membered cyclophosphane, CE 34 eV) with collision-gas pressure at 1 mTorr were also acquired from the diluted lipid extract of mouse plasma in the positive-ion mode. All scans were displayed after normalization to the base peak in individual scan.

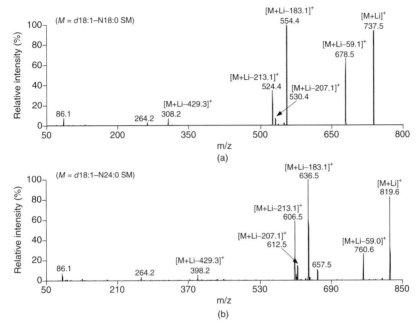

Figure 6.4 Representative product-ion ESI-MS mass spectra of lithiated sphingomyelin species after CID. Product-ion ESI-MS analyses of lithiated d18:1-N18:0 SM at m/z 737.5 (a) and lithiated d18:1-N24:1 SM at m/z 819.6 in the presence of LiOH in the infusion solution were performed on a QqQ mass spectrometer. Collision activation was conducted with collision energy at 32 and 35 eV (for the SM species, respectively) and gas pressure at 1 mTorr. The spectra demonstrated an identical fragmentation pattern from different SM species under different experimental conditions.

The fragmentation pattern of lithiated SM species contains at least four modest to abundant fragment ions (Figure 6.4). These include an abundant fragment ion corresponding to the neutral loss of 59 amu (i.e., trimethylamine), an abundant fragment ion corresponding to the neutral loss of 183 amu (i.e., phosphocholine), an abundant fragment ion corresponding to the neutral loss of 213 amu (i.e., phosphocholine plus methyl aldehyde), and a modest fragment ion corresponding to the neutral loss of 429 amu (i.e., phosphocholine plus long chain sphingoid base) in addition to a few other low-to-modest abundance fragment ions. Therefore, both neutral losses of 59 and 183 amu identify and confirm the head group of SM species; the neutral loss of 213 amu specifically determines the presence of a sphingoid backbone; and the neutral loss of 429 amu defines the structure of backbone as sphingosine. Previous studies have also demonstrated the neutral losses of 431 and 457 amu from sphinganine (d18:0) and d20:1 sphingoid base, respectively, and others [25]. Then, the acyl amide chains of SM species can be derived from the corresponding m/z values once the sphingoid backbones are determined.

Based on this fragmentation pattern, MDMS-SL analysis of SM species present in a biological sample can be designed and performed. The two-dimensional MS mapping is constructed with sequential acquisition of a full MS scan, NLS59, NLS183, NLS213, and a few NL scans corresponding to the sphingoid bases such as NLS429, NLS431, NLS457 (Figure 6.5). The lipid extracts of biological samples for SM analysis with or without alkaline hydrolysis [28, 29] could be used. Alkaline treatment is always recommended for better quantification and accurate identification. Specifically, the full MS scan is usually acquired in the mass range between m/z 650 and 900 (or other appropriate mass ranges) in the positive-ion mode after direct infusion of an alkaline-treated lipid extract in the presence of a small amount of lithium hydroxide. From the identical infusion solution and in the same mass range, other neutral-loss scans specified earlier are also acquired. In the constructed 2D mass spectrum, the fragment ions detected by the tandem MS scans cross with a precursor ion displayed in the full MS scan represent the product ions yielded from the precursor ion after CID. For example, although the signal of the molecular ion at m/z 737.7 is very low, at the baseline level, the broken line in Figure 6.5 highlights the abundant fragment ions detected with NLS59, NLS183, NLS213 and NLS429, but not with NLS431. These results clearly identify the lithiated SM species at m/z 737.7 as d18:1–N18:0 SM.

A few points should be noted for identification of SM species present in biological samples by using the recognized fragmentation pattern.

- The possibility that an artificial ion peak is detected by the combination of NLS59, NLS183, and NLS213 is negligible, particularly after base hydrolysis.

- The ion peaks shown in NLS431 carried the M+2 ^{13}C isotopologues of the ions present in NLS429. Therefore, the abundance of SM species containing sphinganine backbone as determined by NLS431 should be deduced after correction for ^{13}C isotope distribution from NLS429. Similar corrections should also be performed for any other pairs of NLSs of the sphingoid bases with a difference of a double bond.

- When a big number of neutral-loss mass is monitored with NLS, the specificity to the neutrally lost fragment is reduced as the mass value increases since the possibility with a combination of multiple low mass neutral losses is increased.

6.2.2.3 *Triacylglycerol*

TAG species can readily be ionized in the positive-ion ESI-MS as their lithium adducts when lithium hydroxide (or lithium chloride) is used as a modifier for a diluted infusion solution [30, 31]. Tandem MS analysis of lithiated TAG species in the product-ion analysis mode after CID displays abundant and informative fragment ions, but shows a unique fragmentation pattern of fragment ions corresponding to the neutral losses of paired free fatty acid(s) and their lithium salt(s) (Figure 6.6). Therefore, the number of abundant fragment ions present in a product-ion mass spectrum of lithiated TAG species depends on the number of different FA chains present in the TAG species. Specifically, six abundant fragment ions (i.e., three corresponding to the neutral losses of three fatty acids and three from the neutral losses of their lithium salts) are yielded after CID if a TAG species possesses

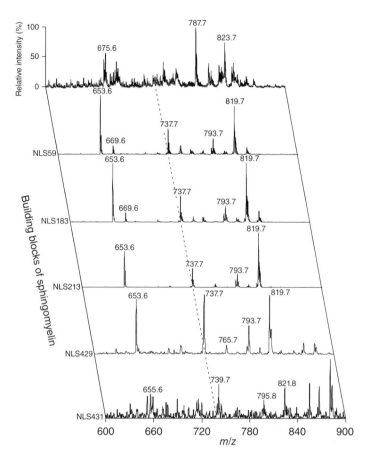

Figure 6.5 Representative two-dimensional mass spectrometric analyses of sphingomyelin molecular species in the alkaline-treated lipid extracts of mouse spinal cord in the positive-ion mode in the presence of LiOH. Lipid extracts from spinal cord of mice at 48 days of age were prepared by using a Bligh–Dyer extraction procedure as previously described [27]. A part of each lipid extract was treated with lithium methoxide as described previously [28] and the residual lipid extract was reconstituted with 50 μL of 1:1 CHCl$_3$/MeOH per milligram of original tissue protein. Each of the reconstituted lipid extracts was diluted 10 times prior to addition of a small amount of LiOH (10 pmol LiOH/μL). Positive-ion ESI mass spectrum in the full MS scan mode was acquired in the mass range from *m/z* 600 to 900 from the diluted lipid extract of mouse spinal cord, which displayed very low abundance ions corresponding to lithiated SM species. Neutral loss scans (NLS) as indicated were also acquired from the diluted lipid extract in the mass range. All scans were displayed after normalization to the base peak in individual scan. The ion at *m/z* 653.6 corresponds to the selected internal standard for quantitative analysis of SM species.

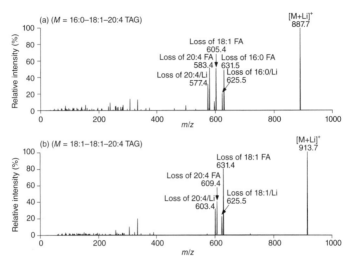

Figure 6.6 Product-ion ESI-MS mass spectra of sodiated triacylglycerol species after CID. Product-ion mass spectra of lithiated 16:0–18:1–20:4 (a) and 18:1–18:1–20:4 TAG (b) were acquired with collision energy at 32 eV and collision gas pressure at 1 mTorr. The tandem MS mass spectra displayed abundant fragment ions corresponding to the neutral losses of either fatty acids or fatty acyl lithium salts from TAG species.

three different FA chains (Figure 6.6a); four fragment ions (two corresponding to the neutral losses of two separate fatty acids and two from the neutral losses of their lithium salts) are produced from a TAG species containing two different FA chains (Figure 6.6b); and only two fragment ions corresponding to the neutral losses of a free fatty acid and its salt, respectively, are generated if the TAG species contains three identical FA chains. This fragmentation pattern is validated for all examined TAG species under various experimental conditions [30–32]. Importantly, either FA chains or their salts can serve as building blocks of TAG species for their identification in biological samples. However, neutral losses of free fatty acids are commonly used for the purpose due to their higher sensitivity and relatively more stable thermodynamics for the resultant fragment ions (i.e., more comparable intensities of these fragment ions) than those of their corresponding salts.

By exploiting this fragmentation pattern, TAG species present in biological samples can be identification by two-dimensional MS analysis in an MDMS-SL approach [31]. The two dimensional MS mapping is constructed with a full MS scan plus all NLS of naturally occurring fatty acids (approximately 30 types of FAs) (i.e., NLS228 for 14:0, NLS256 for 16:0, NLS282 for 18:1, NLS280 for 18:0, NSL304 for 20:4 FA, etc.). In the two dimensional mass spectrum, the full MS scan is acquired in the mass range between m/z 750 and 1000. TAG species present in the majority of biological samples fall in this mass range (see discussion below). From the identical infusion solution and in the same mass range, other NLSs specified earlier are then sequentially acquired. In the constructed two dimensional mass spectrum, the cross peaks of a given precursor ion in the full MS scan with the building blocks present in the second dimension represent the FA chains that constitute the isomeric TAG species corresponding the selected precursor ion. These isomeric TAG species can thus be readily determined from the number and intensities of these underlying neutral-loss fragment peaks in combination with the m/z value of TAG molecular ion [31].

For example, the lipid extract from mouse liver sample ($\sim$20 mg of wet weight) was prepared in the presence of an internal standard for TAG quantification (i.e., 15.0 nmol of tri17:1 TAG/mg of protein) by using a modified Bligh and Dyer extraction procedure [33]. The extract was further diluted with 1:1 chloroform–methanol (v/v) with the addition of a small amount of lithium hydroxide in methanol solution prior to infusion directly into a mass spectrometer with a NanoMate device (Advion Bioscience, Ithaca, NY, USA). The collision energy was optimized for the achievement of essentially identical neutral loss of fatty acids from different glycerol position of TAG species (see below). Two-dimensional MS mapping was obtained with sequential acquisition of the full MS and all NLSs (Figure 6.7, in which only a part of the NLSs are selectively included for the purpose of illustration). From the two dimensional mass spectrum, individual TAG species including isomeric species, except the isomers due to double-bond locations and regiospecific positions, were identified. For example, manual identification of the isomeric TAG species underlying the molecular ion at m/z 865.7 can be conducted as follows. In the two dimensional mass spectrum (Figure 6.7), this ion peak is crossed with the 16:1, 16:0, 18:1, and 18:0 FA building blocks at the modest-to-high abundance and also crossed with 14:0, 18:2, and 20:2 building blocks, as well as a few others (not displayed in the 2D spectrum) at low abundance (as indicated with the broken line). Based on the m/z 865.7 value, the lithiated TAG species underlying this ion peak must contain 52 total carbon atoms with 2 double bonds or 53 total carbon atoms with 9 double bonds in three FA chains, and the ion intensities resulting from the neutral loss of the three acyl chains from a given TAG molecule are nearly equal. Thus, isomeric TAG species of 16:0–18:1–18:1, 16:1–18:0–18:1, 14:0–16:0–20:2 (minor), 16:0–18:0–18:2 (minor), etc. can be identified. Other TAG species corresponding to other molecular ions displayed in the full MS scan can be identified similarly. Unfortunately, manual assignment of individual TAG species as described earlier is time consuming. Recently, a program for identification of TAG species through simulation of TAG ions has been developed [34]. This program allows for not only assignment of individual TAG species but also identification of numerous low to very low abundance TAG species.

A few points should be recognized for identification of TAG species in biological samples by using the recognized fragmentation pattern.

- Sodium adducts of TAG species are present to a certain degree even cautious measures are taken. Minimizing the ion intensities of sodium adducts should always be kept in mind since ionization of TAG species as sodium adducts depends on TAG molecular species, specifically, on the number of double bonds present in individual TAG species as previously demonstrated [31].

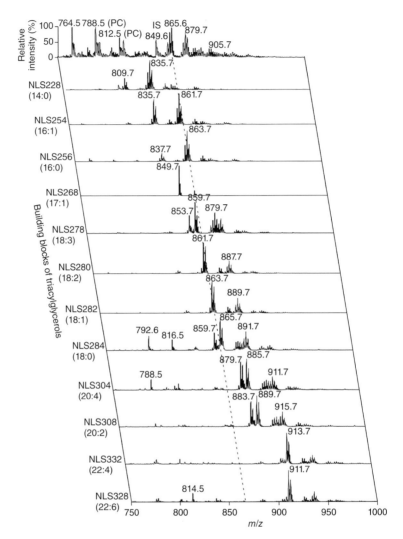

Figure 6.7 Representative two-dimensional mass spectrometric analyses of triacylglycerol species in hepatic lipid extracts of mice. Neutral loss scans (NLS) of all naturally occurring fatty acyl chains (i.e., the building blocks of TAGs) of mouse liver lipid extract were acquired to determine the identities of individual lithiated TAG molecular ion, deconvolute isomeric species, and quantify individual TAG species by comparisons with a selected internal standard (i.e., T17:1 TAG at *m/z* 849.6 as lithium adduct, shown in NLS268). Collision activation was performed with collision energy at 32 eV and collision gas pressure at 1 mTorr on a QqQ-type mass spectrometer (TSQ Vantage, Thermo Fisher Scientific, San Jose, CA, USA). All displayed mass spectral traces are normalized to the base peak in each trace.

- A concern (i.e., whether all the FA chains present in TAG species are detected) always exists when the described MDMS-SL approach is employed for identification of TAG species. Indeed, scanning 10 common fatty acids (even not including any fatty acids containing odd numbers carbon atoms) is enough to cover the major content of TAG species (i.e., >90 mol%) present in the majority of biological samples. It is always advised to scan all possibly occurring fatty acids if sample and resource permit.

- To extend the last point, while the fatty acids containing 14 carbon atoms are usually those at the low mass end to be determined, the relatively short FA chains have been seen present in a large amount [35, 36]. Thus, it should always be paid attention that if NLS228 (i.e.,14:0 FA) shows intense signals, NLS of much shorter FAs should always be examined. Similar strategy should also be employed for determining the existence of fatty acyls containing an odd number of carbon atoms.

- Interfering identification and quantification TAG species due to the neutral loss of fatty acids from PC species is usually negligible as previously discussed [31]. Neutral loss of FAs from PC species only becomes detectable for those FAs that are low abundance in TAG species, but highly abundant in PC species (Figure 6.7). These fragment ions can readily recognized with the even-numbered masses corresponding to the PC ions while those from TAG species are present in odd-numbered masses following the nitrogen rule (Figure 6.7).

- The relative abundance of the fragment ions corresponding to the neutral losses of FA salts from lithiated TAG species is different (Figure 6.6). This difference can be used to determine the regiospecific positions of the FA substituents on the glycerol backbone [30, 37]. The intensity differences between the fragment ions corresponding to the neutral losses of fatty acids from lithiated TAG species are relatively smaller than those corresponding to the losses of their salts. The mechanism leading to these differences is that generation of these fragment ions involves the initial elimination of a free FA in concert with an alpha-hydrogen atom from the adjacent FA chain, followed by formation of a cyclic intermediate that decomposes to yield other characteristic fragment ions as previously elucidated [30]. Under certain collision energy conditions, the peak intensities of the fragment ions corresponding to the losses of FAs can be used for quantification purposes as previously demonstrated [31, 38].

6.2.3 Summary

The previous section has discussed the importance of recognizing the fragmentation pattern of an individual lipid class for identification and quantification by using both LC-MS and shotgun lipidomics approaches. The given examples clearly demonstrated that the fragmentation pattern of a lipid class from a selected ion form is largely independent of collision conditions and instruments. Chapters 7–11 described the fragmentation patterns of the majority of the lipid classes with different ion forms in both positive- and negative-ion modes. Although which ion form is chosen in a particular ionization mode is the researchers' preference, recognition and understanding of the fragmentation patterns resulted from different adducts of a lipid class of interest could facilitate the decision making. Moreover, the use of multiple complementary fragmentation modes can be employed to minimize false discovery rates for abundant species and enhance coverage for extremely low-abundance molecular species. Finally, it is highly recommended that any readers who are interested in understanding the fundamentals and mechanisms of complex lipid fragmentation should further consult a very recently published book entitled *Tandem Mass Spectrometry of Lipids: Molecular Analysis of Complex Lipids* by Dr Murphy [39].

REFERENCES

1. Hsu, F.F., Turk, J., Zhang, K. and Beverley, S.M. (2007) Characterization of inositol phosphorylceramides from Leishmania major by tandem mass spectrometry with electrospray ionization. J. Am. Soc. Mass Spectrom. 18, 1591–1604.

2. Yore, M.M., Syed, I., Moraes-Vieira, P.M., Zhang, T., Herman, M.A., Homan, E.A., Patel, R.T., Lee, J., Chen, S., Peroni, O.D., Dhaneshwar, A.S., Hammarstedt, A., Smith, U., McGraw, T.E., Saghatelian, A. and Kahn, B.B. (2014) Discovery of a class of endogenous mammalian lipids with anti-diabetic and anti-inflammatory effects. Cell 159, 318–332.

3. Jensen, S.M., Brandl, M., Treusch, A.H. and Ejsing, C.S. (2015) Structural characterization of ether lipids from the archaeon Sulfolobus islandicus by high-resolution shotgun lipidomics. J. Mass Spectrom. 50, 476–487.

4. Al-Saad, K.A., Zabrouskov, V., Siems, W.F., Knowles, N.R., Hannan, R.M. and Hill, H.H., Jr. (2003) Matrix-assisted laser desorption/ionization time-of-flight mass spectrometry of lipids: Ionization and prompt fragmentation patterns. Rapid Commun. Mass Spectrom. 17, 87–96.

5. Levery, S.B. (2005) Glycosphingolipid structural analysis and glycosphingolipidomics. Methods Enzymol. 405, 300–369.

6. Yang, K., Zhao, Z., Gross, R.W. and Han, X. (2011) Identification and quantitation of unsaturated fatty acid isomers by electrospray ionization tandem mass spectrometry: A shotgun lipidomics approach. Anal. Chem. 83, 4243–4250.

7. Fitton, E.M., Monaghan, J.J. and Morden, W.E. (1992) Synchronized collision-cell energy ramping. Improving the quality of product-ion spectra. Rapid Commun. Mass Spectrom. 6, 269–271.

8. Hsu, F.-F., Bohrer, A. and Turk, J. (1998) Formation of lithiated adducts of glycerophosphocholine lipids facilitates their identification by electrospray ionization tandem mass spectrometry. J. Am. Soc. Mass Spectrom. 9, 516–526.

9. Yang, K., Zhao, Z., Gross, R.W. and Han, X. (2009) Systematic analysis of choline-containing phospholipids using multi-dimensional mass spectrometry-based shotgun lipidomics. J. Chromatogr. B 877, 2924–2936.

10. Tomer, K.B., Crow, F.W. and Gross, M.L. (1983) Location of double-bond position in unsaturated fatty acids by negative ion MS/MS. J. Am. Chem. Soc. 105, 5487–5488.

11. Crockett, J.S., Gross, M.L., Christie, W.W. and Holman, R.T. (1990) Collisional activation of a series of homoconjugated octadecadienoic acids with fast atom bombardment and tandem mass spectrometry. J. Am. Soc. Mass Spectrom. 1, 183–191.

12. Kind, T., Liu, K.H., Lee do, Y., Defelice, B., Meissen, J.K. and Fiehn, O. (2013) LipidBlast in silico tandem mass spectrometry database for lipid identification. Nat. Methods 10, 755–758.

13. Mitchell, T.W., Pham, H., Thomas, M.C. and Blanksby, S.J. (2009) Identification of double bond position in lipids: From GC to OzID. J. Chromatogr. B 877, 2722–2735.

14. Ma, X. and Xia, Y. (2014) Pinpointing double bonds in lipids by Paternò–Büchi reactions and mass spectrometry. Angew. Chem. Int. Ed. 53, 2592–2596.

15. Griffiths, W.J. (2003) Tandem mass spectrometry in the study of fatty acids, bile acids, and steroids. Mass Spectrom. Rev. 22, 81–152.

16. Murphy, R.C., Fiedler, J. and Hevko, J. (2001) Analysis of nonvolatile lipids by mass spectrometry. Chem. Rev. 101, 479–526.

17. Pulfer, M. and Murphy, R.C. (2003) Electrospray mass spectrometry of phospholipids. Mass Spectrom. Rev. 22, 332–364.

18. Hsu, F.-F. and Turk, J. (2005) Electrospray ionization with low-energy collisionally activated dissociation tandem mass spectrometry of complex lipids: Structural characterization and mechanism of fragmentation. In Modern Methods for Lipid Analysis by Liquid Chromatography/Mass Spectrometry and Related Techniques (Byrdwell, W.C., ed.). pp. 61–178, AOCS Press, Champaign, IL.

19. Hsu, F.F. and Turk, J. (2009) Electrospray ionization with low-energy collisionally activated dissociation tandem mass spectrometry of glycerophospholipids: Mechanisms of fragmentation and structural characterization. J. Chromatogr. B 877, 2673–2695.

20. Murphy, R.C. and Axelsen, P.H. (2011) Mass spectrometric analysis of long-chain lipids. Mass Spectrom. Rev. 30, 579–599.

21. Hsu, F.F. and Turk, J. (2005) Analysis of sulfatides. In The encyclopedia of mass spectrometry (Caprioli, R.M., ed.). pp. 473–492, Elsevier, New York

22. Han, X. and Gross, R.W. (1996) Structural determination of lysophospholipid regioisomers by electrospray ionization tandem mass spectrometry. J. Am. Chem. Soc. 118, 451–457.

23. Hsu, F.-F., Turk, J., Thukkani, A.K., Messner, M.C., Wildsmith, K.R. and Ford, D.A. (2003) Characterization of alkylacyl, alk-1-enylacyl and lyso subclasses of glycerophosphocholine by tandem quadrupole mass spectrometry with electrospray ionization. J. Mass Spectrom. 38, 752–763.

24. Hsu, F.-F. and Turk, J. (2003) Electrospray ionization/tandem quadrupole mass spectrometric studies on phosphatidylcholines: The fragmentation processes. J. Am. Soc. Mass Spectrom. 14, 352–363.

25. Hsu, F.F. and Turk, J. (2000) Structural determination of sphingomyelin by tandem mass spectrometry with electrospray ionization. J. Am. Soc. Mass Spectrom. 11, 437–449.

26. Hsu, F.F. and Turk, J. (2005) Analysis of Sphingomyelins. In The encyclopedia of mass spectrometry (Caprioli, R. M., ed.). pp. 430–447, Elsevier, New York

27. Wang, C., Wang, M., Zhou, Y., Dupree, J.L. and Han, X. (2014) Alterations in mouse brain lipidome after disruption of CST gene: A lipidomics study. Mol. Neurobiol. 50, 88–96.

28. Jiang, X., Cheng, H., Yang, K., Gross, R.W. and Han, X. (2007) Alkaline methanolysis of lipid extracts extends shotgun lipidomics analyses to the low abundance regime of cellular sphingolipids. Anal. Biochem. 371, 135–145.

29. Merrill, A.H., Jr., Sullards, M.C., Allegood, J.C., Kelly, S. and Wang, E. (2005) Sphingolipidomics: High-throughput, structure-specific, and quantitative analysis of sphingolipids by liquid chromatography tandem mass spectrometry. Methods 36, 207–224.

30. Hsu, F.-F. and Turk, J. (1999) Structural characterization of triacylglycerols as lithiated adducts by electrospray ionization mass spectrometry using low-energy collisionally activated dissociation on a triple stage quadrupole instrument. J. Am. Soc. Mass Spectrom. 10, 587–599.

31. Han, X. and Gross, R.W. (2001) Quantitative analysis and molecular species fingerprinting of triacylglyceride molecular species directly from lipid extracts of biological samples by electrospray ionization tandem mass spectrometry. Anal. Biochem. 295, 88–100.

32. Hsu, F.F. and Turk, J. (2010) Electrospray ionization multiple-stage linear ion-trap mass spectrometry for structural elucidation of triacylglycerols: Assignment of fatty acyl groups on the glycerol backbone and location of double bonds. J. Am. Soc. Mass Spectrom. 21, 657–669.

33. Christie, W.W. and Han, X. (2010) Lipid Analysis: Isolation, Separation, Identification and Lipidomic Analysis. The Oily Press, Bridgwater, England. pp 448.

34. Han, R.H., Wang, M., Fang, X. and Han, X. (2013) Simulation of triacylglycerol ion profiles: Bioinformatics for interpretation of triacylglycerol biosynthesis. J. Lipid Res. 54, 1023–1032.

35. Su, X., Han, X., Yang, J., Mancuso, D.J., Chen, J., Bickel, P.E. and Gross, R.W. (2004) Sequential ordered fatty acid a oxidation and D9 desaturation are major determinants of lipid storage and utilization in differentiating adipocytes. Biochemistry 43, 5033–5044.

36. Watkins, S.M., Reifsnyder, P.R., Pan, H.J., German, J.B. and Leiter, E.H. (2002) Lipid metabolome-wide effects of the PPARgamma agonist rosiglitazone. J. Lipid Res. 43, 1809–1817.

37. Herrera, L.C., Potvin, M.A. and Melanson, J.E. (2010) Quantitative analysis of positional isomers of triacylglycerols via electrospray ionization tandem mass spectrometry of sodiated adducts. Rapid Commun. Mass Spectrom. 24, 2745–2752.

38. Duffin, K.L., Henion, J.D. and Shieh, J.J. (1991) Electrospray and tandem mass spectrometric characterization of acylglycerol mixtures that are dissolved in nonpolar solvents. Anal. Chem. 63, 1781–1788.

39. Murphy, R.C. (2015) Tandem Mass Spectrometry of Lipids: Molecular analysis of complex lipids. Royal Society of Chemistry, Cambridge, UK. pp 280.

7

FRAGMENTATION PATTERNS OF GLYCEROPHOSPHOLIPIDS

7.1 INTRODUCTION

Glycerophospholipid (GPL) is a category of lipid classes containing at least one phosphate (or phosphonate) group that is esterified to one of the glycerol hydroxyl groups (Chapter 1). The phosphate or phosphonate together with its linked moiety is commonly called as the head group and defines the lipid class. There exist three subclasses (i.e., phosphatidyl-, plasmanyl-, or plasmenyl-) of a few lipid classes. The common lipid classes classified into subclasses are PC and PE classes. Although some GPL classes such as platelet activation factor and lysoGPL play important roles in biological systems serving as lipid second massagers, the majority of the GPL classes are the major components of cellular membrane, where these lipids provide appropriate matrices for optimal functions of membrane proteins and serve as substrates for releasing signaling lipids.

This category of lipids represents the most extensively characterized ones in comparison to other categories of lipids by MS/MS or even MS^n, particularly after CID. The characterization is conducted in both positive- and negative-ion modes under a variety of experimental conditions. These experimental conditions lead to resulting in a variety of quasimolecular ions including different adducts. Therefore, very different fragment ions are resulted from these different quasimolecular ions, all of which are interesting, particularly from an ion chemistry perspective view. From the extensive characterization, Hsu and Turk concluded that the fragmentation

Lipidomics: Comprehensive Mass Spectrometry of Lipids, First Edition. Xianlin Han.
© 2016 John Wiley & Sons, Inc. Published 2016 by John Wiley & Sons, Inc.

processes are essentially identical for those resulting from positive quasimolecular ions, whereas the fragmentation of all negative quasimolecular ions also follows a general rule [1].

In summary, charge-remote fragmentation processes [2] play a major role in fragmentation of these classes of lipids in the positive-ion mode. This charge-remote fragmentation generally leads to the generation of intense fragment ions yielding from the head groups. In these fragmentation processes, the loss of the fatty acyl (FA) substituent at *sn*-1 position is a more favorable pathway than the counterpart loss of the FA substituent at *sn*-2 position of the glycerol moiety. Through stable isotope labeling and/or MS^n analysis, Hsu and Turk revealed that these differential losses of *sn*-1 and *sn*-2 FA chains are due to the involvement of the α-hydrogen atoms at the FA chain in the elimination of the adjacent FA substituent as an acid and due to that the α-hydrogen atoms of the *sn*-2 FA substituent are more labile than their counterparts at *sn*-1 [1, 3–5]. These differential losses of *sn*-1 and *sn*-2 FA chains are critical for identification of regiospecificities of these GPL species, for example, as their alkaline adducts [6, 7].

In the negative-ion mode, the deprotonated species of GPL classes after CID yield the predominant presence of one or two fragment ions corresponding to the FA carboxylate ions in the mass region of *m/z* 200–350 resulting from the FA chains at the *sn*-1 and *sn*-2 positions of glycerol. In most cases, these carboxylate ions are the base peaks in the spectra. In addition, there exists another set of fragment ions around *m/z* 400, corresponding to the $[M-H-R_xCH_2CO_2H]^-$ and $[M-H-R_xCH=C=O]^-$ (where $x = 1, 2$) via losses of the FA substituents as acids or ketenes, respectively. This second set of fragment ions is usually present in low-to-moderate abundance in the product-ion spectra. Charge-driven fragmentation processes [8] are the major mechanisms leading to the generation of product ions in the negative-ion mode. In this process, the gas-phase basicity of the deprotonated molecular ions determines the loss of the FA substituent as an acid or as a ketene [8, 9]. The gas-phase basicity defines the affinity of a proton to a molecule or a negatively charged molecular ion according to IUPAC Gold Book [10]. The polar head group of individual GPL class determines the gas-phase basicity of their deprotonated ions. Therefore, different molecular species of a class show essentially an identical fragmentation pattern, but distinct mass spectra of different GPL classes are obtained. Generally, loss of the FA chain at *sn*-2 as an acid or as a ketene is more favorable than that at *sn*-1 because the loss of the *sn*-2 FA chain is more sterically favored. This leads to the ion corresponding to $[M-H-R_2CH_2CO_2H]^-$ more abundant than that of $[M-H-R_1CH_2CO_2H]^-$ and the ion corresponding to $[M-H-R_2CH=CO]^-$ more intense than that of $[M-H-R_1CH=CO]^-$. This difference between the losses of the FA chains at *sn*-1 and *sn*-2 positions allows for identification of regioisomers [8, 9, 11–14].

In this chapter, not all, but only the fragmentation patterns of individual lipid classes of GPL under some common experimental conditions are summarized because these experimental conditions can be readily transferred to a practical utilization for a large-scale analysis of lipids in lipidomics. As stated in Chapter 6, any advanced readers should always look for the details from the original studies on the topic and/or study several invaluable review articles as well as a newly published book on the topic [1, 15–20]. The review articles [1, 18, 21] written by Hsu and Turk should always be consulted for a detailed understanding of structural characterization of specific lipid classes.

7.2 CHOLINE GLYCEROPHOSPHOLIPID

7.2.1 Positive Ion Mode

7.2.1.1 *Protonated Species*   Under an acidic condition or in the presence of an ammonium salt as a modifier in a lipid solution, protonated molecular species ($[M+H]^+$) are readily formed in the positive-ion mode for PC class, which contains a quaternary amine in the form of zwitterion with the phosphate. The product-ion spectra yielding from the $[M+H]^+$ of PC species including all subclasses displayed an ion at m/z 184, corresponding to a phosphocholine ion. The structurally informative fragment ions corresponding to the FA chain(s) are generally in low abundance. These ions are essentially buried in baselines when product-ion analysis is conducted with a QqQ-type mass spectrometer although these fragment ions can be well visualized by a QqToF-type instrument. Therefore, structural characterization from $[M+H]^+$ of PC species are less applicable for identification of regioisomers although achievable [4, 17].

Extensive studies have shown that formation of the fragment ion at m/z 184 mainly involves the α-hydrogen atoms of the FA chain at *sn*-2 due to more labile than its counterpart at *sn*-1 [4]. This leads to the more favorable formation of the $[M+H-R_2CH=C=O]^+$ ion than the $[M+H-R_1CH=C=O]^+$ ion, resulting from the losses of the FA chains at the *sn*-2 and *sn*-1 positions as ketenes, respectively. Therefore, the position of the FA moieties at the glycerol backbone can be assigned from these fragment ions to a certain degree.

7.2.1.2 *Alkaline Adducts*   PC species can readily form alkaline (Alk = Li, Na, K) adducts. The formation of alkaline adducts depends largely on the availability or the concentration of the alkaline ions in the lipid solution. Therefore, if lithium adducts would like to be used for the analysis of PC species, lithium hydroxide or a lithium salt should be used as a modifier in the lipid solution. A sodium adduct is always displayed in a mass spectrum if there is no any other modifier added to the lipid solution since sodium ions are essentially present "everywhere."

A common fragmentation pattern yielded from alkaline adducts of PC species after CID as demonstrated [3, 4, 7, 17, 22] can be summarized. This pattern includes the following:

- A fragment ion corresponding to the loss of trimethylamine (i.e. $[M+Alk-59]^+$)
- A fragment ion corresponding to the loss of phosphocholine (i.e., $[M+Alk-183]^+$)
- A product ion corresponding to the loss of alkaline cholinephosphate (i.e., $[M+Alk-(Alk+182)]^+$).

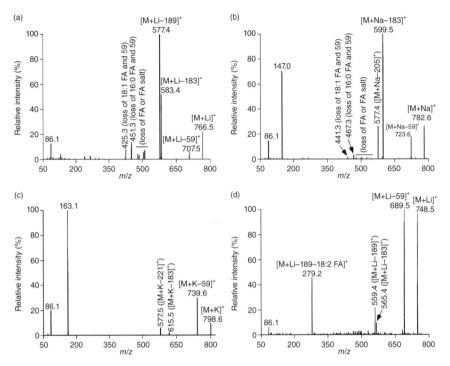

Figure 7.1 Representative product-ion ESI-MS analysis of lithiated, sodiated, and potassiated PC species after CID. Product-ion ESI-MS mass spectra of lithiated, sodiated, and potassiated 16:0–18:1 phosphatidylcholine (dPC) (a–c, respectively), and lithiated 16:0–18:2 plasmenylcholine (pPC) (d) were acquired on a QqQ mass spectrometer (Thermo Fisher TSQ Vantage). Collision activation was carried out at collision energy of 32 eV and gas pressure of 1 mTorr.

Representative examples of MS/MS mass spectra of lithiated, sodiated, and potassiated 16:0–18:1 diacyl PC (dPC) species are displayed in Figure 7.1a–c, respectively.

In addition to these abundant fragment ions corresponding to the phosphocholine head group, fragment ions related to FA chains present in PC species are also present, particularly from those of lithiated PC species. These fragment ions are usually present in relatively low abundance in CID ESI-MS mass spectra of alkaline PC adducts (Figure 7.1a–c). There are three pairs of fragment ions in this category. These include a pair of product ions yielded from the neutral loss of fatty acids (i.e., $[M+Alk-R_xCO_2H]^+$); a pair of fragment ions resulted from the neutral loss of alkaline FA salts (i.e., $[M+Alk-R_xCO_2Alk]^+$); and a pair of fragment ions corresponding to the continuous neutral loss of fatty acids after neutral loss of trimethylamine (59 amu) (i.e., $[M+Alk-(R_xCO_2H+59)]^+$); where $x = 1$ and 2, respectively (Figure 7.1a–c). The loss of the FA substituent at *sn*-1 position is more favored than the counterpart loss of the FA substituent at *sn*-2 position of the glycerol moiety as discussed in the introduction of this chapter. The pair of $[M+Alk-(R_xCO_2H+59)]^+$ ions is usually the most abundant pair of fragment ions among these FA-related ions and commonly used for regiospecific discrimination of dPC species [1, 3, 4, 23].

The fragmentation patterns of alkaline adducts of plamenylcholine (pPC) and plasmanylcholine (aPC) species contain all the fragment ions related to phosphocholine head group as those resulted from dPC species, including $[M+Alk-59]^+$, $[M+Alk-183]^+$, and $[M+Alk-(Alk+182)]^+$. However, the fragmentation patterns of alkaline adducts of pPC and aPC species are also distinct from that of dPC species in three aspects as follows:

- The $[M+Alk-59]^+$ ion is dominated in the product-ion spectra of the alkaline adducts of both pPC and aPC species.
- The fragment ions corresponding to the neutral losses of fatty acids and FA salts (i.e., $M+Alk-R_2CO_2H]^+$ and $[M+Alk-R_2CO_2Alk]^+$) are absent.
- The fragment ion corresponding to $[M+Alk-(R_2CO_2H+59)]^+$ is of low abundance.

These differences of the fragmentation patterns between ether PC subclasses and dPC subclass are widely used to distinguish ether-containing PC species from diacyl ester-containing PC species [4, 7, 23]. This distinct fragmentation pattern of pPC and aPC species from dPC species can be well explained with the absence of labile α-hydrogen atoms of the *sn*-1 aliphatic chains in the pPC and aPC subclasses. Specifically, these labile α-hydrogen atoms are required for elimination of the adjacent FA substituent as an acid (i.e., R_2CO_2H) as discussed in the introduction of this chapter. Therefore, the formation of $[M+Alk-R_2CO_2H]^+$ and $[M+Alk-(R_2CO_2H+59)]^+$ fragments are not favored from the alkaline adducts of both pPC and aPC species. Moreover, the lack of continuous neutral loss of FA chain as a fatty acid from the resultant $[M+Alk-59]^+$ due to the absence of labile α-hydrogen atoms explains this fragment ion as a predominant fragment ion displayed in the product-ion mass spectra of these pPC and aPC subclasses. Figure 7.1d shows a representative product-ion ESI-MS spectrum of lithiated 16:0–18:2 pPC species in comparison to the spectrum of its diacyl counterpart displayed in Figure 7.1a.

Distinction of alkaline adducts of pPC species from aPC species is the presence of a unique feature to the fragmentation pattern of pPC species, that is, the presence of an abundant fragment ion corresponding to $[M+Alk-(182+Alk)-R_2CO_2H]^+$ in the product-ion ESI-MS spectra of pPC alkaline adducts, but not in those of both aPC and dPC counterparts [7]. This prominent ion arises from the further neutral loss of the *sn*-2 FA moiety from the ion corresponding to $([M+Alk-(182+Alk)]^+)$. For example, the *m/z* 279 ion $([M+Li-(182+Li)-R_2CO_2H]^+)$ arises from lithiated 16:0–18:2 pPC species (Figure 7.1d).

7.2.2 Negative-Ion Mode

PC species can readily form various adduct ions with those small anion(s) present in the matrix (i.e., $[M+X]^-$) (where $X=$ Cl, CH_3CO_2, HCO_2, CF_3CO_2, etc.) in the negative-ion mode of ESI-MS [22, 24–26]. These adducts lead to generating an $[M-15]^-$ ion (i.e., $[M+X-CH_3X]^-$) or even showing as the predominant quasimolecular ions of PC species in many cases [25, 27]. However, this quasi-molecular ion can be minimized through tuning the ionization conditions [28], indicating that the neutral loss of CH_3X from the anion adduct is very facile.

The fragmentation pattern of anion adducts of PC species after CID contains three types of fragment ions as follows:

- A predominant $[M-15]^-$ ion is yielded from the neutral loss of CH_3X from quasimolecular ions.

- One or two fragment ions around m/z 300 corresponding to FA carboxylate ions resulted from FA substituent(s) of PC species are dominated.

- A cluster of low-to-modest abundance fragment ions around m/z 450 arising from the neutral losses of FA substituent(s) of PC species are present in the product-ion ESI-MS spectra of anion adducts of PC species.

As discussed in Chapter 6, this fragmentation pattern is independent of collision activation conditions although the ratios of fragment ion intensities may vary with the collision activation conditions employed. Figure 7.2 illustrates such a notion with variation of collision energy levels, which also serves to clearly show the fragmentation pattern of anion adducts of PC species.

Many studies have shown that the peak intensity of the carboxylate anion resulted from the sn-2 FA chain is approximately three times more intense than that arising from the sn-1 FA chain of dPC species [17, 22, 29, 30]. This fact has been used to identify the location of fatty acyls and thus regioisomers of dPC species. However, this ratio becomes smaller than three if the sn-2 FA chain is a polyunsaturated FA substituent [22]. It is now clear that this reduced ratio is due to the fact that the resultant polyunsaturated FA carboxylate anion can go further fragmentation to yield an ion corresponding to the neutral loss of carbon dioxide (i.e., [carboxylate anion−44]$^-$) [31–33]. This continuous loss of carbon dioxide has recently been exploited to identify the location of double bond(s) of fatty acyls [33]. With the recognition of carbon dioxide loss, the combined ion intensity of the carboxylate anion and [carboxylate anion−44] ion arising from sn-2 polyunsaturated FA chain is still approximately three times more intense than that of the sn-1 FA carboxylate ion [32].

In addition to the utility of FA carboxylate anion ratios for identification of regiospecificity, the paired ions resultant from the neutral losses of FA substituents as either fatty acids (i.e., $[M-15-R_xCH_2COOH]^-$) or FA ketenes (i.e., $[M-15-R_xCH=C=O]^-$), respectively, can also be used to determine the regioisomers of dPC species [22, 27]. In this case, 1-acyl demethylated lysoPC fragment ions produced from the loss of sn-2 FA moiety are more intense than those of 2-acyl counterparts arising from the loss of sn-1 FA substituent [22, 27] (Figure 7.2). This is due to the fact that the loss of the sn-2 FA chain is more sterically favored as

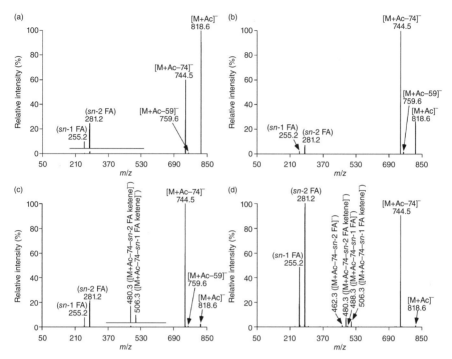

Figure 7.2 Product-ion ESI-MS spectra of 16:0–18:1 phosphatidylcholine acetate adduct after CID at different collision energy. Product-ion ESI-MS analysis of acetate adduct of 16:0–18:1 dPC at *m/z* 818.5 in the presence of ammonium acetate in the infusion solution was performed on a Thermo Fisher TSQ Vantage mass spectrometer. Collision activation was performed at collision energy of 15 (a), 20 (b), 25 (c), and 30 (d) eV, and gas pressure of 1 mTorr. Ac stands for acetate. These results demonstrate two key points. First, while product-ion mass spectra acquired at different collision energy are very different, the fragmentation pattern of dPC species is identical under different experimental conditions. Second, the low-mass fragment ions increase as collision energy.

aforementioned. Moreover, the [M−15−R_2CH=C=O]$^−$ ion is more intense than the [M−15−R_2CH$_2$COOH]$^−$ ion while the [M−15−R_1CH=C=O]$^−$ ion is less abundant than the [M−15−R_1CH$_2$COOH]$^−$ ion [22, 27].

The fragmentation patterns of small anion adducts of pPC and aPC subclasses are essentially identical to that of dPC species containing two identical fatty acyl chains, i.e., only one each of the carboxylate anion, [M−15−RCH$_2$COOH]$^−$, and [M−15−RCH=C=O]$^−$ is present in their product-ion mass spectra. However, the pathways leading to this identical pattern are different. The pattern of pPC and aPC subclasses is due to the absence of the *sn*-1 FA moiety, whereas the pattern from dPC species containing two identical FA chains is due to the resultant fragment ions from *sn*-1 and *sn*-2 FA moieties are identical. Multiple approaches can be employed to distinguish pPC species from aPC species present in biological lipid samples. These include the treatment of acid vapor [34, 35], utilization of the paired rule [32], or a mass shift with iodine in methanol [36].

7.3 ETHANOLAMINE GLYCEROPHOSPHOLIPID

7.3.1 Positive-Ion Mode

7.3.1.1 Protonated Species   PE species in the lipid solution under acidic conditions or in the presence of ammonium ions (e.g., 5 mM) can be readily ionized as
proton adducts ($[M+H]^+$) in the positive-ion mode [25, 37, 38]. It should be recognized that the ionization efficiency of PE species as proton adducts is relatively lower
than that of PC species because the quaternary amine of protonated PC species as
a positive charge site is much stable than the primary amine as the charge site of
protonated PE species.

Fragmentation pattern of protonated PE species contains the following fragment
ions [1, 25, 37–39]:

- An intense fragment ion corresponding to the neutral loss of phosphoethanolamine (i.e., $[M+H-141]^+$).
- One or two low-abundant fragment ions corresponding to fatty acylium ions
  (i.e., R_xCO^+, $x = 1, 2$).

This pattern is very different from that of proteonated PC species (see last
section). The resultant $[M+H-141]^+$ ion, rather than a protonated PE species at *m/z*
142 (equivalent to the *m/z* 184 ion resultant from proteonated PC species), indicates
that the phosphoethanolamine is less competitive for a proton, but easy to lose a
proton from the primary amine. Extensive mechanistic studies have demonstrated
that the fragmentation process leading to the generation of $[M+H-141]^+$ ion also
involves the participation of the α-hydrogen atoms of the FA substituent, mainly that
at the *sn*-2 position [1, 4, 9]. Figure 7.3a shows a representative product-ion ESI-MS
spectrum of protonated 16:0–22:6 diacyl PE (dPE) after CID.

7.3.1.2 Alkaline Adducts   PE species can essentially form adducts with any alkaline (e.g., Li, Na, K) ion in the positive-ion mode when a modifier carrying the alkaline
ion is added to the lipid solution, but sodium adduct is formed if no other alkaline ion
is added. Similar to the formation of proton adducts, the ionization efficiency of PE
alkaline adducts is much lower in comparison to that of PC counterparts. Therefore,
utilization of these adducts for analysis of PE species present in biological samples
is suppressed by PC species by shotgun lipidomics. Application of these adducts for
the analysis of PE species with LC-MS is also minimal.

Fragmentation pattern of PE alkaline adducts has been well characterized [5, 22].
Fragmentation pattern of PE alkaline adducts, which is similar to that of PC alkaline
adducts, contains the following fragment ions:

- A fragment ion corresponding to the alkaline PA adducts arises from the neutral
  loss of aziridine (43 amu) (i.e., $[M+Alk-43]^+$) from the quasimolecular ions.
  This fragment ion is equivalent to $[M+Alk-59]^+$ in PC fragmentation pattern.

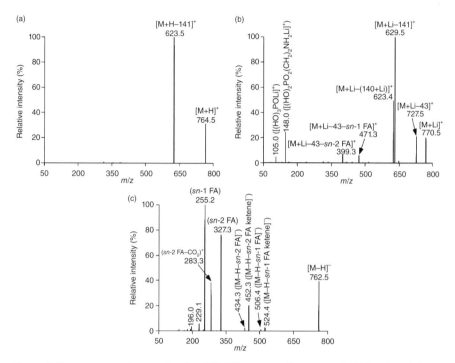

Figure 7.3 Representative product-ion ESI-MS spectra of protonated, lithiated, and deprotonated 16:0–22:6 phosphatidylethanolamine after CID. Product-ion ESI-MS analysis of 16:0–22:6 dPE in the presence of ammonium acetate (as protonated species, a) or lithium hydroxide (as lithiated species, b) in the infusion solution in the positive-ion mode, or either solution in the negative-ion mode (as deprotonated species, c) after CID was performed on a Thermo Fisher TSQ Vantage mass spectrometer. Collision activation was performed at collision energy of 18 (a), 28 (b), and 28 (c) eV, and gas pressure of 1 mTorr.

- Two fragment ions corresponding to lithiated DAG and DAG-like ion arise from the neutral losses of phosphoethanolamine (141 amu) and alkaline ethanolaminephosphate salt (140+Alk amu), respectively, from PE alkaline adducts. These two fragment ions are equivalent to the $[M+Alk-183]^+$ and $[M+Alk-(182+Alk)]^+$ fragment ions in the case of PC adducts. These fragment ions along with the presence of low-abundant alkaline ethanolamine phosphate ($[(HO)_2PO_2(CH_2)_2NH_2Alk]^+$) and phosphoric acid alkaline adduct ($[(HO)_3POAlk]^+$) are characteristic of PE species.

- The third type of fragment ions is those resulted from the neutral losses of the FA substituents from both quasimolecular ion and alkaline PA adduct (i.e., $[M+Alk-R_xCO_2H]^+$ and $[M+Li-43-R_xCO_2H]^+$, where $x = 1, 2$). The loss of the FA substituent at *sn*-1 position is more favored than the counterpart loss of the FA substituent at *sn*-2 position of the glycerol moiety as aforementioned.

Figure 7.3b shows a representative product-ion ESI-MS spectrum of lithiated 16:0–22:6 dPE after CID.

Similar to the case of PC class, product-ion ESI-MS analyses after CID enable researchers to discriminate the molecular species in different sub-classes of PE, for example, as lithium adduct [1]. The ions corresponding to $[M+Li-43]^+$, $[M+Li-141]^+$, $[M+Li-(140+Li)]^+$, and the ion at m/z 148 $([(HO)_2PO_2(CH_2)_2NH_2Li]^+)$ commonly yielded from lithiated dPE species are also observed in the product-ion spectra of the $[M+Li]^+$ ions of plasmenylethanolamine (pPE) and plasmanylethanolamine (aPE) species. However, the product-ion spectra of lithiated pPE species contain unique fragment ions. For example, an ion at m/z 425, yielding from the combined losses of aziridine and the alk-1'-enyl residue at the position sn-1 as an alcohol $(C_{16}H_{33}CH=CHOH)$, and an ion at m/z 307, resulting from further loss of the 18:1 FA substituent at position sn-2 as a free acid from m/z 589, are present in the spectrum of lithiated 18:0–18:1 pPE at m/z 740 [1]. These two ions identify the vinyl ether linkage at position sn-1 and the FA substituent at sn-2, respectively. In contrast, these two counterpart fragment ions are absent in the product-ion ESI-MS spectrum of lithiated 18:0–18:0 aPE at m/z 740 [1].

7.3.2 Negative-Ion Mode

7.3.2.1 Deprotonated Species   PE species can be ionized sensitively by negative-ion ESI-MS and yield deprotonated ions under various experimental conditions [25, 38, 40]. The fragmentation pattern of deprotonated PE species after CID contains three types of fragment ions as follows:

- Abundant one or two intense fragment ions around m/z 300 correspond to FA carboxylate ions depending on whether the FA substituents are identical or not, respectively. The peak intensity of the carboxylate anion resulting from the sn-2 FA substituent is approximately three times more intense due to its sterically favorable loss than that from the sn-1 FA chain of deprotonated dPE species.

- A cluster of low-to-modest abundance fragment ions around m/z 450 arising from the neutral losses of FA substituent(s) as fatty acids (i.e., $[M-H-R_xCH_2COOH]^-$) and FA ketenes (i.e., $[M-H-R_xCH=C=O]^-$) (where $x = 1, 2$ if present) of PE species are present in the product-ion ESI-MS spectra of deprotonated PE species. Similar to the fragmentation pattern of dPC species in the negative-ion mode, 1-acyl lysoPE fragment ions produced by loss of the sn-2 moiety are more intense than those of 2-acyl lysoPE fragment ions resulting from the loss of the sn-1 FA substituent [9, 22, 27]. Again, the $[M-H-R_2CH=C=O]^-$ ion is generally more intense than the $[M-H-R_2CH_2COOH]^-$ ion, while the $[M-H-R_1CH=C=O]^-$ ion is less abundant than the $[M-H-R_1CH_2COOH]^-$ ion [22, 32]. The intensity ratios of these pairs of ions can be used to determine the regioisomers of dPE.

- There also exists a fragment ion at m/z 196, corresponding to a glycerophos-phoethanolamine anion derivative, in the product-ion spectra of all PE species in low abundance, which is characteristic of the phosphoethanolamine head group.

Figure 7.3c shows a representative product-ion ESI-MS spectrum of deprotonated 16:0–22:6 dPE after CID. It should be recognized that the significant loss of CO_2 from polyunsaturated FA (Figure 7.3c) is always present as previously described [32, 33].

It should be noted that, in general, the product-ion ESI-MS spectra of deprotonated PE species consisting of polyunsaturated FA substituents are readily distinguishable from those consisting of saturated ones. This is due to a fact that the polyunsaturated FA carboxylate anions formed after CID in a QqQ-type mass spectrometer undergo vigorous secondary dissociation, while the saturated FA carboxylate anions undergo minimal secondary dissociation after formation [41, 42]. Therefore, the peak intensity of the polyunsaturated FA carboxylate anion is lower than expected.

Both product-ion ESI-MS mass spectra of pPE and aPE species are dominated by a single set of fragment ions corresponding to FA carboxylate anion from *sn*-2 FA, $[M-H-R_2CH_2COOH]^-$, and $[M-H-R_2CH=C=O]^-$. These fragment ions can be used to identify the FA substituent. An ion corresponding to the alkenyl moiety can be detected in low abundance to determine the plasmenyl- or plasmanyl- identity [32, 43]. Confirmation of the structural assignments using a comparison of the two mass spectra of PE species obtained before and after destructive removal of pPE by acid treatment [32, 34, 35] or iodine treatment [36] could be employed. It should be recognized that acid treatment may result in severe sample losses [34].

7.3.2.2 Derivatized Species

By exploring the specific reactivity of the primary amine group, derivative methods including with Fmoc chloride and 4-(dimethylamino)benzoic acid have been developed to enhance the analysis of PE species [44–46]. The enhancement of ionization efficiency in the negative-ion mode is achieved after derivatization through turning weakly anionic lipids (or weakly zwitterionic lipids) into anionic lipids (Chapters 2 and 3).

Fragmentation pattern of Fmoc-derivatized PE species in the negative-ion mode contains a unique, intense fragment ion resulted from the neutral loss of Fmoc and corresponding to deprotonated PE species. The rest of the fragmentation pattern is identical to that of deprotonated PE species as described earlier [44]. Product-ion ESI-MS spectra of 4-(dimethylamino)benzoic acid-derivatized PE species in the negative-ion mode displayed predominant fragment ion(s) corresponding to FA carboxylate ion(s) and low-abundant fragment ion(s) corresponding to the separate neutral loss of individual FA substituent as FA ketene [46]. The derivative of PE species with 4-(dimethylamino)benzoic acid can also readily form a stable positive charge site at dimethylamino moiety under acidic conditions. This charge site can lead to charge-remote fragmentation of the derivatives similar to those recently described [47]. The pattern of charge-remote fragmentation of the derivatized PE species include two intense fragment ions containing the charge site and a fragment ion yielded from the neutral loss of the entire head group (including the derivatized moiety) and corresponding to DAG-like ion [46]. Other derivative method was also employed for the purpose [36].

7.4 PHOSPHATIDYLINOSITOL AND PHOSPHATIDYLINOSITIDES

7.4.1 Positive-Ion Mode

As discussed in Chapter 2, ionization of anionic GPL species in the positive-ion mode is not favorable, but still can be formed in the presence of alkaline ion(s) in the solution or under acidic conditions. Generally, the sensitivity for ionization as alkaline adducts is markedly lower than what is observed as the $[M+H]^+$ ions in the positive-ion mode. Characterization of lithium and dilithium adducts of PI species (i.e., $[M+Li]^+$ and $[M-H+2Li]^+$) or their protonated species has been performed [1].

7.4.2 Negative-Ion Mode

PI species can be readily ionized as deprotonated form. The fragmentation pattern of deprotonated PI species ($[M-H]^-$) is very informative and more complicated than those of other anionic GPLs [41]. The fragmentation pattern of deprotonated PI species contains the following fragments:

- A cluster of fragment ions around m/z 550 in low abundance correspond to those yielded from the losses of FA substituents as fatty acids and FA ketenes (i.e., $[M-H-R_xCH_2CO_2H]^-$ and $[M-H-R_xCH=C=O]^-$, where $x = 1,\ 2$), respectively.

- A cluster of fragment ions around m/z 400 are those arising from the further loss of either inositol or (inositol-H_2O) from the cluster ions around m/z 500 (i.e., $[M-H-R_xCH_2CO_2H-(inositol-H2O)]^-$ or $[M-H-R_xCH=C=O-inositol]^-$). It seems that these fragment ions arising from charge-driven processes are occurring preferentially at the sn-2 position [41] as discussed in Section 7.1. This preference can be used to assign the positions of FA chains linked to the glycerol hydroxyl groups.

- A cluster of fragment ions in abundance correspond to FA carboxylate ions. The intensity of the $R_2CO_2^-$ ion is either relatively lower than or nearly equal to that of the $R_1CO_2^-$ ion. Those showing the lower intensity ones are the polyunsaturated FA substituents due to the further loss of CO_2 from the $R_2CO_2^-$ ion yielding an ion corresponding to $[R_2CO_2-44]^-$. In such a case, the combination of the intensities of both ions is still near to that of the $R_1CO_2^-$ ion.

- A cluster of fragment ions are yielded from the PI head group including those at m/z 315, 297, 279, 259, 241, and 223. Usually, the ion at m/z 241 is the most abundant one. The m/z 297 ion yields from consecutive losses of the FA substituents as fatty acids (i.e., $[M-H-R_1CO_2H-R_2CO_2H]^-$).

Figure 7.4a shows a representative product-ion ESI-MS spectrum of 18:1–20:5 PI species.

The fragmentation pathways of deprotonated phosphatidylinositol phosphate (PIP) and diphosphate (PIP_2) species (i.e., $[M-H]^-$) after low-energy CID are similar to that of PI [41]. However, the doubly charged (i.e., $[M-2H]^{2-}$) ions of PIP and PIP_2 species undergo fragmentation pathways that are similar to that of

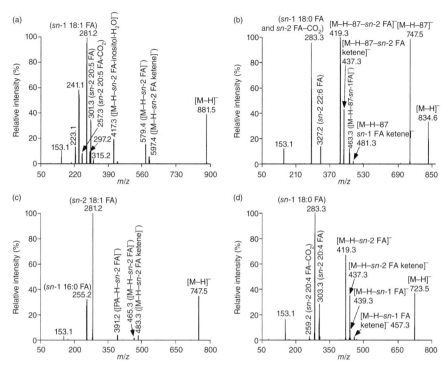

Figure 7.4 Representative product-ion ESI-MS spectra of deprotonated anionic glycerophospholipids (i.e., [M−H]⁻) after CID. Product-ion ESI-MS analyses of 18:1–20:5 PI (a), 18:0–22:6 PS (b), 16:0–18:1 PG (c), and 18:0–20:4 PA (d) in the negative-ion mode after CID were performed on a Thermo Fisher TSQ Vantage mass spectrometer. Collision activation was performed at collision energy of 40 (a), 30 (b), 30 (c), and 32 (d) eV, and gas pressure of 1 mTorr.

deprotonated PE species, which are basic [41]. These results suggest that the further deprotonated gaseous [M−2H]²⁻ ions of PIP and PIP₂ are basic precursors.

7.5   PHOSPHATIDYLSERINE

7.5.1   Positive-Ion Mode

The fragmentation pattern of protonated PS species is similar to that protonated PE species. This suggests that the fragmentation processes and the gas-phase basicities of phosphoserine and phosphoethanolamine are similar. Specifically, the product-ion ESI-MS spectra of protonated PS species display a dominant fragment ion corresponding to [M+H−185]⁺, arising from the elimination of the phosphoserine moiety as that of the PE counterpart [1]. The spectrum also shows the fragment ions arising from further dissociation of [M+H−185]⁺ by losses of the FA substituents as ketenes at positions sn-1 and sn-2, respectively. The acylium ions (R$_x$CO⁺) are observed in low abundance. Intriguingly, in different from the CID fragmentation of protonated PE, the protonated phosphoserine ion at m/z 186 is usually absent.

7.5.2 Negative-Ion Mode

In addition to the formation of deprotonated species as a predominant molecular ion (i.e., $[M-H]^-$) in the negative-ion mode, two other ion forms can also occur under certain experimental conditions. The first one is the $[M-H-87]^-$ form corresponding to a deprotonated PA counterpart resulted from the loss of serine (i.e., 87 amu), which may be formed in the ion source due to its facile loss. To avoid the formation of this ion-source fragmented ion, the ion source conditions for PS ionization should be tuned to minimize its generation. The second molecular ion of PS species that could be formed under basic conditions is the doubly charged format (i.e., $[M-2H]^{2-}$). The fragmentation of this ion form has not been well characterized.

The fragmentation pattern of deprotonated PS species (i.e., $[M-H]^-$) includes an intense fragment ion (i.e., $[M-H-87]^-$) corresponding to a deprotonated PA counterpart resulted from the loss of serine (i.e., 87 amu) (Figure 7.4b). The rest of the product-ion spectra of deprotonated PS species are essentially identical to those resulted from the deprotonated PA counterparts, which are presented below (compare Figure 7.4b with Figure 7.4d). This result suggests that a loss of serine to the formation of $[M-H-87]^-$ is the primary fragmentation process that leads to its further fragmentation. This pathway has been confirmed by MS^n studies utilizing an ion-trap mass spectrometer [12].

7.6   PHOSPHATIDYLGLYCEROL

7.6.1   Positive-Ion Mode

Fragmentation of protonated PG species after CID yields a predominant fragment ion corresponding to the loss of phosphoglycerol (i.e., $[M+H-(HO)_2P(O)OX]^+$, where $X =$ glycerol) [1]. The fragmentation pattern also contains the acylium ions (R_xCO^+) ($x = 1, 2$) as well as the ions arising from the further loss of the FA substituents as ketenes at positions sn-1 and sn-2, respectively, from $[M+H-(HO)_2P(O)OX]^+$. These fragment ions are generally present in low abundance. The FA identities can be determined from these ions, but assigning the regiospecific positions of these FA substituents on the glycerol is impractical. Hsu and Turk made this type of assignment from the MS^3 analysis [12].

7.6.2   Negative-Ion Mode

The fragmentation pattern of deprotonated PG species after CID contains three sets of modest to abundant, informative fragment ions (Figure 7.4c) as follows:

- The FA carboxylate ions ($R_xCO_2^-$, $x = 1, 2$) are usually the most abundant ions in the spectra. The $R_2CO_2^-$ ion peak is more intense than that of the $R_1CO_2^-$ peak.

- A cluster of fragment ions arise from the losses of FA substituents as fatty acids and FA ketenes, corresponding to $[M-H-R_xCH_2CO_2H]^-$ and $[M-H-R_xCH=C=O]^-$ ions, where $x = 1, 2$. The $[M-H-R_2CH_2CO_2H]^-$ and $[M-H-R_2CH=C=O]^-$ ions are more abundant than the counterpart ions at $[M-H-R_1CH_2CO_2H]^-$ and $[M-H-R_1CH=C=O]^-$ in the spectra [11]. This is probably due to the preferential losses of the FA substituent at the sn-2 position as described in Section 7.1. Moreover, the loss of a FA ketene yielding the $[M-H-R_2CH=C=O]^-$ ion is more preferential than the loss of a fatty acid yielding $[M-H-R_2CH_2CO_2H]^-$ ion at sn-2. In contrast, the $[M-H-R_1CH_2CO_2H]^-$ ion is more abundant than the $[M-H-R_1CH=C=O]^-$ ion arising from the analogous losses of the FA substituent at position sn-1 as a free acid or a FA ketene, respectively. This is consistent with the notion that PG is a weakly acidic glycerophospholipid [11], and the gas-phase basicity of the $[M-H]^-$ ion is between that of PE and PA. These observations are also consistent with the notion that the α-hydrogen of the FA substituent at position sn-2 is more labile and undergoes more facile loss of ketene as aforementioned. The presence of the more intense $[M-H-R_2CH=C=O]^-$ ion peak than that of the $[M-H-R_1CH=C=O]^-$ ion along with the finding that the $R_2CO_2^-$ is more abundant than the $R_1CO_2^-$ ion is readily applicable for the structural determination of PG species including regioisomers.

- A set of ions at m/z 227, 209, 171, and 153 in low abundance corresponding to the combined loss of a FA ketene and a fatty acid, the loss of fatty acids, glycerol phosphate, and a phosphoglycerol derivative, respectively, are the indicative of the polar head groups [11].

There exist some occasions that PG fragmentation does not follow the pattern discussed earlier. For example, PG species from *Arabidopsis thaliana* contain an unusual FA chain, 3-*trans*-hexadecenoyl, at the sn-2 position [48]. Product-ion ESI-MS spectra of deprotonated PG species after CID all demonstrate a predominant $[M-H-236]^-$ ion, resulting from the loss of the 3-*trans*-hexadecenoyl moiety as a ketene.

7.7 PHOSPHATIDIC ACID

7.7.1 Positive-Ion Mode

Fragmentation of protonated PA species after CID demonstrates a predominant fragment ion corresponding to the loss of phosphoric acid (i.e., $[M+H-(HO)_2P(O)OH]^+$) [1]. This ion is consistent with the notion that the gas-phase phosphoric acid is less competitive for a proton to form the protonated $(HO)_2P(O)OH$. The acylium ions as well as the ions arising from further dissociation of $[M+H-HO)_2P(O)OH]^+$ via the loss of the FA substituents as ketenes at positions sn-1 and sn-2, respectively, are present in the spectra in low abundance. Although the FA identities can be determined from these ions with efforts, assigning the regiospecific positions of these FA substituents on the glycerol backbone based on the differences in their abundance is impractical.

7.7.2   Negative-Ion Mode

The fragmentation pattern of deprotonated PA species (i.e., $[M-H]^-$) is essentially identical to that arising from $[PS-H-87]^-$ as described in Figure 7.4d. Charge-driven fragmentation processes involve the participation of the exchangeable hydrogen of the phosphate head group [8]. More intense $[M-H-R_2CH_2CO_2H]^-$ and $[M-H-R_2CH=C=O]^-$ ions than the $[M-H-R_1CH_2CO_2H]^-$ and $[M-H-R'1CH=C=O]^-$ ions, respectively, are present in the product-ion ESI-MS spectra of deprotonated PA species as discussed earlier. The differences between these pairs of fragment ions along with the finding that the $R_2CO_2^-$ is more abundant than the $R_1CO_2^-$ ion are commonly used to determine the regiospecific positions of the FA substituents qualitatively. However, the significant loss of CO_2 from sn-2 polyunsaturated FA [32, 33] usually leads to a reversed order of the intensities of $R_1CO_2^-$ and $R_2CO_2^-$ ions (Figure 7.4d).

7.8   CARDIOLIPIN

Cardiolipin (CL) is a unique class of anionic GPL and contains two phosphodiester moieties in each molecule. CL species can yield both singly and doubly charged ions in negative-ion ESI-MS analysis. In most cases, the $[M-2H]^{2-}$ ion is more abundant than the $[M-H]^-$ ion. In contrast to ESI-MS analysis, MALDI-MS analysis in the negative-ion mode shows predominant singly charged deprotonated CL species [49, 50].

Product-ion ESI-MS analyses of deprotonated CL species containing four identical FA chains after low-energy CID are well documented [22, 51]. The fragmentation pattern of $[M-2H]^{2-}$ ions after low-energy CID contains a predominant carboxylate anion (Figure 7.5a). The product-ion spectra also contain a doubly charged fragment ion arising from the loss of the FA substituent as a ketene. However, the fragment ions corresponding to the loss of the FA substituents as free fatty acids are not observed, consistent with the notion that the $[M-2H]^{2-}$ ion is a basic precursor ion and undergoes more facile ketene than acid loss [41]. Moreover, the product-ion spectra of doubly charged CL species containing four identical FA chains also give rise to a singly charged fragment ion, corresponding to the residual ion after splitting a carboxylate anion from the $[M-2H]^{2-}$ ion (Figure 7.5a).

The fragmentation pattern of $[M-H]^-$ ion of CL species containing four identical FA chains is very different from that of their $[M-2H]^{2-}$ ions (Figure 7.5b). Product-ion ESI-MS spectra of $[M-H]^-$ ion of CL species containing four identical FA chains yield a fragment ion in modest to abundant intensity corresponding to the loss of a free fatty acid and contain ions corresponding to a deprotonated PA anion and a deprotonated dehydrated PG anion (i.e., $[M-H-PA]^-$) [8]. The fragment ions resulting from the further loss of the FA substituents as a ketene or a free acid,

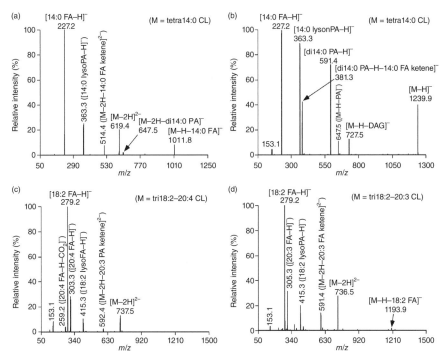

Figure 7.5 Representative product-ion ESI-MS spectra of cardiolipin species in the negative-ion mode after low-energy CID. Product-ion ESI-MS analyses of doubly charged deprotonated tetra14:0 cardiolipin (a), its singly charged counterpart (b), and the doubly charged ion at m/z 737.4 (c) and 736.5 (d) of cardiolipin species present in mouse myocardial lipid extracts were performed on a Thermo Fisher TSQ Vantage mass spectrometer. Collision activation was performed at collision energy of 45 (a), 20 (b), 22 (c), and 25 (d) eV, and gas pressure of 1 mTorr.

respectively, corresponding to lysoPA and lysoPG anion derivatives, are also present [22, 51] (Figure 7.5b).

CL species containing different FA chains are commercially unavailable. Characterization of this type of CL species is directly selected from the full MS analysis of biological samples. Therefore, the selected ions of CL species may represent a mixture of regioisomers, and the fragmentation pattern obtained from such way may not exactly reflect the fragmentation pattern of a single CL species. Han et al. [52] have determined a few CL ions from lipid extracts of mouse myocardium (Figure 7.5c and d), while Hsu et al. [51] studied some isomeric ions of CL species from different origins including bacteria. In general, the fragment ions detected from the synthetic CL species can be found from product-ion spectra of the natural CL samples. However, the peak intensities of the latter vary from sample to sample, reflecting the presence of different regioisomers. It is intriguing that the ratio of the peak intensities of the ions corresponding to the FA substituents is essentially equivalent to the ratio of the number of each FA chain [52].

7.9 LYSOGLYCEROPHOSPHOLIPIDS

7.9.1 Choline Lysoglycerophospholipids

Characterization of choline lysoGPL species as alkaline adducts is well studied in the positive-ion mode after low-energy CID [6, 7, 23, 26]. The fragmentation pattern of alkaline adducted lysoPC species contains three groups of informative fragments as follows:

- The abundant fragment ions corresponding to the neutral losses of polar head group fragments including $[M+Alk-59]^+$, $[M+Alk-183]^+$, and $[M+Alk-(Alk+182)]^+$ are predominant in their product-ion mass spectra (Figure 7.6a). These fragment ions are essentially identical to those arising from PC species.

- An additional ion at $[M+Alk-103]^+$ corresponding to the loss of choline-like molecule is also present in low abundance in the fragmentation pattern (Figure 7.6a). In contrast to the presence of abundant ions reflecting the losses of FA substituents (i.e., $[M+Alk-R_xCO_2H]^+$ and $[M+Alk-59-R_xCO_2H]^+$) from their parent species, these fragment ions are of low abundance or even not present in the product-ion ESI-MS spectra (Figure 7.6a). This is due to the fact that the lack of a FA chain leads to missing the initiation of the α-hydrogen at the FA chain in the elimination of the adjacent FA substituent as an acid.

- The fragment ions resulting from the phosphocholine head group are present at m/z 124+Alk and 104 corresponding to the alkaline adduct of a five-membered ethylene phosphatidic acid and choline ions in the product-ion spectra of lysoPC species (Figure 7.6a). The peak intensity ratio of these fragment ions can be used to determine the position of FA chain linked to the glycerol hydroxyl group [6, 23].

Similar to plasmanylcholine and plasmenylcholine species (see above), due to the lack of α-hydrogen in both the *sn*-1 and *sn*-2 moieties, further dissociation from $[M+Alk-59]^+$ to form $[M+Alk-183]^+$ and $[M+Alk-(Alk+182)]^+$ is significantly reduced [1]. Accordingly, the presence of a prominent $[M+Alk-59]^+$ ion relative to $[M+Alk-183]^+$ and $[M+Alk-(Alk+182)]^+$ ions can be used to discriminate ether subclasses from lysophosphatidylcholine.

The fragmentation pattern of protonated lysoPC species after low-energy CID is very different from those of their alkaline adducts [26]. Fragmentation of all protonated species shows a prominent and distinct ion at m/z 184 corresponding to phosphocholine. Product-ion ESI-MS spectra of protonated lysoPC species also display another very abundant fragment ions corresponding to the loss of H_2O (18 amu) (i.e., $[M+H-18]^+$) (Figure 7.6b). This fragment ion is absent or present in very low abundance in ether lysoPC subclasses. This feature can be used to distinguish lysoPC subclasses.

Although ionization of lysoPC species in the negative-ion mode for lipidomics is not common, these species can be readily ionized as the adducts of small anions available in the media. Characterization of these anionic adducts after low-energy

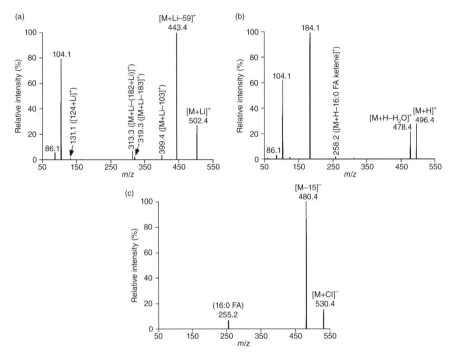

Figure 7.6 Representative product-ion ESI-MS spectra of lysophosphatidylcholine species in both positive- and negative-ion modes after low-energy CID. Product-ion ESI-MS analyses of lithiated (a), protonated (b), and chloride-adducted (c) 1-palmitoyl-*sn*-3-phosphocholine were performed on a Thermo Fisher TSQ Vantage mass spectrometer. Collision activation was performed at collision energy of 22 (a), 21 (b), and 15 (c) eV, and gas pressure of 1 mTorr.

CID yields a prominent ion corresponding to the loss of methyl along with the anion adduct from quasimolecular ions resulting in an $[M-15]^-$ ion [6, 26]. Another abundant fragment ion from lysoPC species is the FA carboxylate ion corresponding to the FA substituent (Figure 7.6c). A few other low-abundant fragment ions could also present in the product-ion ESI-MS spectra of lysoPC anion adducts corresponding to the further loss of FA substituent as a fatty acid or a FA ketene from the $[M-15]^-$ ion (i.e., $[M-15-RCH_2CO_2H]^-$ and $[M-15-RCH=C=O]^-$, respectively). These fragment ions are absent in ether-linked lysoPC subclasses, in which a fragment ion corresponding to the loss of long-chain FA alcohol (i.e., $[M-15-RCOH]^-$) is present in low abundance [6, 26].

7.9.2 Ethanolamine Lysoglycerophospholipids

The fragmentation patterns between deprotonated lysophosphatidylethanolamine (acyl lysoPE) and ether-linked lysoPE species are very different [1, 6]. The former contains a predominent ion around *m/z* 300 corresponding to the sole FA carboxylate anion that identifies the FA substituent of the molecules (Figure 7.7a). There also exist two fragment ions at *m/z* 214 and 196, arising from the losses

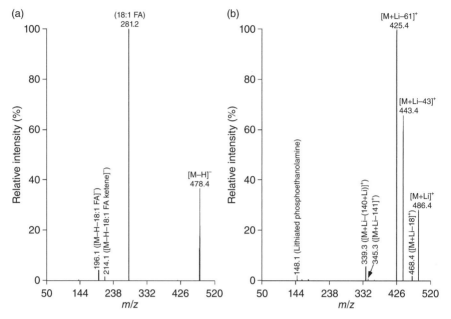

Figure 7.7 Representative product-ion ESI-MS spectra of lysophosphatidylethanolamine species in both negative- and positive-ion modes after low-energy CID. Product-ion ESI-MS analyses of deprotonated (a) and lithiated (b) 1-oleoyl-*sn*-3-phosphoethanolamine were performed on a Thermo Fisher TSQ Vantage mass spectrometer. Collision activation was performed at collision energy of 20 (a) and 18 (b) eV, and gas pressure of 1 mTorr.

of the FA substituent as a FA ketene and acid (i.e., [M−H−RCH=C=O]⁻ and [M−H−RCH$_2$COOH]⁻), respectively. Intriguingly, the *m/z* 214 ion is lower than the *m/z* 196 ion in the product-ion spectra of deprotonated 1-acyl lysoPE (Figure 7.7a), whereas the abundance of these two ions are reversed in the spectra of 2-acyl lysoPE counterparts. This is consistent with the notion that the gas-phase [M−H]⁻ ions of 1- and of 2-acyl lysoPE species are weakly basic ions and undergo more facile loss of the FA substituent as a ketene at *sn*-2 than at *sn*-1. Thus, positional isomers of 1- and 2-acyl lysoPE species can be differentiated.

Ether-linked lysoPE species can be readily recognized in the absence of a FA carboxylate anion, which is replaced by a low-abundant ion corresponding to the *sn*-1-alkenyl or *sn*-1-alkyl chain. An additional product ion (i.e., [M−62]⁻, corresponding to the neutral loss of ethanolamine from precursor ion) is also present in the spectra of ether-linked lysoPE species [6]. Moreover, the fragment ion at *m/z* 196 is usually the most abundant ion present in the product-ion spectra of ether-linked lysoPE species.

The fragmentation pattern of lysoPE species as alkaline adducts contains the following fragments [6] (Figure 7.7b):

- A most abundant ion at [M+Alk−43]⁺ (Alk = alkaline) resulting from the neutral loss of vinylamine.

- A less abundant ion at $[M+Alk-61]^+$ arising from the neutral loss of ethanolamine.
- A low-abundant ion at $[M+Alk-141]^+$ yielding from the neutral loss of phosphoethanolamine.
- A modest ion at $[M+Alk-(140+Alk)]^+$ corresponding to the neutral loss of alkaline ethanolamine phosphate.

The *sn*-1 and *sn*-2 position isomers of acyl lysoPE alkaline adducts can be distinct from the differential intensities of these fragment ions as discussed in Section 7.1. For example, these isomers can be substantiated by the significant difference of the ratio between the fragment ion pair at $[M+Alk-(140+Alk)]^+$ and $[M+Alk-61]^+$ as previously discussed [6]. Finally, a few fragment ions resulting from the phosphoethanolamine head group are present (Figure 7.7b). For example, the *m/z* 148 ion corresponds to the lithiated phosphoethanolamine as previously described [5].

Product-ion spectra of protonated acyl lysoPE species display two fragment ions at $[M+H-141]^+$ and $[M+H-18]^+$ corresponding to the losses of phosphoethanolamine and a water molecule, respectively [53]. Product-ion spectra of protonated alkenyl lysoPE species show three product ions at $[M+H-172]^+$, $[M+H-154]^+$, and $[M+H-18]^+$ [53]. The first two fragment ions correspond to the losses of phosphoethanolamine derivatives and the third fragment ion is the loss of a water molecule. Obviously, the fragment ion corresponding to the loss of phosphoethanolamine from acyl lysoPE species does not occur in the fragmentation pathway of alkenyl lysoPE. This feature can be readily used for distinction of these subclasses.

7.9.3 Anionic Lysoglycerophospholipids

Characterization of anionic lysoGPL species as deprotonated species by product-ion ESI-MS after low-energy CID has extensively been conducted [17, 53–56]. Fragmentation patterns of deprotonated anionic lysoGPL species are essentially identical to their parent diacyl counterparts. However, the fragment ion at *m/z* 153, corresponding to a glycerophosphate derivative, is much more abundant in the spectra of anionic lysoGPL species than their diacyl parents. It should be recognized that the product-ion spectra of deprotonated anionic lysoGPL regioisomers are essentially identical. Therefore, distinction between position isomers of anionic lysoGPL species is impractical.

7.10   OTHER GLYCEROPHOSPHOLIPIDS

7.10.1   *N*-Acyl Phosphatidylethanolamine

Deprotonated *N*-acyl PE species have been characterized in a few studies [57–60]. In summary, the product-ion ESI-MS spectra of these ions after CID display a few extra features in addition to those of its parent PE species (see above). First, a cluster of fragment ions are present, corresponding to the loss of a FA substituent as a FA acid or a FA ketene. Second, the cluster of fragment ions corresponding to the losses of FA substituents as a FA acid or a FA ketene become more complicated with the combination of the loss of the N-acyl as a FA ketene. However, the FA carboxylate ion resulted from N-acyl is absent. This feature can be used to identify the identity of N-acyl.

7.10.2 N-Acyl Phosphatidylserine

The fragmentation pattern of deprotonated N-acyl PS species is essentially identical to that of deprotonated PS species. The only difference is the replacement of a fragment ion corresponding to the loss of serine with N-acyl serine [61].

7.10.3   Acyl Phosphatidylglycerol

The fragmentation pattern of acyl PG is very similar to that of N-acyl PE, but is more complicated than that of N-acyl PE due to the presence of the additional loss of the FA substituent of the third acyl chain as a FA acid [42, 57], which are as follows:

- A cluster of fragment ions at the mass range corresponding to the loss of one FA substituent as a FA acid or a FA ketene, respectively.
- A cluster of fragment ions corresponding to the losses of two FA substituents as FA acids or ketenes or their combinations.
- A cluster of fragment ions corresponding to those of FA carboxylate ions, among which the FA carboxylate ion resulted from the third acyl chain is usually in less abundance likely due to the sterical effect.

7.10.4   Bis(monoacylglycero)phosphate

Bis(monoacylglycero)phosphate (BMP) species are isomeric to PG species. In BMP, two FA chains are acylated to different glycerol molecules, whereas these FA chains are located in one glycerol in PG species. BMP plays an important role in normal lysosomal/endosomal functions in cells [62]. It has become evident that BMP is involved in the pathology of lysosomal storage diseases such as Niemann–Pick C disease (cholesterol accumulation) and certain drug-induced lipidoses [63, 64]. Dysregulation of BMP metabolism and hence of cholesterol homeostasis may also be relevant to atherosclerosis [64].

Characterization of BMP species by ESI-MS has been seldomly conducted, particularly in comparison to PG species. We found that the fragmentation pattern of deprotonated BMP species contains two very different fragment features in comparison to their isomeric PG species (Figure 7.8 compared to Figure 7.4c, unpublished data) as follows:

- The FA carboxylate ions ($R_xCO_2^-$, $x = 1, 2$) are the most abundant ions in the spectra as in both cases of PG and BMP species, but the intensities of these ions

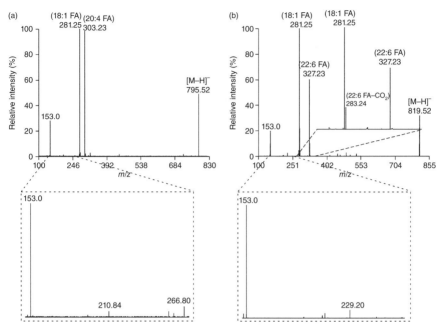

Figure 7.8 Representative product-ion ESI-MS spectra of deprotonated bis(monoacyl-glycero)phosphate (BMP) species (i.e., $[M-H]^-$) after low-energy CID. Product-ion ESI-MS analyses of 18:0–20:4 BMP (a) and 18:0–22:6 BMP (b) in the negative-ion mode after CID were performed on a Thermo Fisher Q-Executive mass spectrometer. Collision activation was conducted with higher energy collisional dissociation (HCD) at 24.0 eV and gas pressure of 1 mTorr.

(if two different FA substituents are present in a BMP species) are essentially identical in the case of BMP, whereas the $R_2CO_2^-$ ion peak is more intense than that of the $R_1CO_2^-$ peak in the case of PG species. It should be pointed out that the relatively low abundance of m/z 327 ion corresponding to docosahexonic acid is largely due to the sequential facile loss of carbon dioxide to yield the ion at m/z 283 (inset of Figure 7.8b) as we recognized [31–33].

- Among the cluster of fragment ions at m/z 227, 209, 171, and 153 corresponding to the combined loss of a FA ketene and a fatty acid, the loss of fatty acids, glycerol phosphate, and a phosphoglycerol derivative, respectively, present in the product-ion spectra of PG species, only the phosphoglycerol derivative ion at m/z 153 can be detected in low abundance in the case of BMP species. All of others are either absent or present in very low intensity relative to the intensity of the ion at m/z 153 (insets of Figure 7.8).

These distinct features can be used to readily identify the isomeric species of PG and BMP as recently conducted [65].

7.10.5 Cyclic Phosphatidic Acid

Characterization of cyclic PA species as deprotonated species by product-ion ESI-MS in the negative-ion mode after low-energy CID has been conducted [66]. The fragmentation pattern of deprotonated cyclic PA species contains two prominent fragment ions, one corresponding to the product ion of FA substituent of the species (equivalent to the loss of 136 amu) and the other at m/z 153 corresponding to a glycerophosphate derivative.

REFERENCES

1. Hsu, F.F. and Turk, J. (2009) Electrospray ionization with low-energy collisionally activated dissociation tandem mass spectrometry of glycerophospholipids: Mechanisms of fragmentation and structural characterization. J. Chromatogr. B 877, 2673–2695.

2. Cheng, C. and Gross, M.L. (2000) Applications and mechanisms of charge-remote fragmentation. Mass Spectrom. Rev. 19, 398–420.

3. Hsu, F.-F., Bohrer, A. and Turk, J. (1998) Formation of lithiated adducts of glycerophosphocholine lipids facilitates their identification by electrospray ionization tandem mass spectrometry. J. Am. Soc. Mass Spectrom. 9, 516–526.

4. Hsu, F.-F. and Turk, J. (2003) Electrospray ionization/tandem quadrupole mass spectrometric studies on phosphatidylcholines: The fragmentation processes. J. Am. Soc. Mass Spectrom. 14, 352–363.

5. Hsu, F.F. and Turk, J. (2000) Characterization of phosphatidylethanolamine as a lithiated adduct by triple quadrupole tandem mass spectrometry with electrospray ionization. J. Mass Spectrom. 35, 595–606.

6. Han, X. and Gross, R.W. (1996) Structural determination of lysophospholipid regioisomers by electrospray ionization tandem mass spectrometry. J. Am. Chem. Soc. 118, 451–457.

7. Hsu, F.-F., Turk, J., Thukkani, A.K., Messner, M.C., Wildsmith, K.R. and Ford, D.A. (2003) Characterization of alkylacyl, alk-1-enylacyl and lyso subclasses of glycerophosphocholine by tandem quadrupole mass spectrometry with electrospray ionization. J. Mass Spectrom. 38, 752–763.

8. Hsu, F.F. and Turk, J. (2000) Charge-driven fragmentation processes in diacyl glycerophosphatidic acids upon low-energy collisional activation: A mechanistic proposal. J. Am. Soc. Mass Spectrom. 11, 797–803.

9. Hsu, F.F. and Turk, J. (2000) Charge-remote and charge-driven fragmentation processes in diacyl glycerophosphoethanolamine upon low-energy collisional activation: A mechanistic proposal. J. Am. Soc. Mass Spectrom. 11, 892–899.

10. IUPAC. (1994) Compendium of Chemical Terminology. Blackwell Scientific Publications, Oxford.

11. Hsu, F.F. and Turk, J. (2001) Studies on phosphatidylglycerol with triple quadrupole tandem mass spectrometry with electrospray ionization: Fragmentation processes and structural characterization. J. Am. Soc. Mass Spectrom. 12, 1036–1043.

12. Hsu, F.F. and Turk, J. (2005) Studies on phosphatidylserine by tandem quadrupole and multiple stage quadrupole ion-trap mass spectrometry with electrospray ionization: Structural characterization and the fragmentation processes. J. Am. Soc. Mass Spectrom. 16, 1510–1522.

13. Nakanishi, H., Iida, Y., Shimizu, T. and Taguchi, R. (2010) Separation and quantification of sn-1 and sn-2 fatty acid positional isomers in phosphatidylcholine by RPLC-ESIMS/MS. J. Biochem. 147, 245–256.

14. Taguchi, R. and Ishikawa, M. (2010) Precise and global identification of phospholipid molecular species by an Orbitrap mass spectrometer and automated search engine Lipid Search. J. Chromatogr. A 1217, 4229–4239.

15. Griffiths, W.J. (2003) Tandem mass spectrometry in the study of fatty acids, bile acids, and steroids. Mass Spectrom. Rev. 22, 81–152.

16. Murphy, R.C., Fiedler, J. and Hevko, J. (2001) Analysis of nonvolatile lipids by mass spectrometry. Chem. Rev. 101, 479–526.

17. Pulfer, M. and Murphy, R.C. (2003) Electrospray mass spectrometry of phospholipids. Mass Spectrom. Rev. 22, 332–364.

18. Hsu, F.-F. and Turk, J. (2005) Electrospray ionization with low-energy collisionally activated dissociation tandem mass spectrometry of complex lipids: Structural characterization and mechanism of fragmentation. In Modern Methods for Lipid Analysis by Liquid Chromatography/Mass Spectrometry and Related Techniques (Byrdwell, W.C., ed.). pp. 61–178, AOCS Press, Champaign, IL

19. Murphy, R.C. and Axelsen, P.H. (2011) Mass spectrometric analysis of long-chain lipids. Mass Spectrom. Rev. 30, 579–599.

20. Murphy, R.C. (2015) Tandem Mass Spectrometry of Lipids: Molecular Analysis of Complex Lipids. Royal Society of Chemistry, Cambridge, UK. pp 280.

21. Hsu, F.F. and Turk, J. (2005) Analysis of sulfatides. In The Encyclopedia of Mass Spectrometry (Caprioli, R.M., ed.). pp. 473–492, Elsevier, New York

22. Han, X. and Gross, R.W. (1995) Structural determination of picomole amounts of phospholipids via electrospray ionization tandem mass spectrometry. J. Am. Soc. Mass Spectrom. 6, 1202-1210.

23. Yang, K., Zhao, Z., Gross, R.W. and Han, X. (2009) Systematic analysis of choline-containing phospholipids using multi-dimensional mass spectrometry-based shotgun lipidomics. J. Chromatogr. B 877, 2924–2936.

24. Weintraub, S.T., Pinckard, R.N. and Hail, M. (1991) Electrospray ionization for analysis of platelet-activating factor. Rapid Commun. Mass Spectrom. 5, 309–311.

25. Kerwin, J.L., Tuininga, A.R. and Ericsson, L.H. (1994) Identification of molecular species of glycerophospholipids and sphingomyelin using electrospray mass spectrometry. J. Lipid Res. 35, 1102–1114.

26. Khaselev, N. and Murphy, R.C. (2000) Electrospray ionization mass spectrometry of lyso-glycerophosphocholine lipid subclasses. J. Am. Soc. Mass Spectrom. 11, 283–291.

27. Houjou, T., Yamatani, K., Nakanishi, H., Imagawa, M., Shimizu, T. and Taguchi, R. (2004) Rapid and selective identification of molecular species in phosphatidylcholine and sphingomyelin by conditional neutral loss scanning and MS3. Rapid Commun. Mass Spectrom. 18, 3123–3130.

28. Han, X., Yang, K., Yang, J., Fikes, K.N., Cheng, H. and Gross, R.W. (2006) Factors influencing the electrospray intrasource separation and selective ionization of glycerophospholipids. J. Am. Soc. Mass Spectrom. 17, 264–274.

29. Ekroos, K., Chernushevich, I.V., Simons, K. and Shevchenko, A. (2002) Quantitative profiling of phospholipids by multiple precursor ion scanning on a hybrid quadrupole time-of-flight mass spectrometer. Anal. Chem. 74, 941–949.

30. Ejsing, C.S., Duchoslav, E., Sampaio, J., Simons, K., Bonner, R., Thiele, C., Ekroos, K. and Shevchenko, A. (2006) Automated identification and quantification of glycerophospholipid molecular species by multiple precursor ion scanning. Anal. Chem. 78, 6202–6214.

31. Ekroos, K., Ejsing, C.S., Bahr, U., Karas, M., Simons, K. and Shevchenko, A. (2003) Charting molecular composition of phosphatidylcholines by fatty acid scanning and ion trap MS3 fragmentation. J. Lipid Res. 44, 2181–2192.

32. Yang, K., Zhao, Z., Gross, R.W. and Han, X. (2007) Shotgun lipidomics identifies a paired rule for the presence of isomeric ether phospholipid molecular species. PLoS One 2, e1368.

33. Yang, K., Zhao, Z., Gross, R.W. and Han, X. (2011) Identification and quantitation of unsaturated fatty acid isomers by electrospray ionization tandem mass spectrometry: A shotgun lipidomics approach. Anal. Chem. 83, 4243–4250.

34. Kayganich, K.A. and Murphy, R.C. (1992) Fast atom bombardment tandem mass spectrometric identification of diacyl, alkylacyl, and alk-1-enylacyl molecular species of glycerophosphoethanolamine in human polymorphonuclear leukocytes. Anal. Chem. 64, 2965–2971.

35. Ford, D.A., Rosenbloom, K.B. and Gross, R.W. (1992) The primary determinant of rabbit myocardial ethanolamine phosphotransferase substrate selectivity is the covalent nature of the sn-1 aliphatic group of diradyl glycerol acceptors. J. Biol. Chem. 267, 11222–11228.

36. Fhaner, C.J., Liu, S., Zhou, X. and Reid, G.E. (2013) Functional group selective derivatization and gas-phase fragmentation reactions of plasmalogen glycerophospholipids. Mass Spectrom. (Tokyo) 2, S0015.

37. Brugger, B., Erben, G., Sandhoff, R., Wieland, F.T. and Lehmann, W.D. (1997) Quantitative analysis of biological membrane lipids at the low picomole level by nano-electrospray ionization tandem mass spectrometry. Proc. Natl. Acad. Sci. U.S.A. 94, 2339–2344.

38. Kim, H.Y., Wang, T.C. and Ma, Y.C. (1994) Liquid chromatography/mass spectrometry of phospholipids using electrospray ionization. Anal. Chem. 66, 3977–3982.

39. Brouwers, J.F., Vernooij, E.A., Tielens, A.G. and van Golde, L.M. (1999) Rapid separation and identification of phosphatidylethanolamine molecular species. J. Lipid Res. 40, 164–169.

40. Han, X. and Gross, R.W. (1994) Electrospray ionization mass spectroscopic analysis of human erythrocyte plasma membrane phospholipids. Proc. Natl. Acad. Sci. U.S.A. 91, 10635–10639.

41. Hsu, F.-F. and Turk, J. (2000) Characterization of phosphatidylinositol, phosphatidylinositol-4-phosphate, and phosphatidylinositol-4,5-bisphosphate by electrospray ionization tandem mass spectrometry: A mechanistic study. J. Am. Soc. Mass Spectrom. 11, 986–999.

42. Hsu, F.F., Turk, J., Shi, Y. and Groisman, E.A. (2004) Characterization of acylphosphatidylglycerols from Salmonella typhimurium by tandem mass spectrometry with electrospray ionization. J. Am. Soc. Mass Spectrom. 15, 1–11.

43. Schwudke, D., Oegema, J., Burton, L., Entchev, E., Hannich, J.T., Ejsing, C.S., Kurzchalia, T. and Shevchenko, A. (2006) Lipid profiling by multiple precursor and neutral loss scanning driven by the data-dependent acquisition. Anal. Chem. 78, 585–595.

44. Han, X., Yang, K., Cheng, H., Fikes, K.N. and Gross, R.W. (2005) Shotgun lipidomics of phosphoethanolamine-containing lipids in biological samples after one-step in situ derivatization. J. Lipid Res. 46, 1548–1560.

45. Berry, K.A. and Murphy, R.C. (2005) Analysis of cell membrane aminophospholipids as isotope-tagged derivatives. J. Lipid Res. 46, 1038–1046.

46. Zemski Berry, K.A., Turner, W.W., VanNieuwenhze, M.S. and Murphy, R.C. (2009) Stable isotope labeled 4-(dimethylamino)benzoic acid derivatives of glycerophosphoethanolamine lipids. Anal. Chem. 81, 6633-6640.

47. Wang, M., Han, R.H. and Han, X. (2013) Fatty acidomics: Global analysis of lipid species containing a carboxyl group with a charge-remote fragmentation-assisted approach. Anal. Chem. 85, 9312–9320.

48. Hsu, F.F., Turk, J., Williams, T.D. and Welti, R. (2007) Electrospray ionization multiple stage quadrupole ion-trap and tandem quadrupole mass spectrometric studies on phosphatidylglycerol from Arabidopsis leaves. J. Am. Soc. Mass Spectrom. 18, 783–790.

49. Sun, G., Yang, K., Zhao, Z., Guan, S., Han, X. and Gross, R.W. (2008) Matrix-assisted laser desorption/ionization time-of-flight mass spectrometric analysis of cellular glycerophospholipids enabled by multiplexed solvent dependent analyte-matrix interactions. Anal. Chem. 80, 7576–7585.

50. Wang, H.Y., Jackson, S.N. and Woods, A.S. (2007) Direct MALDI-MS analysis of cardiolipin from rat organs sections. J. Am. Soc. Mass Spectrom. 18, 567–577.

51. Hsu, F.F., Turk, J., Rhoades, E.R., Russell, D.G., Shi, Y. and Groisman, E.A. (2005) Structural characterization of cardiolipin by tandem quadrupole and multiple-stage quadrupole ion-trap mass spectrometry with electrospray ionization. J. Am. Soc. Mass Spectrom. 16, 491–504.

52. Han, X., Yang, K., Yang, J., Cheng, H. and Gross, R.W. (2006) Shotgun lipidomics of cardiolipin molecular species in lipid extracts of biological samples. J. Lipid Res. 47, 864–879.

53. Chen, S. (1997) Tandem mass spectrometric approach for determining structure of molecular species of aminophospholipids. Lipids 32, 85–100.

54. Xiao, Y., Chen, Y., Kennedy, A.W., Belinson, J. and Xu, Y. (2000) Evaluation of plasma lysophospholipids for diagnostic significance using electrospray ionization mass spectrometry (ESI-MS) analyses. Ann. N. Y. Acad. Sci. 905, 242–259.

55. Lee, J.Y., Min, H.K. and Moon, M.H. (2011) Simultaneous profiling of lysophospholipids and phospholipids from human plasma by nanoflow liquid chromatography-tandem mass spectrometry. Anal. Bioanal. Chem. 400, 2953–2961.

56. Wang, C., Wang, M. and Han, X. (2015) Comprehensive and quantitative analysis of lysophospholipid molecular species present in obese mouse liver by shotgun lipidomics. Anal. Chem. 87, 4879–4887.

57. Holmback, J., Karlsson, A.A. and Arnoldsson, K.C. (2001) Characterization of N-acylphosphatidylethanolamine and acylphosphatidylglycerol in oats. Lipids 36, 153–165.

58. Mileykovskaya, E., Ryan, A.C., Mo, X., Lin, C.C., Khalaf, K.I., Dowhan, W. and Garrett, T.A. (2009) Phosphatidic acid and N-acylphosphatidylethanolamine form membrane domains in *Escherichia coli* mutant lacking cardiolipin and phosphatidylglycerol. J. Biol. Chem. 284, 2990–3000.

59. Kilaru, A., Tamura, P., Isaac, G., Welti, R., Venables, B.J., Seier, E. and Chapman, K.D. (2012) Lipidomic analysis of N-acylphosphatidylethanolamine molecular species in Arabidopsis suggests feedback regulation by N-acylethanolamines. Planta 236, 809–824.

60. Astarita, G., Ahmed, F. and Piomelli, D. (2008) Identification of biosynthetic precursors for the endocannabinoid anandamide in the rat brain. J. Lipid Res. 49, 48–57.

61. Guan, Z., Li, S., Smith, D.C., Shaw, W.A. and Raetz, C.R. (2007) Identification of N-acylphosphatidylserine molecules in eukaryotic cells. Biochemistry 46, 14500–14513.

62. Gallala, H.D. and Sandhoff, K. (2011) Biological function of the cellular lipid BMP-BMP as a key activator for cholesterol sorting and membrane digestion. Neurochem. Res. 36, 1594–1600.

63. Meikle, P.J., Duplock, S., Blacklock, D., Whitfield, P.D., Macintosh, G., Hopwood, J.J. and Fuller, M. (2008) Effect of lysosomal storage on bis(monoacylglycero)phosphate. Biochem. J. 411, 71–78.

64. Hullin-Matsuda, F., Luquain-Costaz, C., Bouvier, J. and Delton-Vandenbroucke, I. (2009) Bis(monoacylglycero)phosphate, a peculiar phospholipid to control the fate of cholesterol: Implications in pathology. Prostaglandins Leukot. Essent. Fatty Acids 81, 313–324.

65. Akgoc, Z., Sena-Esteves, M., Martin, D.R., Han, X., d'Azzo, A. and Seyfried, T.N. (2015) Bis(monoacylglycero)phosphate: A secondary storage lipid in the gangliosidoses. J. Lipid Res. 56, 1006–1013.

66. Shan, L., Li, S., Jaffe, K. and Davis, L. (2008) Quantitative determination of cyclic phosphatidic acid in human serum by LC/ESI/MS/MS. J. Chromatogr. B 862, 161–167.

8

FRAGMENTATION PATTERNS OF SPHINGOLIPIDS

8.1 INTRODUCTION

Sphingolipids are a category of lipid classes in which all the species contain common long-chain sphingoid bases as their core structures (see Chapter 1). These sphingoid bases originate from the condensation of serine (or the other amino acids in some occasion [1]) and a long-chain fatty acyl CoA to yield sphinganine. Acylation of sphinganine with the primary amine at the C2 position produces dihydroceramides, which converts into ceramides, phosphosphingolipids, glycosphingolipids, and other sphingolipid classes through desaturase or transferase activities. Therefore, the polar moieties that are linked to the hydroxy group of sphingoid base at position C1 represent the individual sphingolipid classes.

Although the condensation of serine with palmitoyl CoA is most common due to the enzymatic selectivity, condensation with other fatty acyl CoA species is also present in biological systems. Therefore, sphingolipids represent the most complex category of lipids in cellular lipidomes. Extensive characterization of all these different subcategories is prohibited due to the lack of commercially available authentic standards. The good thing is that the fragmentation patterns of these subcategories of sphingolipids are essentially identical to those of sphinganine or sphingosine analogs.

In the best scenario, information from characterization of sphingolipids could lead us to identify the classes (i.e., polar head groups), subclasses (sphingosine and sphinganine as well as their analogs), and aliphatic chains. Unlike in glycerophospholipids that usually go remodeling after synthesis, the biosynthesis pathways of sphingolipids

Lipidomics: Comprehensive Mass Spectrometry of Lipids, First Edition. Xianlin Han.
© 2016 John Wiley & Sons, Inc. Published 2016 by John Wiley & Sons, Inc.

indicate that the aliphatic chains in sphingolipid species are generally fixed. Therefore, efforts to define the position isomers can be eliminated. Since low-energy CID does not enable us to determine the location of double bond(s), the FA amide chain can be derived once the sphingoid base is identified or vice verse with additional determination of subclasses.

Similar to the category of GPL classes, characterization of sphingolipids can always be conducted in both positive- and negative-ion modes. Very different fragment ions could be generated from these different quasimolecular ions, all of which are interesting, particularly in viewing from the ion chemistry perspective. Selecting one type of quasimolecular ion to characterize a particular sphingolipid class largely depends on the ionization sensitivity and usefulness for lipidomics.

In this chapter, similar to the last one, only the fragmentation patterns of individual lipid classes of sphingolipids under some common experimental conditions are overviewed. Any advanced readers should always look for the details from the original studies on the topic and/or study several invaluable review articles [1–3].

8.2 CERAMIDE

8.2.1 Positive-Ion Mode

Cer species can readily form protonated ions ([M+H]$^+$) under acidic conditions in the positive-ion mode. These ions are very labile and readily lose a water molecule to become the [M+H−18]$^+$ ion in the ion source [4–6]. The fragmentation pattern of protonated Cer species after CID contains the following:

- A predominant fragment ion (usually the base peak), corresponding to the non-specific loss of a water molecule (i.e., [M+H−18]$^+$).
- A moderate to abundant ion at [M+H−36]$^+$ resulted from further loss of a water molecule from the fragment ion of [M+H−18]$^+$.
- The third type of fragment ions corresponding to those characteristic of the sphingoid bases, i.e., m/z 264 and 282 for Cer species containing sphingosine and m/z 266 and 284 for those containing sphinganine.

The m/z 282 and 284 ions arise from the neutral loss of a FA ketene from the predominant [M+H−18]$^+$ of Cer and dihydroceramide species, respectively. The fragment ions at m/z 264 and 266 could result from the further loss of a water molecule from the m/z 282 and 284 ions, respectively, or alternatively are produced from the neutral loss of a FA ketene from [M+H−36]$^+$. More complicated Cer species, especially those found in skin, containing di- and trihydroxy sphingoid bases in combination with hydroxy and nonhydroxy FA amides were also characterized by product-ion ESI-MS analyses of protonated species [7–11].

Cer species can also be ionized as lithium adducts (i.e., $[M+Li]^+$) in modest sensitivity in the positive-ion mode if lithium ions are present in the matrix [4]. The fragmentation pattern of lithiated Cer species ($[M+Li]^+$) has been well characterized [4], which is possibly used to identify the FA amide substituent and the sphingoid bases of the species [4]. Fragment ions specific to each Cer subclass are also observed [4]. Structural information can also be obtained from characterization of the sodium/potassium adducts of Cer species [5–7].

8.2.2 Negative-Ion Mode

Ceramides can form a variety of adducts with small anions present in the matrix (i.e., $[M+X]^-$, where $X=$ a kind of small anion) in the negative-ion mode [8]. Product-ion ESI-MS spectra of these adducted Cer species after low-energy CID display simply an X ion and an ion resulting from the loss of HX. Structural information related to the FA amine and/or sphingoid base is not available [8, 9]. However, it was showed that generation of the fragment ion yielding from the loss of HX is easier from Cer species containing α-hydroxy FA amide than their nonhydroxy-containing counterparts [8]. This feature is very useful for distinction of these Cer subtypes.

Ceramides form deprotonated ($[M-H]^-$) ions under basic conditions in the negative-ion mode with relatively lower sensitivity in comparison to the $[M+X]^-$ ions [10, 11]. Product-ion ESI-MS spectra of deprotonated ceramides after low-energy CID display too many modest to abundant fragment ions to be individually mentioned here [4, 8, 9, 12] (Figure 8.1a and b). Moreover, the fragmentation patterns of deprotonated α-hydroxy and nonhydroxy Cer species are essentially identical. But it can be readily recognized that the fragment-ion intensities resultant from α-hydroxy Cer species relative to the molecular ions are substantially different from those arising from nonhydroxy counterparts (Figure 8.1a and b). These features include the following:

- The most abundant fragment ion at $[M-H-256]^-$ arising from both nonhydroxy and α-hydroxy ceramides is approximately three times more intense in the spectra of the former than the latter when the intensities of both molecular ions are nearly equal.
- An abundant fragment ion at $[M-H-327]^-$, corresponding to acyl carbonyl anion, is only present in the spectra of deprotonated α-hydroxy species.
- However, the fragment ion at $[M-H-240]^-$, corresponding to the loss of 2-*trans*-palmitoleyl alcohol, is nearly equally intense to those ceramide species containing sphingosine.

Therefore, these characteristic features can be used to identify and quantify individual ceramide species containing sphingosine with or without presence of α-hydroxy group in FA amide [13].

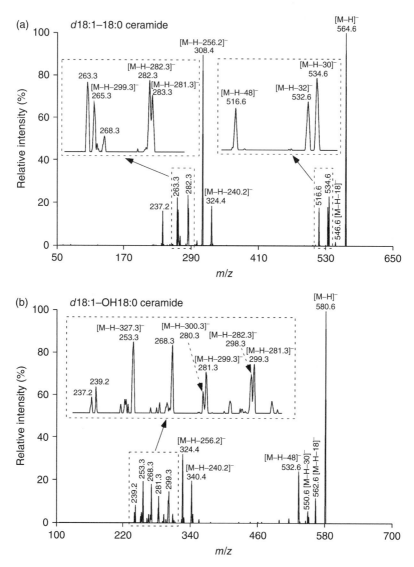

Figure 8.1 Representative product-ion ESI-MS spectra of deprotonated ceramide species in the negative-ion modes. Tandem MS analyses of deprotonated *d*18:1–18:0 (a) and *d*18:1–OH18:0 (b) ceramide species were performed on a QqQ mass spectrometer (Thermo Fisher TSQ Vantage). Collision activation was carried out at collision energy of 32 eV and gas pressure of 1 mTorr.

8.3 SPHINGOMYELIN

8.3.1 Positive-Ion Mode

Product-ion ESI-MS spectra of protonated SM species only display a fragment ion at m/z 184, corresponding to phosphocholine, after low-energy CID. The low-abundant fragment ions, corresponding to the loss of fatty acids present in the product-ion spectra of protonated PC species, are absent in the case of SM species. This is due to a fact that the amide bond in SM species is more stable than the ester bond in PC species.

SM species can be readily ionized as alkaline adducts in the positive-ion mode when alkaline ions are available in the matrix. Again, sodium adducts of SM species are readily formed if no other modifier(s) are added to the lipid solution. The fragmentation pattern of SM alkaline adducts is well studied after low-energy CID (Figure 6.4) [14, 15]. This pattern includes the following:

- A modest fragment ion at $[M+Alk-59]^+$ arising from the neutral loss of trimethylamine.
- An abundant fragment ion at $[M+Alk-183]^+$ arising from the neutral loss of phosphocholine.
- A modest fragment ion at $[M+Alk-(182+Alk)]^+$ arising from the neutral loss of phosphocholine alkaline.
- A modest fragment ion at $[M+Alk-(182+30)]^+$ arising from the neutral loss of phosphocholine plus a methyl aldehyde molecule.
- A low to moderate ion at $[M+Alk-(200+Alk)]^+$ arising from the further loss of a water molecule from the $[M+Alk-(182+Alk)]^+$ ion.
- A modest fragment ion at $[M+Alk-429]^+$ corresponding to the loss of the long-chain sphingoid base as a terminal conjugated diene from the $[M+Alk-(200+Alk)]^+$ ion, which can be used to determine the structure of the sphingoid base of the molecule and then derive the constituent of the FA amide.

8.3.2 Negative-Ion Mode

SM species can readily form anion adducts in the negative-ion mode. Product-ion ESI-MS spectra of anion adducts of SM species after low-energy CID only display a $[M-15]^-$ fragment ion, corresponding to the loss of a methyl group [14].

8.4 CEREBROSIDE

8.4.1 Positive-Ion Mode

In the positive-ion mode, cerebroside (HexCer) species including both GluCer and GalCer can be readily ionized as their proton or alkaline adducts (i.e., $[M+X]^+, X = H$,

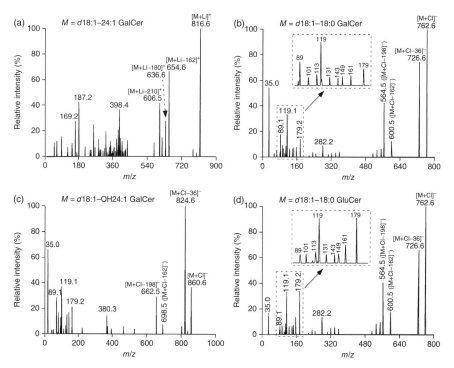

Figure 8.2   Representative product-ion ESI-MS spectra of lithiated and chloride-adducted cerebroside species in the positive- and negative-ion modes, respectively. Product-ion ESI-MS mass spectra of lithiated *d*18:1–24:1 galactosylceramide (GalCer) (a), and chloride-adducted *d*18:1–18:0 GalCer (b), *d*18:1–OH24:1 GalCer (c), and d18:1–18:0 GluCer (d) were acquired on a QqQ mass spectrometer (Thermo Fisher TSQ Vantage). Collision activation was carried out at collision energy of 50 eV in the positive-ion mode and of 30 eV in the negative-ion mode with gas pressure of 1 mTorr.

Li, Na, K), depending on the availability and affinity of the small cation(s) [12, 16, 17]. The fragmentation pattern of alkaline-adducted HexCer species yield three types of fragment ions [12] (Figure 8.2a):

- The ions related to the neutral loss of the hexosyl head group (three fragment ions are present at [M+Alk−162]⁺, [M+Alk−180]⁺, and [M+Alk−210]⁺, corresponding to the neutral loss of different hexosyl derivatives).

- The ions at [180+Alk]⁺ or [162+Alk]⁺ corresponding to the alkaline-adducted hexose derivatives.

- The fragment ions carrying structural information about the FA chains in the cerebrosides (e.g., an abundant ion at *m/z* 399, corresponding to a lithiated FA amide derivative, is present in the product-ion ESI-MS spectrum of the lithium adduct of *d*18:1–N24:1 galactosylceramide. Many other minor fragment ion peaks are also present in this region of the spectrum, corresponding to either the loss of the aliphatic substituent from the sphingoid base or the loss of FA amide derivatives. Yielding of these minor fragment ions was discussed in details in the original study [16].

The fragmentation pattern of alkaline-adducted cerebroside species containing an α-hydroxy group contains abundant fragment ions similar to those of nonhydroxy counterparts [16]. It should be pointed out that MS/MS analysis of GluCer species is unable to be distinguished from their isomeric GalCer counterparts although MS^n analysis of their alkaline adducts can be used to identify the hexose moiety [16].

Fragmentation of protonated cerebroside species (i.e., [M+H]$^+$) undergoes different processes under different CID energy conditions: resulting in ceramide-like fragment ion corresponding to the neutral loss of hexose moiety with a low-energy collision; and yielding protonated cleavage dehydrated sphingoid base ion after the neutral loss of both the hexose head group and the FA chain at higher CID energies [16, 17].

8.4.2 Negative-Ion Mode

Cerebroside species can be readily ionized as small anion adducts (i.e., [M+Y]$^-$, $Y =$ Cl, HCOO, etc.) in the negative-ion mode [12]. Chloride adducts of cerebrosides are formed if no other anionic modifier is present in the lipid solution. The fragmentation pattern of cerebroside chloride adducts (i.e., [M+Cl]$^-$) after low-energy CID yields three unique features (Figure 8.2b–d):

- An abundant fragment ion is present in the product-ion spectrum at [M+Cl−36]$^-$, corresponding to the loss of HCl. The presence of an ion at m/z 35 along with this feature confirms the chloride adduct of the molecular ion. The intensity of this [M+Cl−36]$^-$ fragment ion relative to the quasimolecular ion ([M+Cl]$^-$) resulting from chlorinated HexCer species containing an α-hydroxy moiety is much higher (over threefold) than that of nonhydroxy-containing cerebroside counterparts [12] (comparison between Figure 8.2b and c). This feature can be used to distinct the subclasses of cerebroside.

- An abundant fragment ion at [M+Cl−198]$^-$, corresponding to the loss of a hexosyl moiety, is present. The presence of a cluster of fragment ions between m/z 89 and 179 along with this 198 amu loss ion characterizes the hexosyl group of HexCer species. Intriguingly, the fragment ions between m/z 89 and 179 reflect the fragmentation of galactose or glucose anion at m/z 179, arising from the loss of 30 amu (formaldehyde) or 18 amu (water) or their combination to a different degree (insets of Figure 8.2b and d). It was found that the averaged intensity ratios between m/z 89 and 179 were 0.74 ± 0.10 and 4.8 ± 0.7 for GalCer (Figure 8.2b) and GluCer (Figure 8.2d) species, respectively [12]. Therefore, these ratios can be used to distinguish GalCer and GluCer isomers.

- A unique fragment ion, corresponding to the FA amide anion, is present in the product-ion ESI-MS spectrum of chlorinated HexCer species after low-energy CID, thereby identifying the FA amide substituent of HexCer species. For example, the fragment ion at m/z 282 or 380 is present in the product-ion mass spectra of chlorinated GalCer and GluCer species containing a stearoyl (N-18:0, Figure 8.2b and d) or α-hydroxy nervonoyl (N-OH24:1, Figure 8.2c) amide, respectively.

8.5   SULFATIDE

Sulfatide (3-sulfogalactosylceramide, ST) species can be readily ionized as deprotonated species (i.e., $[M-H]^-$) by ESI-MS in the negative-ion mode. The fragmentation pattern of deprotonated ST species after low-energy CID is well characterized [3, 18] (Figure 8.3a) and contains the following features:

- The product-ion mass spectrum displays a prominent ion at m/z 97, corresponding to the $HOSO_3^-$ ion.
- A cluster of modest to abundant fragment ions reflecting the ST head group (i.e., 3-sulfogalactosyl moiety) at m/z 259, 257, and 241 is present [18]. The m/z 259 and m/z 257 ions may represent the galactose 3-sulfate and galactono-1,5-lactone 3-sulfate anions [3]. Further loss of a water molecule from m/z 259 leads to the ion at m/z 241.
- The fragmentation of deprotonated ST species also yields many product ions containing information about the sphingoid base and FA amide substituent. The direct loss of the FA chain as a ketene from $[M-H]^-$ via the NH–CO bond cleavage results in the m/z 540 ion, which undergoes the loss of a water molecule to yield the m/z 522 ion. The m/z 540 ion can also lead to the m/z 300 ion, corresponding to 1-O-2′-aminoethenyl galactosyl 3-sulfate ion, probably via the combined losses of the sphingoid base as an aldehyde and H_2 as demonstrated previously. These ions are generally present in low abundance.

In addition to the common ions that identify sphingoid base, galactose, and fatty acid moieties, fragmentation of ST species consisting of an α-hydroxy FA substituent and a d18:1 sphingoid base (i.e., d18:1–OHFA ST) after low-energy CID yields some unique fragment ions [3, 18, 19] (Figure 8.3b). For example, prominent ions at m/z 540 and 522 and less abundant ions at m/z 507 are present in product-ion ESI-MS spectra of hydroxy-containing ST species. Moreover, the ion series produced from classical charge-remote fragmentation processes present in nonhydroxy-containing species are not observed in the hydroxy-containing species.

8.6   OLIGOGLYCOSYLCERAMIDE AND GANGLIOSIDES

Oligoglycosylceramides and gangliosides are ceramides with two or more O-linked sugar residues that may be modified by phosphate, sulfate, or various other groups.

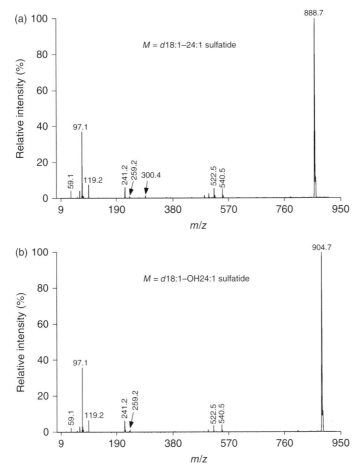

Figure 8.3 Representative product-ion ESI-MS spectra of sulfatide species in the negative-ion modes after low-energy CID. Product-ion ESI-MS mass spectra of deprotonated $d18:1–24:1$ (a) and $d18:1–OH24:1$ (b) ST species were acquired on a QqQ mass spectrometer (Thermo Fisher TSQ Vantage). Collision activation was carried out at collision energy of $65\,eV$ and gas pressure of $1\,mTorr$.

Tandem MS analysis of these glycosphingolipid species in the positive-ion mode after low-energy CID generally yields the neutral loss of the sugar residues with retention of the charge on the ceramide core [2, 20–22]. Therefore, structural information about the FA amide substituent or sphingoid base is unable to be derived from this type of analysis unless multistage tandem MS analysis is performed. Moreover, even when higher energy CID is performed for characterization of these complex glycosphingolipids either in the positive- or negative-ion mode, characterization of FA amide substituent or sphingoid base structures is still prohibited while insights into the glycoside composition can be obtained under the experimental condition. Description of this area of work is beyond the scope of the book. The readers who are interested in the fragmentation patterns of these complex compounds after high-energy CID could consult invaluable review or original research articles [2, 23, 24].

8.7 INOSITOL PHOSPHORYLCERAMIDE

Similar to PI species, inositol phosphorylceramide (IPC) species can be readily ion-ized to form [M−H]− in the negative-ion mode. Ionization of IPC species in the positive-ion mode could form [M+H]+ and [M+Li]+ under certain experimental conditions. However, ionization efficiency to form these ions is much lower in com-parison to that of [M−H]− [25].

The fragmentation pattern of IPC species in the negative-ion mode (i.e., [M−H]−) after low-energy CID is structural informative [25], including the following:

- The fragment ions at m/z 241 (corresponding to an inositol-1,2-cyclic phosphate anion) and at m/z 259 corresponding to an inositol monophosphate anion are prominent in this fragmentation pattern of IPC species, along with the presence of a modest ion at m/z 223 arising from m/z 241 via the further loss of a water molecule.

- Low-abundant fragment ions at [M−H−162]− and [M−H−180]−, arising from the losses of a dehydrated inositol and an inositol residue, respectively, are present.

- A low-abundant ion at [M−H−RCH=C=O]−, corresponding to the loss of the FA substituent as a ketene, is also present in the fragmentation pattern of IPC species. This fragment ion may be used to assign the FA amide constituent in the species as previously demonstrated [25].

8.8 SPHINGOLIPID METABOLITES

8.8.1 Sphingoid Bases

Sphingoid bases (mainly sphingosine and sphinganine) can be readily ionized as proton adducts in the positive-ion mode under acidic conditions (e.g., 0.1% of formic acid in chloroform–methanol (1:1, v/v)) [17, 26]. The fragmentation pattern of the protonated species (i.e., [M+H]+) of sphingosine after low-energy CID (e.g., 10–15 eV) contains three prominent product ions and numerous low-abundant fragment ions (Figure 8.4a).

- An abundant product ions at [M+H−18]+ corresponding to the nonspecific loss of one water molecule.

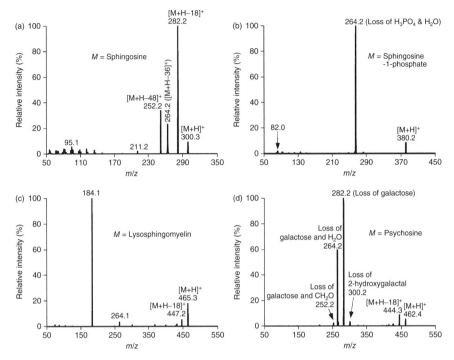

Figure 8.4 Representative product-ion ESI-MS spectra of sphingolipid metabolites in the positive-ion mode after low-energy CID. Product-ion ESI-MS analyses of protonated sphingosine (a), S1P (b), lysoSM (c), and psychosine (d) were performed at collision energy of 15, 24, 22, and 24 eV, respectively, by using a QqQ mass spectrometer (Thermo Fisher TSQ Vantage). Gas pressure of 1 mTorr was employed in the collisional activation.

- An abundant product ions at $[M+H-36]^+$ corresponding to the nonspecific loss of two water molecules.
- An abundant product ion arising from the neutral loss of 48.0 amu, corresponding to the losses of a water molecule and another formaldehyde molecule.

The fragmentation pattern of protonated sphinganine is quite similar to that of protonated sphingosine as obtained under identical experimental conditions as previously mentioned [26]. This is likely due to that the losses of water molecules are facile (occurred at very low collision energy) and nonspecific so that the structural difference between sphingosine and sphinganine does not significantly contribute to their fragmentation patterns. This observation also suggests that the fragmentation pattern of other related sphingoid bases be similar to that of protonated sphingosine under identical experimental conditions. The low-abundant fragment ion clusters arise from the fragmentation of aliphatic chain of sphingoid backbone likely via charge-remote fragmentation processes.

8.8.2 Sphingoid-1-Phosphate

Sphingoid-1-phosphate (S1P) species can be readily ionized as deprotonated ions (i.e., $[M-H]^-$) in the negative-ion mode. The fragmentation pattern of deprotonated S1P species after low-energy CID contains the following fragment ions [17, 27]:

- An intense characteristic ion at m/z 79, corresponding to $[PO_3]^-$.
- A low-abundant fragment ion at m/z 97, corresponding to $[H_2PO_4]^-$.
- A few very low-abundant fragment ions between m/z 200 and 260, corresponding to the losses of H_3PO_4 and one or more water molecules.

S1P species can also form protonated ions (i.e., $[M+H]^+$) in the positive-ion mode under neutral or acidic conditions. The fragmentation pattern of protonated sphingosine-1-phosphate after low-energy CID contains the following fragment ions (Figure 8.4b):

- An abundant ion at m/z 264, arising from the neutral loss of H_3PO_4 and water, which is probably because the resultant allylic cation is stabilized by an adjacent enamine.
- Some low-abundant fragment ions, corresponding to the fragmentation of aliphatic chain of sphingosine, in the low mass range of the product-ion spectrum, which are likely produced via charge-remote fragmentation processes.

The product-ion spectrum of protonated sphinganine-1-phosphate in the positive-ion mode that differs from the protonated sphingosine-1-phosphate is the presence of two abundant ions at m/z 284 and 266, instead of one at m/z 264 in the case of sphingosine-1-phosphate. The ion at m/z 284 probably arises from the neutral loss of H_3PO_4 while the m/z 266 ion results from the further loss of a water molecule from the ion at m/z 284.

8.8.3 Lysosphingomyelin

Lysosphingomyelin (lysoSM) can be readily ionized to form proton or alkaline adducts in the positive-ion mode. As with other choline-containing glycerophospholipids, product-ion mass spectra of protonated lysoSM species display the following fragment ions [26] (Figure 8.4c):

- A prominent fragment ion at m/z 184, corresponding to phosphocholine.
- A modest ion at m/z 447 and a few other very low-abundant ions in the region in the product-ion spectrum of protonated lysoSM containing a sphingosine base (m/z 465), corresponding to the losses of one or more water molecules.
- A low-abundant fragment ion at m/z 264, corresponding to the losses of phosphocholine and a water molecule.

8.8.4 Psychosine

Psychosine (i.e., galactosylsphingosine) can form the proton adduct under acidic conditions (e.g., in the presence of 0.1% formic acid in the infused solution) [26, 27]. Fragmentation of protonated psychosine after low-energy CID contains many structurally informative fragments (Figure 8.4d). These include the fragment ions corresponding to the losses of one (at m/z 444) or more water molecules. Other fragment ions in the spectrum including those at m/z 300 (low abundant), 282 (very abundant), 264 (abundant), and 252 (low abundant) are related to the loss of galactose, corresponding to the losses of 2-hydroxygalactal, galactose, galactose plus a water molecule, and galactose plus formaldehyde, respectively [27].

The sodium adduct of psychosine (i.e., [M+Na]$^+$) can also be formed with relatively low sensitivity compared to the proton adduct [27]. However, fragmentation of sodiated psychosine after low-energy CID yields abundant and informative product ions at m/z 467, 203, 185, 157, and 102 [27]. The ion at m/z 467 arises from the loss of NH$_3$ from [M+Na]$^+$. The one at m/z 203 corresponds to the sodiated galactose, which is probably generated from [M+Na]$^+$ via the formation of a highly favored six-membered transition state between the linking oxygen and the hydrogen on the allylic secondary alcohol. The ion at m/z 185 probably results from the loss of a water molecule from the m/z 203 ion. The ion at m/z 157 arises from the ring opening at the C5–O bond moving two electrons to the C1–O bond, yielding a carbonyl. The ion at m/z 102 probably results from three 1,3-hydrogen shifts followed by the loss of galactosyl sodium and a 1,3-diene, arising from a charge-remote fragmentation process.

REFERENCES

1. Merrill, A.H., Jr. (2011) Sphingolipid and glycosphingolipid metabolic pathways in the era of sphingolipidomics. Chem. Rev. 111, 6387–6422.

2. Levery, S.B. (2005) Glycosphingolipid structural analysis and glycosphingolipidomics. Methods Enzymol. 405, 300–369.

3. Hsu, F.F. and Turk, J. (2005) Analysis of sulfatides. In The Encyclopedia of Mass Spectrometry (Caprioli, R. M., ed.). pp. 473–492, Elsevier, New York

4. Hsu, F.-F., Turk, J., Stewart, M.E. and Downing, D.T. (2002) Structural studies on ceramides as lithiated adducts by low energy collisional-activated dissociation tandem mass spectrometry with electrospray ionization. J. Am. Soc. Mass Spectrom. 13, 680–695.

5. Gu, M., Kerwin, J.L., Watts, J.D. and Aebersold, R. (1997) Ceramide profiling of complex lipid mixtures by electrospray ionization mass spectrometry. Anal. Biochem. 244, 347–356.

6. Levery, S.B., Toledo, M.S., Doong, R.L., Straus, A.H. and Takahashi, H.K. (2000) Comparative analysis of ceramide structural modification found in fungal cerebrosides by electrospray tandem mass spectrometry with low energy collision-induced dissociation of Li+ adduct ions. Rapid Commun. Mass Spectrom. 14, 551–563.

7. Liebisch, G., Drobnik, W., Reil, M., Trumbach, B., Arnecke, R., Olgemoller, B., Roscher, A. and Schmitz, G. (1999) Quantitative measurement of different ceramide species from crude cellular extracts by electrospray ionization tandem mass spectrometry (ESI-MS/MS). J. Lipid Res. 40, 1539–1546.

8. Hsu, F.-F. and Turk, J. (2002) Characterization of ceramides by low energy collisional-activated dissociation tandem mass spectrometry with negative-ion electrospray ionization. J. Am. Soc. Mass Spectrom. 13, 558–570.

9. Zhu, J. and Cole, R.B. (2000) Formation and decompositions of chloride adduct ions. J. Am. Soc. Mass Spectrom. 11, 932–941.

10. Raith, K. and Neubert, R.H.H. (1998) Structural studies on ceramides by electrospray tandem mass spectrometry. Rapid Commun. Mass Spectrom. 12, 935–938.

11. Raith, K. and Neubert, R.H.H. (2000) Liquid chromatography-electrospray mass spectrometry and tandem mass spectrometry of ceramides. Anal. Chim. Acta 403, 295–303.

12. Han, X. and Cheng, H. (2005) Characterization and direct quantitation of cerebroside molecular species from lipid extracts by shotgun lipidomics. J. Lipid Res. 46, 163–175.

13. Han, X. (2002) Characterization and direct quantitation of ceramide molecular species from lipid extracts of biological samples by electrospray ionization tandem mass spectrometry. Anal. Biochem. 302, 199–212.

14. Han, X. and Gross, R.W. (1995) Structural determination of picomole amounts of phospholipids via electrospray ionization tandem mass spectrometry. J. Am. Soc. Mass Spectrom. 6, 1202–1210.

15. Hsu, F.F. and Turk, J. (2000) Structural determination of sphingomyelin by tandem mass spectrometry with electrospray ionization. J. Am. Soc. Mass Spectrom. 11, 437–449.

16. Hsu, F.F. and Turk, J. (2001) Structural determination of glycosphingolipids as lithiated adducts by electrospray ionization mass spectrometry using low-energy collisional-activated dissociation on a triple stage quadrupole instrument. J. Am. Soc. Mass Spectrom. 12, 61–79.

17. Sullards, M.C. (2000) Analysis of sphingomyelin, glucosylceramide, ceramide, sphingosine, and sphingosine 1-phosphate by tandem mass spectrometry. Methods Enzymol. 312, 32–45.

18. Hsu, F.-F., Bohrer, A. and Turk, J. (1998) Electrospray ionization tandem mass spectrometric analysis of sulfatide: Determination of fragmentation patterns and characterization of molecular species expressed in brain and in pancreatic islets. Biochim. Biophys. Acta 1392, 202–216.

19. Hsu, F.-F. and Turk, J. (2004) Studies on sulfatides by quadrupole ion-trap mass spectrometry with electrospray ionization: Structural characterization and the fragmentation processes that include an unusual internal galactose residue loss and the classical charge-remote fragmentation. J. Am. Soc. Mass Spectrom. 15, 536–546.

20. Fuller, M.D., Schwientek, T., Wandall, H.H., Pedersen, J.W., Clausen, H. and Levery, S.B. (2005) Structure elucidation of neutral, di-, tri-, and tetraglycosylceramides from high five cells: Identification of a novel (non-arthro-series) glycosphingolipid pathway. Glycobiology 15, 1286–1301.

21. Tsui, Z.C., Chen, Q.R., Thomas, M.J., Samuel, M. and Cui, Z. (2005) A method for profiling gangliosides in animal tissues using electrospray ionization-tandem mass spectrometry. Anal. Biochem. 341, 251–258.

22. Kaga, N., Kazuno, S., Taka, H., Iwabuchi, K. and Murayama, K. (2005) Isolation and mass spectrometry characterization of molecular species of lactosylceramides using liquid chromatography-electrospray ion trap mass spectrometry. Anal. Biochem. 337, 316–324.

23. Costello, C.E. and Vath, J.E. (1990) Tandem mass spectrometry of glycolipids. Methods Enzymol. 193, 738–768.

24. Guittard, J., Hronowski, X.L. and Costello, C.E. (1999) Direct matrix-assisted laser desorption/ionization mass spectrometric analysis of glycosphingolipids on thin layer chromatographic plates and transfer membranes. Rapid Commun. Mass Spectrom. 13, 1838–1849.

25. Hsu, F.F., Turk, J., Zhang, K. and Beverley, S.M. (2007) Characterization of inositol phosphorylceramides from Leishmania major by tandem mass spectrometry with electrospray ionization. J. Am. Soc. Mass Spectrom. 18, 1591–1604.

26. Jiang, X., Cheng, H., Yang, K., Gross, R.W. and Han, X. (2007) Alkaline methanolysis of lipid extracts extends shotgun lipidomics analyses to the low abundance regime of cellular sphingolipids. Anal. Biochem. 371, 135–145.

27. Jiang, X., Yang, K. and Han, X. (2009) Direct quantitation of psychosine from alkaline-treated lipid extracts with a semi-synthetic internal standard. J. Lipid Res. 50, 162–172.

9

FRAGMENTATION PATTERNS OF GLYCEROLIPIDS

9.1 INTRODUCTION

According to the classification of Lipid MAPS, the glycerolipids are the lipid species that can only yield glycerol, a sugar group, fatty acid(s), and/or alkyl variants after acid/base hydrolysis (see Chapter 1). Glycerolipids can be clarified into the classes of monoglyceride, diglycerides, triglycerides, and their glycosyl derivatives. These glycosyl derivatives are also generally called "glycolipids" as defined by the IUPAC [1].

Ether or vinyl ether-linked species in mono-, di-, and triglycerides are usually present in low abundance [2, 3]. Therefore, unless the fragmentation patterns of these ether-linked glycerides are well characterized, only the patterns of their ester-linked counterparts are presented in this chapter. Similarly, among glycolipid species, the patterns of glycosyl DAG species that are the most characterized are discussed in the chapter.

Mono-, di-, and triglyceride species cannot readily be ionized by protonation from nonaqueous solutions, even in the presence of organic acids (e.g., formic acid, acetic acid) under ESI conditions [4]. However, these species can be readily ionized as their adducts of ammonium, lithium, sodium, etc. (i.e., $[M+X]^+$, $X = NH_4$, Li, Na, ...) in the positive-ion mode [3–11]. Similar to the glycerophospholipids as discussed in Chapter 7, an appropriate modifier has to be added to the lipid solution to form one of these kinds of adducts. If more than one modifiers are present in the lipid solution, ionization as their adducts depends on the concentrations as well as their

Lipidomics: Comprehensive Mass Spectrometry of Lipids, First Edition. Xianlin Han.
© 2016 John Wiley & Sons, Inc. Published 2016 by John Wiley & Sons, Inc.

affinities with the glyceride species. If there is no modifier present in the solution, sodium adducts of these species are always formed due to the common existence of sodium ion in the system. Moreover, ionization of mono-, di-, and triglyceride species as small anion adducts in the negative-ion mode is insensitive relative to that of their alkaline adducts in the positive-ion mode.

Since a prominent charge site is absent in the glyceride species, the structure of FA substituent(s) significantly contributes to their ionization, particularly in the case of sodium adducts [10]. Therefore, the instrument response factors for ionization of these glyceride species as ammonium and alkaline adducts are apparently very different, depending on both the number of carbon atoms and of double bonds in the FA chains [10].

Derivatization has been widely used to enhance the ionization efficiency and selectivity of mono- and diglyceride species in order to be sensitively and specifically analyzed by ESI-MS [12–14]. The fragmentation features of the derivatized mono- and diglyceride species are mentioned in the chapter if they are available.

It should be borne in mind that the majority of glycolipid species with or without derivatization were characterized by many other ionization techniques prior to the utilization of ESI-MS. Those kinds of information are well documented in the books written by Murphy [15, 16], and Christie and Han [17]. The interested readers could consult these books for any further information.

9.2 MONOGLYCERIDE

Monoglyceride species can be ionized as alkaline and ammonium adducts in the positive-ion mode if one of these small cations is present in the lipid solution. Early studies showed that fragmentation of sodiated monoglyceride species is not preferable to that of ammoniated species [4]. Fragmentation of ammoniated monoglyceride species after low-energy CID yields the following fragment ions [4]:

- Two fragment ions at $[M+NH_4-17]^+$ and $[M+NH_4-35]^+$, corresponding to the sequential loss of ammonia and then water. (The absence of $[M+NH_4-18]^+$ fragment ion suggests that a loss of water prior to the loss of ammonia is not preferable in the dissociation process of ammoniated monoglyceride.)
- An abundant fragment ion arising from the FA substituent as an acylium ion (i.e., RCO^+).
- Numerous low-mass hydrocarbon or acylium ions resulting from carbon–carbon cleavages of the FA substituent as well as the ammonium ion at m/z 18.

9.3 DIGLYCERIDE

The fragmentation pattern of lithiated diglyceride species was characterized by ESI-MS/MS in the positive-ion mode [5] and contains the following fragment ions (Figure 9.1a):

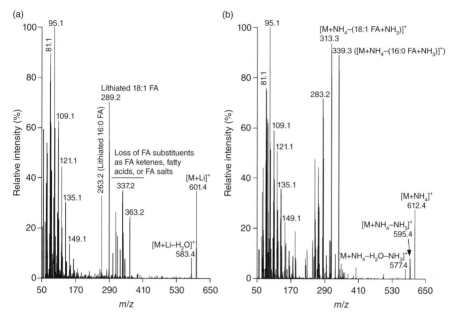

Figure 9.1 Representative product-ion ESI-MS analysis of lithiated and ammoniated dia-cylglycerol species after low-energy CID. Product-ion ESI-MS mass spectra of lithiated and ammoniated 1-16:0–2-18:1 DAG species (a and b, respectively) were acquired on a QqQ mass spectrometer (Thermo Fisher TSQ Vantage). Collision activation was carried out at collision energy of 45 and 38 eV, respectively, and gas pressure of 1 mTorr.

- A most abundant product ion at $[M+Li-18]^+$, corresponding to the loss of H_2O from the molecular ion (i.e., $[M+Li]^+$)
- A complex cluster of abundant product ions around m/z 350, corresponding to the losses of FA substituents as a ketene, as a fatty acid, and as a lithium salt as well as lithiated FA substituents. (The complexity of this cluster of fragment ions depends on the FA constituent(s) of individual DAG species. Three product ions are resulted from DAG species containing identical FA substituents while three different pairs of product ions are yielded from the species containing different FA substituents. Regiospecificity of 1,2- and 1,3-DAG isomers was not elucidated from the fragmentation pattern.)
- Another complex cluster of fragment ions at m/z less than 200, corresponding to the fragments of FA chains.

Product-ion ESI-MS mass spectra of ammonium DAG adducts (i.e., $[M+NH_4]^+$) after low-energy CID display a structurally informative fragmentation pattern (Figure 9.1b) [4, 11]:

- A prominent fragment ion at $[M+NH_4-17]^+$, corresponding to the loss of an ammonia molecule.

- A fragment ion at $[M+NH_4-35]^+$, corresponding to the sequential loss of ammonia and then water.
- A cluster of abundant fragment ions resulting from the loss of free fatty acid(s) plus an ammonia molecule (i.e., $[M+NH_4-(RCOOH+NH_3)]^+$) or other combinations are present. (However, the ammonia-loss ion is generally the base peak. This is likely due to the preferable loss of the proton along with the carboxyl moiety from the α-hydrogen atom of the adjacent FA chain as occurred in the fragmentation process of many lipids in the positive-ion mode including PC and TAG species [8, 18] (see Chapter 7).)

A recent study showed different fragmentation patterns of ammoniated alkyl- and alkenyl-containing diglyceride species between each other and from their diacyl DAG counterparts [3]. These distinct fragmentation features enabled the researchers to identify individual diglyceride species directly from biological lipid extracts (e.g., mouse heart and brain) utilizing an MDMS-SL strategy [3].

The introduction of a quaternary ammonium cation to the hydroxyl moieties of DAG species, using N-chlorobetainyl chloride, was reported [12]. This derivatization not only affords an increased ionization efficiency of two orders compared to their underivatized sodium adducts but also makes the ionization efficiency independent of FA chains present in DAG species because of the existence of a fixed charge site after derivatization. The fragmentation pattern of N-chlorobetainyl-derivatized DAG species is essentially identical to that of PC species (see Chapter 7).

The hydroxyl moieties of DAG species were also derivatized with 2,4-difluorophenyl isocyanate [13] through which a polar head group was introduced. The derivatized DAG species were readily ionized by ESI-MS as ammonium adducts (i.e., $[M+NH_4]^+$) in the positive-ion mode in the presence of ammonium salt (e.g., ammonium acetate or ammonium formate) in the solution [13]. The fragmentation pattern of the derivatized DAG species contains a most abundant ion at $[M+NH_4-190]^+$, corresponding to the neutral loss of difluorophenyl carbamic acid plus ammonia from the derivatized head group. The pattern also shows intense fragment ion(s) arising from the loss of FA substituents plus ammonia (i.e., $[M+NH_4-(R_xCOOH+NH_3)]^+$, $x = 1, 2$). Collisional activation of ammonium adducts of the derivatized 1,3- and 1,2-DAG isomers yielded essentially identical product ions. However, by using a normal-phase column, derivatized 1,2- and 1,3-DAG isomers can be well separated as demonstrated [13], thereby identifying these isomers.

Very recently, the hydroxyl moieties of DAG species have been derivatized with dimethylglycine (DMG) [14] by using an established derivative method [19]. The derivatized DAG species was characterized as both protonated and lithiated adducts in the positive-ion mode [14].

Product-ion mass spectra of protonated DMG–DAG after low-energy CID displayed only a predominant fragment ion peak at $[M+H-103]^+$, corresponding to the neutral loss of DMG from the derivatized group. The product-ion ESI-MS/MS

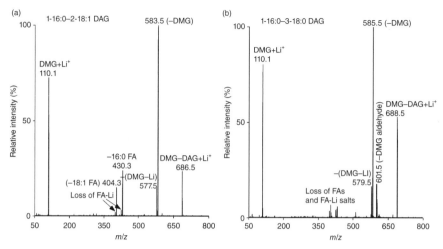

Figure 9.2 Representative product-ion ESI-MS analyses of lithiated DMG–DAG species after low-energy CID. Diacylglycerol species were derivatized with *N,N*-dimethylglycine (DMG) in the presence of *N*-(3-dimethylaminopropyl)-*N'*-ethylcarbodiimide (EDC) hydrochloride and dimethylaminopyridine (DMAP) as previously described [14, 19]. Product-ion mass spectra of DMG-derivatized 1-16:0–2-18:1 (a) and 1-16:0–3-18:0 (b) DAG species as their lithium adducts were acquired on a QqQ mass spectrometer (Thermo Fisher TSQ Vantage). Collision activation was carried out at collision energy of 35 eV and gas pressure of 1 mTorr.

analysis of protonated DMG–DAG did not yield any information about FA substituents and their regioselective positions present in DAG species.

The fragmentation pattern of lithium adducts of DMG–DAG species showed unique features of structural information including 1,2- and 1,3-isomers (Figure 9.2) [14] as follows:

- An abundant fragment ion at *m/z* 110, corresponding to lithiated DMG.

- An abundant product ion, corresponding to the neutral loss of 103 amu (i.e., DMG).

- A moderate abundance fragment ion, corresponding to the neutral loss of 109 amu (i.e., DMG-Li salt) from all examined DMG-DAG species including 1,2- and 1,3-isomers.

- A unique fragment ion at $[M+Li-87]^+$, corresponding to the loss of DMG as an aldehyde, from the DMG-derivatized 1,3-DAG isomers, which allows one to selectively identify 1,3-DAG species through NLS of 87 amu of lithiated DMG–DAG species.

- The fragment ions at $[M+Li-R_xCOOH]^+$ and $[M+Li-R_xCOOLi]^+$ ($x = 1, 2$ or 1, 3, respectively), corresponding to the loss of individual FA substituents as FA acids or their lithium salts, respectively.

The intensity of the fragment ion corresponding to the loss of sn-1 FA was higher than that of sn-2 FA, allowing to identify the regioisomers of 1,2-DAG species. Moreover, the intensities of the fragment ions corresponding to the loss of FA-Li salt from 1,2-DAG ions were in much lower abundance than that resulted from the loss of FA from the molecular ions. However, this pair of fragment ions was present in similar intensities in product-ion mass spectra of lithiated 1,3-DMG–DAG species (Figure 9.2). The differences between the losses of sn-1 and sn-2 FA substituents of 1,2-DAG species and between the 1,2- and 1,3-DAG species is due to the differential fragmentation pathways as previously described [18].

9.4 TRIGLYCERIDE

The fragmentation pattern of the lithium adducts of TAG species after low-energy CID contains the following fragment ions (Figure 6.6):

- A cluster ions corresponding to the losses of fatty acyl substituents as fatty acids (i.e., $[M+Li-R_xCOOH]^+$, where $x = 1$, 2, and 3).
- A cluster ions corresponding to the losses of fatty acyl substituents as their lithium salts (i.e., $[M+Li-R_xCOOLi]^+$, where $x = 1$, 2, and 3).
- A cluster of low-abundant acylium ions (i.e., R_xCO^+, where $x = 1$, 2, and 3) around m/z 300.

Generation of these fragment ions involves the initial elimination of a free fatty acid with an α-hydrogen atom from the adjacent FA chain, followed by formation of a cyclic intermediate that decomposes to yield other characteristic fragment ions. The fragmentation pattern of sodium or silver adducts of TAG species is similar to that of lithiated TAG species [4, 20–22].

Obviously, these fragmentation patterns allow the determination of FA substituents of TAG species. However, determination of the positions of FA substituents from the pattern is not straightforward from lithium or other TAG adducts [20], particularly from those of TAG ions present in the biological samples. Sodium adducts of TAG species were shown to be useful for determination of regioisomers and offered marginally better quantitative results [20].

Product-ion ESI-MS mass spectra of ammoniated TAG species (i.e., $[M+NH_4]^+$) after low-energy CID display multiple abundant ions, which are informative for structural elucidation [4]:

- An abundant fragment ion at $[M+NH_4-17]^+$ (generally not the base peak) corresponding to the loss of ammonia from the ammoniated TAG species.
- The fragment ion(s) at $[M+NH_4-(R_xCOOH+17)]^+$, arising from the loss of free fatty acid(s) plus an ammonia molecule (usually the base peak(s)).

The number of these ions present in the spectra depends on the number of different FA substituents in each species. Mechanistic studies of fragmentation processes demonstrated that the α-hydrogen atom of the adjacent FA chain facilitates the loss of free fatty acid(s) plus an ammonia molecule [8, 11].

Product-ion MS analysis of ether-containing triglyceride as ammonium adducts after low-energy CID was also performed in the positive-ion mode [2]. Essentially an identical fragmentation pattern of ammoniated ether-containing triglyceride species to ammoniated TAG species was obtained. The only difference was the presence of a fragment ion corresponding to the loss of FA alcohol plus ammonia in the former in replacement of the loss of fatty acid plus ammonia.

9.5 HEXOSYL DIACYLGLYCEROL

Large amounts of monoglycoglycerolipids (including hexosyl diacylglycerol (HexDAG)) are present in plants and some bacteria [17, 21]. Similar to cerebroside (see Chapter 8), although HexDAG species do not contain a permanent charge site in the molecule, the presence of hexosyl moiety makes these lipid species readily ionizable in the positive-ion mode as their adducts of ammonium, lithium, and sodium (i.e., $[M+X]^+$, $X = NH_4$, Li, Na) [22–24].

The fragmentation pattern of lithiated HexDAG (as representative by mono-galactosyl diacylglycerol (MGDG)) after low-energy CID is relatively simple (Figure 9.3a), containing two very abundant fragment ions corresponding to the losses of FA substituents (i.e., $[M+Li-R_xCOOH]^+$, $x = 1$, 2). Identification of regioisomers from this pair of ions remains warrant. A few low-abundant ions corresponding to the lithiated hexose derivatives (e.g., m/z 227) are also present.

The fragmentation pattern of ammoniated MGDG after low-energy CID contains much more informative structural information (Figure 9.3b):

- A low fragment ion at $[M+NH_4-17]^+$, corresponding to the loss of an ammonia molecule.
- A modest fragment ion at $[M+NH_4-35]^+$, corresponding to the sequential loss of ammonia and then water.
- A pair of ions arising the further loss of a hexose ring (i.e., 162 amu) from the $[M+NH_4-17]^+$ and $[M+NH_4-35]^+$ ions, respectively.
- A pair of ions at $[M+NH_4-17-Hex-R_xCOOH]^+$ ($x = 1$, 2), respectively, resulting from the further loss of FA substituents from $[M+NH_4-17-Hex]^+$.

Fragmentation of sodiated HexDAG species after low-energy CID by using a triple quadrupole mass spectrometer led to an additional modest fragment ion at m/z 243, corresponding to sodiated hexose [23]. However, by using an ion-trap mass spectrometer, product-ion MS analysis of ammoniated HexDAG displayed fragment ions corresponding to the neutral loss of the hexose moiety and an additional water loss, whereas fragmentation of sodiated counterparts yielded fragment ions corresponding to the neutral loss of the FA substituents [25]. The origin of the fragments from either the *sn*-1 or *sn*-2 position might be deduced from the different signal intensities of the various fragment ions [25]. Fragments resulting from the neutral loss of the FA chain at *sn*-1 position have in general more intense than those resulting from the neutral loss of the FA chain at the *sn*-2 position [26, 27].

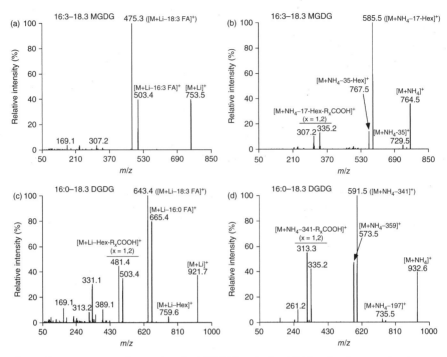

Figure 9.3 Representative product-ion ESI-MS analysis of lithiated and ammoniated mono- and digalactosyl diacylglycerol species after low-energy CID. Product-ion ESI-MS mass spectra of lithiated and ammoniated 1-16:3–2-18:3 MGDG species (a and b, respectively), and lithiated and ammoniated 1-16:0–2-18:3 DGDG species (c and d, respectively) were acquired on a QqQ mass spectrometer (Thermo Fisher TSQ Vantage). Collision activation was carried out at collision energy of 35, 14, 50, and 20 eV, respectively, and gas pressure of 1 mTorr. Hex stands for hexose.

9.6   OTHER GLYCOLIPIDS

Tandem MS characterization of different alkaline adducts of the DGDG species (i.e., $[M+Alk]^+$, Alk = Li, Na, and K) from the bacterium *Bacillus pumilus* was performed by using an ion-trap mass spectrometer to compare the fragmentation patterns [22]. The study indicates that the intensity of fragments and the dissociation pathways depend on the alkaline adduct. Specifically, the basic structures of DGDG species can be obtained from the fragmentation pattern of their sodium adducts, yielding intense fragment ions corresponding to the losses of FA substituents as fatty acids (i.e., $[M+Na-R_xCOOH]^+$, $x = 1, 2$).

The lithium adducts of DGDG species after low-energy CID give rise to an identical fragmentation pattern to that of sodiated DGDG species (Figure 9.3c):

- A moderate abundance fragment ion corresponding to the neutral loss of 162 amu (i.e., hexose derivative) from lithiated DGDG.

- One or two prominent fragment ions (one of them is the base peak) at $[M+Li-R_xCOOH]^+$ $(x = 1, 2)$ corresponding to the losses of FA substituents as FA acids, respectively.
- One or two abundant fragment ions at $[M+Li-Hex-R_xCOOH]^+$ $(x = 1, 2)$ corresponding to the sequential losses of FA substituents as FA acids, respectively, from $[M+Li-Hex]^+$.
- A cluster of fragment ions at the low-mass region corresponding to lithiated hexose, lithiated dihexose, lithiated dihexose glyceride, etc.

Characterization of regioisomers from their sodium or lithium adducts was complicated and not achievable. The product-ion MS analysis of potassium adducts of DGDG species showed a different fragmentation pathway from their lithium or sodium counterparts. Specifically, in addition to the fragment ions corresponding to the losses of FA substituents as fatty acids (i.e., $[M+K-R_xCOOH]^+$, $x = 1, 2$), a new fragment ion corresponding to $[M+K-R_1COOH-CH_2]^+$ was also present. The fragmentation pathway corresponding to this ion was not well understood [22].

The fragmentation pattern of ammonium adducts of digalactosyl DAG (DGDG) species from oat kernels (i.e., $[M+NH_4]^+$) was characterized after low-energy CID [24]. The product-ion spectra of these species display the following fragment ions (Figure 9.3d):

- A pair of low-abundant ions at $[M+NH_4-180]^+$ and $[M+NH_4-197]^+$, corresponding to the loss of ammoniated monogalactose minus a water molecule and ammoniated monogalactose.
- Two very abundant fragment ions (usually the base peak) at $[M+NH_4-341]^+$ and $[M+NH_4-359]^+$, corresponding to the neutral loss of ammoniated digalactose minus a water molecule and ammoniated digalactose, respectively.
- One or two very abundant fragment ions corresponding to the protonated FA substituent(s) esterified to glycerol minus H_2O (resulting from the losses of FA substituents as a FA acid from the ion at $[M+NH_4-341]^+$) depending on whether the DGDG species contain an identical FA chain or two different FA chains, respectively.

These fragment ions determine the FA substituents in the species [24]. Characterization of DGDG regioisomers was not conducted in the study.

Glycolipids can be complicated by the presence of more than two sugar rings as well as FA substituents containing mono-, di-, and triestolides [28]. The ammoniated trigalactosyl and tetragalactosyl species yield the counterparts of ammoniated digalactose minus H_2O and ammoniated digalactose in the case of DGDG [24]. Product-ion spectra of ammoniated estolide-containing glycolipid species display an additional cluster of fragment ions for an estolide moiety, resulting from the neutral loss of either a fatty acid or a FA ketene from the esterified fatty acid in estolide linkage [24].

Sulfonoglycolipids, especially sulfoquinovosyl DAG, are a class of modified glycolipids. This class of lipids is also present in plants and some bacteria. These lipids can be ionized either in the positive-ion mode as $[M-H+2Na]^+$ or in the negative-ion mode as $[M-H]^-$ [29]. Product-ion spectra of the sodiated species display abundant fragment ions corresponding to the neutral loss of either a FA ketene or a fatty acid [30]. The arising fragment ions can further lose a fatty acid or a FA ketene. Product-ion spectra of deprotonated ions in the negative-ion mode after low-energy CID display abundant FA carboxylate anion(s), reflecting the FA substituents in the species [31, 32]. Moreover, the product-ion spectra of all sulfonoglycolipid species in the negative-ion mode show a characteristic fragment ion at m/z 225, corresponding to a dehydrosulfoglycosyl anion.

Characterization of this type of lipid classes and molecular species as well as novel lipid classes with multistage tandem MS was also conducted by using a high-resolution ion trap-Orbitrap mass spectrometer [33]. Any readers who are interested in understanding the fragmentation features of these complex lipids (particularly those novel ones) with high-resolution mass spectrometry should consult the paper.

REFERENCES

1. (a) IUPAC-IUB (1978) Nomenclature of Lipids. Biochem. J. 171, 21–35.
   (b) IUPAC-IUB (1978) Nomenclature of Lipids. Chem. Phys. Lipids 21, 159–173.
   (c) IUPAC-IUB (1977) Nomenclature of Lipids. Eur. J. Biochem. 79, 11–21.
   (d) IUPAC-IUB (1977) Nomenclature of Lipids. Hoppe-Seyler's Z. Physiol. Chem. 358, 617–631.
   (e) IUPAC-IUB (1978) Nomenclature of Lipids. J. Lipid Res. 19, 114–128.
   (f) IUPAC-IUB (1977) Nomenclature of Lipids. Lipids 12, 455–468.
   (g) IUPAC-IUB (1977) Nomenclature of Lipids. Mol. Cell. Biochem. 17, 157–171.

2. Bartz, R., Li, W.H., Venables, B., Zehmer, J.K., Roth, M.R., Welti, R., Anderson, R.G., Liu, P. and Chapman, K.D. (2007) Lipidomics reveals that adiposomes store ether lipids and mediate phospholipid traffic. J. Lipid Res. 48, 837–847.

3. Yang, K., Jenkins, C.M., Dilthey, B. and Gross, R.W. (2015) Multidimensional mass spectrometry-based shotgun lipidomics analysis of vinyl ether diglycerides. Anal. Bioanal. Chem. 407, 5199–5210.

4. Duffin, K.L., Henion, J.D. and Shieh, J.J. (1991) Electrospray and tandem mass spectrometric characterization of acylglycerol mixtures that are dissolved in nonpolar solvents. Anal. Chem. 63, 1781–1788.

5. Hsu, F.F., Ma, Z., Wohltmann, M., Bohrer, A., Nowatzke, W., Ramanadham, S. and Turk, J. (2000) Electrospray ionization/mass spectrometric analyses of human promonocytic U937 cell glycerolipids and evidence that differentiation is associated with membrane lipid composition changes that facilitate phospholipase A2 activation. J. Biol. Chem. 275, 16579–16589.

6. Callender, H.L., Forrester, J.S., Ivanova, P., Preininger, A., Milne, S. and Brown, H.A. (2007) Quantification of diacylglycerol species from cellular extracts by electrospray ionization mass spectrometry using a linear regression algorithm. Anal. Chem. 79, 263–272.

7. Cheng, C., Gross, M.L. and Pittenauer, E. (1998) Complete structural elucidation of triacylglycerols by tandem sector mass spectrometry. Anal. Chem. 70, 4417–4426.

8. Hsu, F.-F. and Turk, J. (1999) Structural characterization of triacylglycerols as lithiated adducts by electrospray ionization mass spectrometry using low-energy collisionally activated dissociation on a triple stage quadrupole instrument. J. Am. Soc. Mass Spectrom. 10, 587–599.

9. Han, X., Abendschein, D.R., Kelley, J.G. and Gross, R.W. (2000) Diabetes-induced changes in specific lipid molecular species in rat myocardium. Biochem. J. 352, 79–89.

10. Han, X. and Gross, R.W. (2001) Quantitative analysis and molecular species fingerprinting of triacylglyceride molecular species directly from lipid extracts of biological samples by electrospray ionization tandem mass spectrometry. Anal. Biochem. 295, 88–100.

11. Murphy, R.C., James, P.F., McAnoy, A.M., Krank, J., Duchoslav, E. and Barkley, R.M. (2007) Detection of the abundance of diacylglycerol and triacylglycerol molecular species in cells using neutral loss mass spectrometry. Anal. Biochem. 366, 59–70.

12. Li, Y.L., Su, X., Stahl, P.D. and Gross, M.L. (2007) Quantification of diacylglycerol molecular species in biological samples by electrospray ionization mass spectrometry after one-step derivatization. Anal. Chem. 79, 1569–1574.

13. Leiker, T.J., Barkley, R.M. and Murphy, R.C. (2011) Analysis of diacylglycerol molecular species in cellular lipid extracts by normal-phase LC-electrospray mass spectrometry. Int. J. Mass Spectrom. 305, 103–109.

14. Wang, M., Hayakawa, J., Yang, K. and Han, X. (2014) Characterization and quantification of diacylglycerol species in biological extracts after one-step derivatization: A shotgun lipidomics approach. Anal. Chem. 86, 2146–2155.

15. Murphy, R.C. (1993) Mass Spectrometry of Lipids. Plenum Press, New York. pp 290.

16. Murphy, R.C. (2015) Tandem Mass Spectrometry of Lipids: Molecular Analysis of Complex Lipids. Royal Society of Chemistry, Cambridge, UK. pp 280.

17. Christie, W.W. and Han, X. (2010) Lipid Analysis: Isolation, Separation, Identification and Lipidomic Analysis. The Oily Press, Bridgwater, England. pp 448.

18. Hsu, F.F. and Turk, J. (2009) Electrospray ionization with low-energy collisionally activated dissociation tandem mass spectrometry of glycerophospholipids: Mechanisms of fragmentation and structural characterization. J. Chromatogr. B 877, 2673–2695.

19. Jiang, X., Ory, D.S. and Han, X. (2007) Characterization of oxysterols by electrospray ionization tandem mass spectrometry after one-step derivatization with dimethylglycine. Rapid Commun. Mass Spectrom. 21, 141–152.

20. Herrera, L.C., Potvin, M.A. and Melanson, J.E. (2010) Quantitative analysis of positional isomers of triacylglycerols via electrospray ionization tandem mass spectrometry of sodiated adducts. Rapid Commun. Mass Spectrom. 24, 2745–2752.

21. Dormann, P. and Benning, C. (2002) Galactolipids rule in seed plants. Trends Plant Sci. 7, 112–118.

22. Wang, W., Liu, Z., Ma, L., Hao, C., Liu, S., Voinov, V.G. and Kalinovskaya, N.I. (1999) Electrospray ionization multiple-stage tandem mass spectrometric analysis of diglycosyldiacylglycerol glycolipids from the bacteria *Bacillus pumilus*. Rapid Commun. Mass Spectrom. 13, 1189–1196.

23. Welti, R., Wang, X. and Williams, T.D. (2003) Electrospray ionization tandem mass spectrometry scan modes for plant chloroplast lipids. Anal. Biochem. 314, 149–152.

24. Moreau, R.A., Doehlert, D.C., Welti, R., Isaac, G., Roth, M., Tamura, P. and Nunez, A. (2008) The identification of mono-, di-, tri-, and tetragalactosyl-diacylglycerols and their natural estolides in oat kernels. Lipids 43, 533–548.

25. Ibrahim, A., Schutz, A.L., Galano, J.M., Herrfurth, C., Feussner, K., Durand, T., Brodhun, F. and Feussner, I. (2011) The alphabet of galactolipids in *Arabidopsis thaliana*. Front. Plant Sci. 2, 95.

26. Guella, G., Frassanito, R. and Mancini, I. (2003) A new solution for an old problem: The regiochemical distribution of the acyl chains in galactolipids can be established by electrospray ionization tandem mass spectrometry. Rapid Commun. Mass Spectrom. 17, 1982–1994.

27. Napolitano, A., Carbone, V., Saggese, P., Takagaki, K. and Pizza, C. (2007) Novel galactolipids from the leaves of Ipomoea batatas L.: Characterization by liquid chromatography coupled with electrospray ionization-quadrupole time-of-flight tandem mass spectrometry. J. Agric. Food Chem. 55, 10289–10297.

28. Fahy, E., Subramaniam, S., Brown, H.A., Glass, C.K., Merrill, A.H., Jr., Murphy, R.C., Raetz, C.R., Russell, D.W., Seyama, Y., Shaw, W., Shimizu, T., Spener, F., van Meer, G., VanNieuwenhze, M.S., White, S.H., Witztum, J.L. and Dennis, E.A. (2005) A comprehensive classification system for lipids. J. Lipid Res. 46, 839–861.

29. Ishizuka, I. (1997) Chemistry and functional distribution of sulfoglycolipids. Prog. Lipid Res. 36, 245–319.

30. Sassaki, G.L., Gorin, P.A., Tischer, C.A. and Iacomini, M. (2001) Sulfonoglycolipids from the lichenized basidiomycete *Dictyonema glabratum*: Isolation, NMR, and ESI-MS approaches. Glycobiology 11, 345–351.

31. Cedergren, R.A. and Hollingsworth, R.I. (1994) Occurrence of sulfoquinovosyl diacylglycerol in some members of the family Rhizobiaceae. J. Lipid Res. 35, 1452–1461.

32. Basconcillo, L.S., Zaheer, R., Finan, T.M. and McCarry, B.E. (2009) A shotgun lipidomics approach in Sinorhizobium meliloti as a tool in functional genomics. J. Lipid Res. 50, 1120–1132.

33. Jensen, S.M., Brandl, M., Treusch, A.H. and Ejsing, C.S. (2015) Structural characterization of ether lipids from the archaeon Sulfolobus islandicus by high-resolution shotgun lipidomics. J. Mass Spectrom. 50, 476–487.

10

FRAGMENTATION PATTERNS OF FATTY ACIDS AND MODIFIED FATTY ACIDS

10.1 INTRODUCTION

Fatty acid (FA) represents a broad family of lipids containing at least one carboxylic acid group and a long aliphatic chain [1], which includes nonesterified and modified FA species. The nonesterified FA species are generally referred as to those species that are not modified through an enzymatic or nonenzymatic process. These species vary in chain length (i.e., number of carbon atoms), the number of double bonds, the locations of these double bonds on the acyl chains, and with or without branched methyl group(s). The different locations of the double bonds form the isomers of an unsaturated FA for which the chain length and the number of double bonds are identical. Modification of these nonesterified FA (NEFA) species through enzymatic or nonenzymatic processes yields a huge number of modified FA species. Oxidized (e.g., eicosanoids, docosanoids, FA species containing hydroxylated or expoxylated group(s)), nitrosylated, and halogenated FAs are some examples of modified FA [1–3].

Although determination of the number of carbon atoms and the number of double bonds of individual NEFA species can be readily achieved once the *m/z* of a FA ion is recorded. Therefore, characterization of FA structure is to determine the location of double bonds for straight chain FA species or the locations of both double bonds and branch chains for FA species containing branch chains. Note that the location of double bonds within a polyunsaturated FA in the majority of those found in mammals is interrelated and always interrupted between each other with a methylene group. For instance, for a FA containing 20 carbon atoms and 4 double bonds, the distribution

Lipidomics: Comprehensive Mass Spectrometry of Lipids, First Edition. Xianlin Han.
© 2016 John Wiley & Sons, Inc. Published 2016 by John Wiley & Sons, Inc.

of the double bonds is either in Δ5,6, Δ8,9, Δ11,12, and Δ14,15 positions (i.e., 5,8,11,14-eicosatetraenoic acid, 20:4(n-6) FA, herein *m:n* was used to denote the FA or FA chain containing *m* of carbon atoms and *n* of double bonds) or in Δ8,9, Δ11,12, Δ14,15, and Δ17,18 positions (i.e., 8,11,14,17-eicosatetraenoic acid, 20:4(n-3) FA). Due to this specific feature of unsaturated FAs, FA double-bond isomers can be distinguished by the location of the first double bond. Although exceptions are present (e.g., conjugated linoleic acids are present in beef and dairy products), these exceptions are only present in a minimal amount [4].

However, definitive determination of the location of each double bond is challenging in the current lipidomics field when the isomers are present as a mixture from biological samples. This is due to that the collision energy, which can be provided with a collision cell in an ESI mass spectrometer, is generally low, although LC separation possesses the power to resolve these isomers under certain conditions [5]. Further challenging is the presence of a large number of modified FA species, the majority of which are present in low to very low abundance and existing in isomers.

To solve these complications, numerous efforts have been made to characterize individual FA species with or without derivatization. In this chapter, these ESI-MS-based efforts are summarized and discussed. As is the case with other lipid classes, prior to the utilization of ESI-MS, FA species including those of modified ones were well characterized by many other ionization techniques after derivatization in general. Those kinds of information are well documented in the book written by Christie and Han [6], which could be consulted if needed.

10.2   NONESTERIFIED FATTY ACID

10.2.1   Underivatized Nonesterified Fatty Acid

10.2.1.1   Positive-Ion Mode   Identifying the position(s) of double bond(s) of unsaturated long-chain FA species was performed in the positive-ion mode from dilithium adducts (i.e., $[M-H+2Li]^+$) [7, 8]. The product-ion ESI mass spectra produced from monounsaturated FA species are relatively simple to interpret with the fragment ions located at the high end of m/z, which are produced from the cleavage of the single bond on the carboxylate side of the C–C double bond. A similar fragmentation pattern is also observed from polyunsaturated FA species with an intense fragment ion (usually the most intense fragment ion) yielded from the chain cleavage adjacent to the terminal (methyl end) double bond while other fragment ions produced from cleavage of each C–C single bond between the terminal double bond and the carboxylate group are also present in the product ion mass spectra of these species.

It should be recognized that following the same line of reasoning, low-energy MS^n analysis has also been used to identify the location of double bond(s) of TAG [7] and glycerophospholipids [9]. However, this type of work currently belongs more to the research laboratory than to the practical lipidomics application.

10.2.1.2   Negative-Ion Mode   The fragmentation pattern of deprotonated NEFA species in the negative-ion mode (i.e., $[M-H]^-$) contains a characteristic ion peak at $[M-H-44]^-$, corresponding to the loss of carbon dioxide, from the majority of unsaturated FA species, and at $[M-H-18]^-$, corresponding to the loss of a water molecule from saturated and some of the monounsaturated FA species after low-energy CID [10, 11]. The intensity of this characteristic ion peak depends on the structures of NEFA species and collision conditions (see Section 4.6) [11]. It is an unfortunate that this fragmentation pattern does not provide any direct information about the location of NEFA double bonds.

However, Yang and colleagues [11] recognized that the variation of the intensity distribution of the characteristic fragment ion corresponding to the loss of carbon dioxide or water with varying CID conditions (i.e., either collision energy or collision gas pressure (see Section 4.6)) and that this distribution pattern can be used to distinct the NEFA isomers. This is due to the fact that the intensity profiles of these fragment ions over collision conditions are distinct for different isomers (Figure 10.1). It was found that the intensity distribution of a characteristic fragment ion over a CID parameter is mainly dependent on the distance of the double bonds to carbonyl

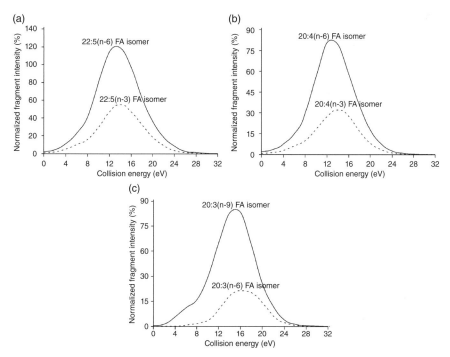

Figure 10.1 Representative intensity distributions of the fragment ions from different paired FA isomers as varied with collision energy. Product-ion ESI analysis of individual FA isomer in the negative-ion mode was performed by varying collision energy at fixed collision gas pressure of 1 mTorr. The intensity distributions of the fragment ions resulting from the loss of CO_2 from 22:5 (a), 20:4 (b), or 20:3 (c) FA isomers as varied with collision energy were displayed after normalization to the intensity of its corresponding molecular ion in the full mass spectrum (i.e., normalized absolute intensity).

as well as on the number of double bonds (Figure 10.2). The distance between the double bonds and carbonyl can be represented with the location of the first double bond from the carboxylate side in polyunsaturated FA species because the double bonds that are present in the majority of the naturally occurring polyunsaturated FA species are almost always interrupted with one methylene group. The underlying mechanism of the altered fragment-ion intensity distribution in FA isomers with varying CID conditions is likely due to the differential interactions between the negative charge carried with the fragment ion and its interactions with electron densities present in the double bonds. The differential interactions are due to the different distance between the terminal charge-carrying carbonyl group and the double bonds in FA isomers and can differentially stabilize the charges in the resultant fragment ions. These differences lead to different fragment-ion intensities, which allow one to distinguish the FA isomers from each other.

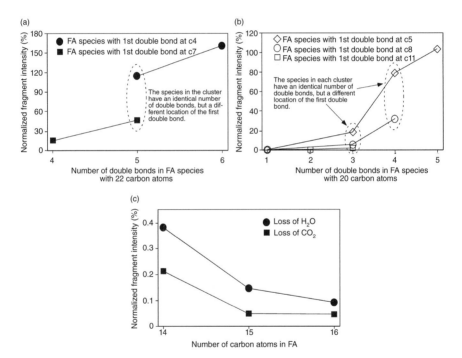

Figure 10.2 The effects of the location of the first double bond at the carboxylate side, the number of the double bonds, and the chain length of FA isomers on the fragment-ion intensities of FA species. The normalized absolute intensities of fragment ions were determined at collision energy of 12 eV and collision gas pressure of 1 mTorr. (a) The effects of the number of double bonds and the location of the first double bond on the fragment-ion intensities of FA species containing 22 carbon atoms; (b) the effects of the number of double bonds and the location of the first double bond on the fragment-ion intensities of FA species containing 20 carbon atoms; and (c) the effects of the FA chain length on the fragment ion intensity of FA species containing one double bond at the $\Delta9,10$ position. Yang et al. [11]. Reproduced with permission of the American Chemical Society.

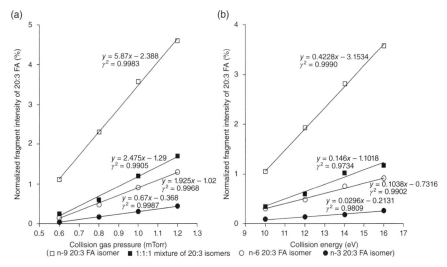

Figure 10.3 Linearity of the intensity distribution of the fragment ion with variation of collision gas pressure or collision energy. Two-dimensional mass spectrometric analysis of 20:3(n-9) FA (open square), 20:3(n-6) FA (open circle), 20:3(n-3) FA (closed circle), or equimolar mixture of 20:3(n-9/6/3) FA isomers (closed square) was performed by varying either collision gas pressure or collision energy in the negative-ion mode. The intensity distribution of the fragment ion resulting from the loss of CO_2 in the low region of either collision gas pressure (a) or collision energy (b) shows similar patterns.

Moreover, by exploiting the intensity distribution difference(s) between the fragment ions, the composition of the identified FA isomers can be quantified with external calibration curves, which are determined with authentic FA isomers [11]. The linear calibration curves could be determined as the relative intensity of a specific fragment ion to the molecular ion with either variation of collision gas pressure at a fixed collision energy or variation of collision energy at a fixed collision gas pressure. The composition of a mixture of FA isomers then can be determined from these linear curves as described [11]. Figure 10.3 shows an example of these kinds of linear curves for determination of the compositions of any 20:3 FA isomeric mixtures. It is important to note that this approach can be extended to identify and quantify the double-bond isomers of FA chains present in individual GPL species of biological samples with multistage tandem mass spectrometry [11].

10.2.2 Derivatized Nonesterified Fatty Acid

10.2.2.1 Off-Line Derivatization MS identification of double-bond location in fatty acyl chains of lipids has also been performed with off-line derivatization followed by product-ion analyses of the derivatized lipids (see [12] for a recent review). For example, Moe and colleagues [13] pretreated GPL species and NEFA species with osmium tetroxide to generate hydroxyl-containing GPL species at the initial site(s) of unsaturation. The dihydroxylated lipids were analyzed with product-ion

MS analyses to locate the position of the initial double bond [14]. The fragmentation of these derivatized fatty acids largely depends on the derivatizing reagents [12].

10.2.2.2   Online Derivatization (Ozonolysis)   Quantitative conversion of olefinic bonds to ozonides after ozone treatment of a dry, thin film of GPL species is achievable [15]. ESI-MS/MS of these adducts in either positive- or negative-ion mode leads to dissociation of the ozonide moiety and thus yields fragment ions uniquely identifying the double-bond position. This success leads to the online ozone treatment, identifying the location of double bonds by ESI-MS/MS. In this approach, lipid species react with ozone that is present in the source gas as part of the electrospray process. The ozone gas could be generated in the ESI source [16] or supplied from an ozone generator [17]. While this technology has its advantages for identification of the double-bond locations present in individual lipid species or simple lipid mixtures, interpretation of the mass spectra resulted from complex mixtures is exceedingly difficult. This is due to the fact that ozone-induced fragments are often isobaric with other lipid ions, and the assignment of fragments to their respective precursor ions becomes ambiguous as the complexity of the mixture increases. To overcome this difficulty, a small amount of ozone could be spiked into the collision cell, thereby identifying individually selected molecular ion of interest from a complex mixture [17]. This approach could be used to identify the double-bond locations of FA chains present in complex lipid species such as TAG species [12].

10.3   MODIFIED FATTY ACID

Numerous modified NEFA species are present in nature, including oxygenated, nitrosylated, halogenated species, etc. All these modifications are generated through enzymatic or nonenzymatic reactions and play many essential roles in biological systems. A well representative and well-studied category of modified NEFA species is those metabolized from arachidonic acid and generally called as eicosanoids. All the eicosanoid species are different in structure. Collision dissociation of underivatized eicosanoid molecular ion species yields unique rearrangement reactions and even breaks carbon–carbon bonds to generate abundant ion products. Therefore, tandem MS analysis of these ions could allow us to elucidate these different structures and identify the species present in biological samples if they are present in a single component, which usually requires effective chromatographic separation.

Since the product-ion ESI mass spectrum of individual eicosanoid species displays its own fragmentation characteristics, a common fragmentation pattern of these eicosanoids does not exists. Therefore, it is too complicated to describe the product-ion mass spectrum of individual eicosanoid species. Moreover, the complication is further increased due to that eicosanoid species can be ionized in both positive- and negative-ion modes as protonated and deprotonated species (i.e., $[M+H]^+$ and $[M-H]^-$), respectively, although the majority of the applications of

ESI for eicosanoid analysis has been in the negative-ion mode [18]. In this section, a few examples of CID mass spectra of eicosanoid species are given, whereas general description of these CID spectra is referred to an excellent review article written by Dr Murphy and colleagues [18].

For example, a panel of eicosanoid species carrying one hydroxyl group (i.e., hydroxyeicosatetraenoic acids (HETEs)) are oxidized metabolites of arachidonic acid, the species of which with the hydroxyl group located at carbon atoms 5, 8, 9, 10, 11, 12, 13, 15, 16, 17, 18, 19, and 20 were determined [19]. All of these species are readily ionized as isomeric carboxylate ions (i.e., $[M-H]^-$) at m/z 319 by ESI. The structural features of individual HETE species (i.e., the location of the hydroxyl group and its relationship to the double bonds) lead to yielding unique fragment ions after CID, which often allows unambiguous determination of these isomers [19, 20]. Figure 10.4 shows a few examples of the product-ion ESI mass spectra of 5-, 8-, 9-, 11-, 12-, and 15-HETE species. These spectra that are essentially identical to those previously reported under the experimental conditions [18, 21] show the differential fragmentation of individual species. More complicated spectra can also be obtained under other conditions [22].

The fragmentation spectrum of deprotonated 5-HETE (i.e., 5-hydroxy-6,8,11,14-eicosatetraenoic acid) displays very abundant fragment ions at m/z 301, 257, 203, and 115 (Figure 10.4a). These ions correspond to the loss of water, water plus carbon dioxide, the fragment ions resulted from the breakage of the C5–C6 bond and rearrangement (i.e., the negative charge resides at the either side of the fragments), respectively, as described [18].

Fragment ions of m/z 301, 257, 163, and 155 are yielded from deprotonated 8-HETE (8-hydroxy-5,9,11,14-eicosatetraenoic acid) after CID (Figure 10.4b). Identically, the fragment ions at m/z 163 and 155 are resulted from the breakage of the C8–C9 bond [18].

The product-ion spectrum of 9-HETE (9-hydroxy-5,7,11,14-eicosatetraenoic acid) also displays the ions due to the loss of either water or carbon dioxide (Figure 10.4c). The typical pair of fragment ions at m/z 151 and 167 due to the breakage of the C9–C10 bond is present in low abundance. Moreover, there exist two low-abundant ions, one at m/z 123 corresponding to the further loss of carbon dioxide from the m/z 167 ion and the other at m/z 179, which likely yields from the cleavage of the C8–C9 bond with rearrangement.

The fragment ions arising from deprotonated 11-HETE (11-hydroxy-5,8,12,14-eicosatetraenoic acid) are simple (Figure 10.4d). There are only two fragment ions at m/z 301 and 167. The former arises from the loss of water from the molecular ion and the latter yields from the cleavage of the C10–C11 bond as previously described [18].

The fragmentation pattern of deprotonated 12-HETE (12-hydroxy-5,8,10,14-eicosatetraenoic acid) contains six abundant product ions at m/z 301, 257, 207, 179, 163, and 135 (Figure 10.4e). The first two ions correspond to the loss of water and water plus carbon dioxide. The m/z 207 ion yields from the cleavage of the C12–C13 bond through formation of aldehyde at the end (C12 site). This ion further loses a

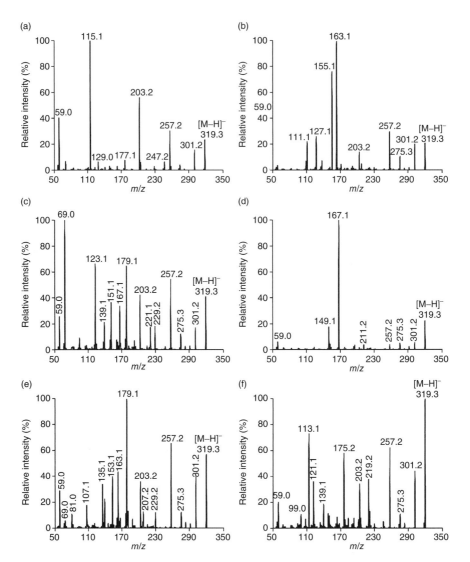

Figure 10.4   Representative product-ion ESI mass spectra of deprotonated HETE species in the negative-ion mode after low-energy CID. Product-ion analyses of 5-HETE (a), 8-HETE (b), 9-HETE (c), 11-HETE (d), 12-HETE (e), and 15-HETE (f) were performed in the negative-ion mode at collision energy of 23 eV and collision gas pressure of 1 mTorr on a QqQ mass spectrometer (Thermo Fisher TSQ Vantage).

carbon dioxide molecule to yield the abundant ion at m/z 163. The very abundant ion at m/z 179 arises from the breakage of C11–C12 bond. This ion further loses a carbon dioxide molecule to yield the abundant ion at m/z 135.

Dissociation of deprotonated 15-HETE (15-hydroxy-5,8,11,13-eicosatetraenoic acid) species after CID yields five abundant product ions at m/z 301, 275 257, 219, and 175 (Figure 10.4f). The first three ions are due to the loss of water, carbon dioxide,

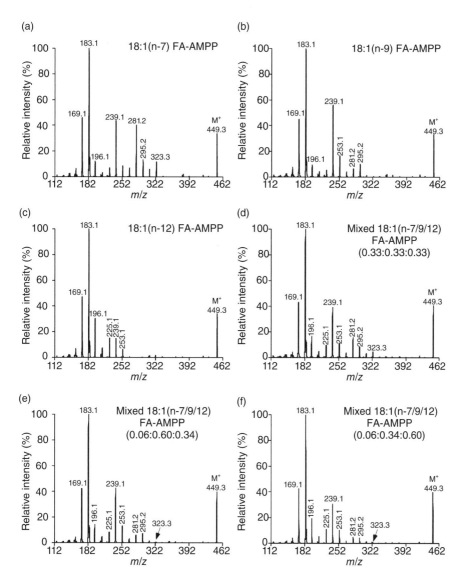

Figure 10.5 Product-ion ESI mass spectra of 18:1 FA isomers and their mixtures after derivatized with AMPP. Product-ion analyses of derivatized 18:1(n-7) (a), 18:1(n-9) (b), and 18:1(n-12) (c) FA isomers, and their 1:1:1 (d), 0.06:0.60:0.34 (e), and 0.06:0.34:0.60 (f) (n-7/9/12) mixtures were performed in the positive-ion mode at collision energy of 40 eV and collision gas pressure of 1 mTorr. The majority of the abundant fragment ions after charge-remote fragmentation with AMPP can be assigned as previously described [28].

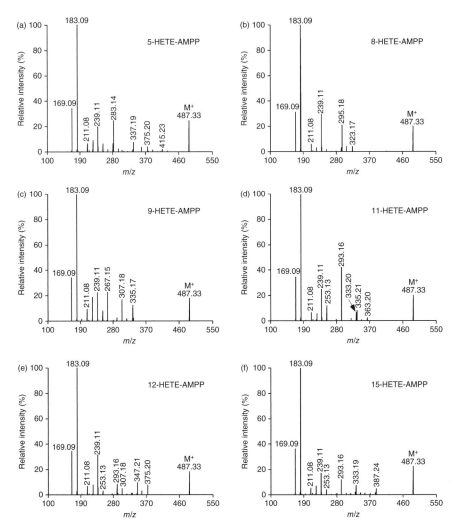

Figure 10.6 Demonstration of fatty acidomics with distinct fragmentation patterns of hydroxyeicosatetraenoic acid (HETE) isomers after derivatization with AMPP for structural determination of these isomers alone or in mixtures.

and water plus carbon dioxide, respectively. The m/z 219 ion arises from the breakage of C14–C15 bond. This ion further loses a carbon dioxide molecule to generate the ion at m/z 175.

10.4   FATTY ACIDOMICS

Charge-remote fragmentation has been well recognized as an effective approach for dissociation of long aliphatic chains [23, 24]. By exploiting this approach, structural identification and quantification of all lipid species containing a carboxylic acid including saturated with or without branch(es), unsaturated, and modified FA species or even other complex species (e.g., retinoic acid and bile acids) by ESI-MS/MS have been achieved after one-step derivatization with a charge-carried reagent through an amidation reaction [25–28]. A variety of charge-carried reagents with variation of hydrophobicity, charge strength, and distance from the charge to the carboxyl group were tested for this purpose.

It was demonstrated that all of these derivatized lipid species containing a carboxylic acid yield a common fragmentation pattern in the positive-ion mode after low-energy CID. This pattern contains a few, very abundant, common fragment ions arising from the derivatized reagent in addition to the signatures generated from the aliphatic component of individual lipid species. These common fragment ions can be used to selectively and sensitively monitor the carboxylic acid-containing lipid species of interest by using either LC-MS [25, 26] or shotgun lipidomics [28, 29]. The enhanced fragment signatures can be used not only to identify the structure of the species but also to determine the composition of isomeric species after simulation of the signature obtained from a mixture with the signatures of individual isomers [28]. Therefore, this success for global identification and quantification of individual lipid species containing a carboxylic acid group was referred as to "fatty acidomics" [28].

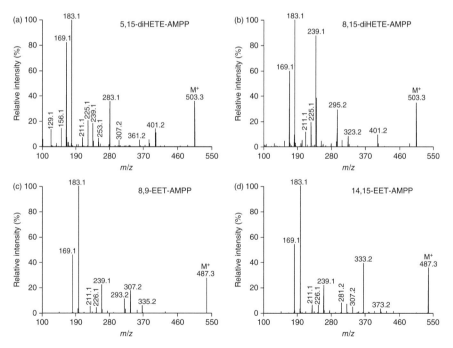

Figure 10.7 Demonstration of fatty acidomics with distinct fragmentation patterns of the dihydroxyeicosatetraenoic acid (diHETE) isomers or of the expoxyeicosatetraenoic acid (EET) isomers after derivatization with AMPP for structural determination of these isomers alone or in mixtures.

For example, distinct fragmentation patterns of 18:1 FA isomers are yielded after derivatized with *N*-(4-aminomethylphenyl)pyridinium (AMPP) or other derivatizing reagents followed by ESI-MS/MS analysis in the positive-ion mode (Figure 10.5a–c). These distinct fragmentation patterns can be used to determine the composition of individual 18:1 isomer in any of their mixtures through simulation of the fragmentation signature resulted from the isomeric ion of a mixture without any chromatographic separation (Figure 10.5d–f). This identification and quantification approach can be readily extended to any other FA isomeric mixtures as long as the distinct fragmentation patterns of isomeric FA species exist after derivatization.

The following are some examples of the distinct fragmentation patterns of different groups of isomeric FA species by using fatty acidomics. Figures 10.6 and 10.7 show some representative, distinct fragmentation patterns of oxidized arachidonic acid (20:4 FA) isomeric species (i.e., isomeric eicosanoids). Very different fragmentation patterns of nitrosylated FA species are shown in Figure 10.8a and b, likely due to the involvement of nitrosyl group in the fragmentation process in addition to the charge-remote fragmentation. By using fatty acidomics, the unique fragmentation pattern of the branched, saturated FA species (e.g., phytanic acid) allows one not

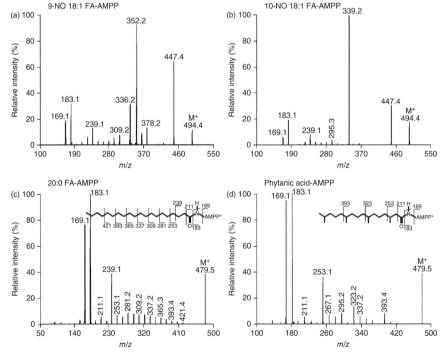

Figure 10.8 Demonstration of fatty acidomics with distinct fragmentation patterns of nitrosylated 18:1 FA isomers and saturated FA species with or without branches after derivatization with AMPP for structural determination.

only to readily identify the location of the methyl branches but also to determine the possible existence of any unbranched isomeric FA species (Figure 10.8c and d). Collectively, as a powerful addition to lipidomics tools, fatty acidomics could be widely used to identify and quantify the lipid species containing a carboxylic acid group, thereby greatly accelerating the identification of the biochemical mechanisms underlying numerous pathological conditions.

REFERENCES

1. Fahy, E., Subramaniam, S., Brown, H.A., Glass, C.K., Merrill, A.H., Jr., Murphy, R.C., Raetz, C.R., Russell, D.W., Seyama, Y., Shaw, W., Shimizu, T., Spener, F., van Meer, G., VanNieuwenhze, M.S., White, S.H., Witztum, J.L. and Dennis, E.A. (2005) A comprehensive classification system for lipids. J. Lipid Res. 46, 839–861.
2. Buczynski, M.W., Dumlao, D.S. and Dennis, E.A. (2009) Thematic review series: Proteomics. An integrated omics analysis of eicosanoid biology. J. Lipid Res. 50, 1015–1038.
3. Guichardant, M., Chen, P., Liu, M., Calzada, C., Colas, R., Vericel, E. and Lagarde, M. (2011) Functional lipidomics of oxidized products from polyunsaturated fatty acids. Chem. Phys. Lipids 164, 544–548.
4. Wahle, K.W., Heys, S.D. and Rotondo, D. (2004) Conjugated linoleic acids: Are they beneficial or detrimental to health? Prog. Lipid Res. 43, 553–587.
5. Borch, R.F. (1975) Separation of long chain fatty acids as phenacyl esters by high pressure liquid chromatography. Anal. Chem. 47, 2437–2439.
6. Christie, W.W. and Han, X. (2010) Lipid Analysis: Isolation, Separation, Identification and Lipidomic Analysis. The Oily Press, Bridgwater, England. pp 448.
7. Hsu, F.-F. and Turk, J. (1999) Structural characterization of triacylglycerols as lithiated adducts by electrospray ionization mass spectrometry using low-energy collisionally activated dissociation on a triple stage quadrupole instrument. J. Am. Soc. Mass Spectrom. 10, 587–599.
8. Hsu, F.-F. and Turk, J. (2008) Elucidation of the double-bond position of long-chain unsaturated fatty acids by multiple-stage linear ion-trap mass spectrometry with electrospray ionization. J. Am. Soc. Mass Spectrom. 19, 1673–1680.
9. Hsu, F.F. and Turk, J. (2008) Structural characterization of unsaturated glycerophospholipids by multiple-stage linear ion-trap mass spectrometry with electrospray ionization. J. Am. Soc. Mass Spectrom. 19, 1681–1691.
10. Kerwin, J.L., Wiens, A.M. and Ericsson, L.H. (1996) Identification of fatty acids by electrospray mass spectrometry and tandem mass spectrometry. J. Mass Spectrom. 31, 184–192.
11. Yang, K., Zhao, Z., Gross, R.W. and Han, X. (2011) Identification and quantitation of unsaturated fatty acid isomers by electrospray ionization tandem mass spectrometry: A shotgun lipidomics approach. Anal. Chem. 83, 4243–4250.
12. Mitchell, T.W., Pham, H., Thomas, M.C. and Blanksby, S.J. (2009) Identification of double bond position in lipids: From GC to OzID. J. Chromatogr. B 877, 2722–2735.
13. Moe, M.K., Anderssen, T., Strom, M.B. and Jensen, E. (2004) Vicinal hydroxylation of unsaturated fatty acids for structural characterization of intact neutral phospholipids by negative electrospray ionization tandem quadrupole mass spectrometry. Rapid Commun. Mass Spectrom. 18, 2121–2130.

14. Moe, M.K., Strom, M.B., Jensen, E. and Claeys, M. (2004) Negative electrospray ion-ization low-energy tandem mass spectrometry of hydroxylated fatty acids: A mechanistic study. Rapid Commun. Mass Spectrom. 18, 1731–1740.

15. Harrison, K.A. and Murphy, R.C. (1996) Direct mass spectrometric analysis of ozonides: application to unsaturated glycerophosphocholine lipids. Anal. Chem. 68, 3224–3230.

16. Thomas, M.C., Mitchell, T.W. and Blanksby, S.J. (2006) Ozonolysis of phospholipid dou-ble bonds during electrospray ionization: A new tool for structure determination. J. Am. Chem. Soc. 128, 58–59.

17. Thomas, M.C., Mitchell, T.W., Harman, D.G., Deeley, J.M., Murphy, R.C. and Blanksby, S.J. (2007) Elucidation of double bond position in unsaturated lipids by ozone electrospray ionization mass spectrometry. Anal. Chem. 79, 5013–5022.

18. Murphy, R.C., Barkley, R.M., Zemski Berry, K., Hankin, J., Harrison, K., Johnson, C., Krank, J., McAnoy, A., Uhlson, C. and Zarini, S. (2005) Electrospray ionization and tan-dem mass spectrometry of eicosanoids. Anal. Biochem. 346, 1–42.

19. Yue, H., Strauss, K.I., Borenstein, M.R., Barbe, M.F., Rossi, L.J. and Jansen, S.A. (2004) Determination of bioactive eicosanoids in brain tissue by a sensitive reversed-phase liquid chromatographic method with fluorescence detection. J. Chromatogr. B 803, 267–277.

20. Puppolo, M., Varma, D. and Jansen, S.A. (2014) A review of analytical methods for eicosanoids in brain tissue. J. Chromatogr. B 964, 50–64.

21. Nakamura, T., Bratton, D.L. and Murphy, R.C. (1997) Analysis of epoxyeicosatrienoic and monohydroxyeicosatetraenoic acids esterified to phospholipids in human red blood cells by electrospray tandem mass spectrometry. J. Mass Spectrom. 32, 888–896.

22. Masoodi, M., Eiden, M., Koulman, A., Spaner, D. and Volmer, D.A. (2010) Comprehen-sive lipidomics analysis of bioactive lipids in complex regulatory networks. Anal. Chem. 82, 8176–8185.

23. Wysocki, V.H. and Ross, M.M. (1991) Charge-remote fragmentation of gas-phase ions: Mechanistic and energetic considerations in the dissociation of long-chain functionalized alkanes and alkenes. Int. J. Mass Spec. 104, 179–211.

24. Cheng, C. and Gross, M.L. (2000) Applications and mechanisms of charge-remote frag-mentation. Mass Spectrom. Rev. 19, 398–420.

25. Bollinger, J.G., Thompson, W., Lai, Y., Oslund, R.C., Hallstrand, T.S., Sadilek, M., Ture-cek, F. and Gelb, M.H. (2010) Improved sensitivity mass spectrometric detection of eicosanoids by charge reversal derivatization. Anal. Chem. 82, 6790–6796.

26. Bollinger, J.G., Rohan, G., Sadilek, M. and Gelb, M.H. (2013) LC/ESI-MS/MS detec-tion of FAs by charge reversal derivatization with more than four orders of magnitude improvement in sensitivity. J. Lipid Res. 54, 3523–3530.

27. Yang, K., Dilthey, B.G. and Gross, R.W. (2013) A shotgun lipidomics approach using charge switch derivatization: Analysis of fatty acid double bond isomers. J. Am. Soc. Mass Spectrom. 24(S1), 228.

28. Wang, M., Han, R.H. and Han, X. (2013) Fatty acidomics: Global analysis of lipid species containing a carboxyl group with a charge-remote fragmentation-assisted approach. Anal. Chem. 85, 9312–9320.

29. Yang, K., Dilthey, B.G. and Gross, R.W. (2013) Identification and quantitation of fatty acid double bond positional isomers: A shotgun lipidomics approach using charge-switch derivatization. Anal. Chem. 85, 9742–9750.

11

FRAGMENTATION PATTERNS OF OTHER BIOACTIVE LIPID METABOLITES

11.1 INTRODUCTION

Metabolism/catabolism of each class of lipids involves multiple steps, and, therefore, generates many corresponding intermediates (or metabolites). The majority of these metabolites play essential roles (e.g., serving as lipid second messengers, being the biosynthesis building blocks, involving energy metabolism) in biological systems. Thus, it is important to accurately identify and quantify these metabolites. However, these lipid intermediates are generally present in low to very low abundance. Moreover, many of them are either nonpolar, or labile, or very active so that these metabolites are either unionizable by ESI-MS, or readily degraded, or easily reacted with other biological compounds once generated, respectively. Therefore, these factors make the analysis of these intermediates in a large scale and high throughput (i.e., in a lipidomics approach) very challenging.

In the last few chapters of this part, fragmentation features of many lipid metabolites are discussed or specified along with their parent molecules. For example, fragmentation patterns of lysophospholipid species are given in Chapter 7, characterization of sphingolipid metabolites is summarized in Chapter 8, DAG and MAG species are discussed in Chapter 9, Chapter 10 is about the metabolites and their modifiers containing a carboxylic acid. In this chapter, only those classes of lipid intermediates that are not mentioned in the earlier chapters and of which characterization by ESI-MS/MS was conducted are summarized. Their biological origins and functions are also overviewed.

Lipidomics: Comprehensive Mass Spectrometry of Lipids, First Edition. Xianlin Han.
© 2016 John Wiley & Sons, Inc. Published 2016 by John Wiley & Sons, Inc.

As pointed out previously, prior to the utilization of ESI-MS for lipid analysis, other ionization techniques were used to characterize many of these metabolites with or without derivatization. These techniques are also briefly mentioned in the chapter. Moreover, numerous review articles summarized the results from the studies of these compounds are available [1–3], which could be consulted for the advanced readers.

11.2   ACYLCARNITINE

Acylcarnitines are essential compounds for the metabolism of fatty acids and represent intermediates of mitochondrial fatty acid β-oxidation. In this process, fatty acids are first activated to form acyl CoAs in the cytosol of cells (see Section 11.3), then the acyl moieties are transferred to carnitine by carnitine palmitoyl transferase I (CPT-I), which is located at the outer mitochondrial membrane. The formed acyl-carnitines are largely and selectively transported into the mitochondria for fatty acid β-oxidation to generate ATP through coordinating activities of CPT-I and CPT-II. The latter is located at the inner mitochondrial membrane and converts acylcarnitines back to acyl CoAs.

The intermediates (medium and short-chain acyl CoAs) produced from β-oxidation could also be reversely transported out through the same machinery to form medium and short-chain acylcarnitines. The unselected long-chain, medium, and short-chain acylcarnitines can be spilled out from cells, leading to the presence of acylcarnitines in blood stream. Therefore, accumulation of unselected long-chain acylcarnitine species represents the increased mitochondrial FA β-oxidation, but accumulation of medium and short-chain acylcarnitines may suggest mitochondrial dysregulation/dysfunction to a certain degree.

Acylcarnitine species can readily form protonated ion (i.e., $[M+H]^+$) in the positive-ion mode. Product-ion ESI mass spectra of protonated acylcarnitine species after low-energy CID display an informative fragmentation pattern for structural elucidation (Figure 11.1a), which includes the following:

- A prominent product ion at m/z 85, corresponding to protonated γ-crotono-lactone resulting from the loss of an acyl group and trimethylamine from the molecular ion.
- A modest ion at m/z 341 corresponding to the loss of trimethylamine (i.e., $[M+H-59]^+$).
- An ion corresponding to protonated fatty acid (usually at very low abundance).
- Acylium ion (i.e., RCO^+).
- An ion arising from the loss of the FA substituent.

When acylcarnitine species contain FA substituents possessing a hydroxyl group (usually at α-position), the product-ion mass spectra are more complicated than those of acylcarnitine counterparts in the absence of a hydroxyl group. In this case, multiple additional fragment ions are present in the tandem mass spectra, arising from the loss of water from different fragment ions as previously demonstrated [4].

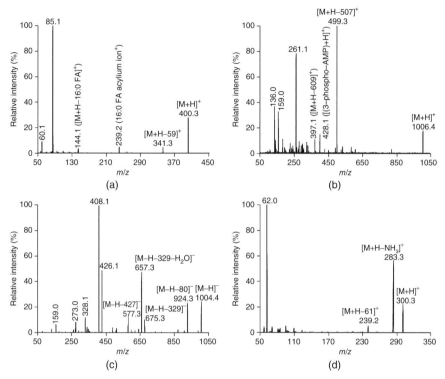

Figure 11.1 Representative product-ion mass spectrometric analyses of protonated acylcarnitine, protonated acyl CoA, deprotonated acyl CoA, and protonated N-acyl ethanolamine. Product-ion ESI-MS analyses of protonated palmitoylcarnitine (a), protonated palmitoyl CoA (b), deprotonated palmitoyl CoA (c), and protonated N-palmitoyl ethanolamine (d) were performed at collision energy of 23, 57, 43, and 18 eV, respectively, by using a QqQ mass spectrometer (Thermo Fisher TSQ Vantage). Gas pressure of 1 mTorr was employed in the collision activation.

11.3 ACYL CoA

Acyl CoA is the activated form of fatty acid in cells with a coenzyme A (CoA) group attached to the end of long-chain fatty acyls and generated through catalysis of acyl-CoA synthetase in the cytosol. Acyl CoA is the essential intermediate of fatty acid metabolism largely through β-oxidation to generate ATP in mitochondria (see Section 11.2). Acyl CoA is also an essential substrate for lipid biosynthesis involving fatty acyl substituents such as the syntheses of complex phospholipids, glycerolipids, and sphingolipids.

Acyl CoA can be ionized in both positive- and negative-ion modes as protonated and deprotonated species, respectively [5]. In addition, [M+Na−2H]⁻ can also be formed in the negative-ion mode. Acyl CoA can also be ionized as doubly charged ions in the negative-ion mode. Acidic conditions are usually required in the formation of protonated ions. Otherwise, sodium or potassium adducts of acyl CoA may be formed in the positive-ion mode [5].

Product-ion ESI mass spectra of protonated acyl CoA species (i.e., [M+H]$^+$) after low-energy CID show the following fragmentation pattern (Figure 11.1b) [5, 6]:

- A very intense product ion (usually the base peak) at [M+H−507]$^+$, corresponding to cleavage between the pantetheinic acid and the ADP residue with charge retention on the FA portion of the molecule.
- An abundant fragment ion at m/z 428, corresponding to protonated 3'-phospho-AMP.
- An abundant ion at [M+H−609]$^+$, corresponding to a part of the CoA with rearrangement.
- A couple of low abundance fragment ions arising from CoA moiety at m/z 508 and 330.

The fragmentation pathways leading to these product ions were previously proposed [5].

The fragmentation pattern of deprotonated acyl CoA species (i.e., [M−H]$^-$) after low-energy CID includes numerous abundant and informative product ions, which can be used for structural elucidation (Figure 11.1c) [5, 6]:

- The one at [M−H−80]$^-$, resulting from the loss of HPO_3.
- The one at [M−H−329]$^-$, arising from the loss of adenosine 3-phosphate.
- The ion at [M−H−347]$^-$, resulting from the further loss of water from the ion at [M−H−329]$^-$.
- An abundant product ion at [M−H−427]$^-$, corresponding to the loss of 3-phospho-AMP.
- Two product ions at m/z 426 and 408, corresponding to 3-phospho-AMP and its counterpart after losing a water molecule, present in very abundance.
- The ion at m/z 328.1, corresponding to adenosine phosphate (i.e., a loss of HPO_3 from the m/z 408 ion).
- A couple of low abundance additional fragment ions in the low mass regions, likely arising from the fragmentation of adenosine phosphate.

11.4 ENDOCANNABINOIDS

Endocannabinoids are defined as endogenously produced compounds that bind to and functionally activate the cannabinoid receptors [7]. Since the discovery of anandamide (N-arachidonoylethanolamine) in 1992 [8], many other endogenous compounds that activate one of the cannabinoid receptors have been identified from mammalian tissues (see Refs [1, 2, 9] for reviews). These endocannabinoids and related compounds generally belonged to the classes of N-acyl ethanolamine, 2-acyl glycerol, and N-acyl amino acid (also called elmiric acids), which are derived from the condensation of fatty acids with amino acids or amino acid derivatives (e.g., N-acyl glycine, N-acyl taurine, N-acyl serotonin, and N-acyl dopamine). These compounds can be ionized and have been characterized in the positive-ion mode as proton, sodium, silver, or ammonium adducts [2].

11.4.1 N-Acyl Ethanolamine

The majority of previous studies on characterization and identification of N-acyl ethanolamine species were conducted by gas chromatography-MS (GC-MS) after derivatization [9, 10]. Lately, ESI-MS has been widely applied for their identification and quantification. Proton adducts of N-acyl ethanolamine species are readily formed in the positive-ion mode. The fragmentation pattern of protonated N-acyl ethanolamine species shows an abundant, unique product ion at $[M+H-61]^+$, corresponding to the loss of ethanolamine, to form an acylium ion (Figure 11.1d) [11].

11.4.2 2-Acyl Glycerol

Characterization of 2-acyl glycerol species as ammonium adducts has been generally described in the section of monoglyceride (Chapter 9).

11.4.3 N-Acyl Amino Acid

A variety of N-acyl amino acid species has been detected in biological samples [12, 13]. N-Acyl amino acid species can be ionized in both positive- and negative-ion modes. Proton adducts or their H_2O-lost forms of N-acyl amino acids (i.e., $[M+H]^+$ and $[M-H_2O+H]^+$) are prominently present in the positive-ion ESI mass spectra under acidic conditions. The ions $[M-H]^-$ and $[M-H_2O-H]^-$ are formed in the negative-ion mode.

The fragmentation pattern of protonated N-acyl amino acids (i.e., $[M+H]^+$) after low-energy CID may vary to a certain degree even from those containing an identical amino acid [14, 15], but largely include the following:

- The one arising from the loss of amino acid or amino acid plus water.
- The one corresponding to protonated amino acid, which is equivalent to the loss of FA substituent as ketene.
- In some cases of protonated N-acyl amino acids, nonspecific loss of water or formic acid may be also present [14].

Dissociation of deprotonated N-acyl amino acids after CID largely shows the following fragmentation pattern:

- The intense fragment ions arising from amino acid (e.g., prominent fragment ions at m/z 80, 107, and 124, corresponding to sulfur trioxide (SO_3^-), vinyl-sulfonic acid, and taurine, respectively, are yielded from deprotonated N-acyl taurine).

- A set of low abundance fragment ions resulted from saturated FA substituent, which are present as a pattern of progressive loss of 14 amu in the mass region between m/z 150–430 [16].

11.5   4-HYDROXYALKENAL

4-Hydroxyalkenal species are a class of peroxidation products of polyunsaturated FA species resulting from a variety of complex enzymatic and nonenzymatic reactions during diverse physiological and pathophysiological processes [17, 18]. These peroxidation pathways could be significantly enhanced under patho(physio)logical conditions in which large amounts of reactive oxygen species (ROS) are present [19]. Therefore, determination of 4-hydroxyalkenal species indicates not only the degree of oxidative stress but also the different oxidative pathways involved under the patho(physio)logical conditions.

It is well known that α,β-unsaturated aldehyde species are metastable, which reactively form covalent adducts with active function groups (e.g., primary amines in proteins and nucleic acids), thereby leading to alterations in protein and DNA functions, dysregulation of enzyme activities, changes of mitochondrial bioenergetics, etc. [17, 20–22]. Moreover, due to their nonpolar properties, the 4-hydroxyalkenal species could easily diffuse from their generating site(s) to propagate the oxidative injury, thereby serving as "toxic second messenger" [23]. Several lines of evidence indicate that accumulation of the modified proteins with these reactive aldehyde species is manifest in cells during aging or under oxidative stress and that such modification and accumulation are linked to the pathogenesis of numerous diseases, such as atherosclerosis, diabetes, muscular dystrophy, rheumatoid arthritis, actinic elastosis, and neurodegenerative diseases (i.e., Alzheimer's disease, Parkinson's disease, and cerebral ischemia) [17, 18, 24].

Many techniques including GC-MS [25–27], HPLC [28–30], HPLC combined with GC-MS [31], ESI-MS after direct infusion [32], and LC-MS/MS [33–37] have been employed for measurement of these species. However, the results are not very satisfied due to the intensive labor, low sensitivity, or severe loss of the compounds during the sample preparation and analysis. Therefore, it is logical to have a method after derivatization to stabilize these reactive metabolites, enhance their ionization efficiency by ESI-MS, and provide characteristic fragments in MS/MS in order to resolve these obstacles. Derivatization of 4-hydroxyalkenal species with carnosine through Michael adduct between C3 of 4-hydroxyalkenals and the imidazole nitrogen of carnosine followed with a natural arrangement to form a hemi-acetal derivative as described previously [38] is perfectly fit to this purpose.

The primary amine in resultant 4-hydroxyalkenal-carnosine adduct can be easily protonated under acidic conditions, which substantially enhances the ionization in comparison to the parent 4-hydroxyalkenal species [39]. It was found that the formed derivatives are stable for long time as examined [39]. Most importantly, product-ion MS analysis of carnosine-adducted 4-hydroxyalkenal species after CID displays many abundant, informative, and characteristic fragment ions which can be exploited to identify and quantify these facile oxidation metabolites in the presence of their stable isotope-labeled counterparts.

Characterization of multiple protonated carnosine adducts of 4-hydroxyalkenal species demonstrates an essentially identical fragmentation pattern including the neutral loss of 17 (ammonia), 71, and 117 amu from the derivatized reagent (Figure 11.2). Other abundant fragment ions include those corresponding to the neutral losses of 46.0 and 63.0 amu and the one at m/z 210.1. All these neutrally lost fragments or those resultant fragment ions represent the building blocks of the adducted 4-hydroxyalkenal species. A combined detection of a few of these fragment ions can be used to specifically identify the presence of carnosine adducts of 4-hydroxyalkenal species, whereas the structures of 4-hydroxyalkenal species can be readily deduced from the detected molecular weight and the knowledge of naturally existing polyunsaturated FA structures.

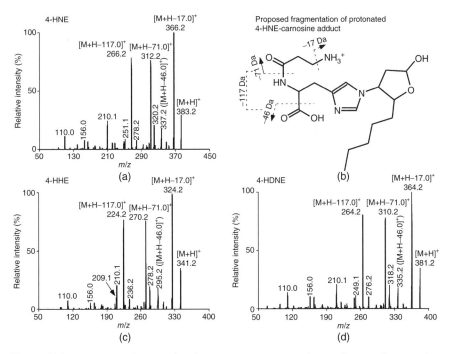

Figure 11.2 Representative product-ion mass spectrometric analyses of carnosine-derivatized 4-hydroxyalkenal species. Carnosine-4-hydroxyalkenal adducts were prepared by incubating individual 4-hydroxyalkenal species with carnosine as previously described [39]. Product-ion ESI-MS analyses of carnosine-derivatized 4-hydroxynonenal (4-HNE, a), 4-hydroxyhexenal (4-HHE, c), and 4-hydroxynondienal (4-HNDE, d) were performed on a QqQ mass spectrometer (Thermo Fisher TSQ Vantage). Collision activation was carried out at collision energy of 25 eV and gas pressure of 1 mTorr. The fragmentation pattern of protonated 4-HNE-carnosine adduct was proposed (b). Wang et al. [39]. Reproduced with permission of the American Chemical Society.

Because the fragment ions corresponding to the neutral losses of 17.0, 63.0, 71.0, and 117.0 amu are abundant and specific to carnosine adducts, it is logical to effectively perform NLS of these neutral fragments to identify the presence of 4-hydroxyalkenal species in biological samples and quantify these identified species in comparison to isotope-labeled internal standard(s) (Figure 11.3).

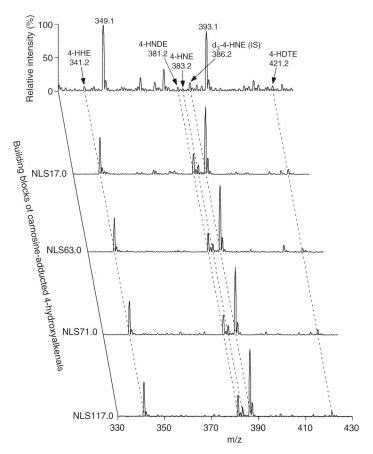

Figure 11.3 Representative tandem mass spectrometric analysis of carnosine-derivatized 4-hydroxyalkenal species from mouse myocardial lipid extracts. Lipid extracts of mouse myocardium were performed by a modified Bligh–Dyer method and derivatized with carnosine as previously described [39]. Neutral-loss scans of 17 (NLS17), 63 (NLS63), 71 (NLS71), and 117 (NLS117) amu were acquired on a QqQ mass spectrometer (Thermo Fisher TSQ Vantage). Collision activation was carried out at gas pressure of 1 mTorr and collision energy of 16, 27, 23, and 28 eV, respectively. The abbreviations of 4-HHE, 4-HNDE, 4-HNE, and 4-HDTE denote 4-hydroxyhexenal, 4-hydroxynondinenal, 4-hydroxynoneal, and 4-hydroxydodecatrienal, respectively. "IS" stands for internal standard.

11.6   CHLORINATED LIPIDS

Cellular plasmalogen species can react with endogenously produced hypochlorous acid, generating 2-chloro fatty aldehydes which further oxidizes or reduces to 2-chloro fatty acids or alcohols, respectively [40, 41]. The resulted 2-chloro fatty acids can accumulate in the activated monocytes and cause cell apoptosis through overproduction of ROS and ER stress [42]. Similarly, plasmalogen species are targeted by endogenously produced hypobromous acid leading to the formation of 2-bromo fatty aldehydes [43, 44], therefore, 2-bromo fatty acids.

Both 2-chloro fatty aldehydes and fatty alcohols can be quantitatively analyzed by classical GC-MS approaches after derivatization with pentafluorobenzoyl hydroxylamine and pentafluorobenzoyl chloride, respectively [40, 41]. Bioactive 2-chloro FA species can be readily ionized in the negative-ion mode of ESI-MS as described in Chapter 2 to form deprotonated molecular ions. Product-ion ESI analysis of these deprotonated molecular ions after CID shows the following fragmentation pattern [45]:

- A very abundant fragment ion at m/z [M−H−36]$^-$, corresponding to the loss of HCl.
- An abundant fragment ion m/z [M−H−82]$^-$, corresponding to the further of a formic acid from the fragment ion at m/z [M−H−36]$^-$.

Therefore, bioactive 2-chloro FA species are readily identified and quantified by LC-MS/MS in the MRM mode utilizing these characteristic fragment ions [40, 41].

11.7   STEROLS AND OXYSTEROLS

Numerous sterols and oxysterols are present in nature [46], produced through enzymatic or nonenzymatic reactions. The majority of these compounds play many essential roles in biological systems. For example, numerous sterols are involved in the biosynthesis and metabolism of cholesterol [47, 48]. GC-MS is used to be the widely used tool for analysis of these sterols and their metabolites after derivatization [48, 49]. Nowadays, LC-MS and LC-MS/MS after ESI become as the popular tool of choice [49].

Although sterols carry a core signature of four fused rings (Figure 1.4a), individual sterol species are different from each other with the additional group(s) or modifications on the core structure. These differences lead to the very different fragmentation patterns after CID. Therefore, there is no common fragmentation pattern(s) available for the sterol species. Individual sterol species has to be characterized and elucidated separately. It is too complicated and beyond the scope of this book to describe each individual product-ion mass spectrum of sterol species. Any advanced readers may find the description of these CID spectra from the excellent review articles written by Dr Griffiths and colleagues [47, 48].

It should be recognized that since the majority of these sterol compounds are relatively hydrophobic, direct ionization of these compounds by ESI is not sensitive. Moreover, many of the sterol species carry at least one hydroxyl group. Detection of quasimolecular ions of these sterol species as the form after loss of water is very common. To enhance the sensitivity and stabilize the original structure of a sterol species, appropriate derivatization is commonly employed for characterization and elucidation of sterol structures. Readers who are interesting in this area of work may consult the work of Dr Griffiths and colleagues [50–53].

11.8 FATTY ACID–HYDROXY FATTY ACIDS

The newest discovered bioactive lipids are the fatty acid–hydroxy fatty acids, which possess anti-diabetic and anti-inflammatory effects as demonstrated [54]. This class of lipids is at the concentration of approximately 50 pmol/ml of serum and 50–100 pmol/mg of white adipose tissue. These bioactive compounds serve as the agonists to G-protein receptors as demonstrated [54].

Essentially, as described in Chapter 2, this class of lipid species as a member among the modified fatty acid family can be readily ionized in the negative-ion mode of ESI-MS to produce deprotonated molecular ions (i.e., $[M-H]^-$) as demonstrated [54]. Product-ion ESI-MS analysis of their deprotonated molecular ions after CID shows the following fragmentation pattern [54]:

- A fragment ion corresponding to the hydroxy FA carboxylate.
- An ion yielded from the loss of a water molecule from hydroxy FA carboxylate.
- The ion of branched FA carboxylate.
- Some of other minor fragments allowing for identification of the location of double bond (if present).

With these fragment ions, the researchers have developed MRM transitions to targetedly determine individual fatty acid–hydroxy fatty acids [54].

The fragmentation of this class of modified fatty acids has also been preliminary characterized with a fatty acidomics approach as described in Chapter 10 in the author's laboratory (Unpublished data). Specifically, we have synthesized a compound of d_4-palmitic acid-12-hydroxy stearic acid and determined its fragmentation pattern after being derivatized with AMPP as described previously [55]. Figure 11.4 shows its product-ion mass spectrum, which displays a very abundant fragment ion at m/z 449.4 corresponding to the loss of branched fatty acid (i.e., d_4-16:0 FA). In addition, the mass spectrum also shows a wealth of fragment fingerprints (see inset), which provide the structural information about the hydroxy fatty acid and the location of the hydroxy moiety. The latter can be derived from the location of the double bond that is predominant between C11 and C12 positions and to a less degree at

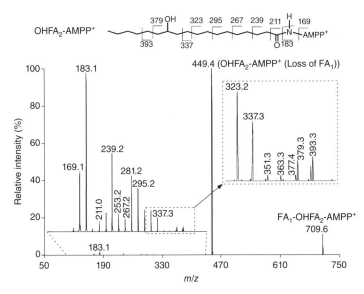

Figure 11.4 Demonstration of fatty acidomics that shows an informative fragmentation pattern of d_4-palmitic acid-12-hydroxy stearic acid after derivatization with AMPP for structural determination.

the C12–C13 position. The lack of any fragment ion carrying the branched fatty acid along with the intense fragment ion at m/z 449.4 resulted from the loss of the branched fatty acid may suggest that the loss of the branched fatty acid after CID is facile and occurs first. However, the presence of relatively abundant peaks corresponding to the C13 and C14 positions is intriguing since such peaks afterward the position of a double bond are absent in the fragmentation pattern of unsaturated fatty acids as previously demonstrated [55]. Moreover, the split peaks with 2 amu differences starting at C11 indicate that at least two fragmentation pathways exist. Clearly, additional studies are necessary to further elucidate the mechanism(s) giving rise to the fragmentation of AMPP-derivatized fatty acid–hydroxy fatty acids.

REFERENCES

1. Murphy, R.C., Barkley, R.M., Zemski Berry, K., Hankin, J., Harrison, K., Johnson, C., Krank, J., McAnoy, A., Uhlson, C. and Zarini, S. (2005) Electrospray ionization and tandem mass spectrometry of eicosanoids. Anal. Biochem. 346, 1–42.

2. Kingsley, P.J. and Marnett, L.J. (2009) Analysis of endocannabinoids, their congeners and COX-2 metabolites. J. Chromatogr. B 877, 2746–2754.

3. Puppolo, M., Varma, D. and Jansen, S.A. (2014) A review of analytical methods for eicosanoids in brain tissue. J. Chromatogr. B 964, 50–64.

4. Su, X., Han, X., Mancuso, D.J., Abendschein, D.R. and Gross, R.W. (2005) Accumulation of long-chain acylcarnitine and 3-hydroxy acylcarnitine molecular species in diabetic

myocardium: Identification of alterations in mitochondrial fatty acid processing in diabetic myocardium by shotgun lipidomics. Biochemistry 44, 5234–5245.

5. Hankin, J.A., Wheelan, P. and Murphy, R.C. (1997) Identification of novel metabolites of prostaglandin E2 formed by isolated rat hepatocytes. Arch. Biochem. Biophys. 340, 317–330.

6. Haynes, C.A., Allegood, J.C., Sims, K., Wang, E.W., Sullards, M.C. and Merrill, A.H., Jr. (2008) Quantitation of fatty acyl-coenzyme As in mammalian cells by liquid chromatography-electrospray ionization tandem mass spectrometry. J. Lipid Res. 49, 1113–1125.

7. Bisogno, T., Ligresti, A. and Di Marzo, V. (2005) The endocannabinoid signalling system: Biochemical aspects. Pharmacol. Biochem. Behav. 81, 224–238.

8. Devane, W.A., Hanus, L., Breuer, A., Pertwee, R.G., Stevenson, L.A., Griffin, G., Gibson, D., Mandelbaum, A., Etinger, A. and Mechoulam, R. (1992) Isolation and structure of a brain constituent that binds to the cannabinoid receptor. Science 258, 1946–1949.

9. Zoerner, A.A., Gutzki, F.M., Batkai, S., May, M., Rakers, C., Engeli, S., Jordan, J. and Tsikas, D. (2011) Quantification of endocannabinoids in biological systems by chromatography and mass spectrometry: A comprehensive review from an analytical and biological perspective. Biochim. Biophys. Acta 1811, 706–723.

10. Kempe, K., Hsu, F.F., Bohrer, A. and Turk, J. (1996) Isotope dilution mass spectrometric measurements indicate that arachidonylethanolamide, the proposed endogenous ligand of the cannabinoid receptor, accumulates in rat brain tissue post mortem but is contained at low levels in or is absent from fresh tissue. J. Biol. Chem. 271, 17287–17295.

11. Markey, S.P., Dudding, T. and Wang, T.C. (2000) Base- and acid-catalyzed interconversions of O-acyl- and N-acyl-ethanolamines: A cautionary note for lipid analyses. J. Lipid Res. 41, 657–662.

12. Tan, B., Bradshaw, H.B., Rimmerman, N., Srinivasan, H., Yu, Y.W., Krey, J.F., Monn, M.F., Chen, J.S., Hu, S.S., Pickens, S.R. and Walker, J.M. (2006) Targeted lipidomics: Discovery of new fatty acyl amides. AAPS J. 8, E461-465.

13. Tan, B., Yu, Y.W., Monn, M.F., Hughes, H.V., O'Dell, D.K. and Walker, J.M. (2009) Targeted lipidomics approach for endogenous N-acyl amino acids in rat brain tissue. J. Chromatogr. B 877, 2890–2894.

14. Bradshaw, H.B., Rimmerman, N., Hu, S.S., Burstein, S. and Walker, J.M. (2009) Novel endogenous N-acyl glycines identification and characterization. Vitam. Horm. 81, 191–205.

15. Huang, S.M., Bisogno, T., Petros, T.J., Chang, S.Y., Zavitsanos, P.A., Zipkin, R.E., Sivakumar, R., Coop, A., Maeda, D.Y., De Petrocellis, L., Burstein, S., Di Marzo, V. and Walker, J.M. (2001) Identification of a new class of molecules, the arachidonyl amino acids, and characterization of one member that inhibits pain. J. Biol. Chem. 276, 42639–42644.

16. Saghatelian, A., Trauger, S.A., Want, E.J., Hawkins, E.G., Siuzdak, G. and Cravatt, B.F. (2004) Assignment of endogenous substrates to enzymes by global metabolite profiling. Biochemistry 43, 14332–14339.

17. Poli, G. and Schaur, R.J. (2000) 4-Hydroxynonenal in the pathomechanisms of oxidative stress. IUBMB Life 50, 315–321.

18. Uchida, K. (2000) Role of reactive aldehyde in cardiovascular diseases. Free Radic. Biol. Med. 28, 1685–1696.

19. Yun, M.R., Park, H.M., Seo, K.W., Lee, S.J., Im, D.S. and Kim, C.D. (2010) 5-Lipoxygenase plays an essential role in 4-HNE-enhanced ROS production in murine macrophages via activation of NADPH oxidase. Free Radic. Res. 44, 742–750.

20. Esterbauer, H., Schaur, R.J. and Zollner, H. (1991) Chemistry and biochemistry of 4-hydroxynonenal, malonaldehyde and related aldehydes. Free Radic. Biol. Med. 11, 81–128.

21. Parola, M., Bellomo, G., Robino, G., Barrera, G. and Dianzani, M.U. (1999) 4-Hydroxynonenal as a biological signal: Molecular basis and pathophysiological implications. Antioxid. Redox. Signal. 1, 255–284.

22. Echtay, K.S. (2007) Mitochondrial uncoupling proteins: What is their physiological role?. Free Radic. Biol. Med. 43, 1351–1371.

23. Uchida, K., Shiraishi, M., Naito, Y., Torii, Y., Nakamura, Y. and Osawa, T. (1999) Activation of stress signaling pathways by the end product of lipid peroxidation. 4-Hydroxy-2-nonenal is a potential inducer of intracellular peroxide production. J. Biol. Chem. 274, 2234–2242.

24. Stadtman, E.R. (2001) Protein oxidation in aging and age-related diseases. Ann. N. Y. Acad. Sci. 928, 22–38.

25. Luo, X.P., Yazdanpanah, M., Bhooi, N. and Lehotay, D.C. (1995) Determination of aldehydes and other lipid peroxidation products in biological samples by gas chromatography–mass spectrometry. Anal. Biochem. 228, 294–298.

26. Bruenner, B.A., Jones, A.D. and German, J.B. (1996) Simultaneous determination of multiple aldehydes in biological tissues and fluids using gas chromatography/stable isotope dilution mass spectrometry. Anal. Biochem. 241, 212–219.

27. Kawai, Y., Takeda, S. and Terao, J. (2007) Lipidomic analysis for lipid peroxidation-derived aldehydes using gas chromatography–mass spectrometry. Chem. Res. Toxicol. 20, 99–107.

28. Goldring, C., Casini, A.F., Maellaro, E., Del Bello, B. and Comporti, M. (1993) Determination of 4-hydroxynonenal by high-performance liquid chromatography with electrochemical detection. Lipids 28, 141–145.

29. Liu, Y.M., Jinno, H., Kurihara, M., Miyata, N. and Toyo'oka, T. (1999) Determination of 4-hydroxy-2-nonenal in primary rat hepatocyte cultures by liquid chromatography with laser induced fluorescence detection. Biomed. Chromatogr. 13, 75–80.

30. Uchida, T., Gotoh, N. and Wada, S. (2002) Method for analysis of 4-hydroxy-2-(E)-nonenal with solid-phase microextraction. Lipids 37, 621–626.

31. Selley, M.L., Bartlett, M.R., McGuiness, J.A., Hapel, A.J. and Ardlie, N.G. (1989) Determination of the lipid peroxidation product trans-4-hydroxy-2-nonenal in biological samples by high-performance liquid chromatography and combined capillary column gas chromatography-negative-ion chemical ionisation mass spectrometry. J. Chromatogr. B Biomed. Sci. Appl. 488, 329–340.

32. Gioacchini, A.M., Calonghi, N., Boga, C., Cappadone, C., Masotti, L., Roda, A. and Traldi, P. (1999) Determination of 4-hydroxy-2-nonenal at cellular levels by means of electrospray mass spectrometry. Rapid Commun. Mass Spectrom. 13, 1573–1579.

33. O'Brien-Coker, I.C., Perkins, G. and Mallet, A.I. (2001) Aldehyde analysis by high performance liquid chromatography/tandem mass spectrometry. Rapid Commun. Mass Spectrom. 15, 920–928.

34. Andreoli, R., Manini, P., Corradi, M., Mutti, A. and Niessen, W.M. (2003) Determination of patterns of biologically relevant aldehydes in exhaled breath condensate of healthy subjects by liquid chromatography/atmospheric chemical ionization tandem mass spectrometry. Rapid Commun. Mass Spectrom. 17, 637–645.

35. Williams, T.I., Lovell, M.A. and Lynn, B.C. (2005) Analysis of derivatized biogenic aldehydes by LC tandem mass spectrometry. Anal. Chem. 77, 3383–3389.

36. Honzatko, A., Brichac, J. and Picklo, M.J. (2007) Quantification of trans-4-hydroxy-2-nonenal enantiomers and metabolites by LC-ESI-MS/MS. J. Chromatogr. B 857, 115–122.

37. Warnke, M.M., Wanigasekara, E., Singhal, S.S., Singhal, J., Awasthi, S. and Armstrong, D.W. (2008) The determination of glutathione-4-hydroxynonenal (GSHNE), E-4-hydroxynonenal (HNE), and E-1-hydroxynon-2-en-4-one (HNO) in mouse liver tissue by LC-ESI-MS. Anal. Bioanal. Chem. 392, 1325–1333.

38. Aldini, G., Carini, M., Beretta, G., Bradamante, S. and Facino, R.M. (2002) Carnosine is a quencher of 4-hydroxy-nonenal: Through what mechanism of reaction? Biochem. Biophys. Res. Commun. 298, 699–706.

39. Wang, M., Fang, H. and Han, X. (2012) Shotgun lipidomics analysis of 4-hydroxyalkenal species directly from lipid extracts after one-step in situ derivatization. Anal. Chem. 84, 4580–4586.

40. Wacker, B.K., Albert, C.J., Ford, B.A. and Ford, D.A. (2013) Strategies for the analysis of chlorinated lipids in biological systems. Free Radic. Biol. Med. 59, 92–99.

41. Wang, W.Y., Albert, C.J. and Ford, D.A. (2013) Approaches for the analysis of chlorinated lipids. Anal. Biochem. 443, 148–152.

42. Wang, W.Y., Albert, C.J. and Ford, D.A. (2014) Alpha-chlorofatty acid accumulates in activated monocytes and causes apoptosis through reactive oxygen species production and endoplasmic reticulum stress. Arterioscler. Thromb. Vasc. Biol. 34, 526–532.

43. Albert, C.J., Crowley, J.R., Hsu, F.F., Thukkani, A.K. and Ford, D.A. (2002) Reactive brominating species produced by myeloperoxidase target the vinyl ether bond of plasmalogens: Disparate utilization of sodium halides in the production of alpha-halo fatty aldehydes. J. Biol. Chem. 277, 4694–4703.

44. Albert, C.J., Thukkani, A.K., Heuertz, R.M., Slungaard, A., Hazen, S.L. and Ford, D.A. (2003) Eosinophil peroxidase-derived reactive brominating species target the vinyl ether bond of plasmalogens generating a novel chemoattractant, alpha-bromo fatty aldehyde. J. Biol. Chem. 278, 8942–8950.

45. Brahmbhatt, V.V., Albert, C.J., Anbukumar, D.S., Cunningham, B.A., Neumann, W.L. and Ford, D.A. (2010) {Omega}-oxidation of {alpha}-chlorinated fatty acids: Identification of {alpha}-chlorinated dicarboxylic acids. J. Biol. Chem. 285, 41255–41269.

46. Fahy, E., Subramaniam, S., Brown, H.A., Glass, C.K., Merrill, A.H., Jr., Murphy, R.C., Raetz, C.R., Russell, D.W., Seyama, Y., Shaw, W., Shimizu, T., Spener, F., van Meer, G., VanNieuwenhze, M.S., White, S.H., Witztum, J.L. and Dennis, E.A. (2005) A comprehensive classification system for lipids. J. Lipid Res. 46, 839–861.

47. Griffiths, W.J. (2003) Tandem mass spectrometry in the study of fatty acids, bile acids, and steroids. Mass Spectrom. Rev. 22, 81–152.

48. Griffiths, W.J. and Wang, Y. (2009) Analysis of neurosterols by GC-MS and LC-MS/MS. J. Chromatogr. B 877, 2778–2805.

49. McDonald, J.G., Smith, D.D., Stiles, A.R. and Russell, D.W. (2012) A comprehensive method for extraction and quantitative analysis of sterols and secosteroids from human plasma. J. Lipid Res. 53, 1399–1409.

50. Griffiths, W.J., Liu, S., Alvelius, G. and Sjovall, J. (2003) Derivatisation for the characterisation of neutral oxosteroids by electrospray and matrix-assisted laser desorption/ionisation tandem mass spectrometry: The Girard P derivative. Rapid Commun. Mass Spectrom. 17, 924–935.

51. Karu, K., Hornshaw, M., Woffendin, G., Bodin, K., Hamberg, M., Alvelius, G., Sjovall, J., Turton, J., Wang, Y. and Griffiths, W.J. (2007) Liquid chromatography-mass spectrometry utilizing multi-stage fragmentation for the identification of oxysterols. J. Lipid Res. 48, 976–987.

52. Griffiths, W.J. and Sjovall, J. (2010) Analytical strategies for characterization of bile acid and oxysterol metabolomes. Biochem. Biophys. Res. Commun. 396, 80–84.

53. Griffiths, W.J., Crick, P.J., Wang, Y., Ogundare, M., Tuschl, K., Morris, A.A., Bigger, B.W. and Clayton, P.T. (2013) Analytical strategies for characterization of oxysterol lipidomes: Liver X receptor ligands in plasma. Free Radic. Biol. Med. 59, 69–84.

54. Yore, M.M., Syed, I., Moraes-Vieira, P.M., Zhang, T., Herman, M.A., Homan, E.A., Patel, R.T., Lee, J., Chen, S., Peroni, O.D., Dhaneshwar, A.S., Hammarstedt, A., Smith, U., McGraw, T.E., Saghatelian, A. and Kahn, B.B. (2014) Discovery of a class of endogenous mammalian lipids with anti-diabetic and anti-inflammatory effects. Cell 159, 318–332.

55. Wang, M., Han, R.H. and Han, X. (2013) Fatty acidomics: Global analysis of lipid species containing a carboxyl group with a charge-remote fragmentation-assisted approach. Anal. Chem. 85, 9312–9320.

12

IMAGING MASS SPECTROMETRY OF LIPIDS

12.1 INTRODUCTION

Imaging mass spectrometry (IMS), combining molecular mass analysis and spatial information determination, directly analyzes intact chemical species *in situ* from the surface they are bound to or the matrix they are embedded in, and provides visualization of molecular distribution on complex surfaces. This is the only technique that generates high-resolution biomolecular images directly from tissue sections and without the need of labeling. Multimodal IMS strategies including MALDI, secondary ion mass spectrometry (SIMS), DESI, and others, have made this technology, a powerful tool for determination of spatial distribution and direct identification of pharmaceuticals, metabolites, lipids, peptides, and proteins in cells and biological tissues.

Lipids play crucial roles in cell and tissue organization and processes. Many human diseases involve the disruption of lipid metabolism pathways (e.g., cancer, diabetes, neurodegenerative, and infectious diseases) [1–10]. Determining the distribution of the lipids and colocalizing the lipids with other pharmaceuticals and biomolecules within whole body sections, organs, biological tissues, or even single cells are always fascinating and important in biology and medicine.

Almost all types of ionization sources used for IMS have been successfully applied for IMS of lipids such as MALDI [11], DESI [12], SIMS [13], and pressurized liquid extraction surface analysis [14]. In general, the majority of the lipids present in tissues ionize easily due to their polar heads (Chapters 1 and 2). PC, SM, and cholesterol species more selectively ionize in the positive-ion mode, while PI, PS, and sulfatide

Lipidomics: Comprehensive Mass Spectrometry of Lipids, First Edition. Xianlin Han.
© 2016 John Wiley & Sons, Inc. Published 2016 by John Wiley & Sons, Inc.

species more readily ionize in the negative-ion mode (Chapter 2). PE species can be analyzed in both positive- and negative-ion modes.

The majority of the biological samples subjected to IMS contain an abundant amount of lipids that can yield a strong ion signal around m/z 800 utilizing soft ionization methods such as MALDI and DESI. The most readily detected and imaged lipid species are those corresponding to PC species due to the presence of a permanent positively charge with the quaternary ammonium moiety of the choline group.

IMS of lipids is a process comprised of four basic steps: sample preparation, desorption/ionization, mass analysis, and imaging processing. As with any analytical technique, sample preparation is a key procedure, which is first extensively discussed in this chapter. After introduction of the prepared sample(s) into a mass spectrometer for IMS, first the biomolecules are desorbed and ionized from the surface through its exposure to a laser beam (MALDI), a primary ion beam (SIMS), or a charged droplet flux (DESI). The produced ions are subsequently separated and detected. Because ion production and mass analysis are generally introduced in Part I of the book, these steps are only briefly mentioned here along with the applications of individual IMS techniques for analysis of lipids. Individual IMS experiment involves data collection of mass spectra at every point, which collectively leads to the generation of ion images. In the chapter, available tools for data and imaging processing are also summarized. However, the interested readers are strongly recommended to extend their readings to the numerous recently published review articles on the topic [15–23].

12.1.1 Samples Suitable for MS Imaging of Lipids

All types of biological samples can be and have been used for IMS, for example, single cells [24], plant sections [25], animal organs [11, 23, 26–30], songbird song systems [31], single zooplankter individuals [32], and rodent whole body sections [15, 33].The mostly used tissue type is the rodent brain due to its suitable size, characteristic structure, and easiness for sectioning. Many other types of animal organs including lungs [34–36], heart [37], kidney [23, 27, 38], colon [28], retina [23], muscle [39, 40], samples with different tumor types [12, 41–43], human biopsies and resected tissue from surgery [41], and cells from cell culture [44, 45]. Fresh, snap-frozen samples are largely used for IMS of lipids, whereas chemically fixed or preserved samples are used much less for lipid imaging than protein analysis due to their possible interference with lipid analysis and perturbation of the location of lipid species [46].

12.1.2 Sample Processing/Preparation

Sample preparation for IMS of lipids involves a few crucial steps including sample collection, storage, embedding, and sectioning. The key during the sample handling is to maintain the integrity and unperturbed spatial distribution of lipids in biological samples. Thus, prevention of lipid degradation is one of the priorities during sample preparation, which implies that samples must be properly and timely collected, processed, and stored prior to the analysis by IMS.

Snap-freezing of the material and storage at –80 °C is the most common step for this purpose to preserve the sample's morphology and minimize lipid degradation through endogenous enzymatic activation (e.g., phospholipases), lipid peroxidation, or changes in cellular metabolism. To avoid sample damage (cracking and fragmentation) during the freezing process, the tissue may be loosely wrapped in aluminum foil and frozen at a temperature below −70 °C by gently lowering the tissue into a liquid (e.g., liquid nitrogen) over a period of 30–60 s as previously described [47]. Properly collected and frozen-stored samples could have a shelf lifetime of a year as previously demonstrated [47].

The embedding of the tissue in a supporting material could be used for easiness of handling and sectioning. To minimize tissue tearing, the use of gelatin or other compounds as the embedding material makes them comparable to the lipid analysis [34, 48]. If biological samples that were usually collected, paraffin-preserved, and stored in tissue banks have to be used for IMS of lipids, paraffin must be removed before tissue analysis [34, 49].

For IMS, the tissue samples are usually sectioned to a thickness of 5–25 μm with an optimal thickness of 10–20 μm (the diameter of a mammalian cell) after considering the pros and cons [47]. This leads to the majority of the cells in the slice cut and exposing the intracellular contents for analysis. For tissue sectioning, a cryomicrotome apparatus is usually used. Tissue samples are mounted to the apparatus cutting stage (which is typically maintained between −5 and −25 °C, depending on the tissue type) and sliced with a stainless steel microtome blade.

The tissue sections are then attached to an electrically conductive steel plate or glass slide. Conductive substrates are used to properly define the electric extraction field that will accelerate the ions produced from the surface. It was recommended to cool the plate by placing it in the cryomicrotome chamber at −15 °C before sectioning [47]. Two approaches are usually used for tissue attachment: either the use of an adhesive double-sided conductive tape or a thaw-mounting method in which the tissue is attached by warming the reverse of the target to produce a localized warm patch. The former requires special care to avoid trapping air bubbles, which can affect the analysis. Although the latter method reduces the risk of sample contamination, it may cause significant variation of the results due to lipid degradation during the attachment process.

12.1.3 Matrix Application

12.1.3.1 Matrix Application It is necessary to add matrix on the surface of the tissue before MALDI or SIMS analysis. On-tissue application of the matrix solution results in *in situ* extraction of biomolecules from the biological sample. Generally, a matrix solution comprises an organic solvent (e.g., methanol or acetonitrile), a matrix (e.g., 2,5-dihydroxy benzoic acid (DHB)), and sometime a modifier (e.g., trifluoroacetic acid). The evaporation of the organic solvent from the matrix solution causes the crystallization of the matrix and incorporation of the analyte molecules into growing crystals. The addition of trifluoroacetic acid increases the amount of available protons for ionization. During desorption/ionization, the energy of the beam from MALDI or SIMS is absorbed by the matrix crystals, which evaporate quickly and release the trapped molecules (desorption), and eventually leads to ionization of the analytes as proton or other cation adducts.

Selection of matrices used for the analysis of cellular glycerophospholipids was previously reviewed [50]. Clearly, DHB is the common choice of matrix for lipid analysis [50]. However, in some applications, neutral matrix (e.g., 9-aminoacridine, 9-AA, 1,5-diaminonathalene, and quercetin) was successfully used for IMS of lipids [51–53]. In order to reduce the chemical background resulting from the matrix, nanoparticles with a size range of 2–10 nm can be used as new-generation matrices for high-resolution imaging [54]. These types of matrices result in desorption and ionization of lipids with minimum background [37, 55, 56].

The concentration of the matrix in solution must be carefully considered before matrix application. If the concentration is too low, the analytes may diffuse from its original position before crystallization, or these will not be sufficient to form proper crystals. If the concentration is too high, rapid crystallization could occur, which makes a limited time for analyte extraction and incorporation. An approximate concentration of 50 mg DHB/mL is usually used for IMS of lipids.

The type of modifier also affects IMS analysis in general as discussed in Chapter 4. For example, mass spectra can be simplified by addition of potassium acetate [57] or LiCl [58] to the matrix solution, which results in the formation of exclusively potassium or lithium adducts of lipids, respectively. By changing the concentration of alkali metal salts in the matrix solution, it is also possible to selectively ionize either polar or nonpolar lipids [59].

12.1.3.2 *Matrix Application Methods*

The matrix solution can be applied to the tissue surface by spotting (as individual droplets) or coating (as a homogeneous layer). Both methods could result inhomogeneous crystallization of matrix with analytes. These methods can be executed either manually or automatically. The latter usually offers better homogeneity and reproducibility, thereby reducing the experimental variations. Moreover, the spotting method needs more time than spraying, which can cause degradation of some molecules during the matrix application process. However, spraying leads to covering the whole surface of the samples with a layer of matrix solution and requires a more careful application protocol than spotting.

A. **Spotting.** The matrix solution can be spotted onto the tissue surface, which limits diffusion of the analytes to the spot size. Manual spotting can be done using a micropipette to deliver microliter droplets generating spots of approximately millimeter size. Robotic spotting produces picoliter droplets and provides a spot size of 100–200 μm, which allows IMS analysis to have a resolving power of approximately 200 μm [60].

B. **Spraying.** Spraying matrix solution to the surface of the sample deposits a fine distribution of matrix mists on the surface. A homogeneous thin film of crystallized matrix layer is formed after solvent evaporation. This approach results in smaller crystal sizes (typically around 20 μm that is comparable with the diameter of the focused laser beam) than robotic spotting, thereby allowing higher image resolution. Spraying can be executed either manually (with a pneumatic sprayer, airbrush, or TLC sprayer) or automatically (with a robotic pneumatic sprayer, vibrational sprayer, or electrospray). Automated spraying generates a more uniform coating and better reproducible results.

C. **Alternative matrix application: Solvent-free and matrix-free methods.**
Sublimation of matrix was successfully developed for matrix application [23, 61, 62]. This approach provided a homogeneous coating of matrix for high-resolution IMS of GPL species from tissue sections. The apparatus used for the method is relatively simple and commercially available. The advantages of sublimation include the elimination of diffusion of the lipid molecules because no solvent is used during the matrix application, the increased purity of the matrix, and the reduction of the crystal size.

A variety of other solvent-free methods with graphite nanoparticles, silver nanoparticles, nanostructural materials, etc., have been developed for IMS and been applied for lipid analysis [25, 37, 54–56, 63, 64]. Finely ground matrix particles that were filtered directly onto the tissue through a 20-μm stainless steel sieve area dry-coating, solvent-free matrix deposition employed for IMS of lipids [65]. This approach provided highly reproducible results and eliminated the variation caused by operator differences. Alternatively, coating solvent-free matrix to the tissue section with a thin layer of seed matrix utilizing a painter's brush was used to distribute the ground matrix on the tissue surface [66] since solvent was necessary for extracting peptides from tissue samples [48].

A matrix-free approach through desorption–ionization on silicon was used for IMS of lipids [67, 68]. In this method, the physical properties of the silicon material (high area surface, UV absorption) are crucial for the desorption/ionization process. The method requires the transfer of analytes to the silicon surface by direct contact with the tissue samples. IMS analysis can be performed with the silicon surface after removal of the tissue.

12.1.4 Data Processing

Lots of mass spectra should be collected during IMS. The higher the MS imaging resolution, the more MS data are required. Up to a few gigabytes that require complex visualization software to process are usually acquired in an experiment of IMS. In this section, a few software tools for processing IMS data are summarized. A detail comparison of these IMS software packages could be found somewhere else [69].

12.1.4.1 Biomap   Biomap (Novartis, Basel, Switzerland, www.maldi-msi.org) is an image processing software tool. It was originally developed for processing the data generated from magnetic resonance imaging, but now is used to process many imaging data including IMS [70]. The imaging is created based on multiplanar reconstruction, allowing extraction of arbitrary slices from a 3D volume. It allows displaying the mass spectrum from any single point. It also allows to display the distribution of any single analyte on the analyzed area [71]. The software provides baseline correction of spectra, spatial filtering, averaging of spectra, etc. [70].

12.1.4.2 FlexImaging   FlexImaging software (Bruker Daltonics GmbH, Bremen, Germany, www.bdal.com) is used for acquisition and evaluation of MALDI-TOF and TOF/TOF imaging data, which allows color-coded visualization of the distribution of any ion detected by IMS. One key feature is its integration of statistical analyses (e.g., hierarchical clustering, PCA, or variance ranking) into data processing, thereby providing comparable determination of analytes between tissue samples.

12.1.4.3 MALDI Imaging Team Imaging Computing System (MITICS)
MITICS software could be used for many types of instruments developed for MALDI imaging [69], which divides into MITICS control and MITICS image. The former is used for setting the acquisition parameters for the imaging sequence, such as creating the raster of acquisition and controlling postacquisition data processing. The latter is for imaging reconstruction.

12.1.4.4 DataCube Explorer   The DataCube Explorer (www.imzml.org>Software Tools) is a visualization software tool for processing IMS data sets. It allows analysts to display both an image-based and a spectrum-based data. The software package possesses features such as spectral analysis at a region of interest, self-organizing map feature for image classification, image smoothing, etc. The software package allows one to convert data sets acquired by a few other software tools.

12.1.4.5 imzML   The software tool of imzML is an extension of MS standard software mzML, which was developed by the Human Proteome Organization-Proteomics Standards Initiative and used to process MS metadata files. The IMS data are stored in a binary format in order to ensure the most efficient storage of these large data sets and controlled by imzML to include parameters that are specific for imaging experiments. These parameters are stored in the imaging MS.obo file.

12.2 MALDI-MS IMAGING

MALDI-MS, particularly with an atmospheric pressure MALDI source, is a powerful method that allows the analysis and detection of a wide variety of biomolecules including lipids directly from tissue sections. The laser serves two purposes: (1) to desorb analytes from the surface material; and (2) induce analyte ionization for MS analysis. The role of the matrix is to absorb the majority of the laser energy, leading to explosive desorption of the matrix crystals, together with incorporation of analyte into the gas phase without degradation of the analyte. Addition of matrix may also aid the ionization of analyte molecules in the gas phase due to the presence of adduct ions from the added modifier(s).

During a MALDI-IMS experiment, the laser is fired across the surface of the matrix covered tissue section, which allows desorption and ionization of lipids.

Nowadays, pulsed, frequency tripled Nd:YAG UV lasers (355 nm) are usually employed for MALDI experiments with a repetition rate of 1000 Hz in commercial instruments for sufficient data acquisition. In MALDI-IMS, the resolving power for application strongly depends on the sample preparation step (e.g., matrix crystal size), stepper motor accuracy, and laser spot sizes. To achieve MALDI-IMS to a practical resolution, the laser spot size of 20 μm is usually used. Therefore, the time needed to obtain images from a sample depends on the number of analyzed spots, the repetition rate of the laser (Hz), and the data collecting and processing speed of computers. For example, imaging a whole-body mouse or rat section with current commercially available MALDI mass spectrometers equipped with lasers operating at 1 kHz would take 2–4 h.

Detection and imaging of lipid molecules were mostly studied using rodent brains. Similar to the analysis of lipid extracts, ion peaks corresponding to PC molecular species are prominent in positive-ion MALDI-IMS analysis of these tissue samples. For example, whole rodent brain sections were determined by MALDI-TOFIMS to examine the distribution of three types of prominent PC species such as PC(32:0), PC(34:1), and PC(36:1) [11] and to construct the atlases of their distributions [57]. Moreover, many lipid species such as PI, PA, PG, PE, PS, ST, and gangliosides were measured from adult mouse brain tissue sections by MALDI-IMS in the negative-ion mode [62]. *In situ* structural analysis by tandem MS analysis is necessary to definitively confirm a lipid species. For example, lithium adducts of PC species were analyzed by MALDI-MS/MS and yielded fragments that allowed for the identification and positional assignment of acyl groups in PC species [58].

From IMS of lipids, alterations in lipid distribution due to changes of physiological or pathological conditions could be determined. For example, age-dependent changes in the distribution and amount of PC species in rat brain have been previously evaluated [57]; and changes in localization of lipids after traumatic brain injury in rat brain have been determined [72]. A study by MALDI-IMS dynamically demonstrated the changes in skeletal muscle lipid composition induced by contraction [40]. The investigators found the reduction of DAG and TAG, as well as the accumulation of PC species in the contracted muscles.

Studies with IMS of lipids can address many specific scientific questions related to the distributions of particular types of lipids. For instance, determination of the distribution of GPL species containing polyunsaturated fatty acids in a particular type of cells in a mouse brain section [57]; of gangliosides in different regions of the brain [73]; of ST in different layers of rat hippocampus [74]; and of PC and GalCer species in rat brain sections [75] has all been well addressed.

It is well known that lipids, particularly those containing a quaternary amine group such as PC and SM species, can readily form adducts with alkali metal ions in the ion sources of soft ionization such as ESI and MALDI (Chapter 2). By exploiting this feature, IMS of lipids could be used to probe the differential distribution of endogenous alkali ions or the changes of these ions under a patho(physio)logical condition such as traumatic brain injury as previously demonstrated [72]. In the study, no signs of trauma on the control hemisphere were revealed by light microscopy. However, MALDI-IMS analysis of the most abundant PC species (i.e., *m/z* 760.6, 16:0–18:1

PC as $[M+H]^+$) showed that this PC species was apparently less intense in the region of injury. Further investigation revealed that this reduction of $[M+H]^+$ intensity is likely due to the changes of local alkali metal ion concentrations in the regions of injury. It was evidenced that the images of both Na^+ and K^+ adduct ions of this PC species were changed dramatically, that is, a significantly increased intensity of its sodium adduct (i.e., $[M+Na]^+$), but depletion of its potassium adduct (i.e., $[M+K]^+$) at the region of injury. The patterns of alkali metal adducts of other PC species also showed the same trend [72]. These observations indicate that IMS of lipids can probe the differential distribution or changes of endogenous alkali metal ions.

New strategies and/or use of new matrices circumvent many problems in tissue mapping [30, 55, 56, 61, 62]. An improved sample preparation protocol by using aqueous washes could significantly enhance signal intensity and thus, increase the number of analytes recorded from adult mouse brain tissue sections [62]. Homogeneous spotting of the matrix on the tissue slice is very critical for a successful and meaningful analysis. Matrix spotting with sublimation has markedly improved this area [61]. However, solutions to the potential disturbance of the spotted matrix to cellular organelles and to the extraction of quantitative results from the analysis are still needed [76].

Matrix implantation of nanoparticles provides advantages for imaging several classes of lipids in tissue sections. An implantation of nanoparticles led to detection of most brain lipids (including neutral lipid species such as GalCer) in a more efficient manner than traditional organic MALDI matrices [75]. A similar implantation of nanoparticles across the entire heart tissue section resulted in a quick, reproducible, solvent-free, uniform matrix layer near the tissue surface. MALDI-IMS analysis of the samples in either positive- or negative-ion mode yielded high-quality images of several heart lipid species. In the negative-ion mode, MS imaging of 24 lipid species (16 PE, 4 PI, 1 PG, 1 CL, and 2 SM species) was obtained. In the positive-ion mode, images of 29 lipid species including 10 PC, 5 PE, 5 SM, and 9 TAG species were obtained from mouse heart sections [37]. These studies clearly demonstrated the advantage and utility of the nanoparticle types of matrices for IMS of lipids.

Introduction of atmospheric pressure (AP) MALDI greatly improves the IMS analysis of lipids. One of the main advantages of AP MALDI relative to conventional vacuum MALDI is the collisional cooling of ions (i.e., the facile removal of ion internal energy by collisions with a neutral gas) during transfer at atmospheric pressure. This results in less fragmentation than conventional MALDI.

Additionally, AP MALDI also possesses the following:

- The ability to easily interchange an AP MALDI source with other ionization sources.
- The capability to analyze compounds and alternative matrices that are not stable under vacuum conditions.
- The ease of introduction and subsequent access to the sample plate resulting in higher throughput and the ability to manipulate the sample during analysis.

Nevertheless, AP MALDI is generally less sensitive than vacuum MALDI due to the ion loss occurred during transferring from the ambient environment into the instrument.

12.3 SECONDARY-ION MASS SPECTROMETRY IMAGING

SIMS is a desorption and ionization technique used for IMS, in which a primary ion beam (e.g., metal ions) is employed to produce secondary ions from the surface of a sample of interest. In SIMS, a sample surface under high vacuum is bombarded with an energetic primary ion beam (typically 1–40 keV). As the primary ion beam strikes the surface of a sample, a collisional cascade involving atoms and fragments within ~10 nm of the surface is initiated. When moving away from the primary collision site, less fragmentation occurs. The released material consists of neutral fragments, electrons, and ionized species, which are the so-called secondary ions. Typically, less than 1% of the total ejected materials are this type of ions. Depending on the electron configuration of the surface molecules, both positive and negative ions can be generated. Monatomic primary ions (Ar^+, Ga^+, In^+, Au^+, Xe^+, Bi^+) or softer cluster primary ion beams such as C_{60}^+, SF_5^+, Bi_3^+, Au_n^+, and Cs_n^+ are usually used [77, 78]. The latter types of primary ion beams allow desorbing secondary ions from the sample surface without extensive fragmentation.

There are two obvious differences of SIMS from MALDI-IMS. The primary ion beam can be focused as sharply as 50 nm, depending on the primary ion beam current and the charge state of the primary ions. Therefore, the spatial resolution of imaging by using SIMS is much higher than that of MALDI-IMS. On the other hand, the energy of the primary ions is typically in the range of 5–25 keV, which is much higher than that of the bond energies of surface molecules. A collision cascade transfers the energy of the primary ions to the sample surface and yields extensive fragmentation of surface molecules.

Samples for SIMS analysis are usually mounted onto steel, glass, or silicon substrates. A wide variety of samples can be analyzed by this surface specific technique. The sample must be stable under high vacuum conditions, ensuing that both the primary and secondary ions without undergoing collisions from their origins. The surface morphology of the sample can affect the generation of secondary ions. Therefore, coating with either metal (e.g., silver, gold, or platinum) or matrix (the same as standard MALDI matrices) to the sample surface has been developed to minimize fragmentation of surface molecules from the primary ion beam impact [79, 80]. It should be recognized that complications could arise from the matrix application, which can lead to analyte delocalization, hot spots associated with matrix crystallization, and a reduction in spatial resolution dependent on the matrix crystal size [81]. Background noise is also a concern in some cases.

Even with those surface modifications and with the developments of cluster ion sources, the high-energy impaction process of SIMS often results in extensive molecular fragmentation. This makes imaging of intact lipids difficult. Thus, SIMS is not the best technique for IMS of lipids present in biological tissue and extracts. Nonetheless, the unique ability of SIMS to acquire high-spatially resolved molecular information at submicron resolution (compared to typical resolutions of 25–200 μm) retains the great interest in SIMS for lipid analysis, in particular imaging of tissue sections and single cells.

Generally, class-specific fragment ions such as the phosphocholine fragment at m/z 184 for PC and SM, an ion at m/z 69 ($C_5H_9^+$, a hydrocarbon fragment characteristic of total GPL species), and an ion at m/z 126 characteristic of 2-aminophosphonolipid are mapped by using monatomic primary ions. The advent of cluster primary ion beams has significantly increased the utility of SIMS for intact lipid analysis [77, 82]. The higher secondary ion yield and lower extent of fragmentation produced by C_{60}^+ ion beams result in much promise for the analysis of intact lipids [83]. Nonetheless, SIMS is currently the only MS-based method providing submicron spatial resolution and allowing the distribution of lipids across a cell surface, including 3D lipid distributions, to be investigated [84], and thus represents an important tool for lipid analysis.

Below are the highlights of a few applications by SIMS for lipid analysis. In a study on rat brain tissue, abundant signals corresponding to cholesterol (m/z 369 and 385) and a range of intact GPL species from m/z 700–800 were detected with a C_{60}^+ ion beam [85]. Similarly, in studying rat cerebellum by SIMS using a Bi_3^+ cluster source, intact ions corresponding to cholesterol, PC, and GalCer lipid species were observed [86]. Other intact lipid species such as glycerophospholipids, glycerolipids, fatty acids, sterols, prenol lipids, and sphingolipids can be detected as intact ions from mammalian tissues [87]. However, lipid species from some classes are seldom detected. For example, intact PE species and cholesterol esters have not been detected by SIMS, which is most likely due to the facile loss of the PE head group and the fatty acyl chains of cholesterol esters, respectively.

12.4   DESI-MS IMAGING

DESI, an ionization technique developed by Cooks and coworkers [88] can be used for IMS analysis under atmospheric pressure with minimal sample preparation [89] (see Section 2.5.2). For DESI imaging, the sample is either placed onto a target (e.g., microscope glass slide) or analyzed *in situ* [12]. A lateral resolution of 40 μm could be achieved by DESI [90].

In typical DESI imaging experiments, the collected tissue sample is flash frozen in liquid nitrogen and subsequently cut in micron thin sections using a cryostat-microtome; the thin tissue slices are thaw mounted onto glass microscope slides for analysis; the ions generated by the DESI (Chapter 2) are transported from surfaces to the gas phase for mass analysis as the surface is being moved in order to cover the entire sample area; a mass spectrum is acquired for each pixel on the surface; and finally, tissue images are constructed to display the spatial intensity distribution of individual selected lipid ion.

For example, DESI-MS imaging was performed to distinguish spatial distributions of lipid profiles present in normal or benign, ductal carcinoma *in situ*, and invasive ductal carcinoma specimens of human breast samples [12]. The distribution of two lipid ions at *m/z* 863 and 818 detected in the negative-ion mode and identified with MS/MS as PI and PS species, respectively, was visualized in breast tissue samples of benign, ductal carcinoma *in situ*, and invasive ductal carcinoma. The intensities of these ions are significantly different among the analyzed tissue samples. Specifically, the distribution obtained from the samples of ductal carcinoma *in situ* demonstrated significant intensities of the PI ion, but virtually absent for the PS ion. These observations indicate that DESI imaging allows the researcher not only to distinguish cancerous *vs.* non-cancerous tissue or regions but also to identify the specific type and stage of the cancer.

A list of studies of IMS of lipids by DESI-MS can be found in a recent review article [18]. Many of the studies by DESI-IMS have focused on identifying the differences of lipid distributions of tissue samples between disease and health states in an attempt to discover lipid biomarkers for disease [41, 91–96]. It indicates that most of the major lipid classes have been successfully detected by DESI. All lipid classes can be detected as intact ions with little fragmentation except cholesterol, which is detected as the $[M+H-H_2O]^+$ ion as usual. In comparison to MALDI-IMS [97], DESI-IMS of human lens [98] indicated that DESI allows detecting a wider variety of lipid classes in a single acquisition under typical conditions. For example, MALDI-IMS only detected SM, Cer, and cholesterol species, whereas DESI-IMS detected SM, Cer, ceramide-1-phosphate, PE, lysoPE, PS, LacCer, and cholesterol.

The ability to use the same spray solution for analysis of lipid species in both positive-and negative-ion modes is an advantage of DESI for imaging with aid of a modifier in comparison to MALDI-IMS, which often needs different matrices for optimal desorption/ionization of different lipids. For example, in the presence of sodium or ammonium salts, TAG, PC, and SM species can be readily detected in the positive-ion mode [99], whereas PC and SM as well as other anionic lipids can be detected in the negative-ion mode [100].

Through manipulation of the composition of the spray solution by adding a reagent that selectively reacts with a target analyte and enhances its detection, detection of cholesterol and other hydroxyl group-containing nonpolar lipids such as steroids and some vitamins [101–103]. For example, inclusion of betaine aldehyde into the spray solution, which reacts with cholesterol led to production of a hemiacetal with a fixed-positively charged trimethylammonium group, thereby greatly enhancing the detection sensitivity for analysis of cholesterol [103]. In the study, quantitation was also achieved by the addition of a standard of d_7-cholesterol. This resulted in a relative standard deviation of 1.2–6.4% and is in a good agreement with other analysis methods as compared.

A drawback of DESI-IMS analysis of lipids in tissue sections in comparison to that of MALDI or SIMS is its tissue damage to a much larger degree. This consequence makes the tissue samples after DESI analysis difficult to acquire the complementary information such as histological data for direct comparison between data sets. To resolve this issue, Eberlin et al. developed a method [29, 104] by using binary solvent systems containing dimethylformamide and either ethanol or acetonitrile to acquire high-quality mass spectral data with minimal tissue damage.

12.5   ION-MOBILITY IMAGING

Ion mobility (IM) is a gas-phase separation technology (see Chapter 4) that adds a new dimension of separation to MS analysis. Therefore, when IM is coupled to an ionization source (e.g., DESI or MALDI), IM-MS results in improving the resolving power for MS imaging and allows better characterization of detected biomolecules of interest. Due to the rapid gas-phase separation of molecules in ion mobility, this technology offers unique advantages for IMS of tissue sections as demonstrated in some recent studies [75, 105, 106]. It should be pointed out that the coupling of IM apparatus to an MS instrument does not lead to enhancing, indeed reducing in some cases, the sensitivity of the instruments, since it does not modify the ionization conditions.

The coupling of MALDI with IM-MS allows researchers to directly probe tissue samples, map the distribution of lipids, and elucidate molecular structure with minimal preparation and within an interval of a few hundred microseconds between the applications of each focused laser desorption pulse to the sample [107]. For example, the distribution of PC and GalCer species was mapped from 16-µm-thick coronal rat brain sections using MALDI-IM-oTOF-MS [75]. Although GalCer species are enriched in brain tissue, detection of these molecular ions in the positive-ion mode is generally prohibited due to the ion suppression of the coexisting lipid species such as PC and SM that are readily ionized, indicating the great advantage of IM in IMS analysis of GalCer species.

MALDI-IM IMS has the ability to improve the imaging of some lipids by separating such ions from endogenous or matrix-related isobaric ions. One such application is the demonstration of selective lipid imaging from rat brain [105]. In the study, analyte ions were separated on the basis of both ion–neutral collision cross section and m/z, which provided rapid separation of isobaric, but structurally distinct ions. The study demonstrated that IM-MS provides three primary benefits: (1) qualitative identification of the analyte molecular class, (2) suppression of chemical noise, and (3) potential for high mass accuracy measurement by using internal calibrants that do not interfere with the analyte of interest on the basis of structure. Thus, in imaging IM-MS, multiple images can be selectively obtained for different types of ions that are isobaric, or for structural/conformational differences for the same type of ion (e.g., GluCer $vs.$ GlaCer for lipid analysis).

12.6   ADVANTAGES AND DRAWBACKS OF IMAGING MASS SPECTROMETRY FOR ANALYSIS OF LIPIDS

12.6.1   Advantages

IMS offers unique features for sample surface analysis and possess many advantages in comparison to other analytical techniques. Most of the available IMS methods offer high spatial resolution for analysis of biomolecules present in tissue section surface. Typically, MALDI-IMS could achieve probing spots as small as 25 µm in diameter [108], while an SIMS ion beam can be focused to the size of 50 nm in diameter [109]. Considering a typical mammalian cell size of 10 µm, MALDI-based IMS could acquire individual mass spectrum to cover approximately four cells per image point while SIMS can probe subcellular structures [110].

The majority of the modern mass spectrometers are capable of separating and detecting ions with very high mass resolution. Resolving power in excess of 10^5 is routinely available on commercial instruments. High mass resolving power means that the isobaric ions hidden with low-resolution technologies can be detected [111]. Ion mobility separation combined with mass spectrometry allows the gas-phase separation of isobaric ions with similar nominal mass during an imaging experiment [106].

IMS analysis takes full advantage of high sensitivity, which is available for the majority of the mass spectrometers commercially available nowadays. Such high sensitivity allows us to image cellular lipids present at low to very low concentrations in biological samples. Many instruments possess the sensitivity to detect analytes at the concentration of low femtomolar to high attomolar levels, which facilitates detection of the components from a single cell [112].

Currently, the technologies utilized in many types of commercially available mass spectrometers are mature enough to detect a variety of lipid classes from an identical biological sample (Chapter 2). This provides a foundation for IMS to visualize different classes of lipids present inside the cells. MS instruments can also be used for imaging unknown lipid species present in the biological sample without any *a priori* knowledge or labels. This is a key advantage of IMS, as unknown molecules can subsequently be identified using MS/MS techniques or in combination with the IM technology. For this purpose, an ion of interest (i.e., a molecular ion) is fragmented and its fragment ions are mass analyzed. This type of analysis can be performed to the MS^n nowadays.

Another advantage of IMS is that endogenous lipid analysis directly from tissue sections can be performed with minimal sample preparation, particularly in the case of IMS by DESI. Usually, sample preparation for IMS only requires a short tissue wash followed by matrix application (e.g., for MALDI and SIMS).

Lastly, the time of sample analysis by IMS is relatively short. The data acquisition of an average imaging experiment lasts from minutes to several hours depending on the instrumentation utilized. The process is fully automated and usually does not require any supervision.

12.6.2 Limitations

Generally, an IMS experiment lasts for a few hours, thus requires that the samples remain stable at room temperature and/or in high vacuum, under which some lipids might degrade. The sample preparation process must be carried out as fast as possible without exposing the samples to the air or room temperature for too long. Moreover, the risk of sample contamination and molecular diffusion, which can affect the reproducibility of the data, complicate their analysis, or affect the quality of the image, must be recognized during sample preparation, in which cutting, washing, and matrix application are generally involved as aforementioned. In the case of IMS analysis of lipids by MADLI-MS, the spatial resolution can be affected by the matrix crystal size, and the quality of the images can be compromised by the presence of matrix clusters and their alkali metal ion adducts. On the other hand, although sample preparation in IMS analysis by SIMS is relatively simple in comparison to MALDI-MS, its extensive in-source fragmentation eliminates lots of information about the spatial distributions at the molecular levels.

Biological tissue represents an extremely complex and challenging sample for direct analysis of lipids by IMS. The coexisting biomolecules may result in ion suppression for lipid analysis. The presence of ion-suppressing phenomena apparently limits the number of detected lipid species. Therefore, great care has to be taken that proper control experiments are conducted before drawing any conclusion from a given IMS experiment.

REFERENCES

1. Adibhatla, R.M. and Hatcher, J.F. (2010) Lipid oxidation and peroxidation in CNS health and disease: From molecular mechanisms to therapeutic opportunities. Antioxid. Redox. Signal. 12, 125–169.

2. DeFronzo, R.A. (2010) Insulin resistance, lipotoxicity, type 2 diabetes and atherosclerosis: the missing links. The Claude Bernard Lecture 2009. Diabetologia 53, 1270–1287.

3. Jones, L., Harold, D. and Williams, J. (2010) Genetic evidence for the involvement of lipid metabolism in Alzheimer's disease. Biochim. Biophys. Acta 1801, 754–761.

4. Narayan, S. and Thomas, E.A. (2011) Sphingolipid abnormalities in psychiatric disorders: A missing link in pathology? Front. Biosci. 16, 1797–1810.

5. O'Donnell, V.B. and Murphy, R.C. (2012) New families of bioactive oxidized phospholipids generated by immune cells: Identification and signaling actions. Blood 120, 1985–1992.

6. Takahashi, T. and Suzuki, T. (2012) Role of sulfatide in normal and pathological cells and tissues. J. Lipid Res. 53, 1437–1450.

7. Thomas, C.P. and O'Donnell, V.B. (2012) Oxidized phospholipid signaling in immune cells. Curr. Opin. Pharmacol. 12, 471–477.

8. Aldrovandi, M. and O'Donnell, V.B. (2013) Oxidized PLs and vascular inflammation. Curr. Atheroscler. Rep. 15, 323.

9. Murphy, S.A. and Nicolaou, A. (2013) Lipidomics applications in health, disease and nutrition research. Mol. Nutr. Food Res. 57, 1336–1346.

10. Puppolo, M., Varma, D. and Jansen, S.A. (2014) A review of analytical methods for eicosanoids in brain tissue. J. Chromatogr. B 964, 50–64.

11. Mikawa, S., Suzuki, M., Fujimoto, C. and Sato, K. (2009) Imaging of phosphatidylcholines in the adult rat brain using MALDI-TOF MS. Neurosci. Lett. 451, 45–49.

12. Dill, A.L., Ifa, D.R., Manicke, N.E., Ouyang, Z. and Cooks, R.G. (2009) Mass spectrometric imaging of lipids using desorption electrospray ionization. J. Chromatogr. B 877, 2883–2889.

13. Malmberg, P., Nygren, H., Richter, K., Chen, Y., Dangardt, F., Friberg, P. and Magnusson, Y. (2007) Imaging of lipids in human adipose tissue by cluster ion TOF-SIMS. Microsc. Res. Tech. 70, 828–835.

14. Almeida, R., Berzina, Z., Arnspang, E.C., Baumgart, J., Vogt, J., Nitsch, R. and Ejsing, C.S. (2015) Quantitative spatial analysis of the mouse brain lipidome by pressurized liquid extraction surface analysis. Anal. Chem. 87, 1749–1756.

15. Brunelle, A. and Laprevote, O. (2009) Lipid imaging with cluster time-of-flight secondary ion mass spectrometry. Anal. Bioanal. Chem. 393, 31–35.

16. Chughtai, K. and Heeren, R.M. (2010) Mass spectrometric imaging for biomedical tissue analysis. Chem. Rev. 110, 3237–3277.

17. Gode, D. and Volmer, D.A. (2013) Lipid imaging by mass spectrometry: A review. Analyst 138, 1289–1315.

18. Ellis, S.R., Brown, S.H., In Het Panhuis, M., Blanksby, S.J. and Mitchell, T.W. (2013) Surface analysis of lipids by mass spectrometry: More than just imaging. Prog. Lipid Res. 52, 329–353.

19. Mclean, J.A., Schultz, J.A. and Woods, A.S. (2010) Ion mobility – mass spectrometry. In Electrospray and MALDI Mass Spectrometry: Fundamentals, Instrumentation, Practicalities, and Biological Applications (Cole, R. B., ed.). pp. 411–439, John Wiley & Sons, Inc., Hoboken, NJ

20. Woods, A.S. and Jackson, S.N. (2010) The application and potential of ion mobility mass spectrometry in imaging MS with a focus on lipids. Methods Mol. Biol. 656, 99–111.

21. Ellis, S.R., Bruinen, A.L. and Heeren, R.M. (2014) A critical evaluation of the current state-of-the-art in quantitative imaging mass spectrometry. Anal. Bioanal. Chem. 406, 1275–1289.

22. Berry, K.A., Hankin, J.A., Barkley, R.M., Spraggins, J.M., Caprioli, R.M. and Murphy, R.C. (2011) MALDI imaging of lipid biochemistry in tissues by mass spectrometry. Chem. Rev. 111, 6491–6512.

23. Murphy, R.C., Hankin, J.A., Barkley, R.M. and Zemski Berry, K.A. (2011) MALDI imaging of lipids after matrix sublimation/deposition. Biochim. Biophys. Acta 1811, 970–975.

24. Fletcher, J.S. (2009) Cellular imaging with secondary ion mass spectrometry. Analyst 134, 2204–2215.

25. Cha, S., Zhang, H., Ilarslan, H.I., Wurtele, E.S., Brachova, L., Nikolau, B.J. and Yeung, E.S. (2008) Direct profiling and imaging of plant metabolites in intact tissues by using colloidal graphite-assisted laser desorption ionization mass spectrometry. Plant J. 55, 348–360.

26. Delvolve, A.M., Colsch, B. and Woods, A.S. (2011) Highlighting anatomical sub-structures in rat brain tissue using lipid imaging. Anal. Methods 3, 1729–1736.

27. Murphy, R.C., Hankin, J.A. and Barkley, R.M. (2009) Imaging of lipid species by MALDI mass spectrometry. J. Lipid Res. 50 Suppl, S317–S322.

28. Brulet, M., Seyer, A., Edelman, A., Brunelle, A., Fritsch, J., Ollero, M. and Laprevote, O. (2010) Lipid mapping of colonic mucosa by cluster TOF-SIMS imaging and multivariate analysis in cftr knockout mice. J. Lipid Res. 51, 3034–3045.

29. Eberlin, L.S., Liu, X., Ferreira, C.R., Santagata, S., Agar, N.Y. and Cooks, R.G. (2011) Desorption electrospray ionization then MALDI mass spectrometry imaging of lipid and protein distributions in single tissue sections. Anal. Chem. 83, 8366–8371.

30. Chen, Y., Allegood, J., Liu, Y., Wang, E., Cachon-Gonzalez, B., Cox, T.M., Merrill, A.H., Jr. and Sullards, M.C. (2008) Imaging MALDI mass spectrometry using an oscillating capillary nebulizer matrix coating system and its application to analysis of lipids in brain from a mouse model of Tay-Sachs/Sandhoff disease. Anal. Chem. 80, 2780–2788.

31. Amaya, K.R., Sweedler, J.V. and Clayton, D.F. (2011) Small molecule analysis and imaging of fatty acids in the zebra finch song system using time-of-flight-secondary ion mass spectrometry. J. Neurochem. 118, 499–511.

32. Ishida, Y., Nakanishi, O., Hirao, S., Tsuge, S., Urabe, J., Sekino, T., Nakanishi, M., Kimoto, T. and Ohtani, H. (2003) Direct analysis of lipids in single zooplankter individuals by matrix-assisted laser desorption/ionization mass spectrometry. Anal. Chem. 75, 4514–4518.

33. Shahidi-Latham, S.K., Dutta, S.M., Prieto Conaway, M.C. and Rudewicz, P.J. (2012) Evaluation of an accurate mass approach for the simultaneous detection of drug and metabolite distributions via whole-body mass spectrometric imaging. Anal. Chem. 84, 7158–7165.

34. Berry, K.A., Li, B., Reynolds, S.D., Barkley, R.M., Gijon, M.A., Hankin, J.A., Henson, P.M. and Murphy, R.C. (2011) MALDI imaging MS of phospholipids in the mouse lung. J. Lipid Res. 52, 1551–1560.

35. Desbenoit, N., Saussereau, E., Bich, C., Bourderioux, M., Fritsch, J., Edelman, A., Brunelle, A. and Ollero, M. (2014) Localized lipidomics in cystic fibrosis: TOF-SIMS imaging of lungs from Pseudomonas aeruginosa-infected mice. Int. J. Biochem. Cell. Biol. 52, 77–82.

36. Sparvero, L.J., Amoscato, A.A., Dixon, C.E., Long, J.B., Kochanek, P.M., Pitt, B.R., Bayir, H. and Kagan, V.E. (2012) Mapping of phospholipids by MALDI imaging (MALDI-MSI): realities and expectations. Chem. Phys. Lipids 165, 545–562.

37. Jackson, S.N., Baldwin, K., Muller, L., Womack, V.M., Schultz, J.A., Balaban, C. and Woods, A.S. (2014) Imaging of lipids in rat heart by MALDI-MS with silver nanoparticles. Anal. Bioanal. Chem. 406, 1377–1386.

38. Ruh, H., Salonikios, T., Fuchser, J., Schwartz, M., Sticht, C., Hochheim, C., Wirnitzer, B., Gretz, N. and Hopf, C. (2013) MALDI imaging MS reveals candidate lipid markers of polycystic kidney disease. J. Lipid Res. 54, 2785–2794.

39. Touboul, D., Piednoel, H., Voisin, V., De La Porte, S., Brunelle, A., Halgand, F. and Laprevote, O. (2004) Changes in phospholipid composition within the dystrophic muscle by matrix-assisted laser desorption/ionization mass spectrometry and mass spectrometry imaging. Eur. J. Mass Spectrom. 10, 657–664.

40. Goto-Inoue, N., Manabe, Y., Miyatake, S., Ogino, S., Morishita, A., Hayasaka, T., Masaki, N., Setou, M. and Fujii, N.L. (2012) Visualization of dynamic change in contraction-induced lipid composition in mouse skeletal muscle by matrix-assisted laser desorption/ionization imaging mass spectrometry. Anal. Bioanal. Chem. 403, 1863–1871.

41. Eberlin, L.S., Norton, I., Dill, A.L., Golby, A.J., Ligon, K.L., Santagata, S., Cooks, R.G. and Agar, N.Y. (2012) Classifying human brain tumors by lipid imaging with mass spectrometry. Cancer Res. 72, 645–654.

42. Jones, E.E., Powers, T.W., Neely, B.A., Cazares, L.H., Troyer, D.A., Parker, A.S. and Drake, R.R. (2014) MALDI imaging mass spectrometry profiling of proteins and lipids in clear cell renal cell carcinoma. Proteomics 14, 924-935.

43. Cimino, J., Calligaris, D., Far, J., Debois, D., Blacher, S., Sounni, N.E., Noel, A. and De Pauw, E. (2013) Towards lipidomics of low-abundant species for exploring tumor heterogeneity guided by high-resolution mass spectrometry imaging. Int. J. Mol. Sci. 14, 24560–24580.

44. Passarelli, M.K., Ewing, A.G. and Winograd, N. (2013) Single-cell lipidomics: characterizing and imaging lipids on the surface of individual Aplysia californica neurons with cluster secondary ion mass spectrometry. Anal. Chem. 85, 2231–2238.

45. Li, L., Garden, R.W. and Sweedler, J.V. (2000) Single-cell MALDI: A new tool for direct peptide profiling. Trends Biotechnol. 18, 151–160.

46. Carter, C.L., McLeod, C.W. and Bunch, J. (2011) Imaging of phospholipids in formalin fixed rat brain sections by matrix assisted laser desorption/ionization mass spectrometry. J. Am. Soc. Mass Spectrom. 22, 1991–1998.

47. Schwartz, S.A., Reyzer, M.L. and Caprioli, R.M. (2003) Direct tissue analysis using matrix-assisted laser desorption/ionization mass spectrometry: Practical aspects of sample preparation. J. Mass Spectrom. 38, 699–708.

48. Chen, R., Hui, L., Sturm, R.M. and Li, L. (2009) Three dimensional mapping of neuropeptides and lipids in crustacean brain by mass spectral imaging. J. Am. Soc. Mass Spectrom. 20, 1068–1077.

49. Lemaire, R., Desmons, A., Tabet, J.C., Day, R., Salzet, M. and Fournier, I. (2007) Direct analysis and MALDI imaging of formalin-fixed, paraffin-embedded tissue sections. J. Proteome Res. 6, 1295–1305.

50. Kim, Y., Shanta, S.R., Zhou, L.H. and Kim, K.P. (2010) Mass spectrometry based cellular phosphoinositides profiling and phospholipid analysis: A brief review. Exp. Mol. Med. 42, 1–11.

51. Cerruti, C.D., Benabdellah, F., Laprevote, O., Touboul, D. and Brunelle, A. (2012) MALDI imaging and structural analysis of rat brain lipid negative ions with 9-aminoacridine matrix. Anal. Chem. 84, 2164–2171.

52. Thomas, A., Charbonneau, J.L., Fournaise, E. and Chaurand, P. (2012) Sublimation of new matrix candidates for high spatial resolution imaging mass spectrometry of lipids: Enhanced information in both positive and negative polarities after 1,5-diaminonapthalene deposition. Anal. Chem. 84, 2048–2054.

53. Wang, X., Han, J., Pan, J. and Borchers, C.H. (2014) Comprehensive imaging of porcine adrenal gland lipids by MALDI-FTMS using quercetin as a matrix. Anal. Chem. 86, 638–646.

54. McLean, J.A., Stumpo, K.A. and Russell, D.H. (2005) Size-selected (2–10 nm) gold nanoparticles for matrix assisted laser desorption ionization of peptides. J. Am. Chem. Soc. 127, 5304–5305.

55. Patti, G.J., Shriver, L.P., Wassif, C.A., Woo, H.K., Uritboonthai, W., Apon, J., Manchester, M., Porter, F.D. and Siuzdak, G. (2010) Nanostructure-initiator mass spectrometry (NIMS) imaging of brain cholesterol metabolites in Smith-Lemli-Opitz syndrome. Neuroscience 170, 858–864.

56. Hayasaka, T., Goto-Inoue, N., Zaima, N., Shrivas, K., Kashiwagi, Y., Yamamoto, M., Nakamoto, M. and Setou, M. (2010) Imaging mass spectrometry with silver nanoparticles reveals the distribution of fatty acids in mouse retinal sections. J. Am. Soc. Mass Spectrom. 21, 1446–1454.

57. Sugiura, Y., Konishi, Y., Zaima, N., Kajihara, S., Nakanishi, H., Taguchi, R. and Setou, M. (2009) Visualization of the cell-selective distribution of PUFA-containing phosphatidyl-cholines in mouse brain by imaging mass spectrometry. J. Lipid Res. 50, 1776–1788.

58. Jackson, S.N., Wang, H.Y. and Woods, A.S. (2005) In situ structural characterization of phosphatidylcholines in brain tissue using MALDI-MS/MS. J. Am. Soc. Mass Spectrom. 16, 2052–2056.

59. Sun, G., Yang, K., Zhao, Z., Guan, S., Han, X. and Gross, R.W. (2008) Matrix-assisted laser desorption/ionization time-of-flight mass spectrometric analysis of cellular glyc-erophospholipids enabled by multiplexed solvent dependent analyte-matrix interactions. Anal. Chem. 80, 7576–7585.

60. Franck, J., Arafah, K., Barnes, A., Wisztorski, M., Salzet, M. and Fournier, I. (2009) Improving tissue preparation for matrix-assisted laser desorption ionization mass spec-trometry imaging. Part 1: Using microspotting. Anal. Chem. 81, 8193–8202.

61. Hankin, J.A., Barkley, R.M. and Murphy, R.C. (2007) Sublimation as a method of matrix application for mass spectrometric imaging. J. Am. Soc. Mass Spectrom. 18, 1646–1652.

62. Angel, P.M., Spraggins, J.M., Baldwin, H.S. and Caprioli, R. (2012) Enhanced sensitiv-ity for high spatial resolution lipid analysis by negative ion mode matrix assisted laser desorption ionization imaging mass spectrometry. Anal. Chem. 84, 1557–1564.

63. Zhang, H., Cha, S. and Yeung, E.S. (2007) Colloidal graphite-assisted laser desorp-tion/ionization MS and MS(n) of small molecules. 2. Direct profiling and MS imaging of small metabolites from fruits. Anal. Chem. 79, 6575–6584.

64. Patti, G.J., Woo, H.K., Yanes, O., Shriver, L., Thomas, D., Uritboonthai, W., Apon, J.V., Steenwyk, R., Manchester, M. and Siuzdak, G. (2010) Detection of carbohydrates and steroids by cation-enhanced nanostructure-initiator mass spectrometry (NIMS) for biofluid analysis and tissue imaging. Anal. Chem. 82, 121–128.

65. Puolitaival, S.M., Burnum, K.E., Cornett, D.S. and Caprioli, R.M. (2008) Solvent-free matrix dry-coating for MALDI imaging of phospholipids. J. Am. Soc. Mass Spectrom. 19, 882–886.

66. Aerni, H.R., Cornett, D.S. and Caprioli, R.M. (2006) Automated acoustic matrix depo-sition for MALDI sample preparation. Anal. Chem. 78, 827–834.

67. Wei, J., Buriak, J.M. and Siuzdak, G. (1999) Desorption-ionization mass spectrometry on porous silicon. Nature 399, 243–246.

68. Liu, Q., Guo, Z. and He, L. (2007) Mass spectrometry imaging of small molecules using desorption/ionization on silicon. Anal. Chem. 79, 3535–3541.

69. Jardin-Mathe, O., Bonnel, D., Franck, J., Wisztorski, M., Macagno, E., Fournier, I. and Salzet, M. (2008) MITICS (MALDI Imaging Team Imaging Computing System): A new open source mass spectrometry imaging software. J. Proteomics 71, 332–345.

70. Sanchez, J.C., Corthals, G.L. and Hochstrasser, D.F. (2004) Biomedical Applications of Proteomics. Weinheim, Wiley-VCH Verlag Gmbh. pp 373.

71. Rohner, T.C., Staab, D. and Stoeckli, M. (2005) MALDI mass spectrometric imaging of biological tissue sections. Mech. Ageing Dev. 126, 177–185.

72. Hankin, J.A., Farias, S.E., Barkley, R.M., Heidenreich, K., Frey, L.C., Hamazaki, K., Kim, H.Y. and Murphy, R.C. (2011) MALDI mass spectrometric imaging of lipids in rat brain injury models. J. Am. Soc. Mass Spectrom. 22, 1014–1021.

73. Sugiura, Y., Shimma, S., Konishi, Y., Yamada, M.K. and Setou, M. (2008) Imaging mass spectrometry technology and application on ganglioside study; visualization of age-dependent accumulation of C20-ganglioside molecular species in the mouse hippocampus. PLoS One 3, e3232.

74. Ageta, H., Asai, S., Sugiura, Y., Goto-Inoue, N., Zaima, N. and Setou, M. (2009) Layer-specific sulfatide localization in rat hippocampus middle molecular layer is revealed by nanoparticle-assisted laser desorption/ionization imaging mass spectrometry. Med. Mol. Morphol. 42, 16–23.

75. Jackson, S.N., Ugarov, M., Egan, T., Post, J.D., Langlais, D., Albert Schultz, J. and Woods, A.S. (2007) MALDI-ion mobility-TOFMS imaging of lipids in rat brain tissue. J. Mass Spectrom. 42, 1093–1098.

76. Jones, J.J., Borgmann, S., Wilkins, C.L. and O'Brien, R.M. (2006) Characterizing the phospholipid profiles in mammalian tissues by MALDI FTMS. Anal. Chem. 78, 3062–3071.

77. Weibel, D., Wong, S., Lockyer, N., Blenkinsopp, P., Hill, R. and Vickerman, J.C. (2003) A C60 primary ion beam system for time of flight secondary ion mass spectrometry: its development and secondary ion yield characteristics. Anal. Chem. 75, 1754–1764.

78. Klerk, L.A., Lockyer, N.P., Kharchenko, A., MacAleese, L., Dankers, P.Y., Vickerman, J.C. and Heeren, R.M. (2010) C60+ secondary ion microscopy using a delay line detector. Anal. Chem. 82, 801–807.

79. Altelaar, A.F., van Minnen, J., Jimenez, C.R., Heeren, R.M. and Piersma, S.R. (2005) Direct molecular imaging of Lymnaea stagnalis nervous tissue at subcellular spatial resolution by mass spectrometry. Anal. Chem. 77, 735–741.

80. McDonnell, L.A., Piersma, S.R., MaartenAltelaar, A.F., Mize, T.H., Luxembourg, S.L., Verhaert, P.D., van Minnen, J. and Heeren, R.M. (2005) Subcellular imaging mass spectrometry of brain tissue. J. Mass Spectrom. 40, 160–168.

81. Heeren, R.M.A., McDonnell, L.A., Amstalden, E., Luxembourg, S.L., Altelaar, A.F.M. and Piersma, S.R. (2006) Why don't biologists use SIMS?: A critical evaluation of imaging MS. Appl. Surf. Sci. 252, 6827–6835.

82. Touboul, D., Kollmer, F., Niehuis, E., Brunelle, A. and Laprevote, O. (2005) Improvement of biological time-of-flight-secondary ion mass spectrometry imaging with a bismuth cluster ion source. J. Am. Soc. Mass Spectrom. 16, 1608–1618.

83. Fletcher, J.S. and Vickerman, J.C. (2010) A new SIMS paradigm for 2D and 3D molecular imaging of bio-systems. Anal. Bioanal. Chem. 396, 85–104.

84. Postawa, Z., Czerwinski, B., Szewczyk, M., Smiley, E.J., Winograd, N. and Garrison, B.J. (2003) Enhancement of sputtering yields due to C60 versus Ga bombardment of Ag[111] as explored by molecular dynamics simulations. Anal. Chem. 75, 4402–4407.

85. Jones, E.A., Lockyer, N.P. and Vickerman, J.C. (2007) Mass spectral analysis and imaging of tissue by ToF-SIMS—The role of buckminsterfullerene, C60+, primary ions. Int. J. Mass Spec, 260, 146–157.

86. Nygren, H., Borner, K., Hagenhoff, B., Malmberg, P. and Mansson, J.E. (2005) Localization of cholesterol, phosphocholine and galactosylceramide in rat cerebellar cortex with imaging TOF-SIMS equipped with a bismuth cluster ion source. Biochim. Biophys. Acta 1737, 102–110.

87. Passarelli, M.K. and Winograd, N. (2011) Lipid imaging with time-of-flight secondary ion mass spectrometry (ToF-SIMS). Biochim. Biophys. Acta 1811, 976–990.

88. Takats, Z., Wiseman, J.M., Gologan, B. and Cooks, R.G. (2004) Mass spectrometry sampling under ambient conditions with desorption electrospray ionization. Science 306, 471–473.

89. Ifa, D.R., Wiseman, J.M., Song, Q. and Cooks, R.G. (2007) Development of capabilities for imaging mass spectrometry underambient conditions with desorption electrospray ionization (DESI). Int. J. Mass Spectrom. 259, 8–15.

90. Kertesz, V. and Van Berkel, G.J. (2008) Improved imaging resolution in desorption electrospray ionization mass spectrometry. Rapid Commun. Mass Spectrom. 22, 2639–2644.

91. Wiseman, J.M., Puolitaival, S.M., Takats, Z., Cooks, R.G. and Caprioli, R.M. (2005) Mass spectrometric profiling of intact biological tissue by using desorption electrospray ionization. Angew. Chem. Int. Ed. Engl. 44, 7094–7097.

92. Dill, A.L., Ifa, D.R., Manicke, N.E., Costa, A.B., Ramos-Vara, J.A., Knapp, D.W. and Cooks, R.G. (2009) Lipid profiles of canine invasive transitional cell carcinoma of the urinary bladder and adjacent normal tissue by desorption electrospray ionization imaging mass spectrometry. Anal. Chem. 81, 8758–8764.

93. Eberlin, L.S., Dill, A.L., Golby, A.J., Ligon, K.L., Wiseman, J.M., Cooks, R.G. and Agar, N.Y. (2010) Discrimination of human astrocytoma subtypes by lipid analysis using desorption electrospray ionization imaging mass spectrometry. Angew. Chem. Int. Ed. Engl. 49, 5953–5956.

94. Dill, A.L., Eberlin, L.S., Zheng, C., Costa, A.B., Ifa, D.R., Cheng, L., Masterson, T.A., Koch, M.O., Vitek, O. and Cooks, R.G. (2010) Multivariate statistical differentiation of renal cell carcinomas based on lipidomic analysis by ambient ionization imaging mass spectrometry. Anal. Bioanal. Chem. 398, 2969–2978.

95. Masterson, T.A., Dill, A.L., Eberlin, L.S., Mattarozzi, M., Cheng, L., Beck, S.D., Bianchi, F. and Cooks, R.G. (2011) Distinctive glycerophospholipid profiles of human seminoma and adjacent normal tissues by desorption electrospray ionization imaging mass spectrometry. J. Am. Soc. Mass Spectrom. 22, 1326–1333.

96. Eberlin, L.S., Dill, A.L., Costa, A.B., Ifa, D.R., Cheng, L., Masterson, T., Koch, M., Ratliff, T.L. and Cooks, R.G. (2010) Cholesterol sulfate imaging in human prostate cancer tissue by desorption electrospray ionization mass spectrometry. Anal. Chem. 82, 3430–3434.

97. Deeley, J.M., Hankin, J.A., Friedrich, M.G., Murphy, R.C., Truscott, R.J., Mitchell, T.W. and Blanksby, S.J. (2010) Sphingolipid distribution changes with age in the human lens. J. Lipid Res. 51, 2753–2760.

98. Ellis, S.R., Wu, C., Deeley, J.M., Zhu, X., Truscott, R.J., in het Panhuis, M., Cooks, R.G., Mitchell, T.W. and Blanksby, S.J. (2010) Imaging of human lens lipids by desorption electrospray ionization mass spectrometry. J. Am. Soc. Mass Spectrom. 21, 2095–2104.

99. Gerbig, S. and Takats, Z. (2010) Analysis of triglycerides in food items by desorption electrospray ionization mass spectrometry. Rapid Commun. Mass Spectrom. 24, 2186–2192.

100. Manicke, N.E., Wiseman, J.M., Ifa, D.R. and Cooks, R.G. (2008) Desorption electrospray ionization (DESI) mass spectrometry and tandem mass spectrometry (MS/MS) of phospholipids and sphingolipids: Ionization, adduct formation, and fragmentation. J. Am. Soc. Mass Spectrom. 19, 531–543.

101. Song, Y. and Cooks, R.G. (2007) Reactive desorption electrospray ionization for selective detection of the hydrolysis products of phosphonate esters. J. Mass Spectrom. 42, 1086–1092.

102. Nyadong, L., Late, S., Green, M.D., Banga, A. and Fernandez, F.M. (2008) Direct quantitation of active ingredients in solid artesunate antimalarials by noncovalent complex forming reactive desorption electrospray ionization mass spectrometry. J. Am. Soc. Mass Spectrom. 19, 380–388.

103. Wu, C., Ifa, D.R., Manicke, N.E. and Cooks, R.G. (2009) Rapid, direct analysis of cholesterol by charge labeling in reactive desorption electrospray ionization. Anal. Chem. 81, 7618–7624.

104. Eberlin, L.S., Ferreira, C.R., Dill, A.L., Ifa, D.R., Cheng, L. and Cooks, R.G. (2011) Nondestructive, histologically compatible tissue imaging by desorption electrospray ionization mass spectrometry. Chembiochem 12, 2129–2132.

105. McLean, J.A., Ridenour, W.B. and Caprioli, R.M. (2007) Profiling and imaging of tissues by imaging ion mobility-mass spectrometry. J. Mass Spectrom. 42, 1099–1105.

106. Trim, P.J., Henson, C.M., Avery, J.L., McEwen, A., Snel, M.F., Claude, E., Marshall, P.S., West, A., Princivalle, A.P. and Clench, M.R. (2008) Matrix-assisted laser desorption/ionization-ion mobility separation-mass spectrometry imaging of vinblastine in whole body tissue sections. Anal. Chem. 80, 8628–8634.

107. Jackson, S.N. and Woods, A.S. (2009) Direct profiling of tissue lipids by MALDI-TOFMS. J. Chromatogr. B 877, 2822–2829.

108. Stoeckli, M., Chaurand, P., Hallahan, D.E. and Caprioli, R.M. (2001) Imaging mass spectrometry: A new technology for the analysis of protein expression in mammalian tissues. Nat. Med. 7, 493–496.

109. Slaveykova, V.I., Guignard, C., Eybe, T., Migeon, H.N. and Hoffmann, L. (2009) Dynamic NanoSIMS ion imaging of unicellular freshwater algae exposed to copper. Anal. Bioanal. Chem. 393, 583–589.

110. Goodwin, R.J., Pennington, S.R. and Pitt, A.R. (2008) Protein and peptides in pictures: Imaging with MALDI mass spectrometry. Proteomics 8, 3785–3800.

111. Taban, I.M., Altelaar, A.F., van der Burgt, Y.E., McDonnell, L.A., Heeren, R.M., Fuchser, J. and Baykut, G. (2007) Imaging of peptides in the rat brain using MALDI-FTICR mass spectrometry. J. Am. Soc. Mass Spectrom. 18, 145–151.

112. Northen, T.R., Yanes, O., Northen, M.T., Marrinucci, D., Uritboonthai, W., Apon, J., Golledge, S.L., Nordstrom, A. and Siuzdak, G. (2007) Clathrate nanostructures for mass spectrometry. Nature 449, 1033–1036.

PART III

QUANTIFICATION OF LIPIDS IN LIPIDOMICS

13

SAMPLE PREPARATION

13.1 INTRODUCTION

Sample preparation is one of the key steps to the successful analyses of cellular lipidomes by mass spectrometry. Different platforms for analysis of lipids may exploit different methods for sample preparation. For example, the methods employing direct infusion require much cleaner lipid extracts than those based on LC-MS since any contamination could yield significant ion suppression for lipid analysis after direct infusion, whereas the column could get rid of most of the contaminants. Moreover, the methods that use external calibration curves require highly possible extraction recovery of lipids, whereas any method utilizing internal standard(s) for analysis of the species of individual lipid class demands less on the extraction recovery since the internal standard(s) compensate the loss of the lipids of the class. Any differential losses of individual lipid species of the class are only a matter of secondary effects of fatty acyl chain(s) on extraction. Regardless of these differences between the methods used for lipid analysis, the essential procedures of sample preparation are virtually identical, including sample collection, sample storage, lipid extraction (except the analysis by imaging mass spectrometry (Chapter 12)), utilization of standards, etc. In this chapter, the principles of sample preparation from sampling to extraction and the methods of extraction for lipidomics are discussed in great detail.

Lipidomics: Comprehensive Mass Spectrometry of Lipids, First Edition. Xianlin Han.
© 2016 John Wiley & Sons, Inc. Published 2016 by John Wiley & Sons, Inc.

13.2   SAMPLING, STORAGE, AND RELATED CONCERNS

13.2.1   Sampling

Similar to the conventional sample preparation for lipid analysis [1], contamination of other materials in the interested sample should be avoided. For example, in lipid analysis with mammalian materials, blood is always present in many organs, particularly those from the peripheral system (e.g., the liver and the heart) if perfusion of the organs is not conducted at the time of harvest. Different amount of residual blood could cause a significant variation in the results of lipid analysis, depending on the interests of the analysis, since blood is enriched with many types of lipids such as TAG, cholesterol, cholesteryl esters, and PC species. Accordingly, thorough perfusion or rinsing of these organs to eliminate the blood content at sampling is vital. After this procedure, it is important to keep a similar, if not possibly identical, degree of dryness of the samples if one would like to use wet sample weight as a normalizer.

In the lipidomics era, how to collect a sample for lipid analysis is one of the important issues to be considered. Unlike the old-fashioned lipid extraction where a large portion or the entire organ is needed, with current highly sensitive analysis methods, only a very small amount of source material is needed to prepare a sample for analysis. It would be a big waste if the entire available sample (e.g., an organ) or a large part of it would still be used for lipid extraction as the old manner. When taking a small amount of samples from the source material, two points should be kept in mind. First, the small amount of samples taken must be representative to the entire source material. Second, enough materials, but not too much, should be sampled for all of the planned analyses.

To address the first concern regarding the representation, it is necessary to understand the source of the problem. As an example of mammalian organs, blood vessels/capillaries, epidermis, fat pads, and other inhomogeneously distributed tissues always exist in the organs. A piece of small sample randomly taken from such an organ hardly represents the entire organ. Similar cases are present in sampling materials from plants. Therefore, the correct approach is to freeze-clamp the entire source material (the organ or a plant) and further pulverize the clamped wafers into a fine powder with a stainless steel mortar and pestle at the temperature of liquid nitrogen prior to taking a small amount of the powder materials. For a study only focusing on a particular section instead of the entire organ, careful dissection to avoid cross-contamination from different regions is essential. For example, the brain is very complicated and elegant. It contains a variety of very different cell populations (e.g., neurons and glia), and thus different lipid classes and compositions in different regions. Therefore, dissecting tissues from the brain requires precision and care. With a large brain, such as a human brain, representative sampling is even more difficult. Criteria for determination of sample representation must be established before large-scale sample analyses are performed. To this end, the PE profile was demonstrated to be a useful criterion to determine the cross-contamination between gray and white matter samples (Figure 13.1) [2, 3].

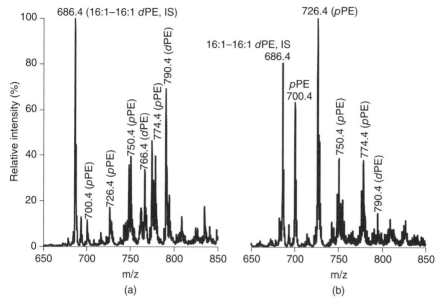

Figure 13.1 Distinct profiles of ethanolamine glycerophospholipid molecular species in lipid extracts of cognitively normal human occipital gray matter and white matter. Brain samples were obtained from the brain bank of the Washington University ADRC Neuropathology/Tissue Resource Core, and brain lipids were extracted by a modified procedure of Bligh–Dyer [1]. Negative-ion ESI mass spectra of lipid extracts of occipital gray matter (a) and white matter (b) were acquired in the presence of a small amount of LiOH as previously described [33]. Individual molecular species corresponding to each ion peak were identified using MDMS-SL analysis as previously described [34]. Plasmenylethanolamine and phosphatidylethanolamine are abbreviated as "pPE" and "dPE," respectively. "IS" denotes internal standard. Source: Han [3]. Adapted with permission of Elsevier.

Regarding the second concern, the answer is dependent on the sensitivity of an instrument and the platform employed. For example, as in the author's laboratory using MDMS-SL with a triple quadrupole instrument (i.e., a TSQ Vantage mass spectrometer, Thermo Fisher Scientific) with a TriVersa NanoMate device (Advion Bioscience), the commonly used samples include those of approximately 10 mg of wet tissue, a million of cells, 100 μL of plasma, or 200 μg of protein of a membrane fraction. With such a sample size, the platform allows analyzing approximately 40 lipid classes and hundreds to thousands of individual lipid species following multiplexed sample processing [4, 5]. The shotgun lipidomics approach based on multiple tandem MS scans using high mass accuracy/high mass resolution instruments uses a similar amount of starting materials for this purpose [6].

In general, larger amounts (>10-fold) of material are necessary when employing any LC-MS-based method to achieve a similar coverage of lipid classes to MDMS-SL [7]. However, the necessary sample size with an LC-MS method for a targeted lipid class depends on the abundance of the lipids in that class. If it is an abundant lipid class, one column step may be able to analyze the species of the class. Approximately 10 mg or less of tissue sample should be adequate for the experiment. However, if the interest is the analysis of a low- or very-low-abundant lipid class, multiple enrichment steps are required prior to identification and quantification of individual species of the class. Accordingly, a much larger sample size (e.g., many grams of tissue) could be required [8–10].

13.2.2 Sample Storage Prior to Extraction

Whenever possible, lipids should be extracted from the samples immediately after removal, so that minimal changes of lipids could occur during the preservation of the samples. When possible, the samples should be frozen rapidly, for example, with dry ice or liquid nitrogen, and stored in sealed glass containers at $-20\,°C$ or a lower temperature in an atmosphere of nitrogen. A storage temperature of as low as $-60\,°C$ was recommended for plasma samples of clinical origin [11]. The process of freezing tissues damages them irreversibly, because the osmotic shock together with formation of ice crystals disrupts the cell membranes. The lipids present in the samples then minimize the degradation catalyzed by enzymes. Lipolytic enzymes, which can hydrolyze lipids on prolonged standing even at $-20\,°C$, are especially troublesome, and contact with organic solvents can facilitate the process. Therefore, tissues should be homogenized and extracted with solvent at the lowest temperature applicable and certainly without being allowed to thaw.

The presence of an unusual amount of lipid intermediates such as NEFA, DAG, PA, or lysoGPL species in lipid extracts generally indicates that some permanent damage to the tissues and hence to the lipids has occurred. For example, it was found that phosphatidylmethanol could be produced by phospholipase D (PLD)-catalyzed transphosphatidylation during extraction of developing soybean seeds with chloroform–methanol [12]. Phosphatidylethanol can be detected if the samples are preserved under an atmosphere of ethanol (Han, unpublished observation). Other alterations in lipids can occur that are more subtle and are discerned less easily. For example, losses of galactolipids can occur without any obvious accumulation of partially hydrolyzed intermediates [13]. Moreover, lipoxygenases can cause artifactual formation of oxygenated fatty acids, and autoxidation can be troublesome for analysis of oxidized polyunsaturated fatty acids. It should keep in mind that the aforementioned lipid intermediates should not be in high concentrations in the biological samples since the majority of these types of lipids are powerful surfactants and enzyme inhibitors.

Often these changes are marginal in their overall importance, since alterations in the abundant lipid classes may be small. On the other hand, they can make a crucial difference to the concentrations of some important lipid intermediates. Generally, the precise concentrations of NEFA and 1,2-DAG species present in tissue samples are recognized to be key metabolic parameters. In an especially thorough study, Kramer and Hulan [14] observed very-low-NEFA concentrations when heart tissue was frozen rapidly and pulverized at dry ice temperatures before extraction. Values obtained were only about 15% of those when similar tissues were extracted by more widely used techniques, i.e., by extracting directly with a homogenizer of the rotating blade type at 0 °C. With the latter, autolysis was presumed to occur during extraction. The DAG levels were also threefold higher when the latter technique was used. Similarly, lysoPC, which had earlier been reported to be a major constituent of chromaffin granules in the adrenal gland, was found to be absent when the tissues were frozen in liquid nitrogen immediately after dissection [15].

It is recommended to keep the tissues dry in an atmosphere of nitrogen in all-glass containers or in bottles with Teflon-lined caps at low temperatures. Endogenous tissue antioxidants generally provide sufficient protection against oxidation under these conditions. Bags, vials, or other containers made of plastic materials should be avoided for storage purposes, as plasticizers leach out and contaminate the biological samples.

13.2.3 Minimizing Autoxidation

Polyunsaturated fatty acids and plasmalogen species can be readily oxidized in air if they are not protected, resulting in confused results if such samples are analyzed. Autoxidation (i.e., nonenzymatically catalyzed oxidative process) of lipids is common if biological samples (particularly the prepared lipid extracts) are exposed to air for a certain time. This process can be exacerbated by strong light and in the presence of metal ions. Once initiated, the oxidative reaction proceeds autocatalytically. The more double bonds the unsaturated fatty acids contain, the more rapidly the oxidative reaction conducts. For example, linoleic acid is autoxidized 20 times as rapidly as oleic acid, and each additional double bond in a fatty acid can increase the rate of oxidative process by twofold to threefold. Autoxidation of arachidonic and docosahexaenoic acids can yield different isomers of eicosanoids and docosanoids, respectively. It was demonstrated that plasmalogen species can be oxidized more rapidly than those species containing polyunsaturated fatty acids [16], suggesting that autoxidation-caused consequences could be more severe than what we previously understood. Collectively, wherever possible, lipids should be handled under an atmosphere of nitrogen by using nitrogen lines to flush the air out of glass containers.

In addition to keep lipid extracts under an atmosphere of nitrogen, it is also advisable to add synthetic antioxidants, such as butylated hydroxytoluene (BHT) to storage solvents at a level of 50–100 mg/L. This compound generally does not interfere with ESI-MS analysis of lipids and can be readily removed as it is relatively volatile. However, it should be noted that excessive amounts of antioxidants can sometimes act as pro-oxidants!

Because the sample size is usually small, the solvents used to prepare lipid samples can be easily evaporated under a nitrogen stream. However, when it is necessary to use a heated bath to facilitate removal of solvents, the temperature of the bath, in general, should not exceed about 40 °C to avoid any accelerated autoxidation of lipids. Moreover, the more volatile lipids may also be lost by evaporation.

13.3   PRINCIPLES AND METHODS OF LIPID EXTRACTION

Lipid extraction is one of the key steps for the successful analyses of cellular lipidomes by ESI-MS, in general, and by approaches using direct infusion, in particular. Traditionally, lipid samples from biological sources are extracted using a mixture of chloroform and methanol based on the Folch method [17] or the modified method of Bligh and Dyer [18] or other solvent combinations [19, 20].

For any method of analysis of lipids, if external standard(s) are used for quantitation, extraction recovery of individual species is always a concern. Therefore, this parameter has to be determined at an early stage during the development of a method. Different solvent systems may have different extraction recoveries for different lipid classes and molecular species. In contrast to the methods utilizing external standards, any methods utilizing internal standards should have less concerns for the extraction recovery since an incomplete extraction recovery, if present, is only a secondary effect on the quantitative analysis of a lipid class caused by the differential extraction recovery from different molecular species of the class relative to that of the internal standard(s) of the class.

As lipidomic analysis is classified into the LC-MS-based approach and shotgun lipidomics, the requirement for sample preparation is also different. The former is more tolerant with the presence of inorganic ions in lipid extracts than the latter since the pre-column or even the analytical column can get rid of these extraction contaminants. However, the presence of inorganic residues may affect substantially the ionization stability as well as ionization efficiencies of different lipid classes and individual species of a class in shotgun lipidomics. Therefore, minimizing the inorganic residues, particularly eliminating any aqueous phase contamination, becomes crucial.

By using classical extraction methods such Folch or Bligh and Dyer procedures, regardless of how carefully the extractions are conducted, a small amount of water and therefore inorganic residues (e.g., salts and glucose) are carried into the solvent extracts. If a high-salt concentration (e.g., in the case of extracting lipids from cerebrospinal fluid or urine) is present in the aqueous phase during extraction, the residual contamination of the solvent extract by a small amount of aqueous phase results in a high chemical noise level due to the presence of inorganic ions and other compounds of low molecular weight. Therefore, it is advisable that an additional solvent extraction should be conducted against an aqueous phase with a lower salt concentration to further "clean up" the lipid extracts. This issue for those recently developed extraction methods are discussed in the next section.

A type of salt is usually used during lipid extraction to enhance the phase separation and extraction efficiency. It is advisable that it is better to add the salt that matches with the adduct preferred for analysis of lipids in both positive- and negative-ion modes by shotgun lipidomics. For example, if proton or ammonium is the preferred adduct for lipid analysis in the positive-ion mode and acetate adduct is the choice in the negative-ion mode, ammonium acetate is the preferred salt used for lipid extraction. If an inorganic salt is not introduced during the extraction step, sodium chloride is always the most abundant salt in nature. Therefore, sodium and chlorine adducts of zwitterionic lipids are always predominant in mass spectra acquired in the positive- and negative-ion modes, respectively (see Chapter 4 for a detail discussion). Although acidic conditions are favorable for improving the extraction efficiency of anionic lipids, particularly those soluble in aqueous phase such as lysoGPL species to a great degree, acidic extraction conditions can lead to degradation of plasmalogen species and resulting in corresponding *sn*-2 lysolipids.

Detergents can cause severe ion suppression in shotgun lipidomics and may affect the column separation of lipids. Therefore, the existence of detergents can complicate the MS analysis of lipids. Thus, utilization of detergents should be avoided in sample preparation when possible. If present, great efforts should be made to remove them as much as possible.

A low-concentration of lithium chloride (e.g., 50 mM in aqueous phase) is preferable for lipid extraction for MDMS-SL since the lithium adduct of lipid species can yield unique and informative fragmentation patterns after CID (Part II). Moreover, the weakly acidic conditions resulted from a weak Lewis base of lithium ion and a strong Lewis acid of chloride could improve the extraction efficiency of anionic lipids to a certain degree and could lead to increased intrasource separation of molecular specie of anionic lipid classes from PE species in negative-ion ESI-MS (Chapter 3). Under such weakly acidic conditions, degradation of plasmalogen species does not occur.

13.3.1 Principles of Lipid Extraction

A variety of lipid classes that possess very different physical properties are present in tissues. Although the majority of lipids should be easily extractable, many lipids that are closely associated with proteins and polysaccharides in cellular membranes are not extracted so readily. These lipids do not, in general, form covalent bonds with those proteins and polysaccharides, although covalently linked fatty acids do occur in nature. Generally, these lipids are associated with those cellular components by weak hydrophobic or van der Waals' forces, by hydrogen bonds, and by ionic bonds. For example, the hydrophobic aliphatic moieties of lipids interact with the nonpolar regions of amino acids, such as valine, leucine, and isoleucine, of proteins to form weak associations. Hydroxyl, carboxyl, and amino groups in lipid molecules, on the other hand, can interact strongly with biopolymers via hydrogen bonds. Finally, the strongest bonds of all are the ionic linkages between acidic phosphate or sulfate groups on lipids and metal ions, which may in turn be bound similar to cellular proteins or polysaccharides.

Pure lipids dissolve in a variety of solvents, depending on the relative strengths of the interactions between the solvent and either the hydrophobic or the hydrophilic regions of the lipids. Nonpolar lipids such as TAG or cholesteryl esters are very soluble in hydrocarbon solvents such as hexane, cyclohexane, or toluene, and in solvents of somewhat higher polarity, such as chloroform or ethers. These nonpolar lipids tend to be rather insoluble in polar solvents such as alcohols (e.g., methanol in particular). Their solubility decreases as the chain lengths of the fatty acid moieties in these lipids increase or as the chain lengths of the solvent alcohol decrease. Unsaturated lipids tend to dissolve in most solvents more readily than saturated and higher melting analogs. On the other hand, polar complex lipids dissolve readily in more polar solvents such as chloroform, methanol, and ethanol, but are only sparingly soluble in hydrocarbon solvents. Acetone is a good solvent for glycolipids, but not for GPL species, and it can even be used to precipitate the latter from their solutions.

In order to extract lipids from tissues, it is necessary to find solvents that will not only dissolve the lipids readily but also break the interactions of lipids with the tissue matrices. Moreover, some lipids can be physically trapped within a tissue matrix. For example, starch macromolecules in cereal grains could trap lysoGPL species as an inclusion complex. The cell walls of some organisms are less permeable than others to solvents; water then assists the extraction by causing swelling of the biopolymers and it is an essential component of any extractant. Methanol may play a compromised role in aiding lipid solubility and cell wall breakage in some cases. In some circumstances, it may be necessary to conduct a denaturation of the other constituents of the cell walls by some means such as sonication before a thorough extraction of the lipids is possible.

The potential toxicity of the solvents to the users is another important factor for consideration. Therefore, it is always better to employ safer extraction conditions if possible. For example, the isopropanol-hexane (3:2 by volume) system was used in many cases due to their relatively low toxicity [21, 22]. Unfortunately, this system is unable to recover some lipid classes such as gangliosides quantitatively.

In addition to toxicity consideration, high throughput is another factor that is usually considered in development of a lipid extraction method. In the method developed utilizing chloroform, the lipid extracts (i.e., the chloroform layer) are present in the bottom layer. This outcome leads to two potential problems for lipidomic analysis. First, taking the bottom layer out is not straightforward and not easy for automation. Thus, it is not suitable for achieving high throughput. Second, taking the bottom layer out certainly has the contacts of the pipette with the protein precipitated middle layer and the aqueous top layer. Thus, potential contamination from these layers is inevitable. Accordingly, many novel extraction methods for lipidomic analysis were developed, which are summarized in next section.

In classical lipid extraction, most lipid analysts use chloroform–methanol (2:1, v/v), with the endogenous water in the tissue as a ternary component of the system, to extract lipids from animal, plant, and bacterial samples. Usually, the biological sample is homogenized in the presence of both solvents. However, better results could be obtained if the sample is first mixed with methanol alone before the chloroform is added to the solution. In aid of homogenization, low-power sonication (e.g., bath sonication) at 0 °C could be employed at this stage. With the samples difficult to extract cellular lipids, more than one extraction may be performed to enhance the lipid extraction recovery. As aforementioned, making the hardly homogenized tissue (e.g., muscle) into fine powder at the temperature of liquid nitrogen not only aids the homogenization to minimize heterogeneous but also could enhance the extraction recovery due to the increased contact surface. With the lyophilized tissue samples, it may be necessary to rehydrate them prior to performing lipid extraction.

There are many cases in which alternative or modified procedures must be used. For example, butanol saturated with water appears to be the very useful solvent mixture to disrupt the inclusion complexes of lipids in starch and give the best extraction recoveries of lipids from cereals [23]. This solvent combination was recommended for the quantitative recovery of lysoGPL classes and acylcarnitines [23]. Recently, modified versions of this system for lipid extraction were developed for global lipid analysis [20] (see below). If quantitative recoveries of the highly complex glycosphingolipids of intestinal cells are required, it was recommended that the tissue be partially digested by alkali, RNAase, DNAase, and a protease prior to extraction with chloroform–methanol [24]. Similar procedures are sometimes suggested for preparation of bacterial lipids. Acidic extraction conditions may be required for quantitative recovery of particular lipid classes, but they can cause destruction of plasmalogens. Solvent mixtures containing acetonitrile are sometimes recommended for CoA esters and acylcarnitines. Some plant tissues should be pre-extracted with isopropanol to minimize artifactual degradation of lipids by tissue enzymes [1].

Lipid extracts obtained from biological samples, as aforementioned, tend to contain significant amounts of nonlipid contaminants, such as sugars, amino acids, urea, and salts. These must be removed before the lipids are analyzed. A common and classical approach is to use a simple washing procedure devised by Folch, Lees and Sloane Stanley [17], in which a chloroform–methanol (2:1, v/v) extract is shaken and equilibrated with one-fourth its volume of saline solution (i.e., 0.88% potassium chloride in water). The mixture partitions into two layers, of which the lower phase is comprised of chloroform–methanol–water in a proportion of 86:14:1 (by volume) and contains virtually all of the lipids, while the upper phase consists of the same solvents in the proportion of 3:48:47 (by volume), respectively, and contains much of the nonlipid contaminants. It is important that the proportion of chloroform, methanol, and water in the combined phases should be as close as possible to 8:4:3 (by volume), otherwise selective losses of lipids may occur. If a second wash of the lower phase is needed to remove any remaining contaminants, a mixture of similar composition to that of the upper phase should be used, i.e., methanol–saline solution (1:1, v/v).

The gangliosides present in the sample will partition into the aqueous layer, together with varying amounts of oligoglycosphingolipids, in conventional lipid extraction procedures. However, they can be recovered from this layer by dialyzing out most of the impurities of low molecular weight, and then lyophilizing the residue [25], or by diisopropyl ether–butanol extraction followed by desalting with gel filtration [26] or another type of SPE column [27]. Moreover, SPE columns packed with reversed-phase materials of the octadecylsilyl type are used for the quantitative recovery of gangliosides [28], prostaglandins [29], and isotopically labeled GPL species also [30].

It should be recommended that in any extraction procedure, it is important that the weight of fresh tissue extracted is recorded, together with the weight of lipid obtained from it. For many purposes, it could be desirable to normalize the lipid content to the amount of dry materials or protein content in the tissue later.

It should be emphasized that it is difficult to achieve a complete recovery of every lipid class by any known method of extraction. Any incomplete recovery will lead to an inaccurate measurement of the lipid content in a sample or an inconsistency in the results from different laboratories if the analytical methods are based on external calibration curves. To avoid this potential problem, internal standards may be added for quantification of lipid classes during the extraction procedure. Therefore, the effects of any incomplete recovery from extraction on quantitative analysis can be minimized. The differential recovery of individual molecular species of a lipid class of interest relative to the selected internal standard is only a minor secondary effect in the most cases. This point is particularly important for the accurate analysis of individual lipid molecular species employing a lipidomic approach since internal standards are essential for MS-based lipid analysis.

13.3.2   Internal Standards

No rule exists in the relationship of the absolute intensity (i.e., counts) of a molecular ion as determined by ESI-MS with the concentration in solution of the analyte that yields the ion. The ion intensity of an analyte determined by ESI-MS could be affected by many factors related to sample preparation, ionization conditions, tuning conditions, the analyzer and detector used in the mass spectrometer, etc. Minor changes of these factors could lead to significant changes of ion intensity from one state to another. As MS instruments become more and more sensitive, the influences of these factors on determination of ion intensities by ESI-MS become more and more evident. Thus, it would be difficult for scientists to virtually repeat a measurement of absolute ion counts for an analyte in a biological sample from time to time, from an instrument to another instrument, or from one laboratory to the others.

Accordingly, accurate quantification of any compound by ESI-MS has to be made by comparisons to either an internal or external standard similar to the compounds of interest (e.g., their stable isotopologues). An internal standard is preferably added during sample preparation at the earliest step possible and analyzed at the same time as the sample. The external standard is analyzed separately, but should be analyzed under "identical" conditions with the sample of interest. A calibration curve has to be established by using the external standard also under "identical" conditions. Both methods have advantages and disadvantages. The former is known for its simplicity and accuracy if the internal standard is within a linear dynamic range of the measurement with the sample (see below). However, selection of an internal standard may be difficult and the dynamic range of the measurement must be predetermined. For the latter, control of the measurements being conducted under identical conditions is the key. It becomes particularly difficult for this requirement when many steps of sample preparation, separation, and quantification are involved. Global lipidomic analysis exactly represents such a complicated case in which a method using external standards alone is almost impossible. Employing internal standards (or in combination with external standards in the case of LC-MS-based methods for lipid analysis) should be employed to achieve accurate quantification of complex cellular lipidomes. Therefore, any unexpected changes of ion counts during analysis can be controlled internally or normalized. Results obtained from sample analysis without any internal control can only be used for qualitative comparisons.

The lipid species that can be used as internal standards for the analysis of a sample depend on the existence of endogenous lipid species in the sample. In theory, accurate quantification of any compound by MS can only be achieved by comparison of its peak intensity (or area) with that of a chemically identical internal standard with stable isotopes incorporated within a linear dynamic range. However, it is impractical to use thousands of such isotopically labeled internal standards for quantitative analyses of a cellular lipidome, although quantification of a few known lipids can be achieved by employing such a method [31]. Fortunately, it has been demonstrated that the response factors of individual species of a polar lipid class by ESI-MS in the full MS mode depend predominantly on the charge properties of the polar head group in the concentration range of picomoles per microliter or lower after correction for different ^{13}C isotopologue distributions [32–35]. This finding becomes a foundation for quantification of individual lipid species possessing an identical polar head group employing a species in the class as an internal standard. With such an internal standard, a reasonable accuracy of measurement (approximately 95%) could be achieved if determination is made in the low concentration region and correction for different ^{13}C isotopologue distributions is conducted. Correction for different ^{13}C isotopologue distributions is also called hereafter as "^{13}C de-isotoping," which refers to a process in which the intensities of all the isotopologues of a species are converted to that of their monoisotopic counterpart (see further discussion later).

A primary condition for selection of an internal standard is the absence or very low abundance (e.g., ≪ 1% of the most abundant species of the class in the lipid extract) of an isobaric ion peak. Such a condition of any isobaric peaks from the endogenous cellular lipids coexisting with the (quasi)molecular ions of those selected species has to be predetermined for both shotgun lipidomics and LC-MS analyses.

For accurate quantification (e.g., >90% accuracy) of lipid species in a biological sample, or comparison of lipid profiles between lipid extracts of biological samples, appropriate amounts of suitable internal standards are added to each homogenate prior to extraction. In most cases, the amount of internal standard added and thereafter the quantified lipid content (that could be normalized to) should be based on a parameter that is stable from sample to sample and can be determined relatively easily. The protein, DNA, or RNA content in tissue or cell samples, the phosphorus content in the lipid extract, the tissue wet or dry weight, the cell number, and the volumes of the biofluids are the parameters frequently used by investigators in the field. Each of these "normalization" parameters has benefits and detriments depending on the physiological or pathological system studied. In terms of detriments, for example, determination of phosphorus content may carry a large experimental error, and the phosphorus content in lipid extract may also vary under different physiological or pathological conditions. Tissue samples may carry different amounts of water in preparation, making the ratio of lipids to tissue wet weight inconsistent, while it takes too long (at least overnight) to prepare dry tissue samples. The volume of biofluid may be influenced by the fluid and/or food intake, whereas counting cell numbers becomes difficult when the cells are clustered. Therefore, protein, DNA, or RNA content is relatively stable and can be determined in a high-throughput manner. Employing one of these contents as a normalization parameter is highly recommended. Note that although the levels of many proteins may change from one state to the other, the amounts of the structural proteins, which account for the majority of the protein content of a sample, are usually quite stable.

Below is a list of typical internal standards used in the author's laboratory for analysis of lipid species present in most biological samples by MDMS-SL: di15:0 PG (3), 17:0–20:4 PI (4.5), di14:0 PS (19), tetra14:0 CL (1.5), di14:0 PA (0.5), di16:1 PE (26), di14:1 PC (37.5), 17:0 lysoPC (1.5), 14:0 lysoPE (1.2), 17:1 lysoPG (0.004), lysoPI (0.08), 17:1 lysoPS (0.08), lysoPA (0.045), N12:0 SM (3), N15:0 GalCer (8, Matreya), N16:0 ST (3, Matreya), N17:0 Cer (1), 17:0 sphingoid base (0.1), 17:0 S1P (0.05), N,N-dimethyl psychosine (0.05, laboratory-synthesized), tri17:1 TAG (1, NU-CHEK-PREP, Inc.), di17:1 DAG (0.1, NU-CHEK-PREP, Inc.), 17:1 MAG (0.1, NU-CHEK-PREP, Inc.), $^2H_4$-16:0 NEFA (5, Cambridge Stable Isotope Laboratories), $^{13}C_4$-16:0 acylcarnitine (0.05, Sigma Chemical Co.), 17:0 acyl CoA (0.05, Sigma Chemical Co.), $^2H_3$-4-hydroxynonenal (2, Cayman Chemical), $^2H_4$-cholesterol (170, Cambridge Stable Isotope Laboratories), and $^2H_5$-16:0 cholesteryl ester (1, Cambridge Stable Isotope Laboratories). The levels used for mouse cortical lipid extracts are given in parentheses in nanomoles per microgram protein, which can be adjusted appropriately for different tissue, cell type, or other biological samples. All these standards can be purchased from Avanti Polar Lipids, Inc., except those specified.

The amount of individual internal standard should be optimized to make the relative intensity of the internal standard peak in the range of >20–<500% in comparison to the ion peak corresponding to the most abundant species in the class. When the relative peak intensity of a selected internal standard is lower than 20% in comparison to the base peak, the experimental error is greatly amplified. Addition of too much internal standard could lead to an ion suppression effect, making the endogenous lipid species close to the baseline. Accordingly, the optimal amounts of internal standards necessary for lipid quantification could vary largely for different kinds of samples.

The given amounts for mouse cortical lipid analysis (see above) can also be used as references for estimating the internal standards used for a particular class of cellular membrane lipids present in other biological samples. Here, the assumption is that the levels of cellular membrane lipids are similar for different cell types. For example, the aforementioned levels of internal standards are not suitable for quantification of lipid species in the liver. But we can estimate the levels of the internal standards for quantification of hepatic lipids based on the levels of internal standards for the cortex. Specifically, the content of an internal standard for quantification of PC species in the liver can be estimated as follows: the total content of lipids in the liver is ~400 nmol/mg of protein; the total content of lipids in the cortex is ~800 nmol/mg of protein; the internal standard for quantification of PC species in the cortex is 37.5 nmol/mg of protein (see above); thus the level of the internal standard for PC species in the liver should be $\sim 37.5 \times 400/800 = 18.75$ nmol/mg of protein. This number is consistent with that used in experiments [36]. These estimated levels of internal standards could then serve as a starting point for a pilot experiment. Moreover, it should also be recognized that the different lipid compositions among lipid classes in different samples are present. For example, only minimal amounts of ST and GalCer, as well as a greatly reduced amount of Cer species are present in non-neuronal samples such as the heart or the liver.

Although the requirement of the internal standard levels for LC-MS analysis of lipids is not as strict as shotgun lipidomics, recognizing the effects of internal standard levels on accurate quantification is still useful and should be integrated into the method development or validation. For example, the level of an internal standard in LC-MS analysis also should not be too low since a lower level of an internal standard makes the experimental error inherited with the internal standard to be amplified as in shotgun lipidomics.

Alternative approaches exist for semiquantitative or qualitative comparisons between biological samples. In those approaches, only a limited set of or no internal standards are used during the analysis and/or only the peak intensities (or areas) of the detected ions are compared. These ions may or may not be well characterized and/or identified. A normalized composition relative to a selected ion among the detected ions is usually used for comparison. These approaches are generally referred to as lipid profiling. Although lipid profiling does not provide direct information on the stoichiometric relationship between lipid species and may be prone to poor analytical reproducibility, it can provide a comprehensive comparison between biological samples. High-throughput analysis is a major goal for such profiling, whereas statistical analysis (e.g., PCA) is always required. In this section, these approaches are not discussed at length.

13.3.3   Lipid Extraction Methods

13.3.3.1   Folch Extraction   Many modifications of the basic extraction procedure [17] were devised for use in particular circumstances, and the analyst must decide what is required of a method to make some additional modifications such as addition of salts or acids, change of sample size, etc. The modified "Folch" procedure [37] could serve as the basic method.

> The tissue (1 g) is homogenized with methanol (10 mL) for 1 min in a blender, then chloroform (20 mL) is added, and then the homogenization continued for additional 2 min. The mixture is filtered when the solid remaining is resuspended in chloroform–methanol (2:1 by volume, 30 mL) and homogenized for 3 min. The mixture is filtered again and rewashed with fresh solvent. The combined filtrates are transferred to a measuring cylinder, then one-fourth of the total volume of 0.88% potassium chloride in water is added, and then the mixture is shaken thoroughly before being allowed to settle. The aqueous (upper) layer is drawn off by aspiration, then one-fourth the volume of lower layer of methanol–saline solution (1:1 by volume) is added, and then the washing procedure is repeated. The bottom layer, containing the extracted lipid, is filtered before the solvent is removed on a rotary film evaporator. The lipid is stored in a small volume of chloroform or hexane under an atmosphere of nitrogen at −20 °C until it can be analyzed.

Obviously, a reduction in sample and solvent for this procedure of lipid extraction of biological samples for lipidomics analysis is necessary. The key in the procedure is to use methanol to homogenize the tissue sample and use chloroform–methanol (2:1, v/v) to dissolve the lipids. As used only for small sample materials in the case of lipid extraction for lipidomics, the water content present in the original samples can essentially be neglected in the phase separation procedure.

13.3.3.2 Bligh–Dyer Extraction   In contrast, a method by Bligh and Dyer [18] was devised originally for the extraction of GPL species from fish muscle tissue in a relatively economical manner. In other words, it is recommended for large samples with a high proportion of endogenous water.

It is assumed that 100 g of the wet tissue to be extracted contains 80 g of water. A hundred gram of the tissue is homogenized for 4 min in a blender with a solvent mixture consisting of chloroform (100 mL) and methanol (200 mL). If the mixture forms two phases, more solvent should be added until a single phase is achieved. The mixture is filtered through a sintered glass funnel and the tissue residue is rehomogenized with chloroform (100 mL) and filtered once more. The two filtrates are combined and transferred to a graduated cylinder, then aqueous potassium chloride solution (0.88%, 100 mL) is added, and then the mixture is shaken thoroughly before being allowed to settle. The mixture should be biphasic (or more aqueous phase should be added). The upper layer with any interfacial material is removed by aspiration; the lower phase contains the lipids, which are recovered as in the previous method.

The authors of the method noted that quantitative recovery of TAG was not always achieved when TAG species were the major components of a tissue, and in this instance, it was recommended a re-extraction with chloroform alone, to be combined with the first extract. The method is much misunderstood and therefore misused.

13.3.3.2.1 A Modified Bligh and Dyer Procedure for Lipidomics   To succeed in lipidomic analysis with ESI-MS using direct infusion or with MALDI-MS, a key point is to have a lipid extract carrying only a minimal amount of inorganic salts. Although solid-phase extraction cartridges can be used to eliminate the salt contaminants, a careful solvent wash is recommended as routinely used in the author's laboratory [36, 38].

Individual tissue sample (10–50 mg) is cut into small pieces followed by homogenization in 1 mL of ice-cold phosphate-buffered saline (i.e., 13.7 mM NaCl, 1 mM Na_2HPO_4, 0.27 mM KCl, pH 7.4) using a disposable tissue grinder. Brief bath sonication if available could be performed at 0 °C to enhance the homogenization. Protein concentration of each homogenate is then determined. A small volume of homogenate containing 1–2 mg of protein is transferred to a glass test tube. Methanol and dichloromethane (or chloroform) (4 mL, 1:1, v:v) are added. The solution is titrated with a volume of LiCl solution to a final concentration of 50 mM in 1.8 mL aqueous solution. Internal standards as a pre-mixture are also added to each test tube based on the protein content in the tube. Thus, the quantified lipid content can be normalized to the protein content. At least one internal standard for each class of lipids of interest is recommended.

The extraction mixture is centrifuged at 2500 rpm for 10 min. The dichloromethane layer (bottom layer) is carefully removed and saved. Into the MeOH and aqueous layer of each test tube, an additional 2 mL of dichloromethane is added and the dichloromethane layer is separated as above. The dichloromethane extracts from each identical sample are combined and dried under a nitrogen stream. Each individual residue is then resuspended in 4 mL of dichloromethane–methanol (1:1), re-extracted against 1.8 mL of 10 mM LiCl aqueous solution, and the extract is dried as described earlier. The dichloromethane extracts are subsequently dried under a nitrogen stream and each individual residue is resuspended with a volume of 100 μL/mg of protein in 1:1 dichloromethane–methanol. The lipid extracts are finally flushed with nitrogen, capped, and stored at −20 °C for lipidomic analyses (typically within 1 week).

The lithium chloride solution in the extraction can be replaced with other salts such as ammonium acetate or a weak acid if necessary [33]. However, an acidic environment may lead to plasmalogen degradation, thus it should be very cautious. It should be pointed out that although the aqueous phase is usually discarded after extraction, it could be used for analysis of many lipid classes, which largely disperse to it as reported [39].

13.3.3.3 MTBE Extraction

Chloroform extraction of lipids from biological samples is widely used for sample preparation in lipid analysis. However, in addition to solvent toxicity, difficulty in collecting the chloroform layer makes sample preparation difficult for automated, high-throughput analysis of lipids. Moreover, the presence of potential contamination of salts from the aqueous phase during solvent collection is a big concern for shotgun lipidomics. To accommodate these difficulties, an alternative extraction method using methyl-*tert*-butyl ether (MTBE) that is present in the top layer against a lower aqueous phase was developed [19] and widely adopted in the lipidomics field.

Methanol (1.5 mL) is added to a 200-μL sample aliquot, which is placed into a glass tube with a Teflon-lined cap, and the tube is vortexed. Then, 5 mL of MTBE is added and the mixture is incubated for 1 h at room temperature in a shaker. Phase separation is induced by adding 1.25 mL of de-ionized water. After 10 min of incubation at room temperature, the sample is centrifuged at 1000g for 10 min. The upper (organic) phase is collected, and the lower phase is re-extracted with 2 mL of a solvent mixture, whose composition is equivalent to the expected composition of the upper phase (obtained by mixing MTBE–methanol–water (10:3:2.5 by volume) and collecting the upper phase). Combined organic phases are dried in a vacuum centrifuge. To speed up sample drying, 200 μL of MS grade methanol is added to the organic phase after 25 min of centrifugation. Extracted lipids are dissolved in 200 μL of chloroform–methanol–water (60:30:4.5 by volume) for storage.

The MTBE extraction method was further improved for extraction of lipids from brain samples with mechanical homogenization utilizing ceramic beads to enhance high throughput and automation [40]. Another development based on this method is to utilize an intermediate fraction as well as the aqueous phase for total analysis of lipids and metabolites [41].

It seems that the extraction recoveries of many lipid classes examined by the MTBE method are comparable to those using chloroform. The 1-h incubation period may be shortened by using low-power sonication [42] or a microwave oven [43]. One drawback of the MTBE extraction method is that the organic phase contains quite a large amount of aqueous component. This may prolong the evaporation of the solvents prior to reconstituting the lipid solution. Moreover, the contamination of aqueous phase may cause complication of lipid analysis due to the presence of inorganic salts and other small molecules.

13.3.3.4 BUME Extraction
Following the same line of reasoning as the development of MTBE extraction method and considering the automation utilizing any standard 96-well robot, an extraction method using butanol:methanol (BUME) as initial extraction solvents was developed [20]. Specifically, 300 μL of butanol:methanol (BUME) mixture (3:1, v/v) is first mixed with 10–100 μL of plasma sample to form an initial phase 1 extraction; 300 μL of heptane:ethyl acetate (3:1, v/v) is added to the vessel containing BUME mixture; and finally, 300 μL of 1% acetic acid is added to the extraction solution to form phase 2 separation. Again, it should be cautious that an acidic environment might lead to plasmalogen hydrolysis.

This method may compensate the MTBE method with less salts/inorganic contaminants carried over in the organic phase. However, the organic phase contains the butanol component that is hard to be evaporated under a stream of nitrogen. This method was further improved in a few aspects including replacing acetic acid solution with a salt solution; applying for tissue homogenates; and diluting the organic phase directly for shotgun lipidomics in the author's laboratory. The results after these improvements are well comparable to those from the modified Bligh–Dyer extraction if not considering the analysis of lipids present in the aqueous phase.

13.3.3.5 Extraction of Plant Samples
With plant tissues, it is necessary to extract first with isopropanol in order to deactivate the enzymes, and a procedure devised by Nichols is usually recommended [44]. The plant tissues are macerated with 100 parts by weight of isopropanol. The mixture is filtered, the solid is extracted again in a similar manner and finally is shaken overnight with 199 parts of chloroform–isopropanol (1:1, v/v). The combined filtrates are taken up in chloroform–methanol (2:1, v/v) and given a "Folch" wash as described earlier. The purified lipids are recovered from the lower layer as described in Folch extraction.

This method is modified for lipid extraction for lipidomic analysis. For example, Welti et al. omitted the filtering step and directly extracted the lipids from plant tissues by Folch extraction method [45].

> Plant tissues (10–50 mg dry weight) are cut at the sampling time and transferred to 3 mL of isopropanol with 0.01% butylated hydroxytoluene at 75 °C. After 15 min, 1.5 mL of chloroform and 0.6 mL of water are added. The tubes are shaken for 1 h, followed by removal of the extract. The plants are re-extracted with chloroform–methanol (2:1, v/v) with 0.01% butylated hydroxytoluene five times with 30 min of agitation each time, until all of the remaining plant tissue appears white. The combined extracts are washed once with 1 mL of 1 M KCl and once with 2 mL of water. The solvent is evaporated under nitrogen, and the lipid extract is dissolved in 1 mL of chloroform.

13.3.3.6 Special Cases
Special methods are required for quantitative extraction of some lipid classes such as gangliosides and polyphosphoinositides, and to minimize artifactual formation of NEFA, DAG, and polyphosphoinositides when

accurate analyses are required. Methods for extraction of gangliosides are mentioned in Section 13.3.1. Due to the low concentration of polyphosphoinositides present in biological samples and their strongly acidic nature, special methods for extraction of these lipids have to be employed. For quantitative extraction of these lipids, it is essential that an acid should be incorporated into the extraction medium to ensure that the inositides are effectively solubilized, but the acid must then be neutralized as quickly as possible to prevent hydrolysis of the lipids. It appears that two procedures that are based on established methods (see above) are favored. For example, a "Folch" extraction has been recommended with chloroform–methanol–0.4% hydrochloric acid (2:1:0.05 by volume) as extraction medium [46], or a modified "Bligh and Dyer" method can be used in a two-stage extraction process, first with a neutral solvent and then with it acidified [47].

Effective recovery of lysoGPL species by using conventional extraction methods is not ideal due to their high solubility in an aqueous phase. Addition of an acid can enhance the recovery, but leads to hydrolysis of plasmalogen species. Therefore, some special extraction procedures are developed for this purpose. For instance, a portion of the mixture of methanol and plasma could be directly loaded to the column for efficient LC-MS analysis of these compounds in plasma [48]. Recovery of lysoGPL species from the aqueous phase after extraction of biological samples with a modified Bligh–Dyer procedure (see above) by using a HybridSPE column can be achieved [39].

Recently, a monophase strategy by using chloroform–methanol–water was practiced for broader lipid analysis including gangliosides [49]. Analysis of lipids present in a single rat retina is achieved in the study.

13.3.4 Contaminants and Artifacts in Extraction

The production of NEFA, DAG, PA, or lysoGPL species during faulty storage of tissue samples prior to extraction is mentioned earlier. Extraneous substances can also be possibly introduced into lipid extracts from innumerable sources. All solvents, including from time to time those grades that are nominally of high purity, can contain contaminants, and as large volumes of solvent may be used to obtain small amounts of lipids, any such impurities can be troublesome. The higher quality grades of solvent may have to be checked periodically with background runs to ensure that they meet the required standards, while those of poorer quality of solvents should be redistilled before use. Buffers prepared for use in mobile phases and stored for lengthy periods in refrigerators may gradually accumulate a substantial microbial population. Some extraneous substances, for example, antioxidants, are added deliberately by manufacturers to minimize peroxide formation in ethers, chloroform, etc. Their presence should be recognized and integrated into data interpretation.

Other extraneous lipid-like materials can be introduced accidentally into lipid samples from a variety of sources. Plastic ware of all kinds (other than that made from Teflon) can be especially troublesome and is best avoided, since plasticizers (diesters of phthalic acid usually) very easily leach out by organic solvents. They tend to co-chromatograph with some lipids or cause ion suppression, so they may spread confusion and obscure compounds of interest in lipidomic analysis. They are especially troublesome in GC analyses of fatty acid methyl esters. Conversely, it has been shown that lipids can themselves dissolve in some plastics, leading to selective losses of less polar constituents [50].

Manufacturers of fine chemicals are fallible, and all laboratory reagents can on occasion contain impurities that may cause problems in analytical procedures. It is necessary to exercise vigilance to detect and eliminate these at an early stage. Further lipid contaminants can arise from such obscure sources as fingerprints, and from a host of materials in everyday use in laboratories, including cosmetics, hair preparations, hand creams, soaps, polishes, the exhausts from vacuum pumps, lubricants, and greases, if they are used carelessly.

Under optimum conditions, lipids should not change in composition and structure during extraction or storage. However, there are some combinations of conditions that can give rise to unwanted changes. For example, it is well recognized that trans-esterification of lipids can occur under certain conditions [51]. The same problem may arise if the pH is too low, and the reaction may even be catalyzed enzymatically by transacylases. Similarly, acetone can cause some dephosphorylation of polyphosphoinositides, and it can react with PE species to form an imine derivative. Some rearrangement of plasmalogen species was found to occur when they were stored in methanol for long periods, especially if acidic conditions were employed during extraction, leading to accumulation of lysoGPL species as artifacts.

Artifactual enzymatic hydrolysis of lipids, catalyzed by tissue enzymes, can be promoted by the solvents used for extraction. This is especially troublesome with plant tissues, in which PLD activity (both hydrolytic and transphosphatidylase) is stimulated by some solvents. The problem is usually circumvented by a pre-extraction with isopropanol, which deactivates the enzyme. It is also possible to obtain an artifactual enzyme-catalyzed acylation of some lipids (such as glycosyldiacylglycerols) in certain circumstances [52].

13.3.5 Storage of Lipid Extracts

Autoxidation is always a concern for storage of lipid extracts. They should not be stored in a dry state, but dissolved in a small volume of a relatively nonpolar (aprotic) solvent. The lipid samples should be stored at $-20\,°C$ at least in a glass (never plastic) container. Air should be excluded from the container by flushing with a stream of nitrogen. Antioxidants should be used for prolonged storage.

REFERENCES

1. Christie, W.W. and Han, X. (2010) Lipid Analysis: Isolation, Separation, Identification and Lipidomic Analysis. The Oily Press, Bridgwater, England. pp 448.
2. Han, X., Holtzman, D.M. and McKeel, D.W., Jr. (2001) Plasmalogen deficiency in early Alzheimer's disease subjects and in animal models: Molecular characterization using electrospray ionization mass spectrometry. J. Neurochem. 77, 1168–1180.

3. Han, X. (2010) Multi-dimensional mass spectrometry-based shotgun lipidomics and the altered lipids at the mild cognitive impairment stage of Alzheimer's disease. Biochim. Biophys. Acta 1801, 774–783.

4. Yang, K., Cheng, H., Gross, R.W. and Han, X. (2009) Automated lipid identification and quantification by multi-dimensional mass spectrometry-based shotgun lipidomics. Anal. Chem. 81, 4356–4368.

5. Han, X., Yang, K. and Gross, R.W. (2012) Multi-dimensional mass spectrometry-based shotgun lipidomics and novel strategies for lipidomic analyses. Mass Spectrom. Rev. 31, 134–178.

6. Stahlman, M., Ejsing, C.S., Tarasov, K., Perman, J., Boren, J. and Ekroos, K. (2009) High throughput oriented shotgun lipidomics by quadrupole time-of-flight mass spectrometry. J. Chromatogr. B 877, 2664–2672.

7. Quehenberger, O., Armando, A.M., Brown, A.H., Milne, S.B., Myers, D.S., Merrill, A.H., Bandyopadhyay, S., Jones, K.N., Kelly, S., Shaner, R.L., Sullards, C.M., Wang, E., Murphy, R.C., Barkley, R.M., Leiker, T.J., Raetz, C.R., Guan, Z., Laird, G.M., Six, D.A., Russell, D.W., McDonald, J.G., Subramaniam, S., Fahy, E. and Dennis, E.A. (2010) Lipidomics reveals a remarkable diversity of lipids in human plasma. J. Lipid Res. 51, 3299–3305.

8. Guan, Z., Li, S., Smith, D.C., Shaw, W.A. and Raetz, C.R. (2007) Identification of N-acylphosphatidylserine molecules in eukaryotic cells. Biochemistry 46, 14500–14513.

9. Astarita, G. and Piomelli, D. (2009) Lipidomic analysis of endocannabinoid metabolism in biological samples. J. Chromatogr. B 877, 2755–2767.

10. Tan, B., Yu, Y.W., Monn, M.F., Hughes, H.V., O'Dell, D.K. and Walker, J.M. (2009) Targeted lipidomics approach for endogenous N-acyl amino acids in rat brain tissue. J. Chromatogr. B 877, 2890–2894.

11. Naito, H.K. and David, J.A. (1984) Laboratory considerations: Determination of cholesterol, triglyceride, phospholipid, and other lipids in blood and tissues. Lab. Res. Methods Biol. Med. 10, 1–76.

12. Roughan, P.G., Slack, C.R. and Holland, R. (1978) Generation of phospholipid artefacts during extraction of developing soybean seeds with methanolic solvents. Lipids 13, 497–503.

13. Sastry, P.S. and Kates, M. (1964) Hydrolysis of monogalactosyl and digalactosyl diglycerides by specific enzymes in Runner-Bean leaves. Biochemistry 3, 1280–1287.

14. Kramer, J.M. and Hulan, H.W. (1978) A comparison of procedures to determine free fatty acids in rat heart. J. Lipid Res. 19, 103–106.

15. Arthur, G. and Sheltawy, A. (1980) The presence of lysophosphatidylcholine in chromaffin granules. Biochem. J. 191, 523–532.

16. Messner, M.C., Albert, C.J., Hsu, F.F. and Ford, D.A. (2006) Selective plasmenylcholine oxidation by hypochlorous acid: Formation of lysophosphatidylcholine chlorohydrins. Chem. Phys. Lipids 144, 34–44.

17. Folch, J., Lees, M. and Sloane Stanley, G.H. (1957) A simple method for the isolation and purification of total lipides from animal tissues. J. Biol. Chem. 226, 497–509.

18. Bligh, E.G. and Dyer, W.J. (1959) A rapid method of total lipid extraction and purification. Can. J. Biochem. Physiol. 37, 911–917.

19. Matyash, V., Liebisch, G., Kurzchalia, T.V., Shevchenko, A. and Schwudke, D. (2008) Lipid extraction by methyl-tert-butyl ether for high-throughput lipidomics. J. Lipid Res. 49, 1137–1146.

20. Lofgren, L., Stahlman, M., Forsberg, G.B., Saarinen, S., Nilsson, R. and Hansson, G.I. (2012) The BUME method: A novel automated chloroform-free 96-well total lipid extraction method for blood plasma. J. Lipid Res. 53, 1690–1700.

21. Hara, A. and Radin, N.S. (1978) Lipid extraction of tissues with a low-toxicity solvent. Anal. Biochem. 90, 420–426.

22. Radin, N.S. (1981) Extraction of tissue lipids with a solvent of low toxicity. Methods Enzymol. 72, 5–7.

23. Morrison, W.R., Tan, S.L. and Hargin, K.D. (1980) Methods for the quantitative analysis of lipids in cereal grains and similar tissues. J. Sci. Food Agric. 31, 329–340.

24. Slomiany, A. and Slomiany, B.L. (1981) A new method for the isolation of the simple and highly complex glycosphingolipids from animal tissue. J. Biochem. Biophys. Methods 5, 229–236.

25. Kanfer, J.N. (1969) Preparation of gangliosides. Methods Enzymol. 14, 660–664.

26. Ladisch, S. and Gillard, B. (1985) A solvent partition method for microscale ganglioside purification. Anal. Biochem. 146, 220–231.

27. Popa, I., Vlad, C., Bodennec, J. and Portoukalian, J. (2002) Recovery of gangliosides from aqueous solutions on styrene-divinylbenzene copolymer columns. J. Lipid Res. 43, 1335–1340.

28. Williams, M.A. and McCluer, R.H. (1980) The use of Sep-Pak C18 cartridges during the isolation of gangliosides. J. Neurochem. 35, 266–269.

29. Powell, W.S. (1980) Rapid extraction of oxygenated metabolites of arachidonic acid from biological samples using octadecylsilyl silica. Prostaglandins 20, 947–957.

30. Figlewicz, D.A., Nolan, C.E., Singh, I.N. and Jungalwala, F.B. (1985) Pre-packed reverse phase columns for isolation of complex lipids synthesized from radioactive precursors. J. Lipid Res. 26, 140–144.

31. Harrison, K.A., Clay, K.L. and Murphy, R.C. (1999) Negative ion electrospray and tandem mass spectrometric analysis of platelet activating factor (PAF) (1-hexadecyl–2-acetyl-glycerophosphocholine). J. Mass Spectrom. 34, 330–335.

32. Han, X. and Gross, R.W. (1994) Electrospray ionization mass spectroscopic analysis of human erythrocyte plasma membrane phospholipids. Proc. Natl. Acad. Sci. U. S. A. 91, 10635–10639.

33. Han, X. and Gross, R.W. (2005) Shotgun lipidomics: Electrospray ionization mass spectrometric analysis and quantitation of the cellular lipidomes directly from crude extracts of biological samples. Mass Spectrom. Rev. 24, 367–412.

34. Han, X. and Gross, R.W. (2005) Shotgun lipidomics: Multi-dimensional mass spectrometric analysis of cellular lipidomes. Expert Rev. Proteomics 2, 253–264.

35. Han, X., Gubitosi-Klug, R.A., Collins, B.J. and Gross, R.W. (1996) Alterations in individual molecular species of human platelet phospholipids during thrombin stimulation: Electrospray ionization mass spectrometry-facilitated identification of the boundary conditions for the magnitude and selectivity of thrombin-induced platelet phospholipid hydrolysis. Biochemistry 35, 5822–5832.

36. Han, X., Yang, J., Cheng, H., Ye, H. and Gross, R.W. (2004) Towards fingerprinting cellular lipidomes directly from biological samples by two-dimensional electrospray ionization mass spectrometry. Anal. Biochem. 330, 317–331.

37. Ways, P. and Hanahan, D.J. (1964) Characterization and quantification of red cell lipids in normal man. J. Lipid Res. 5, 318–328.

38. Cheng, H., Guan, S. and Han, X. (2006) Abundance of triacylglycerols in ganglia and their depletion in diabetic mice: Implications for the role of altered triacylglycerols in diabetic neuropathy. J. Neurochem. 97, 1288–1300.

39. Wang, C., Wang, M. and Han, X. (2015) Comprehensive and quantitative analysis of lysophospholipid molecular species present in obese mouse liver by shotgun lipidomics. Anal. Chem. 87, 4879–4887.

40. Abbott, S.K., Jenner, A.M., Mitchell, T.W., Brown, S.H., Halliday, G.M. and Garner, B. (2013) An improved high-throughput lipid extraction method for the analysis of human brain lipids. Lipids 48, 307–318.

41. Chen, S., Hoene, M., Li, J., Li, Y., Zhao, X., Haring, H.U., Schleicher, E.D., Weigert, C., Xu, G. and Lehmann, R. (2013) Simultaneous extraction of metabolome and lipidome with methyl tert-butyl ether from a single small tissue sample for ultra-high performance liquid chromatography/mass spectrometry. J. Chromatogr. A 1298, 9–16.

42. Ametaj, B.N., Bobe, G., Lu, Y., Young, J.W. and Beitz, D.C. (2003) Effect of sample preparation, length of time, and sample size on quantification of total lipids from bovine liver. J. Agric. Food Chem. 51, 2105–2110.

43. Virot, M., Tomao, V., Colnagui, G., Visinoni, F. and Chemat, F. (2007) New microwave-integrated Soxhlet extraction. An advantageous tool for the extraction of lipids from food products. J. Chromatogr. A 1174, 138–144.

44. Nichols, B.W. (1963) Separation of the lipids of photosynthetic tissues: Improvements in analysis by thin-layer chromatography. Biochim. Biophys. Acta 70, 417–422.

45. Welti, R., Li, W., Li, M., Sang, Y., Biesiada, H., Zhou, H.-E., Rajashekar, C.B., Williams, T.D. and Wang, X. (2002) Profiling membrane lipids in plant stress responses. Role of phospholipase Da in freezing-induced lipid changes in Arabidopsis. J. Biol. Chem. 277, 31994–32002.

46. Singh, A.K. (1992) Quantitative analysis of inositol lipids and inositol phosphates in synaptosomes and microvessels by column chromatography: Comparison of the mass analysis and the radiolabelling methods. J. Chromatogr. 581, 1–10.

47. Vickers, J.D. (1995) Extraction of polyphosphoinositides from platelets: Comparison of a two-step procedure with a common single-step extraction procedure. Anal. Biochem. 224, 449–451.

48. Zhao, Z. and Xu, Y. (2010) An extremely simple method for extraction of lysophospholipids and phospholipids from blood samples. J. Lipid Res. 51, 652–659.

49. Lydic, T.A., Busik, J.V. and Reid, G.E. (2014) A monophasic extraction strategy for the simultaneous lipidome analysis of polar and nonpolar retina lipids. J. Lipid Res. 55, 1797–1809.

50. Lee, K.Y. (1971) Loss of lipid to plastic tubing. J. Lipid Res. 12, 635–636.

51. Lough, A.K., Felinski, L. and Garton, G.A. (1962) The production of methyl esters of fatty acids as artifacts during the extraction or storage of tissue lipids in the presence of methanol J. Lipid Res. 3, 476–478.

52. Heinz, E. and Tulloch, A.P. (1969) Reinvestigation of the structure of acyl galactosyl diglyceride from spinach leaves. Hoppe Seylers Z. Physiol. Chem. 350, 493–498.

14

QUANTIFICATION OF INDIVIDUAL LIPID SPECIES IN LIPIDOMICS

14.1 INTRODUCTION

Quantification in omics generally falls into two categories, i.e., relative and absolute quantifications. The former measures the pattern change of the lipid species in a lipidome, which can be used as a tool for readout after stimulation or for biomarker discovery. The latter determines the mass levels of individual lipid species, and then each individual lipid subclass and class of a lipidome. Measurement of the changed mass levels of individual lipid class, subclass, and molecular species is critical for elucidation of biochemical mechanism(s) responsible for the changes and for pathway/network analysis in addition to serving as a tool for readout after a perturbation or for biomarker discovery. Thus, only the latter case is extensively discussed in this and the following chapters.

Many modern technologies (including MS, nuclear magnetic resonance spectroscopy, fluorescence spectroscopy, chromatography, and microfluidic devices) have been used in lipidomics for quantification of lipid species in biological systems [1]. Clearly, ESI-MS has evolved to be one of the most popular and powerful technologies for quantitative analyses of individual lipid species [2–5].

Modern mass spectrometry with an ESI source possesses many advantages for structural characterization and identification of analytes with high sensitivity as discussed in the last two parts. Moreover, the majority of mass spectrometers can not only very accurately measure the mass-to-charge ratio of an ion but also accurately determine the molar ratios of "isotopologues" of an analyte. Here, *isotopologues*

Lipidomics: Comprehensive Mass Spectrometry of Lipids, First Edition. Xianlin Han.
© 2016 John Wiley & Sons, Inc. Published 2016 by John Wiley & Sons, Inc.

(i.e., isotope-labeled analogs) are molecules that differ only in their isotopic composition. "Isotopologues" are different from "*isotopomers*," which are the isomers having the same number of each isotopic atom but differing in their positions.

The accurate determination of the molar ratios of isotopologues of an analyte by MS indicates that this powerful tool can be used for accurate quantification of a compound if an isotopologue of the compound in a known amount (which should be within a linear dynamic range relative to the compound) is present in the same solution with the compound. This statement specifies the multiple requirements for quantification by ESI-MS as follows:

- Accurate quantification should be conducted using an internal standard.
- It is better to use a stable isotopologue as the internal standard.
- Accurate quantification by ESI-MS can only be carried out in a relative measurement manner to the internal standard.
- The compound and its isotopologue should be within a linear dynamic range of quantification.

These points are extensively discussed in this chapter and in Chapter 15. However, the conclusion is that ESI-MS can be used as a tool for quantification under certain conditions. In practice, ESI-MS has become one of the most popular methods for quantitative analyses of individual lipid species in lipidomics, thereby greatly facilitating the development of the field.

Why can accurate quantification by ESI-MS only be conducted in a relative measurement manner with an isotopologue as the internal standard? A mass spectrometer is unlike an UV–vis spectrophotometer in which the relationship between the optical intensity and the concentration of an analyte follows the Beer–Lambert law. There is no direct relationship in MS between the determined ion counts and the concentration of an analyte yielding the ion due to numerous reasons, which are as follows:

- The differences of ion source designs, ion transmission, and ion detection devices between instruments
- The differences of the response factors of an analyte to the instruments
- An instrument responding to the changes of solution matrices
- Changes of the operational conditions of an instrument.

Therefore, very different ion intensities (i.e., ion counts) of an identical solution of an analyte can be obtained from different laboratories by ESI-MS by using different types of instruments, or even in the same laboratory under apparently "identical" experimental conditions, since numerous minor variations such as the degree of ion source clearance, moisture levels, vacuum stability, and electronic noise levels can affect the ion intensity of the analyte.

Regarding the use of a stable isotopologue of the analyte as an internal standard, unfortunately, it is impractical to prepare thousands of stable isotopologues to quantify individual lipid species in lipidomics. Then the question is, what is the minimal number of internal standards required in different methods? These questions are largely discussed in Chapter 15, but these methods and their potential caveats are discussed in this chapter.

It should be recognized that the word "quantification" leads to very different expectations for analytical chemists and bioanalytical chemists. To an analytical chemist, quantification must be very "accurate." All attempts in each step of a quantitative analysis from sampling to data processing should achieve the highest degree of accuracy and/or precision possible. Therefore, error propagation can be pre-estimated and controlled. Using a stable isotopologue as an internal standard for quantification of its counterpart somewhat falls into this category. However, to a bioanalytical chemist, the expectation for quantification is relatively loose since many uncertainties or variations are present in the entire process of analysis from sampling, sample preparation, and analysis, which are as follows:

- The variations present in sampling of biological samples could be substantial, which could be bigger than any analytical errors (Chapter 13).
- The variation of the protein assays (if the protein content is used as a normalizer) is usually at the level of 10%.
- The variation for the other normalizers such as RNA and DNA contents, which is similar to that protein assay, is also present.

Therefore, any method in lipidomics employing some kind of compromise way or correction factors for quantification of a particular category of compounds to achieve a variation around 10% should be acceptable and practical.

It should also be recognized that due to the presence of the variations in biological samples, a statistic analysis of the data obtained is usually essential for quantification of a group of samples and comparison between the sample groups. Unfortunately, different statistical methods and personally favored selection of parameters could lead an analyst to have different conclusions, particularly if the accuracy and/or reproducibility of the employed lipidomics approach for acquiring analytical data are also relatively low. Therefore, the statistical method and parameters should always be clearly reported.

Quantification of lipids by ESI-MS through a lipidomic approach is an interdisciplinary task that largely determines the amounts of intact individual species in a biological sample based on the selected internal standard(s) and a normalizer (although other relative measurements are also used (see below)). Unfortunately, the measurements of the contents of both the internal standard and the normalizer contain experimental errors. These errors impact the lipid quantification. Collectively, by this approach for lipidomic analysis, the amounts of individual lipid species in a selected sample size can be determined if appropriate internal standards are added prior to extraction with a tolerant experimental error, in which correction for any bias in extraction recovery, molecular species-dependent ionization efficiencies, and other factors is considered within a variation of 10%.

14.2   PRINCIPLES OF QUANTIFYING LIPID SPECIES BY MASS SPECTROMETRY

Quantification of the concentration of an analyte with MS analysis, in principle, employs a correlation between the concentration and the ion intensity of the analyte, which is linear within a predetermined linear dynamic range:

$$I = I_{app} - b = a * c \qquad (14.1)$$

where c is the concentration of the analyte; I_{app} is the apparent ion intensity of the analyte measured by MS; b is the mass spectral baseline resulting from both the baseline drift if present and the chemical noise, which can be determined as described recently [6], or other methods [7, 8], or ignored if the baseline is indeed low; I is the baseline-corrected ion intensity of the analyte (i.e., the actual ion intensity); and a is the response factor of the employed mass spectrometer. When $I_{app} \gg b$ (e.g., $S/N > 10$), $I \approx I_{app}$; otherwise, correction for the spectral baseline is required to obtain the actual ion intensity I from the measured apparent ion intensity I_{app} of the analyte. Apparently, if a constant response factor of a could be determined for the analyte from a mass spectrometer or mass spectrometers, absolute quantification of its concentration (within linear dynamic range) of the analyte would be simply obtained from its baseline-corrected ion intensity with Equation 14.1. This is similar to the Beer–Lambert law in the spectroscopic analysis. Unfortunately, as mentioned in Section 14.1, the ion intensity of an analyte measured by MS could be easily affected with even minor alterations in the conditions of the solution, analyte ionization, and instrumentation. This suggests that the response factor a is hardly to be a constant in MS. Moreover, most of the alterations could not be controlled or might not even be noticed. Accordingly, it would be difficult to determine the constant response factor for an analyte of interest, thus performing absolute quantification based on Equation 14.1 would be mostly impossible.

Therefore, quantification of an analyte by MS analysis usually requires comparisons to either an external standard that is itself in most cases or an internal standard that is an analog to the analyte (e.g., its stable isotopologue). When an external standard is used, a calibration curve of the standard is established at a series of concentrations each of which should be analyzed under identical conditions that will be applied to the MS analysis of the analyte of interest. When an internal standard is used, the standard is added at the earliest step possible during sample preparation and its concentration should be in an appropriate ratio to the analyte (see Section 13.3.2).

The advantage of using an external standard is that there is no concern of the potential overlapping of extraneously added standards with endogenous molecular species. However, control of the analyses of external standard and analyte of interest under identical conditions is generally difficult. For example, the multiple steps involved in sample preparation (including separation) may lead to differential recovery and carryover from sample to sample; the varied composition of the analyzed solution due to the use of gradients or the presence of co-eluents during chromatographic separation may contribute to differential ionization conditions from run to run; and the varied spray stability during ESI-MS analysis and other factors may lead to differential ionization efficiency from time to time. Therefore, use of external standards alone is usually not the best choice for the analysis of a complex system, which is generally associated with a complicated process such as the global analyses of the cellular lipidome.

The advantage of using an internal standard is its simplicity and accuracy resulting from its being processed and analyzed simultaneously with the analyte of interest. However, selection of an appropriate internal standard might be difficult because different systems may need different internal standards and specifically synthesized standards may be necessary to avoid any potential overlap with endogenous species in the analyzed system. Moreover, addition of an appropriate amount of internal standard is not straightforward, but needs some expertise and predetermination (see Section 13.3.2).

When a standard is used for quantification of the concentration of an analyte, many researchers simply use a formula of *ratiometric comparison*:

$$\frac{I_u}{I_s} = \frac{c_u}{c_s} \tag{14.2}$$

where I_u and I_s are the actual or baseline-corrected ion peak intensities (or areas) of the analyte and the selected standard, respectively; c_u and c_s are the unknown concentration of the analyte and the known concentration of the standard, respectively. This equation is naturally obtained in analogs to that used in spectroscopic analysis. Unfortunately, this equation holds true only under some limited conditions for MS analysis (see below for discussions).

Based on Equation 14.1, the ratio of the ion peak intensity of an analyte *vs.* that of a standard should be as follows:

$$\frac{I_u}{I_s} = \frac{(a_u * c_u + b_u)}{(a_s * c_s + b_s)} \tag{14.3}$$

Therefore, Equation 14.2 can only be true when two conditions are met. First, the background terms of b_u and b_s are either so small that these terms can be ignored or corrected as aforementioned. In this case, Equation 14.3 turns into

$$\frac{I_u}{I_s} = \left(\frac{a_u}{a_s}\right)\left(\frac{c_u}{c_s}\right) \tag{14.4}$$

Obviously, the second condition that has to be met is that the response factors of the analyte and the standard are identical (i.e., $a_u/a_s = 1$).

So the question is, when these factors are identical? To answer this question, we have to recognize the individual components leading to the response factor. Generally, the response factor could be divided into multiple subterms considering different components as

$$a = a_1 * a_2 * a_3 * a_4 * \ldots \tag{14.5}$$

where a_1, a_2, a_3, a_4, and more are the different components that influence the response factor. These components include, but are not limited to, the following factors: ionization efficiency of the analyte (ionization factor); an aggregated state of concentration affecting the response factor (concentration factor); the factor of tandem MS that influences the response factor in a molecular species-dependent manner (tandem MS factor); different matrices leading to different ionization response (matrix factor); etc. These factors are discussed in the following paragraphs and are further discussed in Chapter 15.

In the field of lipidomics, it was validated that individual lipid species of a polar lipid class could possess nearly identical ionization efficiencies in the low-concentration region [9–12]. This is due to these ionization efficiencies being predominantly dependent on their identically charged head group while their differential acyl chains including the length and unsaturation only minimally affect the ionization under low-concentration conditions. The reason why this only holds at the low-concentration region is discussed in the concentration factor (the following paragraph). Accordingly, the requirements of high polarity and low concentration are critical to minimize the differential response factors between the different lipid species of a class. These two requirements will suffice $a_u/a_s = 1$ if other conditions are identical, thereby facilitating the accurate quantitation of lipid species with comparison to an analog of the class serving as an internal standard. However, since identical response factors are not valid for individual lipid species of nonpolar lipid classes (e.g., TAG, DAG, and cholesteryl ester) even in the low-concentration region, the response factors for these species or a correlation between response factors and acyl chain length and unsaturation have to be predetermined for accurate quantification [13, 14]. Alternatively, a nonpolar lipid class could be converted to a polar lipid class through appropriate derivatization as demonstrated in many cases such as 4-hydroxyalkenal [15], DAG [16], oxysterol [17], and modified and unmodified fatty acids [18]. The fact that individual species of a polar lipid class possess nearly identical ionization efficiencies at the low-concentration region also indicates that any species among the lipid class of interest could serve as an internal standard of the class for quantification within a certain degree of accuracy.

A unique physical property of lipids is that they tend to form aggregates at a high concentration (even in an organic solvent like chloroform) or in a polar environment (e.g., methanol, acetonitrile, and water and particularly in the presence of salt(s)). The formed aggregates are poorly ionizable and/or detectable. The former is due to that the aggregates possess very different polarity, geometry, size, etc., from the single lipid species, and thus lead to very different ionization process from the individual molecular species. The latter is due to the aggregates (which could contain very different numbers of lipid species and be adducted with different numbers of small ions) possessing a different distribution of m/z, and some of them could be out of the mass detection range of the instrument. Those if they are within the mass detection range are in the noise levels of a broad mass range due to a wide combination of the number of lipid species and the adducted ions when an aggregate is ionized. Another complication resulting from aggregation is that the formation of lipid aggregates is acyl chain-dependent, i.e., the species containing long fatty acyl chain(s), and a fewer number of double bonds are much easier to form aggregates than those containing short and highly unsaturated fatty acyl chain(s). This physical property of lipid aggregates makes the response factors of individual lipid species at the aggregated state in a molecular species-dependent manner. Such a dependency of response factors on individual lipid molecular species has been well demonstrated in early studies of lipidomics [10, 19]. Therefore, analysis of individual lipid species of a polar lipid class at a low concentration is one of the key factors for accurate quantification of these species by ESI-MS.

Although individual species of a lipid class yield an essentially identical fragmentation pattern after collision-induced dissociation (CID) (Chapter 6), the intensities of the resultant individual fragments of these species could be very different depending on the collision energy due to their different structures of fatty acyl chains and their differential spatial configurations (Chapter 4). This fact has been extensively demonstrated in previous studies [20], indicating that tandem MS mass spectra (e.g., precursor-ion scan (PIS) and neutral-loss scan (NLS)) acquired from a lipid solution could be very different from each other and from the spectra acquired in the full MS mode in a collision-dependent manner (i.e., tandem MS factor). These results suggest that any quantification methods developed based on tandem MS including PIS, NLS, and SRM/MRM should consider this tandem MS factor. A few strategies could be employed to minimize the effects of this factor on quantification including the following:

- To optimize the CID energy making the tandem MS of a lipid class essentially identical or comparable to that acquired in the full MS mode.
- To loop the collision energy at a fine step (e.g., 1, 2, or 5 eV for a QqQ-type instrument) in a certain range to balance all of the possible fragmentation patterns.
- To be corrected in the presence of two or more internal standards.

It has already been extensively discussed that different matrices could affect the ionization efficiency of a lipid species (Chapter 4). This matrix factor should be minimal if analysis is carried out under an identical matrix condition such as direct infusion. However, for any method developed based on LC-MS, this factor should be recognized and well addressed since the matrix including solvent(s), eluent(s), and other constituents could be different from one elute time of an analyte to another elute time corresponding to the other analyte. Although the influence of different matrix conditions could be tested, it is hard to predict all possibilities during an LC-MS analysis. It is advisable to use two or more internal standards to correct those unidentified effects.

Many other factors leading to the differential response factors of different molecular species of a class (see Chapter 15 for more details) could also be present. For instance, if the instrument is not correctly tuned to equally transmit the ions generated from these different lipid species, the apparent ionization response factors of these species could be different. Anyone who is developing a quantitative method by using ESI-MS including tandem MS should recognize and make great efforts to minimize all these potential differences of the response factors. If the efforts are unable to be made to achieve such a goal, inclusion of additional internal standard(s) would be the ultimate solution to correct the differences of individual molecular species from the selected internal standard.

It should be emphasized again that selection of the stable isotopologue of the analyte as the internal standard for its quantification would perfectly satisfy the requirement of having identical response factors. This is due to the stable isotopologue having the same structure and property as the analyte (e.g., the same recovery and same ionization efficiency), and the internal standard is processed and analyzed at the same time as the analyte. However, this approach is impractical if not impossible to analyze thousands of species of interest in a complex system such as a cellular lipidome [21].

14.3   METHODS FOR QUANTIFICATION IN LIPIDOMICS

14.3.1   Tandem Mass Spectrometry-Based Method

As extensively described in Part II of this book, the species of an individual lipid class contain one or more characteristic fragments after CID. These diagnostic fragments can be used to filter the presence of individual species of the class through PIS, NLS, or both after direct infusion. Based on these special PIS and NLS, a method for quantitative analyses after ESI has been initially developed by Brugger et al. at the very early developmental stage of ESI-MS analysis of lipids [22]. Utilization of this technique was followed by Welti et al. for plant lipidomics [23–25] and many others [26–28]. In this methodology, at least two species of a lipid class of interest are used as internal standards, which are added during extraction of biological samples. A tandem MS mass spectrum is acquired in either neutral-loss or precursor-ion mode to detect the presence of the individual species of the class after direct infusion of a lipid solution through monitoring the characteristic neutral fragment or fragment ion, respectively (Chapter 2), of the lipid class of interest. The included standards are then used to correct many of the experimental factors (see discussed in the previous section and Chapter 15) for accurate quantification.

It should be recognized that at least two criteria should be considered in selection of the internal standards in this approach:

- To be absent or present in insignificant amounts in the original lipid extracts. This can be examined with a lipid extract of the biological sample without addition of any internal standards.
- To be representative of the physical properties (e.g., acyl chain length and unsaturation) of the entire lipid class. Therefore, those analogs of the lipid class located at the both ends of the mass range or isotope-labeled species are frequently used.

In this approach, the specific PIS or NLS to a lipid class is acquired in the profile mode with a stable ion current for a certain time period (commonly a few minutes) after direct infusion with a syringe pump or a NanoMate device (e.g., TriVersa NanoMate, Advion BioScience, Ithaca, NY). A mass spectrum averaged from all these acquired scans is obtained to reduce the noise level in proportion to $1/N^{1/2}$ (where N is the number of acquired mass spectra) and minimize any signal fluctuation, thereby increasing the signal-to-noise ratio. This acquired mass spectrum should then display "all the species of the class" including the selected internal standards at the sensitivity of the instrument employed. It should be noted that the detected ion peaks do not necessarily represent a lipid species of the class. Some artificial peaks could be displayed in the acquired mass spectrum (see discussion below). However, any detected ion peak at an m/z value corresponding to a species of the lipid class under the experimental conditions should have a very high probability (e.g., >90%) of representing this species.

As aforementioned, the peak intensities of the fragment ions depend on the chemical structures (i.e., the number of carbon atoms, and the number and location of double bonds) of each individual species of the class in addition to the conditions for CID. This is because the intense fragment ions represent those thermodynamically stable ions or because these fragment ions produced are favored in dissociation kinetics, both of which depend on the chemical structures (i.e., the number of carbon atoms, and the number and location of double bonds) of individual species of the class. The presence of this species-dependent process was demonstrated in a very early study reporting such a method [22] and confirmed in other studies [23, 29].

For example, a tandem MS analysis through PIS of m/z 184 was performed for analysis of PC species present in a lipid extract of Chinese hamster ovary cells in which an equimolar mixture of four PC species was used as internal standards [22]. Since this fragment ion represents the protonated phosphocholine under experimental conditions, the PIS detects all the protonated choline-containing GPL species in the solution. However, the spectrum clearly showed the decreases in the ion peak intensities of these internal standards as their molecular weights increased. A calibration curve from the ion peak intensities of the standards was derived and used to correct the experimental ion abundances to quantify individual PC species in this approach [22]. Accordingly, a built-in (non)linear calibration curve from two or more internal standards can be determined from their peak intensities. The concentration of each species of the class can then be derived from its ion peak intensity of the tandem MS mass spectrum by comparison with the calibration curve.

The authors of the paper pointed out that the net effect of the molecular weight on signal intensity is the sum of the following physical phenomena [22]:

- Variation of ionization efficiency
- Variation in fragmentation efficiency
- Variation in ion transmission
- Variation in detector response.

In their discussion, the authors did not mention that the effects of the number of double bonds as a factor in chemical structure as well as the concentration of the lipid

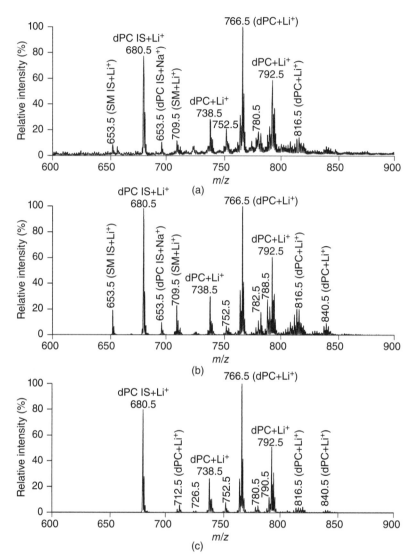

Figure 14.1 Comparisons between the molecular species profiles of choline glycerophospholipids in the lipid extracts of CHO cells acquired in the full MS scan mode and neutral-loss mode. Lipid extracts of CHO cells were prepared by using a Bligh–Dyer extraction procedure. Each individual lipid extract was diluted to a lipid concentration of <50 pmol total lipids/μL. A small amount of LiOH (25 pmol LiOH/μL) was added to the diluted solution. A positive-ion ESI mass spectrum in the full MS scan mode was acquired in the mass range from m/z 600 to 900 from the diluted lipid extract of CHO cells (a). Neutral-loss scans of 183.1 amu (i.e., phosphocholine) (b) and 189.1 amu (i.e., lithium cholinephosphate) (c) as indicated were also acquired from the diluted lipid extract in the mass range with a collision energy of 35 eV. IS denotes internal standards; SM stands for sphingomyelin; and dPC abbreviates diacyl phosphatidylcholine. Source: Yang et al. [20]. Reproduced with permission of Elsevier.

solution could also lead to the apparent differential ionization responses of different species as aforementioned. After inclusion of these two factors, virtually identical factors affecting the ionization response factor of individual species of a class to what described in the last section are discussed.

It should be recognized that differential ^{13}C isotope distributions of individual species containing different numbers of carbon atoms are present (this will be extensively discussed in Chapter 15). Although the authors of the study [22] did not recognize this factor, this difference could be corrected by the calibration curve with different species containing different fatty acyl chains (i.e., different numbers of carbon atoms). However, another factor related to ^{13}C isotopologues was not addressed in the study [22], which is caused by an overlap between the isotopologues containing two ^{13}C atoms and the species containing one less double bond. The contribution of this factor to the accuracy of quantification of the species containing one less double bond could be varied from a minimal effect to a significant amount depending on the concentration ratio of these species.

The effects of different molecular structures of a lipid class on accurate quantification of the species in comparison to a selected internal standard were demonstrated with an alternative approach by Yang et al. [20]. In the study, the investigators determined the effects of different structures of PC species on their profiles acquired through different neutral-loss scans, which are characteristics of this lipid class including the losses of choline (NLS59), phosphocholine (NLS183), and lithium cholinephosphate (NLS189) (Figure 14.1). These observations clearly indicate that the fragmentation process of a lipid class significantly depends on the structure of different species of the class, especially those containing polyunsaturated fatty acyl substituents.

This quantification approach possesses many advantages:

- Simple and efficient.
- Suitable for analysis of lipid classes that possess one or more characteristic fragments.
- Analysis can be carried out in a high-throughput manner.
- Using the selected standards, many experimental factors causing differential response factors are essentially eliminated.

The advantages of this methodology are obvious and the performance of the method is straightforward. Therefore, it has become increasingly popular for analyses of individual lipid species of many lipid classes.

However, some limitations of this method should be recognized, which the users should make great efforts to minimize them in order to improve the accuracy of quantification. These include, but are not limited to, those as follows:

- Selection of the standards to meet the two criteria aforementioned is not easy. Thus, the selected standards may only partially represent the species of a lipid class in the sample. The effects of differential fragmentation kinetics and/or thermodynamics of different chemical structures as discussed in the last section on accurate quantification would not be able to be fully ignored.

- The method uses a characteristic fragment of a class to filter the potential species present in the lipid class of interest. The possibility in the presence of artifactual ions in the so-called specific spectrum is relatively high. This possibility was demonstrated with the PIS of m/z 241 (corresponding to inositol phosphate) (Chapter 7) to filter the PI species [29]. It was shown that an approximately 10% of artifactual ions at the modest or higher intensities are present in the spectrum.

- The method is not able to provide any information about the fatty acyl substituents of the filtered ions that match with lipid species of the class. Obviously, identification of the fatty acyl substituents could substantially eliminate the artifacts present in the species [29], which is one of the principles of MDMS-SL technology.

- The linear dynamic range for quantification may be limited for some lipid classes when a sensitive tandem MS profiling is lacking, and quantitative analysis of many lipid classes in which a characteristic fragment does not exist is unable to be performed.

- As discussed earlier, the presence of isobaric species containing two ^{13}C atoms is unable to be corrected from the species containing one less double bond. Such overlapping could contribute a large component to the overlapping mono-isotopic peak depending on the abundance of the species containing one more double bond than itself. This factor can be corrected through de-isotoping or as described elsewhere [13, 30].

Taken together, this approach may be more suitable for semiquantification or profiling of different lipid classes than for accurate quantification.

An improved version of this approach is to use mass spectrometers possessing high mass accuracy and high mass resolution, for example, a Q-ToF–type mass spectrometer [21, 31–34]. By using this type of instrument, a product-ion spectrum of each ion in a unit mass or from a data-dependent acquisition after direct infusion can be rapidly and efficiently acquired in a small mass range (e.g., m/z 200–350 or other mass range) in which the most abundant and informative product ions corresponding to fatty acyl carboxylates are present [34]. After performing product-ion analyses of a mass range of interest, any PIS and/or NLS spectra of interest can be extracted from the product ions detected to "isolate" the lipid species of a class of interest. Many limitations present in the original tandem MS approach can be resolved in this new version of the method, including identification of fatty acyl substituents, correction for ^{13}C isotopologue overlaps, elimination of artifactual ions by using multiple PIS and/or NLS, and broadening linear dynamic range.

Many lipid classes can be quantified by this improved shotgun lipidomics approach. Quantification of individual species of a lipid class is conducted based on the summed abundance of its major fragment(s) in comparison to the counterpart of the spiked internal standard of the class. The effects of different ^{13}C isotopologue distributions between the species and the internal standard on quantification are considered. The sum of the fragment abundance may contribute to an increased sensitivity of detection.

In addition to the increased sensitivity, this shotgun lipidomics approach has many advantages in comparison to the profiling method based on a single specific tandem MS spectrum as described earlier. For example, for those lipid classes that do not possess a class-specific fragment (e.g., TAG and ceramide), this approach is still able to quantify the species through the use of many other fragments for identification. Moreover, the linear dynamic range of the method for quantification is 4 orders of magnitude for most lipid classes [34].

Note that this shotgun lipidomics approach is based on tandem MS analysis. As discussed earlier, due to the differential kinetics and/or thermodynamics of different species of a class during CID, two or more internal standards of a class spiked during lipid extraction are necessary to minimize any effects of differential fragmentation patterns on quantification. This point is particularly important to those species containing polyunsaturated fatty acyl substituents [20]. Moreover, identification and quantification of low-abundant isomeric/isobaric species overlapped with an abundant species may be missed due to the limitation of a relatively narrow dynamic range.

14.3.2 Two-Step Quantification Approach Used in MDMS-SL

After separation of different lipid classes in the ESI source (i.e., intrasource separation) and MDMS identification of individual species (see Section 3.2.3.3), quantification of the identified individual species of a lipid class of interest is performed in a two-step procedure in MDMS-SL [11, 35]. This procedure can be conducted automatically [29].

Briefly, this approach first determines whether ion peaks of interest are present in low abundance, or whether ion peaks overlapping with those of other lipid classes exist [29]. The approach performs the first step of quantification (see below) for those species that do not overlap with the species of other lipid class(es) and are present in abundance. Then, a second step of quantification is conducted for those species that either overlap with the species of other lipid class(es) or are present in low abundance. In this two-step quantification approach, the abundant and nonoverlapping peaks are first quantified with ratiometric comparison to the selected internal standard of the class with the mass spectrum acquired in the full MS mode and displaying all the species of the class after baseline correction and ^{13}C de-isotoping. The determined contents of these nonoverlapping and abundant species plus the preselected exogenously added internal standard of the class are the candidate standards for the second step of quantification. It should be recognized that peak intensities of the ions present in the class-specific MS/MS scan(s) might vary with individual subclasses or subtypes of species of the class [20]. Therefore, selection of different standards from the candidate standard list obtained from the first step of quantification for individual subclasses or subtypes of species for the second step of quantification should also be considered [20].

The second step of quantification is carried out to quantify the remaining species of the class, which are identified as overlapping and/or low-abundant species in the first step of quantification. An algorithm is generated based on two variables (i.e., the differences in the number of total carbon atoms and the number of total double bonds present in fatty acyl chains of each individual species) from those of the selected standards with multivariate least-square regression. By using this algorithm, the correction factors for each individual species of the class can be determined [29]. The corrected ion peak intensities of the overlapping and/or low-abundant species from the class-specific PIS or NLS are used to quantify these species with ratiometric comparison (i.e., Equation 14.4 with $a_u/a_s = 1$) to the ion peak intensities of the internal standards. With this second step of quantification, the linear dynamic range is extended dramatically for quantification of those species that are present in overlapping with the species of other lipid class(es) and/or in low abundance utilizing one or more PISs and NLSs. This is due to the significant reduction of background noise and increases in S/N ratios of low-abundant species [30]. Moreover, after filtering the overlapping species utilizing class-specific MS/MS scan(s), the effects of any presence of isobaric/isomeric species from other lipid class(es) on quantification are minimized in the approach [30].

Although this second step of quantification in the approach is similar to tandem MS-based shotgun lipidomics for quantification of individual lipid species (see Section 14.3.1) in some aspects, they are different in a key point. The MDMS-SL uses both an exogenously added internal standard and endogenously determined standards from the first step, whereas the tandem MS-based shotgun lipidomics exclusively uses exogenously added internal standards. One of the big advantages utilizing this second step in comparison to that of the tandem MS approach is that the use of endogenous species as standards can generally provide a more comprehensive representation of the physical properties of structurally similar but low-abundant species in the class. This is due to that the choices of those externally added standards in tandem MS-based shotgun lipidomics are generally limited in order to eliminate any potential overlapping with endogenous species.

It is worthy to note that when only two species, including the preselected internal standard, meet the criteria for selection as the standards for the second step of quantification in MDMS-SL, this second step of quantification is virtually identical to that used in tandem MS-based shotgun lipidomics with a linear standard curve [24]. In this case, the presence of different numbers of double bonds might affect the accurate quantification of those overlapping and/or low-abundant species performed in the second step of quantification. This effect is relatively small in MDMS-SL, especially for the total content of the class, with the following reasons: (1) the selected, endogenously obtained standard usually contains a certain number of double bonds; and (2) the species determined in the full MS mode in the first step of quantification largely contributes to the total content of the lipid class of interest.

There is another big difference between the second step of quantification in MDMS-SL and that used in tandem MS-based shotgun lipidomics. All quantified individual species in the two-step quantification approach are pre-identified with MDMS; therefore, any artifactual peaks present in the tandem MS spectrum that is used for quantification in the second step have been eliminated [29].

Finally, MDMS-SL possesses other advantages in comparison to the tandem MS-based shotgun lipidomics. For example, the second step of quantification in MDMS-SL can apply any or all head group-related PIS and/or NLS of the lipid class, if present and sensitive enough, for quantification of individual species in the second step. This redundant process is very useful to refine the data and serves as an internal validation of the determined results. Moreover, with the second step of quantification, an over 5000-fold linear dynamic range for many lipid classes can be readily achieved [36] due to the second step of quantification serving as a relay for the dynamic range.

This approach has been used to quantify individual species of over 40 lipid classes directly from lipid extracts of biological samples with or without derivatization [16, 18, 29, 37]. For example, in the analysis of SM species present in lipid extracts of mouse cortices, which were treated with lithium methoxide as previously described [38], the quantities of lithiated N18:1 SM at m/z 735.5, N18:0 SM at m/z 737.5, and N24:1 SM at m/z 819.6 were accurately measured with ratiometric comparisons to the selected internal standard (i.e., N14:0 SM at m/z 653.5) after ^{13}C de-isotoping of the mass spectrum acquired in the full MS mode (Trace a of Figure 14.2). The contents of all other low-abundant SM species were determined with the second step of quantitation utilizing the mass spectrum of NLS183 (Trace b of Figure 14.2) or NLS213 (as previously demonstrated [20]) with N14:0, N18:1, and N24:1 SM as internal standards.

In the same analysis of SM species above, a large number of nonoverlapping and abundant HexCer species (e.g., ions at m/z 806.6, 816.6, 820.6, and 834.6) could be quantified in the first step of the quantitation. This procedure was conducted utilizing the mass spectrum acquired in the full MS mode (Trace a of Figure 14.2) with ratiometric comparisons of the ion intensities to that of the selected internal standard (N15:0 GalCer at m/z 692.6) after ^{13}C de-isotoping. It should be noted that the class of HexCer contains two subclasses based on the presence or absence of an α-hydroxy moiety in the chain of fatty acyl amide. The presence of such an α-hydroxy moiety can be sensitively determined with NLS36.0 analyses of the identical lipid solution in the negative-ion mode as previously described [39]. Selection of the standards for quantification of the subclass species containing α-hydroxy moiety in the second step has to be considered to include the ions at m/z 806.6 and 820.6. Accordingly, the contents of the HexCer species in low abundance or overlapping with those of SM species were determined from the NLS162 mass spectrum (Trace c of Figure 14.2) by using the ions at m/z 692.6, 806.6, 816.6, 820.6, and 834.6 as standards. Unfortunately, this approach is unable to separately determine the contents of GalCer and GluCer species (thus, a general term of HexCer was used) as aforementioned. A conventional TLC approach [40] or an LC method [41] can be employed to determine the composition of these different subclasses.

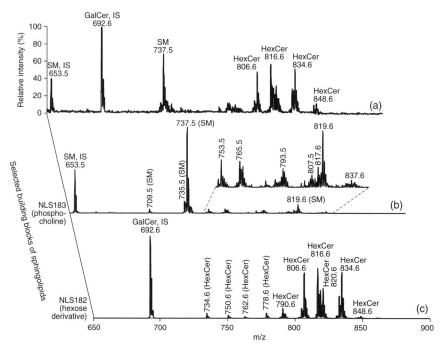

Figure 14.2 Two-step quantification of sphingomyelin and galactosylceramide species from lipid samples of mouse cortices. Lipid samples were prepared after treatment of mouse cortex lipid extracts with lithium methoxide as previously described [38]. MDMS-SL analysis of lipid samples was conducted in the positive-ion mode after addition of a small amount of LiOH. The MS trace "a" was the mass spectrum of the lipid samples acquired in the full MS mode. The MS trace "b" displayed a mass spectrum of NLS183, corresponding to phosphocholine, characteristic of sphingomyelin (SM) species in the solution under the conditions. The MS trace "c" showed a mass spectrum of NLS162, corresponding to a hexose derivative that represents a characteristic building block of HexCer species in the solution under the experimental conditions. IS denotes internal standard. Each spectrum displayed is normalized to the base peak in the spectrum.

Two caveats of this two-step quantification approach should be recognized.

- This two-step quantification approach cannot be applied to any lipid class for which a class-specific and sensitive PIS or NLS is not present, e.g., TAG, Cer, PE, and CL. Special quantification approaches for these lipid classes have been developed in MDMS-SL with or without derivatization [42–45].
- The experimental error for the species determined in the second step of quantification is larger than that in the first step due to the propagation and amplification of experimental errors from the first step. To reduce the amplified errors in the second step, it is critical to minimize any potential experimental error in the first step.

To this end, the key is to use the species in high abundance determined from the first step as standards for quantification of other species of the class in the second step to guarantee accuracy. The good thing is that since the species quantified in the second step only account for a small amount of the total contents of the class, the propagated experimental error in the second step affects the accuracy of total content of the class only to a small degree.

It should be pointed out that a misconception that has been consistently stated in the literature and in symposia is that *ion suppression* present in the analysis of complex lipid mixtures precludes quantification in all methods that use direct infusion. Thus, it has been erroneously argued that lipids cannot be quantified with a shotgun lipidomics approach, which can only provide a profile comparison between the different states. In fact, this concept is entirely incorrect if appropriate conditions for MS analysis are employed in the recommended low-concentration region for lipid analysis. Although it is possible to misuse any analytical procedure in its nonlinear range, careful consideration and use of conditions resulting in a linear dynamic range for analyte quantitation can be easily employed. If one uses concentrations outside of the linear dynamic range of a mass spectrometer, corrupted data are gathered due to competition for ionization that exceeds the 5 orders of magnitude linearity present in many mass spectrometers. Moreover, at high concentrations, the formation of lipid aggregates precludes meaningful quantitation as discussed in Section 14.2.

14.3.3 Selected Ion Monitoring Method

Selected ion monitoring (SIM) is an MS scanning mode in which only a single m/z or a limited number of m/z ratios is transmitted and/or detected by the instrument [46]. This is a classic method for analysis of particular compound(s) or fragment(s) of interest when mass spectrometers are predominantly constructed with a mass analyzer of single quadrupole or Fourier transform ion cyclotron resonance [47]. In the method, one monitors a single m/z at maximum sensitivity. The enhanced sensitivity is solely due to the high duty cycle in the technique. Specifically, in the method, the dwell times at a given m/z can typically be 10–100 times longer than when scanning a range of m/z, thus yielding a statistical advantage on to one to 2 orders of magnitude higher, and ultimately improving the S/N ratio [46].

For quantification, the monitored ion intensity can be directly compared to that of the selected internal standard after direct infusion. In the case of LC-MS analysis, the reconstituted ion peak area of each m/z corresponding to a species of interest can be compared to either a standard curve of the molecular species or to the reconstituted ion peak area of an internal standard under identical experimental conditions.

Unfortunately, the monitored m/z in the SIM technique could represent the combination of isobaric/isomeric species leaving us without unique identifiers. This is typical for the SIM analysis of a very complicated sample such as a serum or plasma sample where many compounds have the same m/z including multiply charged m/z of the species or stable isotopologues. Therefore, the technique is a kind of trading specificity for sensitivity. This technique is seldom used in modern lipidomics analysis due to its limitations.

However, an alternative method of SIM is still commonly used in lipidomics, particularly in the platforms associated with LC-MS, where high duty cycle instruments such as Q-ToF-type mass spectrometers are employed. In this case, a product-ion analysis at any moment of elution time could be performed for certain ions above a preset threshold for identification of these species (i.e., *data-dependent acquisition*), while a mass spectrum in the full MS mode, which detects both m/z values and intensities of the ions between the mass ranges of interest at the eluent time, is acquired over the entire elution time period for quantification. Owing to the very high scan rate, high sensitivity, and very fast and efficient acquisition of full product-ion mass spectra with the Q-ToF-type instruments over QqQ-type mass spectrometers, multiple acquisitions can be recorded at an elution time for identification of the relatively abundant species. The combination of elution time, m/z value, and a number of product-ion mass spectra provides reasonably accurate information about the chemistry of lipid species.

Importantly, the combination of ESI-MS detection with high-performance liquid chromatography (HPLC) separation makes this approach a potential choice for lipid quantification in lipidomics if one of the following conditions can be met.

- An appropriate standard curve of an individual lipid species of interest is carefully established under "identical" experimental conditions to the analysis of a sample [48]
- A stable isotopic labeled internal standard of each lipid species is available. While this is impractical for quantification in a lipidomic approach, a case study was reported [21]
- Ionization efficiencies of each individual species of a polar lipid are identical under the selected experimental conditions after considering the appropriate correction factors [12, 49, 50].

Among these three conditions, only the first approach or its variations was used widely in practice. Accordingly, a total ion current chromatogram of each individual molecular ion can be extracted (i.e., *selected-ion monitoring*) from the recorded mass spectra in the full MS mode from an HPLC run. Indeed, many ions of interest can be extracted simultaneously in this way [51]. To perform quantitative analysis of lipids, the linear dynamic range, limit of detection, and calibration curves of the molecular species of interest are generally predetermined. Thus, the reconstituted ion peak area of each species can be compared to a standard curve obtained under "identical" experimental conditions.

Note that at least one control compound (more are preferred) should be included in each sample. If this sample contains many classes of lipids, one control species for each class should be included in consideration of the differential ionization efficiencies between lipid classes. Accordingly, each of the reconstituted ion peak areas of the species are then normalized to the built-in control compound of the class prior to comparison with the appropriate standard curve for quantification. By taking this step, any effects of the variations due to HPLC separation conditions and/or ESI-MS conditions on the detected ion counts of a particular species, thereby influencing the quantification, can be diminished if not entirely eliminated.

In one example, Masukawa and colleagues studied the quantification of Cer species in human stratum corneum by LC-MS [52]. The authors employed a normal-phase column (Inertsil SIL 100A-3; 150×1.5 mm; GL Science, Tokyo, Japan) with a nonlinear gradient from mobile phase A (hexane–isopropanol–formic acid, 95:5:0.1) to B (hexane–isopropanol–ammonium formate (50 mM) aqueous solution, 25:65:10) for 80 min at a flow rate of 0.1 mL/min. The researchers prepared numerous Cer analogs used as standards. They determined the limit of detection ($S/N > 3$), limit of quantification ($S/N > 10$), and linearity of calibration curves using these compounds. Moreover, they employed a Cer species (d18:1-N17:0) as an internal standard for normalization of all analyzed samples. By using these controls, they achieved a maximum interday reproducibility of 14.3 RSD% and a systematic error of $\pm 21.4\%$. By using SIM, as many as 182 molecular-related ions derived from the diverse Cer species in the stratum corneum were measured.

Although this approach is commonly used for the analysis of a small number of lipids (e.g., a prepurified class) [52, 53], quantification of lipids on a global lipidomics scale is quite limited [12, 51]. This is because of the following parameters:

- Generation of the necessary standard curves for all species in a lipidome is impractical.
- The presence of multiple isomeric and isobaric molecular species may complicate the analysis
- It is very inefficient if each individual lipid class has to be pre-isolated.

To achieve identification of the quantified species, an approach with data-dependent product-ion analysis would be useful. However, an increased duty cycle of the instrument employed is required as the number of the analyzing lipids is increased. Alternatively, a high mass accuracy/resolution mass spectrometer would help to resolve the isobaric molecular ions from different lipid classes although isomeric species resulting from the regiospecificity and/or the double-bond location still cannot be resolved.

It should be kept in mind that the standard curves are established with synthetic lipid species in a solution containing a single species or a simple mixture, which may not mimic the analysis conditions for a biological extract. Specifically, the effects of differential matrices between biological samples and the standard solution on quantification are unknown. "Dynamic ion suppression" (Chapter 15) resulting from the interactions between any unresolved lipid species during a column run occurs.

Furthermore, when a normal-phase HPLC column is employed for separation of different lipid classes, species are not uniformly distributed in the eluted peak (i.e., each individual species of a class may possess its own distinct retention time and peak shape due to differential interactions with the stationary phase). "Dynamic ion suppression" is then a major factor affecting the accuracy in quantification.

If a reversed-phase HPLC column is used to resolve individual species of a pre-isolated class, then the relatively polar mobile phase that is commonly employed causes difficulties with solubility in a molecular species-dependent manner, leading to different apparent ionization efficiencies for individual species. If a solvent gradient is employed to resolve species by a reversed-phase method, changes in the components of the mobile phase may also cause an ionization stability problem (i.e., matrix effects [54]). Moreover, differential loss of lipid species on the column is not unusual [55]. Finally, while a reversed-phase column can eliminate lipid–lipid interactions of one lipid with another in most cases, there is up to a 1000-fold increase in the amount of lipid–lipid interactions with the same species (homodimer formation) since reversed-phase HPLC is typically used to concentrate samples, and lipids in sufficient concentrations tend to aggregate (see Section 14.2). These are some potentially problems, among many others, regarding quantification of lipid species by ESI-MS coupled to HPLC, which should be recognized.

14.3.4   Selected Reaction Monitoring Method

To improve the specificity while retaining the high sensitivity, a new method (i.e., selected or multiple reaction monitoring (SRM/MRM, Chapter 2)) is evolved from the SIM technique, which is only focused on one m/z of a compound to the SRM that is focused on both a molecular (or precursor) ion and a fragment-ion resultant from the precursor ion. The specific experiment in known as a "transition" and is usually written as "precursor-ion mass $\rightarrow$ fragment-ion mass." The only requirement to perform this technique is that the mass spectrometer employed has to possess the capability to perform MS/MS.

In comparison to the SIM methods, both specificity and sensitivity of detection by the SRM/MRM method are much higher. High specificity is due to the specific monitoring of a pair of ion transitions, while high sensitivity is due to the significant reduction of the noise level through MS/MS monitoring. For example, a study carried out by Kingsley and Marnett [56] clearly demonstrated the different results detected by using both SIM and SRM methods (Figure 14.3). In the study, a lipid sample from rodent brain tissue was analyzed using a C18 column with a methanol/water gradient in which both methanol and water contained $70\,\mu M$ silver acetate ($70–100\%$ methanol in 15 min, followed by 10-min hold at 100% methanol). The analysis of the sample was conducted twice under identical experimental conditions, including identical injection volumes, elution gradients, and ionization parameters, with detection by the two procedures (Figure 14.3b). In the SRM experiments, the precursor-/product-ion pairs for the analyses of 2-arachidonoylglycerol, d_8-2-arachidonoylglycerol, anandamide, and d_8-anandamide were included as indicated in Figure 14.3. The higher sensitivity yielded from the SRM technique is apparently evidenced with the analysis of anandamide. Specifically, the peak intensity of the anandamide ion is barely distinguishable from the noise level in the SIM experiment, whereas the peak corresponding to this lipid is clearly demonstrated with an S/N ratio of >5 by the SRM method.

Unlike the SIM method where analysis is largely dependent on the m/z of the compound, the identity of each individual species can be virtually defined in the SRM experiment by the elution time, the m/z of the ion, and the characteristic fragment

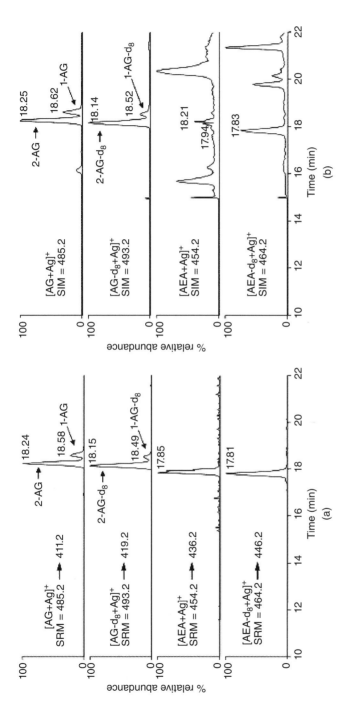

Figure 14.3 Comparison of LC-MS detection via selected reaction monitoring (SRM, a) and selected ion monitoring (SIM, b) of anandamide and 2-arachidonoylglycerol as silver adducts in extracted rodent brain samples. Chromatograms from top to bottom are for the analyses of 2-arachidonoylglycerol, d₈-2-arachidonoylglycerol, anandamide and d₈-anandamide, respectively. Source: Kingsley and Marnett [56]. Reproduced with permission of Elsevier.

monitored as shown successfully for oxysterols [57], positional isomer analysis of oxidized GPL species [58], and an in-depth analysis of PIP species [59]. Of course, if the LC method is unable to resolve the regioisomers or isomers due to different double-bond locations, the identities of these isomers cannot be determined by this approach since the characteristic fragment is largely resulted from the head group of the species.

Generally, in LC-MS analysis, a particular pair or pairs of precursor-/product-ion transitions are monitored at a specified elution time. Of course, these transitions at such an elution time should be predetermined utilizing authentic compounds or close analogs. Alternatively, a data-dependent acquisition approach could be set up with a certain type of instruments [60]. In either case, some degree of preknowledge about the individual lipid species present in the samples is required since currently available instruments are still unable to perform an infinite number of transitions at an elusion time due to the limitation of instrumental duty cycle and/or sensitivity. Moreover, similar to the SIM method, the linear dynamic range, limit of detection, and calibration curves of the species of interest should generally be predetermined for quantitative analysis of lipid species. Thus, the constructed ion peak area of each species can be compared to a standard curve of the species under "identical" experimental conditions.

At least one control compound should be included in each sample analyzed. If the sample contains many classes of lipids, one control species for each class should be included in consideration of the differential ionization efficiencies between classes (see Chapter 15 for further discussion). Each of the reconstituted ion peak areas of the lipid species is then normalized to the control compound of the class prior to comparison with a standard curve for quantification. For instance, Merrill and colleagues have described this approach in detail for the analysis of sphingolipids [41, 61].

The LC-ESI-MS/MS approach is commonly used for the analysis of a small number of lipids (e.g., a preseparated lipid class). This is because only a few pairs of ion transitions can be monitored at any elution time due to the limitation of the duty cycle as aforementioned. Moreover, generation of the necessary standard curves for all species in a cellular lipidome is impractical. However, because of the increased specificity and sensitivity of detection, this SRM/MRM method is particularly useful for quantitative analysis of those lipid classes that are present in low or very low abundance in the cellular lipidomes after a few steps of prechromatographic enrichment. One such typical example is the quantitative analysis of fatty acyl amino acids [62].

In addition to a few aforementioned limitations such as the requirement of an instrument possessing a high duty cycle capability, establishment of numerous standard curves, and preseparation of a lipid class of interest, the requirement of predetermination of the elution time of individual species to set up the pairs of transitions indicates that the SRM/MRM method is only suitable for targeted lipidomics analysis. Moreover, the effects of differential matrices between biological samples and the standard solution on quantification are unknown. Finally, "dynamic ion suppression" (see Chapter 15) resulting from the interactions between unresolved lipid species during an LC run should also be recognized.

14.3.5 High Mass Accuracy Mass Spectrometry Approach

With the increased mass accuracy and resolution in commercialized mass spectrometers, applications of this type of instruments for lipidomics analysis are remarkably increased. With the aid of this type of instruments, analysis in a high-throughput manner could be achieved [34, 63]. Among the instruments in this category, the Q-ToF-type mass spectrometers are the most commonly used ones in the field. This type of instrumentation offers mass resolution of up to 40,000 and mass accuracy of >5 ppm, which is sufficient for pinning down many of the elemental compositions encountered in lipidomics analysis, particularly with the very fast and efficient acquisition of full product-ion spectra. The drawback of this type of instruments is its limited dynamic range of the detector. Quantitation by using this type of instruments can only be performed in a narrow concentration range.

FT-ICR-MS offers literally unlimited mass resolution and sub-ppm mass accuracy. Its combination with a linear ion trap (LTQ-FT) also becomes a high-end standard instrumentation in lipidomics research. The instrument's hybrid character holds the possibility to run the linear ion trap and the FT-ICR-MS as two instruments in parallel, resulting in high-resolution precursor spectra and low-resolution product-ion spectra at an increased duty cycle. Therefore, when it couples to HPLC, the established experimental platform provides retention time, sub-ppm precursor masses, and product-ion spectra for both identification and quantification of individual lipid species [63]. In this particular study, MS/MS spectra were acquired in a data-dependent manner on the four most intensive ions present in each mass spectrum acquired in the full MS mode, leading to an MS/MS coverage of 66%. The investigators recognized that owing to the ultrahigh resolving power and mass accuracy, the platform could allow them to detect lipid species present in crude lipid extracts at very low quantities and those species from which no MS/MS spectra are available for the precursor ions. Other successful applications with a similar setup include quantitation of GPL species and TAG in plasma samples [64] and identification of GPL species in yeast [65]. The drawbacks of FT-ICR-MS are its relatively slow duty cycle ($\sim$3 s at 200,000 mass resolution) and its low-mass cutoff in the linear ion trap. This latter limitation might cause the loss of many low mass diagnostic fragment ions for the analysis of lipid species.

Instrumentation with an Orbitrap analyzer begins replacing the LTQ-FT. This new type of instrument offers a lot of advantages for lipidomic applications, especially when it is hybrid with a linear ion trap or quadrupole. For example, although resolving power and mass accuracy are less than the LTQ-FT, they provide unambiguous elemental compositions for most applications. Mass accuracy can be increased into the sub-ppm range by the use of internal lock mass calibrants. The inclusion of an HCD quadrupole alleviates the low mass cutoff limitations of the LTQ and allows the acquisition of high-resolution MS/MS spectra, although at a much slower speed than a Q-ToF-type instrument. Moreover, the most recent LTQ-Orbitrap setup in comparison to the LTQ-FT offers a faster scan rate of the linear ion trap at uncompromised sensitivity, allowing acquisition of up to 20 low-resolution MS/MS spectra per cycle. Bearing these advantages, this type of instruments has been broadly applied for a variety of lipidomics studies with or without coupling to LC [66–70].

Quantification of the well-resolved species of a polar lipid class is straightforward by ratiometric comparison of the peak intensity of the ion acquired in full MS mode with that of the selected internal standard of the class [35] after ^{13}C de-isotoping [42] assuming that any overlap with the species of other lipid class(es) or isotopologues from the identical class does not exist. If the ion contains isomeric species, the composition of these isomeric species can then be determined from the intensity ratio of fragments resulting from the isomeric species. So the question is, what is the minimal mass resolution of an instrument required for such a purpose? Wang et al. have recently investigated this requirement and found that at this moment, the mass resolution of the majority of the commercially available mass spectrometers is not high enough to totally resolve the overlaps present in the analysis of lipids after direct infusion [71]. The investigators discovered that the major issue is the partial overlap between the two ^{13}C atom-containing isotopologue of a species M (i.e., M+2 isotopologue) and the ion of a species with one less double bond than M

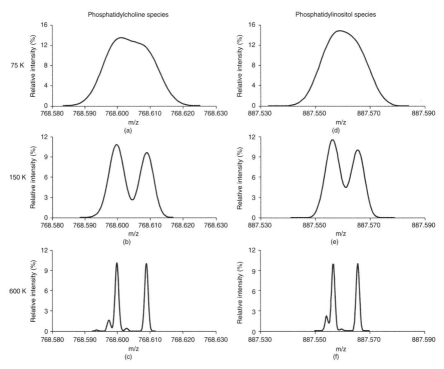

Figure 14.4 Simulation of the mass spectra displaying the M+2 ^{13}C isotopologue and the ion (L), which contains one double bond less than the species M with different instrumental mass resolution. An equal intensity ratio of the M+2 isotopologue and the L ion from the classes of both phosphatidylcholine (a–c) and phosphatidylinositol species (d–f) was used. The mass spectra of these ions with instrumental mass resolution of 75 K (a and d), 150 K (b and e), and 600 K (c and f) were simulated as described in the study [71]. Source: Wang et al. [71]. Reproduced with permission of John Wiley and Sons, Inc.

(assigned here as L) (referred to as the *double bond overlapping effect* hereafter). This issue was not problematic in the case of unit mass resolution MS where these ions are completely overlapped over each other, and the intensity of the overlapped peak is additive. Therefore, the intensity of each individual ion can be extracted after de-isotoping [13, 28, 72].

However, when simulated the double bond overlapping effect-induced mass shift, it was found that the resolution and locations of these ion peaks depend on the mass resolution of the instrument and the relative concentration of the overlapped species. For example, the two equally abundant ions essentially collapse into a single, broad peak at an instrumental mass resolution of 75,000 at the mass range (which corresponds to the mass resolution of the Q-Exactive mass spectrometer setting at 140,000 at m/z 200) (Figure 14.4a and d). Although these ions are partially resolved at the mass resolution of 150,000 (Figure 14.4b and e), the m/z values of the ion peaks of the M+2 ^{13}C isotopologue and the ion of the species L are not affected as much as at the mass resolution of 75,000 by the totally overlapping. All of the isotopologues ion peaks coming from ^{13}C, D, ^{14}N, etc. could only be totally resolved with an instrumental mass resolution of 600,000 at this mass range, which is unachievable with the majority of the commercially available instruments at the current time (Figure 14.4c and f).

This partial overlap alone could cause a mass shift of the species L to the lower mass end up to 12 ppm (at m/z 750 where the majority of lipid ions are detected) depending on the molar ratio of the species M and L and the mass resolution of the instrument used. However, if a mass-searching window of -15 to $+3$ ppm (a 12 ppm mass shift plus ± 3 ppm mass accuracy) is used for accurate mass searching, a substantial number of false positive and nonspecific hits could be included. Therefore, any attempt to conduct direct mass searching of a high accuracy mass spectral data set for lipidomics analysis should be precautious. This overlap could be solved according to the researchers by using M+1 isotopologues [71].

REFERENCES

1. Feng, L. and Prestwich, G.D., eds. (2006) Functional Lipidomics. CRC Press, Taylor & Francis Group, Boca Raton, FL

2. Wenk, M.R. (2010) Lipidomics: New tools and applications. Cell 143, 888–895.

3. Blanksby, S.J. and Mitchell, T.W. (2010) Advances in mass spectrometry for lipidomics. Annu. Rev. Anal. Chem. 3, 433–465.

4. Ivanova, P.T., Milne, S.B., Myers, D.S. and Brown, H.A. (2009) Lipidomics: A mass spectrometry based systems level analysis of cellular lipids. Curr. Opin. Chem. Biol. 13, 526–531.

5. Han, X., Yang, K. and Gross, R.W. (2012) Multi-dimensional mass spectrometry-based shotgun lipidomics and novel strategies for lipidomic analyses. Mass Spectrom. Rev. 31, 134–178.

6. Yang, K., Fang, X., Gross, R.W. and Han, X. (2011) A practical approach for determination of mass spectral baselines. J. Am. Soc. Mass Spectrom. 22, 2090–2099.

7. Coombes, K.R., Tsavachidis, S., Morris, J.S., Baggerly, K.A., Hung, M.C. and Kuerer, H.M. (2005) Improved peak detection and quantification of mass spectrometry data

acquired from surface-enhanced laser desorption and ionization by denoising spectra with the undecimated discrete wavelet transform. Proteomics 5, 4107–4117.

8. Xu, Z., Sun, X. and Harrington Pde, B. (2011) Baseline correction method using an orthogonal basis for gas chromatography/mass spectrometry data. Anal. Chem. 83, 7464–7471.

9. Han, X. and Gross, R.W. (1994) Electrospray ionization mass spectroscopic analysis of human erythrocyte plasma membrane phospholipids. Proc. Natl. Acad. Sci. U. S. A. 91, 10635–10639.

10. Koivusalo, M., Haimi, P., Heikinheimo, L., Kostiainen, R. and Somerharju, P. (2001) Quantitative determination of phospholipid compositions by ESI-MS: Effects of acyl chain length, unsaturation, and lipid concentration on instrument response. J. Lipid Res. 42, 663–672.

11. Han, X. and Gross, R.W. (2005) Shotgun lipidomics: Multi-dimensional mass spectrometric analysis of cellular lipidomes. Expert Rev. Proteomics 2, 253–264.

12. Hermansson, M., Uphoff, A., Kakela, R. and Somerharju, P. (2005) Automated quantitative analysis of complex lipidomes by liquid chromatography/mass spectrometry. Anal. Chem. 77, 2166–2175.

13. Han, X. and Gross, R.W. (2001) Quantitative analysis and molecular species fingerprinting of triacylglyceride molecular species directly from lipid extracts of biological samples by electrospray ionization tandem mass spectrometry. Anal. Biochem. 295, 88–100.

14. Bowden, J.A., Shao, F., Albert, C.J., Lally, J.W., Brown, R.J., Procknow, J.D., Stephenson, A.H. and Ford, D.A. (2011) Electrospray ionization tandem mass spectrometry of sodiated adducts of cholesteryl esters. Lipids 46, 1169–1179.

15. Wang, M., Fang, H. and Han, X. (2012) Shotgun lipidomics analysis of 4-hydroxyalkenal species directly from lipid extracts after one-step in situ derivatization. Anal. Chem. 84, 4580–4586.

16. Wang, M., Hayakawa, J., Yang, K. and Han, X. (2014) Characterization and quantification of diacylglycerol species in biological extracts after one-step derivatization: A shotgun lipidomics approach. Anal. Chem. 86, 2146–2155.

17. Jiang, X., Ory, D.S. and Han, X. (2007) Characterization of oxysterols by electrospray ionization tandem mass spectrometry after one-step derivatization with dimethylglycine. Rapid Commun. Mass Spectrom. 21, 141–152.

18. Wang, M., Han, R.H. and Han, X. (2013) Fatty acidomics: Global analysis of lipid species containing a carboxyl group with a charge-remote fragmentation-assisted approach. Anal. Chem. 85, 9312–9320.

19. Zacarias, A., Bolanowski, D. and Bhatnagar, A. (2002) Comparative measurements of multicomponent phospholipid mixtures by electrospray mass spectroscopy: Relating ion intensity to concentration. Anal. Biochem. 308, 152–159.

20. Yang, K., Zhao, Z., Gross, R.W. and Han, X. (2009) Systematic analysis of choline-containing phospholipids using multi-dimensional mass spectrometry-based shotgun lipidomics. J. Chromatogr. B 877, 2924–2936.

21. Ekroos, K., Chernushevich, I.V., Simons, K. and Shevchenko, A. (2002) Quantitative profiling of phospholipids by multiple precursor ion scanning on a hybrid quadrupole time-of-flight mass spectrometer. Anal. Chem. 74, 941–949.

22. Brugger, B., Erben, G., Sandhoff, R., Wieland, F.T. and Lehmann, W.D. (1997) Quantitative analysis of biological membrane lipids at the low picomole level by nano-electrospray ionization tandem mass spectrometry. Proc. Natl. Acad. Sci. U. S. A. 94, 2339–2344.

23. Welti, R., Li, W., Li, M., Sang, Y., Biesiada, H., Zhou, H.-E., Rajashekar, C.B., Williams, T.D. and Wang, X. (2002) Profiling membrane lipids in plant stress responses. Role of phospholipase Da in freezing-induced lipid changes in Arabidopsis. J. Biol. Chem. 277, 31994–32002.

24. Welti, R., Wang, X. and Williams, T.D. (2003) Electrospray ionization tandem mass spectrometry scan modes for plant chloroplast lipids. Anal. Biochem. 314, 149–152.

25. Welti, R., Shah, J., Li, W., Li, M., Chen, J., Burke, J.J., Fauconnier, M.L., Chapman, K., Chye, M.L. and Wang, X. (2007) Plant lipidomics: Discerning biological function by profiling plant complex lipids using mass spectrometry. Front. Biosci. 12, 2494–2506.

26. Mitchell, T.W., Turner, N., Hulbert, A.J., Else, P.L., Hawley, J.A., Lee, J.S., Bruce, C.R. and Blanksby, S.J. (2004) Exercise alters the profile of phospholipid molecular species in rat skeletal muscle. J. Appl. Physiol. 97, 1823–1829.

27. Deeley, J.M., Mitchell, T.W., Wei, X., Korth, J., Nealon, J.R., Blanksby, S.J. and Truscott, R.J. (2008) Human lens lipids differ markedly from those of commonly used experimental animals. Biochim. Biophys. Acta 1781, 288–298.

28. Liebisch, G., Lieser, B., Rathenberg, J., Drobnik, W. and Schmitz, G. (2004) High-throughput quantification of phosphatidylcholine and sphingomyelin by electrospray ionization tandem mass spectrometry coupled with isotope correction algorithm. Biochim. Biophys. Acta 1686, 108–117.

29. Yang, K., Cheng, H., Gross, R.W. and Han, X. (2009) Automated lipid identification and quantification by multi-dimensional mass spectrometry-based shotgun lipidomics. Anal. Chem. 81, 4356–4368.

30. Han, X. and Gross, R.W. (2005) Shotgun lipidomics: Electrospray ionization mass spectrometric analysis and quantitation of the cellular lipidomes directly from crude extracts of biological samples. Mass Spectrom. Rev. 24, 367–412.

31. Ejsing, C.S., Duchoslav, E., Sampaio, J., Simons, K., Bonner, R., Thiele, C., Ekroos, K. and Shevchenko, A. (2006) Automated identification and quantification of glycerophospholipid molecular species by multiple precursor ion scanning. Anal. Chem. 78, 6202–6214.

32. Schwudke, D., Oegema, J., Burton, L., Entchev, E., Hannich, J.T., Ejsing, C.S., Kurzchalia, T. and Shevchenko, A. (2006) Lipid profiling by multiple precursor and neutral loss scanning driven by the data-dependent acquisition. Anal. Chem. 78, 585–595.

33. Schwudke, D., Liebisch, G., Herzog, R., Schmitz, G. and Shevchenko, A. (2007) Shotgun lipidomics by tandem mass spectrometry under data-dependent acquisition control. Methods Enzymol. 433, 175–191.

34. Stahlman, M., Ejsing, C.S., Tarasov, K., Perman, J., Boren, J. and Ekroos, K. (2009) High throughput oriented shotgun lipidomics by quadrupole time-of-flight mass spectrometry. J. Chromatogr. B 877, 2664–2672.

35. Han, X., Cheng, H., Mancuso, D.J. and Gross, R.W. (2004) Caloric restriction results in phospholipid depletion, membrane remodeling and triacylglycerol accumulation in murine myocardium. Biochemistry 43, 15584–15594.

36. Han, X., Yang, K. and Gross, R.W. (2008) Microfluidics-based electrospray ionization enhances intrasource separation of lipid classes and extends identification of individual molecular species through multi-dimensional mass spectrometry: Development of an automated high throughput platform for shotgun lipidomics. Rapid Commun. Mass Spectrom. 22, 2115–2124.

37. Wang, C., Wang, M. and Han, X. (2015) Comprehensive and quantitative analysis of lysophospholipid molecular species present in obese mouse liver by shotgun lipidomics. Anal. Chem. 87, 4879–4887.

38. Jiang, X., Cheng, H., Yang, K., Gross, R.W. and Han, X. (2007) Alkaline methanolysis of lipid extracts extends shotgun lipidomics analyses to the low abundance regime of cellular sphingolipids. Anal. Biochem. 371, 135–145.

39. Han, X. and Cheng, H. (2005) Characterization and direct quantitation of cerebroside molecular species from lipid extracts by shotgun lipidomics. J. Lipid Res. 46, 163–175.

40. Abe, T. and Norton, W.T. (1974) The characterization of sphingolipids from neurons and astroglia of immature rat brain. J. Neurochem. 23, 1025–1036.

41. Merrill, A.H., Jr., Sullards, M.C., Allegood, J.C., Kelly, S. and Wang, E. (2005) Sphingolipidomics: High-throughput, structure-specific, and quantitative analysis of sphingolipids by liquid chromatography tandem mass spectrometry. Methods 36, 207–224.

42. Han, X., Yang, J., Cheng, H., Ye, H. and Gross, R.W. (2004) Towards fingerprinting cellular lipidomes directly from biological samples by two-dimensional electrospray ionization mass spectrometry. Anal. Biochem. 330, 317–331.

43. Han, X. (2002) Characterization and direct quantitation of ceramide molecular species from lipid extracts of biological samples by electrospray ionization tandem mass spectrometry. Anal. Biochem. 302, 199–212.

44. Han, X., Yang, K., Cheng, H., Fikes, K.N. and Gross, R.W. (2005) Shotgun lipidomics of phosphoethanolamine-containing lipids in biological samples after one-step in situ derivatization. J. Lipid Res. 46, 1548–1560.

45. Han, X., Yang, K., Yang, J., Cheng, H. and Gross, R.W. (2006) Shotgun lipidomics of cardiolipin molecular species in lipid extracts of biological samples. J. Lipid Res. 47, 864–879.

46. Watson, J.T. (1990) Selected-ion measurements, Methods Enzymol. 193, 86–106.

47. Murphy, R.C. (1993) Mass Spectrometry of Lipids. Plenum Press, New York. pp 290.

48. Lieser, B., Liebisch, G., Drobnik, W. and Schmitz, G. (2003) Quantification of sphingosine and sphinganine from crude lipid extracts by HPLC electrospray ionization tandem mass spectrometry. J. Lipid Res. 44, 2209–2216.

49. Sommer, U., Herscovitz, H., Welty, F.K. and Costello, C.E. (2006) LC-MS-based method for the qualitative and quantitative analysis of complex lipid mixtures. J. Lipid Res. 47, 804–814.

50. Sparagna, G.C., Johnson, C.A., McCune, S.A., Moore, R.L. and Murphy, R.C. (2005) Quantitation of cardiolipin molecular species in spontaneously hypertensive heart failure rats using electrospray ionization mass spectrometry. J. Lipid Res. 46, 1196–1204.

51. Shui, G., Bendt, A.K., Pethe, K., Dick, T. and Wenk, M.R. (2007) Sensitive profiling of chemically diverse bioactive lipids. J. Lipid Res. 48, 1976–1984.

52. Masukawa, Y., Narita, H., Sato, H., Naoe, A., Kondo, N., Sugai, Y., Oba, T., Homma, R., Ishikawa, J., Takagi, Y. and Kitahara, T. (2009) Comprehensive quantification of ceramide species in human stratum corneum. J. Lipid Res. 50, 1708–1719.

53. Liebisch, G., Drobnik, W., Reil, M., Trumbach, B., Arnecke, R., Olgemoller, B., Roscher, A. and Schmitz, G. (1999) Quantitative measurement of different ceramide species from crude cellular extracts by electrospray ionization tandem mass spectrometry (ESI-MS/MS). J. Lipid Res. 40, 1539–1546.

54. Cappiello, A., Famiglini, G., Palma, P., Pierini, E., Termopoli, V. and Trufelli, H. (2008) Overcoming matrix effects in liquid chromatography-mass spectrometry. Anal. Chem. 80, 9343–9348.

55. DeLong, C.J., Baker, P.R.S., Samuel, M., Cui, Z. and Thomas, M.J. (2001) Molecular species composition of rat liver phospholipids by ESI-MS/MS: The effect of chromatography. J. Lipid Res. 42, 1959–1968.

56. Kingsley, P.J. and Marnett, L.J. (2003) Analysis of endocannabinoids by Ag^+ coordination tandem mass spectrometry. Anal. Biochem. 314, 8–15.

57. Karu, K., Turton, J., Wang, Y. and Griffiths, W.J. (2011) Nano-liquid chromatography-tandem mass spectrometry analysis of oxysterols in brain: Monitoring of cholesterol autoxidation. Chem. Phys. Lipids 164, 411–424.

58. Nakanishi, H., Iida, Y., Shimizu, T. and Taguchi, R. (2009) Analysis of oxidized phosphatidylcholines as markers for oxidative stress, using multiple reaction monitoring with theoretically expanded data sets with reversed-phase liquid chromatography/tandem mass spectrometry. J. Chromatogr. B 877, 1366–1374.

59. Pettitt, T.R., Dove, S.K., Lubben, A., Calaminus, S.D. and Wakelam, M.J. (2006) Analysis of intact phosphoinositides in biological samples. J. Lipid Res. 47, 1588–1596.

60. Tan, B., Bradshaw, H.B., Rimmerman, N., Srinivasan, H., Yu, Y.W., Krey, J.F., Monn, M.F., Chen, J.S., Hu, S.S., Pickens, S.R. and Walker, J.M. (2006) Targeted lipidomics: Discovery of new fatty acyl amides. AAPS J. 8, E461–E465.

61. Merrill, A.H., Jr., Stokes, T.H., Momin, A., Park, H., Portz, B.J., Kelly, S., Wang, E., Sullards, M.C. and Wang, M.D. (2009) Sphingolipidomics: A valuable tool for understanding the roles of sphingolipids in biology and disease. J. Lipid Res. 50, S97–S102.

62. Tan, B., Yu, Y.W., Monn, M.F., Hughes, H.V., O'Dell, D.K. and Walker, J.M. (2009) Targeted lipidomics approach for endogenous N-acyl amino acids in rat brain tissue. J. Chromatogr. B 877, 2890–2894.

63. Fauland, A., Kofeler, H., Trotzmuller, M., Knopf, A., Hartler, J., Eberl, A., Chitraju, C., Lankmayr, E. and Spener, F. (2011) A comprehensive method for lipid profiling by liquid chromatography-ion cyclotron resonance mass spectrometry. J. Lipid Res. 52, 2314–2322.

64. Hu, C., van Dommelen, J., van der Heijden, R., Spijksma, G., Reijmers, T.H., Wang, M., Slee, E., Lu, X., Xu, G., van der Greef, J. and Hankemeier, T. (2008) RPLC-ion-trap-FTMS method for lipid profiling of plasma: Method validation and application to p53 mutant mouse model. J. Proteome Res. 7, 4982–4991.

65. Hein, E.M., Blank, L.M., Heyland, J., Baumbach, J.I., Schmid, A. and Hayen, H. (2009) Glycerophospholipid profiling by high-performance liquid chromatography/mass spectrometry using exact mass measurements and multi-stage mass spectrometric fragmentation experiments in parallel. Rapid Commun. Mass Spectrom. 23, 1636–1646.

66. Ogiso, H., Suzuki, T. and Taguchi, R. (2008) Development of a reverse-phase liquid chromatography electrospray ionization mass spectrometry method for lipidomics, improving detection of phosphatidic acid and phosphatidylserine. Anal. Biochem. 375, 124–131.

67. Sato, Y., Nakamura, T., Aoshima, K. and Oda, Y. (2010) Quantitative and wide-ranging profiling of phospholipids in human plasma by two-dimensional liquid chromatography/mass spectrometry. Anal. Chem. 82, 9858–9864.

68. Schuhmann, K., Herzog, R., Schwudke, D., Metelmann-Strupat, W., Bornstein, S.R. and Shevchenko, A. (2011) Bottom-up shotgun lipidomics by higher energy collisional dissociation on LTQ Orbitrap mass spectrometers. Anal. Chem. 83, 5480–5487.

69. Schuhmann, K., Almeida, R., Baumert, M., Herzog, R., Bornstein, S.R. and Shevchenko, A. (2012) Shotgun lipidomics on a LTQ Orbitrap mass spectrometer by successive switching between acquisition polarity modes. J. Mass Spectrom. 47, 96–104.

70. Yamada, T., Uchikata, T., Sakamoto, S., Yokoi, Y., Nishiumi, S., Yoshida, M., Fukusaki, E. and Bamba, T. (2013) Supercritical fluid chromatography/Orbitrap mass spectrometry based lipidomics platform coupled with automated lipid identification software for accurate lipid profiling. J. Chromatogr. A 1301, 237–242.

71. Wang, M., Huang, Y. and Han, X. (2014) Accurate mass searching of individual lipid species candidates from high-resolution mass spectra for shotgun lipidomics. Rapid Commun. Mass Spectrom. 28, 2201–2210.

72. Eibl, G., Bernardo, K., Koal, T., Ramsay, S.L., Weinberger, K.M. and Graber, A. (2008) Isotope correction of mass spectrometry profiles. Rapid Commun. Mass Spectrom. 22, 2248–2252.

15

FACTORS AFFECTING ACCURATE QUANTIFICATION OF LIPIDS

15.1 INTRODUCTION

The conditions for "accurate" quantification of lipid species by ESI-MS or ESI-MS/MS are discussed in Chapter 14. It is recognized that quantification of lipid species by MS is different from that of spectroscopic analysis in which the Beer–Lambert law can be followed. As discussed in Chapter 14, two essential parameters need to be considered in quantification of lipids by MS, i.e., the baseline of a mass spectrum used for the purpose and the response factor of an individual lipid species in comparison to that of the selected internal standard. The contribution of the baseline to the intensity of an ion should be minimal or a correction for the baseline has to be conducted as previously described [1]. Regarding the response factor of a lipid species of interest, the essential condition is that it has to be comparable to that of the selected internal standard or a correction factor has to be predetermined.

The response factor a is a combination of many different variables that affect the intensity of an ion of interest (see Equation 14.5). We briefly discussed some of these factors in Chapter 14, which are extensively reiterated in this chapter. Moreover, all of the other potential factors that may affect the accurate quantification of lipid species are discussed.

Lipidomics: Comprehensive Mass Spectrometry of Lipids, First Edition. Xianlin Han.
© 2016 John Wiley & Sons, Inc. Published 2016 by John Wiley & Sons, Inc.

15.2 LIPID AGGREGATION

Lipids readily form aggregates (e.g., dimers, oligomers, or micelles) as the lipid concentration increases or the solvent of a lipid solution becomes more polar. This is due to the nature of lipids that possess unique high hydrophobicity as well as a certain degree of hydrophilicity. The higher the hydrophobicity of a lipid species (e.g., longer acyl chain or less unsaturation), the lower the concentration at which the lipids aggregate. Lipids in aggregated forms cannot be ionized efficiently (Section 14.2). This tendency and the ionization nature of the aggregates make the lipid species containing short and/or polyunsaturated acyl chains show higher apparent response factors than those in the same class containing long and/or saturated acyl chains at a concentration that lipid aggregates form [2]. Therefore, lipid aggregation could substantially affect ionization response factor of an individual lipid species depending on the physical property of the species. Subsequently, ionization of individual lipid species in a polar lipid class becomes not only charged head group-dependent but also species-dependent. Under such conditions, it is obvious that the equation used for quantification of lipid species by using an internal standard (i.e., Equation 14.2) does not hold true. This process is further complicated by the dependence on the concentration of the solution, thereby making the use of calibration curves ineffective. It is, therefore, critical to keep the total lipid concentration lower than the concentration that favors aggregate formation.

The maximum lipid concentrations at which lipid aggregation is negligible also depend on the solvent system of the lipid solution in addition to the physical properties of individual lipid class. The recommended upper limit of total lipid concentration for direct infusion-based approaches (i.e., shotgun lipidomics) is approximately 100 pmol/µL in a 2:1 (v/v), 50 pmol/µL in a 1:1 (v/v), and 10 pmol/µL in a 1:2 (v/v) chloroform–methanol solvent system. However, when an extract contains a large amount of nonpolar lipids such as TAG, and cholesterol and its esters, this upper limit of lipid concentration should be substantially reduced. Alternatively, the upper limit remains for the polar lipid quantification after a prefractionation with hexane or with other nonpolar solvent to remove most of the nonpolar lipids from polar lipids. The total lipid concentration of a lipid extract could be estimated with the following knowledge: approximately 300–500 nmol total lipids/mg of protein for organs, such as the heart, skeletal muscle, liver, kidney, and for some cultured cell types, and 1000–2000 nmol total lipids/mg of protein for brain samples. Trial experiments could be performed to estimate the content of a lipid extract when analyzing an unknown sample.

The effects of lipid aggregation on quantification in shotgun lipidomics have been appreciated by many investigators; however, the effects of lipid aggregation on quantification by LC-MS-based approaches still have not been broadly recognized. A species eluted from a column is substantially concentrated at its peak time in comparison to the loaded solution in the majority of the cases. At such a condition of elution, formation of aggregates (i.e., homo-aggregates from same species) could potentially occur. Moreover, the mobile phase used in a reversed-phase HPLC column typically contains polar solvents (e.g., water, acetonitrile, high percentage of methanol, or salts) that favor lipid aggregation in a relatively low concentration. These factors potentially affect the response factors of the lipid species eluted at different times. For this reason, isotope-labeled internal standards for LC-MS are recommended in order to accurately quantify individual species of a lipid class. Alternatively, multiple calibration curves in combination with one internal standard or even multiple internal standards for semiquantification could be used as previously described [3–5]. In such cases, the internal standard(s) are used to normalize the variations between runs.

The concentration at which aggregates form could be estimated through determination of the linear dynamic range of concentration (see below). The upper concentration of a dynamic range at which the linear regression begins deviating from the log plot line represents the concentration of a lipid species at which aggregates begin to form.

15.3   LINEAR DYNAMIC RANGE OF QUANTIFICATION

Dynamic range is one of the critic factors to be dealt with for quantitative analysis by any method. In the modern era, the detectors used in mass spectrometers generally possess a very wide dynamic range although some types of mass detectors (e.g., the detector with multiple-channel plates commonly used in time-of-flight mass spectrometers) have dynamic ranges that are relatively smaller than others. Therefore, the detectors are usually not a factor to limit the dynamic range of a method for quantitative analysis of lipids.

In the quantitative analysis of lipids, the upper limit of the dynamic range corresponds to the concentration at which lipid aggregates begin to form. This concentration depends on the lipid class of interest, the solvents used, and the method (direct infusion or LC-MS) employed. The lower limit of the dynamic range of a method is the lowest concentration that the method is capable of quantifying individual species (which is generally higher than the limit of detection). This lower concentration is dependent on the sensitivity of the instrument, the sensitivity of the method, the effects of matrices, and others. For example, LC-MS/MS enhances the S/N through increases in duty cycle and selectivity and typically possesses an extended dynamic range in comparison to LC-MS (see Chapter 14).

There are at least two different measures of a dynamic range of a method due to the nature of MS analysis as discussed in Chapter 14. One is the linear range of the concentration of the analyte of interest, which can be accurately quantified. This measure of the dynamic range defines the linear relationship between absolute ion counts and the concentration of a species. This linear range is over 10,000-fold in the low concentration region and has been validated by many independent studies [2, 5–11]. As aforementioned, measurement of the absolute ion counts of an analyte may not be very meaningful in quantitative analysis of lipids by ESI-MS. However, this relationship has to be linear under a particular experimental condition. This concentration dynamic range is commonly measured by plotting the peak intensity ratio of the species of interest and an internal standard in a solution *vs.* the concentration of the solution that spans a wide range of concentrations [12, 13]. A horizontal line is expected within the linear dynamic range of concentration [12–15].

Another measure of the dynamic range of a method is the linear range of the ratio of the species of interest to an internal standard. This is the ion peak intensity (or area) ratio between ion intensities in a spectrum in the case of direct infusion or is the extracted ion peak area ratio in the case of LC-MS analysis. Due to the presence of background noise (e.g., chemical noise) and possible baseline drift (i.e., instrument stability), only an approximately 100-fold dynamic range (from 0.1 to 10 of the ratio) of this measure can be obtained for direct comparison between ion intensities in a mass spectrum [16]. This ratio is to guarantee an $S/N > 10$ (if the noise level is $<1\%$ of the base peak) for Equation 14.4 to be true. It should be noted that although the background noise and baseline drift cannot be viewed directly in the extracted ion current chromatogram, the contribution of these factors to accurate quantification of individual lipid species, especially those with low abundance, must be recognized. However, with the help of a two-step procedure in many cases, as discussed earlier, a 1000-fold dynamic range can be achieved, as long as the concentration measures of dynamic range are linear over 1000-fold [16, 17].

For both shotgun lipidomics and LC-MS-based approaches, the dynamic range should be examined in the presence of sample matrices instead of using a pure standard. Under such conditions, the matrix effects (e.g., ion suppression) that become more severe in analysis of minor species (or classes) in the presence of abundant species (or classes) can be eliminated for sample analysis.

For example, for a pure lipid species or a very abundant species in a mixture, a method can be developed to have over a 10,000-fold linear dynamic range using a two-step procedure of multidimensional MS-based shotgun lipidomics [18]. However, if this species is only a component of a minor lipid class in an extract, the first step of quantification is not achievable. The linear dynamic range in this case is reduced at least 100-fold. This represents the effects of "steady-state ion suppression" on the linear dynamic range for quantification of minor lipid classes in multidimensional MS-based shotgun lipidomics (see below). Quantification of sphingosine-1-phosphate directly from lipid extracts using shotgun lipidomics represents one such case. Since sphingosine-1-phosphate only accounts for $\ll 1\%$ of the total sphingolipids and is only present at noise level in the full MS spectrum, only an approximately 100-fold dynamic range for its quantification using precursor-ion scanning of m/z 79 is obtained [19]. As discussed earlier, a reduced dynamic range of a method for analysis of a minor component in the presence of abundant species is also evident in methods with LC-MS. Again, the minor species can be analyzed quantitatively with a pre-isolated fraction or with a saturated concentration of the abundant species to increase the dynamic range for quantification of the minor components.

Most studies report their linear dynamic range with a straight line ($y = ax + b$) to best fit the data with a good apparent correlation coefficient (γ^2). Unfortunately, by using this model, the data do not contribute equally to the straight-line fitting. The smaller the data point, the smaller the contribution to the fitting. Therefore, as long as a few data points with big numbers are linear, the data set can be fit into a straight line very well. By using this approach, the linear dynamic range may contain some artifacts and mislead the linearity on data points at the low concentration end. To solve this issue, it is advisable to use a least-square linear regression with a weighting factor as previously described [20].

There are many other ways to test the linear dynamic range. One is to perform linear least-square regression to fit a line

$$y = a + bx + cx^2 + \ldots \tag{15.1}$$

to find out whether the third or later terms can be ignored. If these terms cannot be ignored, the data point(s) at either the low or high concentration end should be excluded to finally achieve the stage when they can be ignored. This final data set defines the linear dynamic range of the method.

Alternatively, one can formulate the straight-line equation

$$y = ax + b \tag{15.2}$$

to the following format:

$$\log(y - b) = \log(a) + \log(x) \tag{15.3}$$

With the parameters a and b obtained from Equation 15.2, it is possible to test whether the original data set that creates the parameters a and b still holds as a straight line when plotted by $\log(y - b)$ vs. $\log(x)$. If the data set does not hold a straight line with this plot, it is indicated that the original fitting contains some degree of artifacts. Then the data points from either the low or high concentration end should be deleted one by one to test the linearity to finally achieve a straight line for both formulae. If preferred, Equation 15.3 can be employed to obtain parameters a and b as well as the correlation coefficient (γ^2) [21]. This is due to the fact that the contribution of each point (large or small) in the data set to the fitting of a straight line with Equation 15.3 is more equivalent. Deviation of the straight line at the upper concentration indicates the concentration at which lipid aggregation starts (see above).

15.4   NUTS AND BOLTS OF TANDEM MASS SPECTROMETRY FOR QUANTIFICATION OF LIPIDS

As discussed in Section 14.3, the use of tandem MS for quantification of lipids possesses numerous advantages including enhancing detection sensitivity, thereby greatly extending the linear dynamic range of quantification. Thus, the utility of tandem MS for quantification of lipid species becomes very popular. It is unfortunate that many of the researchers utilizing this technique for quantification of lipids do not recognize the potential problems inherited with this approach. The most severe concern comes from the nature of tandem MS for analysis of lipids, i.e., the fragmentation of individual species of a lipid class depends on the structure of individual lipid species. Therefore, unless an isotopologue of an individual lipid species is employed as an internal standard, the potential concerns leading to inaccurate quantification of such a species in comparison to a selected internal standard have to be recognized. In this section, these concerns are discussed in addition to what were mentioned in Chapter 14.

First, it is always better to optimize the collision energy that can balance the fragment intensities of all the ions of the entire class of interest for quantification of lipid species by tandem MS in shotgun lipidomics even more than one internal standards are employed in the method. Similarly, optimization of the SRM/MRM conditions for individual species in an LC-MS/MS method is not recommended when interest is to quantify all the species of a class, unless a calibration curve for each individual species is established under the identical conditions, since optimization of MRM conditions for individual lipid species leads to an incomparable response factor of the species of interest to that of the selected internal standard. In both cases of shotgun lipidomics and LC-MS/MS analyses, different collision energies applied for different species could lead to substantial errors in quantitative analysis, as discussed previously [22]. Careful attention to CID energy must be exercised if accurate quantification is a goal.

Second, the fragment ion in each pair of transition ions for quantification in an LC-based SRM method should be highly specific to the lipid class. To this end, a fragment ion corresponding to a fatty acyl chain (i.e., fatty acyl carboxylate anion) should be avoided as a fragment, as this negates the purpose of SRM to enhance the specificity for the analysis of species in a given class, as fatty carboxylate anions are obtained with all lipids. Moreover, it is well known that the fragment ion intensity of a fatty acyl chain yielded from different regioisomeric positions is very different [23].

Third, the mass resolution of the ion peaks in either precursor-ion scans or the SRM experiments is always lower than those observed in full mass spectra, product-ion spectra, or neutral-loss scans. This phenomenon is probably due to a minor time-lag resulting from a different free path experienced by the interrogated ions in the collision cell [22]. Reduction of the collision gas pressure, which can be compensated with an increase in collision energy to achieve an optimized fragmentation, is advisable if peak-broadening becomes an issue in either precursor-ion scans or SRM analyses of lipids.

Fourth, a distorted tuning and calibration file of a QqQ-type mass spectrometer could lead to a very different profile of a lipid class. It is advisable to examine the consistency for analysis of lipids with the previous conditions after a new file is established. It is particularly important to make such a comparison when performing MS/MS analysis since two tuning and calibration files (i.e., for both Q1 and Q2) are involved in this case.

Fifth, the mass accuracy during precursor-ion or neutral-loss scanning or SRM analysis depends not only on the mass accuracy of both the first and second mass analyzers but also on the mass difference between these two analyzers. In other words, during the neutral-loss or precursor-ion scanning, the quadrupoles must be correctly calibrated to transmit the desired ions at the appropriate time. Another issue comes from inaccurately approximating either the mass loss in the neutral-loss scanning mode or the selected ion *m/z* value in the precursor-ion scanning mode. For example, artifacts can result from setting the neutral loss of stearic acid at an approximation of 284 amu instead of 284.3 amu or setting the precursor ion of stearate at the approximation of m/z 283 instead of m/z 283.3. It was demonstrated that when the neutral-loss mass was intentionally offset by as little as 0.4 amu, an error greater than 50% in quantification as well as a dramatic reduction in overall detection sensitivity was observed [22].

Lastly, establishment of CID conditions for tandem MS analysis (i.e., product ion, precursor-ion, neutral loss, and SRM/MRM) are instrument dependent. It is always advisable to optimize these parameters for each instrument in a laboratory utilizing the parameters available in many review papers as a starting point [22, 24, 25].

15.5   ION SUPPRESSION

The term "ion suppression" is widely used to describe a phenomenon that ionization efficiency or intensity of a compound or a group of compounds is significantly reduced due to the presence of other compounds, changes in the matrix components, or simply due to the dramatic changes of the concentration of the compound itself as in the case of lipid aggregation. Such a reduction of ionization efficiency is largely due to the changes in the efficiency of droplet formation or droplet desolvation due to the presence of these compounds. This process is species and/or concentration dependent. Therefore, ion suppression may affect ion formation (thereby dynamic range and the limitation of detection), detection precision, and quantification accuracy.

Ion suppression is present in both shotgun lipidomics and LC-MS analysis although the mechanism(s) are likely different. In the case of shotgun lipidomics, low abundance and/or less ionizable species are always suppressed in the analysis by the full mass scan due to the coexistence of abundant and/or readily ionizable species. In the case of LC-MS, the varied matrix effect always affects its analysis, regardless of the sensitivity or selectivity of the mass analyzer used [26]. The origin and/or mechanism of ion suppression are still a mystery.

There are many possible sources leading to ion suppression, including endogenous compounds from the sample matrices as well as exogenous substances from contamination during sample preparation, such as polymers extracted from plastic tubes. Some factors that are known to induce ion suppression include high concentration, mass overlapping, basicity, and elution in the same retention window as the analyte of interest (in the case of LC-MS analysis) [27]. The latter case is similar to what occurs in shotgun lipidomics.

In samples comprised of multiple components at high concentrations, competition for either space or charge in the process of ionization most likely occurs. Such competition leads to a lower signal of each component (i.e., they suppress each other). This appears to be the case for lipid analysis with direct infusion where there exist many components, and the ionization efficiency depends on the physical properties of each lipid class. Therefore, the current dogma is that ion suppression is present in all methods employing direct infusion. In turn, the presence of ion suppression leads to inaccurate quantification of lipid species. Thus, it is sometimes argued that lipids cannot be quantified by using a direct infusion approach, which can only provide a profile comparison between the different states. In fact, this is an incorrect concept for the analysis of lipids at low concentration. However, this statement does hold true in the high-concentration region, where competition for either space or charge occurs. Indeed, in this high-concentration region, lipids have aggregated already and cannot be quantified as extensively discussed in Chapter 14 and in Section 15.2.

Two phenomena that are related to ion suppression should be noted here. First, molecular species of different lipid classes in a lipid extract show different ionization efficiencies after direct infusion even in the low-concentration region. This difference largely reflects the charge properties of these lipid classes (Chapter 3). This phenomenon might be called a type of "ion suppression". However, in MDMS-SL, this phenomenon has been exaggerated to selectively ionize only a certain lipid class or a certain category of lipid classes under specified conditions to achieve a separation without chromatography (i.e., intrasource separation) [28]. Second, although a particular lipid class may be "suppressed" in the presence of the other lipids in the analysis of full mass scan, the intensity ratios of individual species of this class is essentially kept unchanged as long as the analysis is performed in the low-concentration region, as aforementioned. Accordingly, the effects of this so-called ion suppression, which may actually be called "**steady-state ion suppression,**" on the analysis of this particular lipid class are only on the linear dynamic range due to the reduced detection limit in the first-quantification step rather than on accurate quantification of individual species with internal standard(s) of the class. The linear dynamic range after direct infusion can always be improved through tandem MS analysis (i.e., the second-step quantification). Therefore, this "steady-state ion suppression" is not a severe issue with shotgun lipidomics approaches, at least for MDMS-SL when the analysis is performed in the low-concentration region, if the developed methods for analysis of those species can be extended to the low or very low abundance region. Moreover, the rapid development of sensitive mass spectrometers further improves this aspect.

Actually, a similar phenomenon to this "steady-state ion suppression" in shotgun lipidomics is also present in any method developed with LC-MS for quantitative analysis of lipid mixtures. For example, if it is intended to quantify a species of a minor lipid class in the presence of other abundant species [24], the amount of total lipids that can be loaded onto a column are capped by the upper limit of the linear dynamic range of the most abundant species in the mixture under the experimental conditions. The loaded amount of total lipids to expand the linear dynamic range of the minor component in the method cannot be increased greatly if there is a need for quantification of major components as well. Of course, the minor species can be analyzed separately with a pre-isolated fraction or with a saturated concentration of the abundant species to increase the dynamic range for quantification of the minor components.

In addition to the steady-state ion suppression, there exists another complication of ion suppression for any method based on LC-MS and LC-MS/MS if the species either within a class or between classes cannot be completely resolved. Since the concentration of each individual species constantly changes during elution from the column, competition for either space or charge during ionization most probably occurs, particularly for the other species at the same retention time. To distinguish it from "steady-state ion suppression," this type of ion suppression can be called as **"dynamic ion suppression**." The real harm of "dynamic ion suppression" is that it is unpredictable and varies all the time in an analysis, while "steady-state ion suppression" leads to reduction of linear dynamic range as aforementioned, which is only a concern for the analysis of lipids less ionizable and/or at the very low abundance for which the desired sensitivity of the mass spectrometer is not achievable.

Diluting the sample for injection, especially in the case of shotgun lipidomics, or reducing the volume and concentration of a sample loaded to a system in the case of an LC-MS method is one way to reduce ion suppression efficiently. This measure is able to reduce the amount of interfering compounds, although this approach might not be appropriate for the analysis of low- or very-low-abundant lipid classes and/or molecular species. However, with the development of mass spectrometers, the sensitivity is substantially improved and is still improving. Therefore, it is strongly recommended that analysis of lipids, by either an LC-MS-based method or shotgun lipidomics, should be conducted in a concentration range as low as possible. It has been shown also that nanospray ESI-MS analysis leads to reduced signal suppression due to generating smaller, more highly, charged droplets that are more tolerant of nonvolatile salts [29].

In general, reducing matrix ions in the infused solution or mobile phase and improving chromatographic resolving power are the other two effective ways of circumventing ion suppression, particularly for MS/MS- or LC-MS-based methods. An easy and effective way to change chromatographic separation is to modify mobile phase strength or gradient conditions once the type of column is selected. The use of additives and/or buffers to aid in separation and improve the chromatographic performance is common. However, such additives can cause the suppression of electrospray ionization or contamination of the mass spectrometer. It is important to use them at a concentration as low as possible. It is also advisable that establishing the standard curves should be performed in the environment of a sample of interest, e.g., through spiking standards in the prepared samples, in order to minimize the matrix effects.

In order to pursue high-throughput analyses, more and more investigators have been minimizing the time for sample preparation, for instance, adding plasma sample direct to a loading solution prior to injection. Unfortunately, eliminating the sample cleanup step could lead to severe ion suppression, and poor reproducibility and performance, especially when complex matrices are involved. Careful consideration must be given to evaluating and eliminating matrix effects when developing any assay.

15.6 SPECTRAL BASELINE

Baseline noise resulting from either chemical or electronic noise is always present in MS analysis particularly by using a type of QqQ mass spectrometer [14]; however, it is continuing to improve with instrument development. The contribution of the noise to ion intensity may be negligible for quantification of the abundant lipid species since the degree of the baseline ion current (or ion counts) is only a few percent or less of the ion current of the molecular ion of interest. However, when a quantitative analysis is to cover the low-abundant species, the degree of the baseline noise that contributes to the ion peak becomes severe. In this case, elimination of such a noise contribution has to be considered.

The presence of baseline noise is obvious when mass spectra are examined in the case of shotgun lipidomics. Therefore, this can be easily corrected as demonstrated [1]. However, such noise may not be directly visible in the case of LC-MS analysis of lipids, particularly when a mobile-phase gradient is employed. It is difficult to access the magnitude of this noise level. Many researchers use the extracted ion intensity at neighboring elution times to evaluate the baseline; however, it does not entirely represent the level of noise at the real elution time.

For accurate quantification of lipid species, it is important to set a high signal/noise ratio to guarantee that only real ions corresponding to the species of interest are quantified, but not artifacts. Tandem MS-based methods in both shotgun lipidomics and LC-MS/MS can enhance the signal-to-noise ratios (Figure 14.3), but it is impossible to reduce the noise levels by tandem MS to a negligible level for the analysis of low abundance species.

15.7   THE EFFECTS OF ISOTOPES

Each lipid class in a cellular lipidome comprises a variety of lipid species that contain an identical head group but various fatty acyl chains containing a differential number of carbon atoms and differential degrees of unsaturation. Therefore, if an equal molar mixture of lipid species of a class is analyzed by MS, the monoisotopic peak intensities of these species displayed in the mass spectrum decrease as the number of carbon atoms in the species increases. This is due to the differential distribution of isotopologues in those species.

This differential isotopologue distribution of lipid species can affect their quantification by ratiometric comparison with a selected internal standard of the class if only the monoisotopic peak intensities are compared. The general approach to correct this differential isotopologue distribution is to convert all the isotopologues of a species, including the internal standard, to its monoisotopologue to make a summarized peak intensity of each ion (i.e., de-isotoping) prior to comparison. De-isotoping prior to quantification is widely used in proteomics, and programs for de-isotoping in lipidomics have also been developed [30, 31]. Some commercial instruments may carry such software for data processing. In this section, only a simplified approach is discussed in detail, which can be easily implemented in any data processing program.

In general, the isotopologue distribution of each species of a class mainly depends on the number of total carbon atoms in the species. This is due to the following facts. First, all the species of a class of interest should contain identical numbers of O, N, P, or other atoms if the lipid species of interest is not a modified one. This means that these atoms equally contribute their isotopologues to all the species of the class. Therefore, their contribution to the ion intensities does not cause any difference for relative comparison to the selected internal standard(s). Second, the natural abundance of deuterium (^{2}H) (0.0115%) is much lower than that of ^{13}C isotope (1.07%) although there is a larger number of hydrogen atoms compared to carbon atoms of lipid species. This suggests that the contribution due to the different number of ^{2}H atoms to differential isotopologue distribution can be negligible relative to that of ^{13}C isotopes. Therefore, only the correction for differential ^{13}C isotopologue distribution is generally sufficient and is discussed below. It should be noted that when the atoms such as Cl or S, whose isotopologues have large natural abundances, are present in the species of a class, the effects of their isotopologue distribution on quantification are not negligible and have to be taken into account carefully.

Quantification by ratiometric comparison with an internal standard is based on the ratio of the sum of the isotopologue intensities of a species to that of the internal standard. The fact is that the monoisotopic peak is the most intense peak in the isotopologue cluster of a lipid species for almost all lipids, and its intensity can therefore be determined more accurately compared to the intensities of other isotopic peaks of the species. Therefore, correcting for differential isotopologue distribution is based on the deduction of the intensity of each isotopologue of a species from the determined monoisotopic peak intensity.

Taken together, the total ion intensity $I_{\text{total}}(n)$ of an isotopologue cluster of a lipid species is

$$I_{\text{total}}(n) = I_n(1 + 0.0109n + 0.0109^2 n(n - 1)/2 + \ldots) \tag{15.4}$$

where I_n is the monoisotopic peak intensity of the species containing n carbon atoms and 0.0109 is the abundance of ^{13}C in nature when the abundance of ^{12}C is defined as 1. For quantification of this species with an internal standard containing s carbon atoms, we have when conditions of Equation 14.2 are satisfied:

$$
\begin{aligned}
c_n &= I_{\text{total}}(n)/I_{\text{total}}(s) * c_s \\
&= (1 + 0.0109n + 0.0109^2 n(n - 1)/2 + \ldots)I_n/ \\
&\qquad (1 + 0.0109s + 0.0109^2 s(s - 1)/2 + \ldots)I_s * c_s \\
&= Z_1 * (I_n/I_s) * c_s
\end{aligned}
\tag{15.5}
$$

where

$$
\begin{aligned}
Z_1 &= (1 + 0.0109n + 0.0109^2 n(n - 1)/2 + \ldots)/ \\
&\qquad (1 + 0.0109s + 0.0109^2 s(s - 1)/2 + \ldots)
\end{aligned}
\tag{15.6}
$$

Z_1 has previously been called the type I ^{13}C isotope correction factor [22]; n and s are the numbers of total carbon atoms in the species of interest and the selected internal standard, respectively; I_n and I_s are the monoisotopic peak intensities of the species and the internal standard, respectively; c_n and c_s are the concentration of the species of interest and the internal standard, respectively. The dots represent the contribution of other isotopologues which contain more than two ^{13}C atoms. These terms can be ignored in most cases without affecting the accuracy of quantification.

It should be mentioned that, following this line of reasoning, a method based on comparison of the intensities between cardiolipin $M + 1$ isotopologues, which exploits the uniqueness of doubly charged cardiolipin ions, was developed and is very powerful for quantification of individual cardiolipin molecular species [14, 32].

Different from the correction for differences of the number of carbon atoms between individual species of a lipid class, correction for another isotope effect on quantification due to the double-bond overlapping effects (see below) has to be performed for those utilizing unit-resolution mass spectrometers. For those who utilize high-resolution mass spectrometers for quantification, another approach to extract ion peak intensity from a partially overlapped ion peak can be consulted as described [33]. Herein, **the double-bond overlapping effect** refers to the overlap or partial overlap between the two ^{13}C atom-containing isotopologue of a species M (i.e., M+2 isotopologue) and the ion of a species with one less double bond than M.

The double-bond overlapping effect is common and severe when low-to-moderate resolution mass spectrometers are used for analysis of lipids by either shotgun lipidomics or an LC-MS approach. In the latter case, this effect is present unless resolving individual species of a class can be achieved. Due to the presence of this overlapping effect, the measured monoisotopic peak intensity of a species may not represent the true monoisotopic peak intensity of the species. Therefore, prior to performing de-isotoping as represented by Equation 15.6, correction for the double-bond overlapping effects has to be performed as follows:

$$I_n = I_{n'} - I_N * (0.0109^2 n(n-1)/2)$$
$$= (1 - (I_N/I'_n)(0.0109^2 n(n-1)/2)) * I_{n'}$$
$$= Z_2 * I_{n'} \tag{15.7}$$

where
$$Z_2 = 1 - (I_N/I_{n'})(0.0109^2 n(n-1)/2) \tag{15.8}$$

n is the number of total carbon atoms in the species of interest; I_n is the corrected monoisotopic peak intensity of the species of interest; $I_{n'}$ is the apparent monoisotopic peak intensity of the species that can be determined by MS; and I_N is the monoisotopic peak intensity of the species that differs from the species of interest with only one more double bond or two mass units less (I_N can also be determined by MS in the experiment). This correction factor can be negligible if $I_N \ll I_{n'}$. Z_2 has previously been called type II ^{13}C correction factor in distinguishing from Z_1.

It should be specifically pointed out that when a tandem MS spectrum is used for quantification of lipid species using Equation 15.5 in which I_n and I_s are corrected using Equations 15.6 and 15.8, respectively, both types of correction factors (i.e., Z_1 and Z_2) may need to be modified because the fragment monitored in tandem MS (i.e., the fragment ion in PIS or the neutral fragment in NLS) is the monoisotopic peak and therefore contains ^{12}C atoms only. Accordingly, the number of total carbon atoms in Equations 15.6 and 15.8 should be deduced by subtraction of the number of the carbon atoms in the monitored fragment that do not contribute ^{13}C isotopologue effects.

It should also be noted that if a calibration curve using two or more internal standards covering a wide mass range is used (e.g., in the class-specific tandem MS-based shotgun lipidomics), the first type of correction factor Z_1 can largely be covered by using their corresponding calibration curves. But the second type of correction factor Z_2 still needs to be performed. In LC-MS-based approaches, if the chromatographic separation can totally resolve individual lipid species in a class and a calibration curve is established for each individual species, neither correction factors are needed. Otherwise, these corrections or other alternative types of de-isotoping should always be taken into account. **Unfortunately, the majority of lipidomics studies based on LC-MS or MS/MS and the corresponding software tools have not paid attention on the effects of isotopes on quantification**. This factor alone could lead to a large experimental error.

15.8   MINIMAL NUMBER OF INTERNAL STANDARDS FOR QUANTIFICATION

In Section 13.3.2, the importance of the mass levels of internal standards used for quantification is extensively discussed. In fact, the number of internal standards used for quantification of lipid species of a class is equally critical. In this section, a minimal number of internal standards that should be used for quantification of the species of a class by different approaches are discussed. In summary, the number of internal standards that should be used really depends on the number of the variables present in the employed approach.

In Chapter 14, it has been extensively discussed that the ionization response factors of individual lipid species of a polar class are essentially identical after ^{13}C de-isotoping, as described earlier, if the analysis is conducted in a concentration range less than that at which lipid aggregates begin to form. Therefore, if quantification is conducted by shotgun lipidomics in the full mass scan mode, one internal standard of a class should be adequate. However, since tandem MS process depends on both the subclass linkage and individual molecular species (two variables), this adds a variable to the analysis of individual species of each subclass in comparison to that conducted in the full mass scan mode. Therefore, any method developed based on tandem MS in shotgun lipidomics should employ an extra internal standard to cover each variable. In other words, two or more internal standards are needed for relatively accurate quantification of individual species of a subclass. However, as discussed in Chapter 14, if ramping the collision energy or optimization of the collision energy to balance the fragmentation process of all individual molecular species of a (sub)class is conducted and validated, one less internal standard employed may be sufficient.

In the SIM method after LC-MS with isocratic elution, the additional variable in comparison to that conducted in the full mass scan mode after direct infusion is the changes in the concentration of individual species. Therefore, two or more internal standards at different elution times are necessary in this case. If a gradient of mobile phase is used for elution, an extra variable is introduced and at least one additional internal standard should be employed. In the case of an SRM/MRM method after

TABLE 15.1 Summary of Variables Present in the Methods and Their Required Minimal Number of Internal Standards for Quantification of Individual Species of a Polar Lipid Class

Platforms[a]	Variables	Minimal Number of Internal Standards
MDMS-SL	No variable	1
MS/MS-based SL	MS/MS	2
SL with HMR mass spectrometer	No variable	1
SL with HMR mass spectrometer utilizing MS/MS	MS/MS	2
LC-MS with solvent gradient	Conc., solvent gradient	3 (or 1 with external calibration curves)
LC-MS/MS with solvent gradient	Conc., solvent gradient, MS/MS	4 (or 1 with external calibration curves)

[a]MDMS, SL, and HMR stand for multidimensional mass spectrometry, shotgun lipidomics, and high mass resolution, respectively.

LC-MS, one additional variable exists in comparison to that of the SIM method after LC-MS, that is, the tandem MS process. Therefore, an extra internal standard should be employed relative to the SIM method. In both SIM and SRM/MRM methods after LC-MS, the necessary number of internal standards employed for accurate quantification could be compensated with a combination of an internal standard with a few external calibration curves.

Taken together, the minimal number of internal standards that should be used for accurate quantification is varied from method to method, largely depending on the existence of the variables in each method. Table 15.1 summarizes the easily recognizable variables present in each method and therefore the minimal number of internal standards necessary for the methods. As aforementioned, some alternative approaches could be used to reduce the minimal number of internal standards for quantification. It should be pointed out that any method employing a number of internal standards much less than the variables present in the method is unlikely able to provide accurate quantification, but it can still be used for relative comparison between the samples. If this is the case, it is advisable that the investigators should not overstate the results as quantitative.

So far, all the discussions are for quantification of individual species of a polar lipid class. For quantification of those of a nonpolar lipid class due to the differential ionization response factors of individual molecular species (a variable), more internal standards than those of a polar lipid class should be used. Alternatively, a correction factor for this variable could be predetermined and implemented in the developed method [12].

15.9 IN-SOURCE FRAGMENTATION

In-source fragmentation, as named, is a fragmentation process that occurs in the ion source during an MS analysis. The occurrence of this process is largely due to an artifact in a method development. For instance, the researcher might not recognize such a process or the severe consequence of this process on quantification, or the investigator might try to push an optimal condition for best ionization of a certain category of lipids. In both cases, a harsh ionization condition (e.g., high ionization temperature, high ionization voltage, very short distance between the spray tip and the inlet, leading to an extremely high electrical capacity, inappropriate setting of the gate voltages) could be introduced in the method during its development.

The presence of in-source fragmentation could have the following consequences:

- Reduced ionization sensitivity due to the loss of a certain amount of lipids in this process.
- Addition of a variable for quantification since in-source fragmentation, which is similar to CID and dependent on the molecular species.
- Complication in both quantitative and qualitative analyses (see below).

Accordingly, elimination of the in-source fragmentation process should be a task of a researcher who develops a method in lipidomics.

Here are some tips that should aid in elimination of this process. A standard solution of lipids should be further used to fine-tune the instrument after an instrument is tuned with a standard solution as recommended by the instrumental vendor(s) since those standard solutions are generally in favor of proteomics. This measure alone could enhance the ionization efficiency by an order of magnitude in many cases in the experiences of author's laboratory. Alternatively, a less harsh ionization condition (temperature, voltage, etc.) can be purposely set when one begins a tune/calibration experiment.

A suitable ionization condition can be examined with the minimal existence of some in-source fragments in the negative-ion mode. These fragments include the ions corresponding to the PA species generated from their PS counterparts and the ions corresponding to the dimethyl PE species that result from their PC counterparts. It should be kept in mind that a particular standard of PS species should be used in order to see the in-source resulted PA species, which should not overlap with that endogenously present in the lipid extracts. The amount of endogenous dimethyl PE species is usually in very low abundance. Under the ionization conditions, chlorinated PC species should be the dominant molecular ions if a modifier (see Chapter 2) is not added in the spray solution. To this end, observation of lysoPA species resulted from in-source fragmentation of lysoPC species was reported [34], which could be similarly minimized with careful tuning/calibration during a method development.

15.10   QUALITY OF SOLVENTS

Multiple types of solvents are employed in the entire process of lipid analysis, from extraction, storage, to solvent gradients in the case of LC-MS or a solvent mixture in the case of direct infusion. Different manufacturers use different types of stabilizer reagents for solvent storage. For example, a variety of stabilizers (from amylenes to ethanol) are used to eliminate the accumulation of phosgene for storage of chloroform. These different types of stabilizers may not only affect the extraction efficiency as previously recognized [35] but also may influence the ionization efficiency.

Solvents from different manufacturers may contain different amounts of residues, which may vary from neutral lipids (free fatty acids and TAG), acidic contaminants, cosmetics, detergents, to plasticizers. These residues could significantly affect the results of analysis. Therefore, it is advised that the solvents used for lipidomics should be carefully selected, and we should make efforts to purchase the highest quality of solvents possible. To this end, the solvents in the author's laboratory are obtained from Honeywell Burdick and Jackson (Muskegon, MI).

15.11   MISCELLANEOUS IN QUANTITATIVE ANALYSIS OF LIPIDS

No matter what delivery tool (i.e., HPLC, syringe, or loop injection) is used for analysis of lipids, it is advisable to use a metal needle or a metal-coated capillary as the spray tip instead of a silica capillary tube. The latter is frequently used in proteomics. This is because a high content of certain solvents (e.g., chloroform) is frequently used for lipid analysis, and these solvents can readily damage silica tubing under a high voltage, leading to the loss of ionization stability.

Currently, a flow rate ranging between hundred nanoliters to a few microliters per minute from an HPLC system can be coupled directly to the ion source chamber. However, a splitter must be utilized if an HPLC system can only deliver at a higher flow rate. To a similar setting, a postsource delivery device could be utilized to introduce a type of particular modifier if this modifier might interfere with the HPLC separation with or without a solvent gradient.

REFERENCES

1. Yang, K., Fang, X., Gross, R.W. and Han, X. (2011) A practical approach for determination of mass spectral baselines. J. Am. Soc. Mass Spectrom. 22, 2090–2099.
2. Koivusalo, M., Haimi, P., Heikinheimo, L., Kostiainen, R. and Somerharju, P. (2001) Quantitative determination of phospholipid compositions by ESI-MS: effects of acyl chain length, unsaturation, and lipid concentration on instrument response. J. Lipid Res. 42, 663–672.
3. Sparagna, G.C., Johnson, C.A., McCune, S.A., Moore, R.L. and Murphy, R.C. (2005) Quantitation of cardiolipin molecular species in spontaneously hypertensive heart failure rats using electrospray ionization mass spectrometry. J. Lipid Res. 46, 1196–1204.
4. Blom, T.S., Koivusalo, M., Kuismanen, E., Kostiainen, R., Somerharju, P. and Ikonen, E. (2001) Mass spectrometric analysis reveals an increase in plasma membrane

polyunsaturated phospholipid species upon cellular cholesterol loading. Biochemistry 40, 14635–14644.

5. Hermansson, M., Uphoff, A., Kakela, R. and Somerharju, P. (2005) Automated quantitative analysis of complex lipidomes by liquid chromatography/mass spectrometry. Anal. Chem. 77, 2166–2175.

6. DeLong, C.J., Baker, P.R.S., Samuel, M., Cui, Z. and Thomas, M.J. (2001) Molecular species composition of rat liver phospholipids by ESI-MS/MS: The effect of chromatography. J. Lipid Res. 42, 1959–1968.

7. Han, X. and Gross, R.W. (1994) Electrospray ionization mass spectroscopic analysis of human erythrocyte plasma membrane phospholipids. Proc. Natl. Acad. Sci. U. S. A. 91, 10635–10639.

8. Kim, H.Y., Wang, T.C. and Ma, Y.C. (1994) Liquid chromatography/mass spectrometry of phospholipids using electrospray ionization. Anal. Chem. 66, 3977–3982.

9. Lehmann, W.D., Koester, M., Erben, G. and Keppler, D. (1997) Characterization and quantification of rat bile phosphatidylcholine by electrospray-tandem mass spectrometry. Anal. Biochem. 246, 102–110.

10. Han, X., Yang, K. and Gross, R.W. (2008) Microfluidics-based electrospray ionization enhances intrasource separation of lipid classes and extends identification of individual molecular species through multi-dimensional mass spectrometry: Development of an automated high throughput platform for shotgun lipidomics. Rapid Commun. Mass Spectrom. 22, 2115–2124.

11. Stahlman, M., Ejsing, C.S., Tarasov, K., Perman, J., Boren, J. and Ekroos, K. (2009) High throughput oriented shotgun lipidomics by quadrupole time-of-flight mass spectrometry. J. Chromatogr. B 877, 2664–2672.

12. Han, X. and Gross, R.W. (2001) Quantitative analysis and molecular species fingerprinting of triacylglyceride molecular species directly from lipid extracts of biological samples by electrospray ionization tandem mass spectrometry. Anal. Biochem. 295, 88–100.

13. Cheng, H., Sun, G., Yang, K., Gross, R.W. and Han, X. (2010) Selective desorption/ionization of sulfatides by MALDI-MS facilitated using 9-aminoacridine as matrix. J. Lipid Res. 51, 1599–1609.

14. Han, X., Yang, K., Yang, J., Cheng, H. and Gross, R.W. (2006) Shotgun lipidomics of cardiolipin molecular species in lipid extracts of biological samples. J. Lipid Res. 47, 864–879.

15. Wang, M., Han, R.H. and Han, X. (2013) Fatty acidomics: Global analysis of lipid species containing a carboxyl group with a charge-remote fragmentation-assisted approach. Anal. Chem. 85, 9312–9320.

16. Han, X. and Cheng, H. (2005) Characterization and direct quantitation of cerebroside molecular species from lipid extracts by shotgun lipidomics. J. Lipid Res. 46, 163–175.

17. Han, X., Yang, J., Cheng, H., Ye, H. and Gross, R.W. (2004) Towards fingerprinting cellular lipidomes directly from biological samples by two-dimensional electrospray ionization mass spectrometry. Anal. Biochem. 330, 317–331.

18. Han, X., Yang, K., Cheng, H., Fikes, K.N. and Gross, R.W. (2005) Shotgun lipidomics of phosphoethanolamine-containing lipids in biological samples after one-step in situ derivatization. J. Lipid Res. 46, 1548–1560.

19. Jiang, X. and Han, X. (2006) Characterization and direct quantitation of sphingoid base-1-phosphates from lipid extracts: A shotgun lipidomics approach. J. Lipid Res. 47, 1865–1873.

20. Almeida, A.M., Castel-Branco, M.M. and Falcao, A.C. (2002) Linear regression for calibration lines revisited: Weighting schemes for bioanalytical methods. J. Chromatogr. B 774, 215–222.

21. Jiang, X., Yang, K. and Han, X. (2009) Direct quantitation of psychosine from alkaline-treated lipid extracts with a semi-synthetic internal standard. J. Lipid Res. 50, 162–172.

22. Han, X. and Gross, R.W. (2005) Shotgun lipidomics: Electrospray ionization mass spectrometric analysis and quantitation of the cellular lipidomes directly from crude extracts of biological samples. Mass Spectrom. Rev. 24, 367–412.

23. Han, X. and Gross, R.W. (1995) Structural determination of picomole amounts of phospholipids via electrospray ionization tandem mass spectrometry. J. Am. Soc. Mass Spectrom. 6, 1202–1210.

24. Merrill, A.H., Jr., Sullards, M.C., Allegood, J.C., Kelly, S. and Wang, E. (2005) Sphingolipidomics: High-throughput, structure-specific, and quantitative analysis of sphingolipids by liquid chromatography tandem mass spectrometry. Methods 36, 207–224.

25. Bielawski, J., Szulc, Z.M., Hannun, Y.A. and Bielawska, A. (2006) Simultaneous quantitative analysis of bioactive sphingolipids by high-performance liquid chromatography-tandem mass spectrometry. Methods 39, 82–91.

26. Cai, S.S., Short, L.C., Syage, J.A., Potvin, M. and Curtis, J.M. (2007) Liquid chromatography-atmospheric pressure photoionization-mass spectrometry analysis of triacylglycerol lipids--effects of mobile phases on sensitivity. J. Chromatogr. A 1173, 88–97.

27. Annesley, T.M. (2003) Ion suppression in mass spectrometry. Clin. Chem. 49, 1041–1044.

28. Han, X., Yang, K., Yang, J., Fikes, K.N., Cheng, H. and Gross, R.W. (2006) Factors influencing the electrospray intrasource separation and selective ionization of glycerophospholipids. J. Am. Soc. Mass Spectrom. 17, 264–274.

29. Gangl, E.T., Annan, M.M., Spooner, N. and Vouros, P. (2001) Reduction of signal suppression effects in ESI-MS using a nanosplitting device. Anal. Chem. 73, 5635–5644.

30. Liebisch, G., Lieser, B., Rathenberg, J., Drobnik, W. and Schmitz, G. (2004) High-throughput quantification of phosphatidylcholine and sphingomyelin by electrospray ionization tandem mass spectrometry coupled with isotope correction algorithm. Biochim. Biophys. Acta 1686, 108–117.

31. Eibl, G., Bernardo, K., Koal, T., Ramsay, S.L., Weinberger, K.M. and Graber, A. (2008) Isotope correction of mass spectrometry profiles. Rapid Commun. Mass Spectrom. 22, 2248–2252.

32. Han, X., Yang, J., Yang, K., Zhao, Z., Abendschein, D.R. and Gross, R.W. (2007) Alterations in myocardial cardiolipin content and composition occur at the very earliest stages of diabetes: A shotgun lipidomics study. Biochemistry 46, 6417–6428.

33. Wang, M., Huang, Y. and Han, X. (2014) Accurate mass searching of individual lipid species candidates from high-resolution mass spectra for shotgun lipidomics. Rapid Commun. Mass Spectrom. 28, 2201–2210.

34. Zhao, Z. and Xu, Y. (2009) Measurement of endogenous lysophosphatidic acid by ESI-MS/MS in plasma samples requires pre-separation of lysophosphatidylcholine. J. Chromatogr. B 877, 3739–3742.

35. Fuhrmann, A., Gerzabek, M.H. and Watzinger, A. (2009) Effects of different chloroform stabilizers on the extraction efficiencies of phospholipid fatty acids from soils. Soil Biol. Biochem. 41, 428–430.

16

DATA QUALITY CONTROL AND INTERPRETATION

16.1 INTRODUCTION

After collection of a data set from MS analysis, data processing (including identification and quantification) can be conducted by using one of the programs introduced in Chapter 5 or those home-made or other strategies [1, 2]. At this stage, it is important to know whether the processed data set is self-consistent, and that there are no artifacts due to improper operation or any hidden experimental error(s) prior to making efforts and devoting resources to bioinformatically analyze the data, to interpret the observations, and even to publish the results. The possible factors that may affect data quality are extensively discussed in Chapter 15. In this chapter, the first part is to discuss how to learn the quality of the data set (i.e., data quality control).

After the data set passes the quality control, the standard procedure is to apply a commercially available software package (e.g., SAS, NCSS, IBM SPSS Statistics, SIMCA-P) for statistical analysis, and clustering of the changed lipid species is often visualized by multivariate analysis such as PCA [3–6]. Significant changes of an individual lipid species, a lipid subclass, a lipid class, a category of lipids, or a desired parameter derivable from the analysis between the sample sets can be revealed from the statistical analysis. It is worth noting that typical PCA is usually variance based and when absolute quantitation values are used as input data for the analysis, low-abundant lipid species are usually ignored in the analysis. Thus, it is better to use composition of individual lipid species in the class for PCA or use decomposition of the cross-correlation matrix instead of the covariance matrix. PCA can also be used as a tool for outlier detection [4].

Lipidomics: Comprehensive Mass Spectrometry of Lipids, First Edition. Xianlin Han.
© 2016 John Wiley & Sons, Inc. Published 2016 by John Wiley & Sons, Inc.

Then, the difficulty for the majority of the investigators, particularly for those with a background in analytical chemistry, is how to further study the mechanisms, which lead to these lipid changes and how to interpret the significance of these determined changes. Owing to this difficulty, many of the publications have only reported the determined significant differences without an in-depth explanation of these changes. Unfortunately, this makes those papers only descriptive.

In fact, the underlying biochemical mechanism(s) responsible for the changes of these lipids are reflected by the changes themselves to a great degree since the pathways and networks of lipid metabolism were extensively studied previously and well documented [7]. Therefore, the determined changes can reveal the possibly changed enzymatic levels and/or activities, or gene expressions based on these pathways and networks. To this end, targeted analysis of these possibly changed protein levels (e.g., Western blot analysis) and gene expression (e.g., real-time PCR analysis) could be effectively performed. A combination of lipidomics data with those obtained from the analysis of genes and proteins could provide deep insights into the molecular mechanism(s) underpinning the uncovered lipid changes. Thus, a large part of this chapter is to introduce these pathways and networks and briefly discuss their application for the studies on biochemical mechanism in lipidomics.

In order to accurately interpret the importance of the revealed lipid changes, it is essential for the researchers to understand the functions of individual lipid classes, subclasses, and individual molecular species. Therefore, an important part of this chapter is to describe the functions of these lipid classes and individual species with examples.

16.2 DATA QUALITY CONTROL

For an established method or platform, the variation of an analysis should be extensively determined and validated. For example, a reproducibility of analyzing a prepared sample by MDMS-SL is approximately 95%, and the precision of analyzing a group of animal samples with the platform is approximately 90% largely due to the variation of protein assay in determining protein content for normalizing lipid levels [8]. Thus, the data quality from the studies with animal samples by MDMS-SL should first be examined to see whether the variation of one (or a few) lipid classes is larger than that of the established method while the variations of other lipid classes are comparable to those of the platform. If this is the case, the bigger variation largely comes from the analysis of this class of lipids. Unstable ionization of this lipid class could be the causal factor, which could be examined from the ion chromatograph. Alternatively, if the variations of all determined lipid classes are larger than those of the established platform, either the variation of samples self or sample preparation should be considered. For example, if the levels of all determined lipid species of a particular sample are lower (or higher) than the average of other samples in the group, it is very likely that an error of sample preparation, protein assay, or addition of internal standards to this particular sample is introduced during sample preparation. Reanalysis of this sample, in this case, is necessary. It should be recognized that the variation of analyzing animal samples is exemplified herein since the variations of human samples are usually much larger than that of the analysis method itself, and thus, it is not straightforward to utilize the variations to carry out the data quality control.

It is well known that membrane lipids are manifest in the majority of the cells and organs and follow a specific pattern. For example, the levels of PC and PE classes are comparable; the content of anionic GPL species is approximately 10% of total lipids; SM only counts for approximately 5% of total lipids; the ratio of PC + SM *vs.* PE + anionic lipids is roughly one; etc. [7, 9, 10]. Obviously, this knowledge should be followed and can generally be used for data quality control if global lipidomics analysis is the goal of a study.

Under normal physiological conditions, lipid metabolites are only present in a small amount (usually < 5%) in comparison to their parent lipid classes. Therefore, it is hard to believe that, for instance, the levels of lysoGPLs are over 10% of those of GPLs. Moreover, lysoGPLs are potent detergents, and a high content of these lipids can lysate membrane structure or lead to membrane leakage. Thus, it is impossible that high levels of these lipids exist in normal cells. This knowledge should always be kept in mind in any targeted or global lipidomics studies.

Another important approach to conduct data quality control is to compare the determined data with those reported in the literature. Although most of the previous studies with conventional methods only report the total amount of individual lipid classes without details of molecular species, these data at least can provide a rough comparison to the current studies.

Many lipidomics studies are carried out with human plasma (or serum) samples. A couple of facts about plasma samples should be useful for data quality control. First, plasma lipoproteins (e.g., VLDL, LDL, and HDL) are enriched with lipids [7]. The lipid compositions of these lipoproteins depend on the type of lipoproteins. The mass levels of cholesteryl esters and TAG species decrease from VLDL to HDL while the contents of phospholipids (including SM) and free cholesterol are in a reversed order in these lipoproteins. Second, the concentrations of plasma phospholipids are in an order of PC > SM ~ PE > lysoPC > PI ~ lysoPE > PS ~ PG > PA [11].

It should be recognized that different cell types possess different types of lipids (classes, subclasses, and individual molecular species). For example, plants lack plasmalogen species; cholesterol is also unique to mammalian; mammalian does not possess the machinery to synthesize essential fatty acids; and the numbers and locations of double bonds in fatty acid species in eukaryotes are not randomly present. This nature of lipids in biological samples can be used as a criterion for data quality control. Violation of such a rule indicates something wrong in MS identification of lipid species.

16.3 RECOGNITION OF LIPID METABOLISM PATHWAYS FOR DATA INTERPRETATION

One of the tasks of lipidomics as originally described [12] is to identify the molecular mechanism(s) responsible for the altered lipid levels induced by a stimulus.

To delineate the underlying mechanism(s) of the altered lipid levels due to the perturbation, it is crucial to recognize all the pathways of the altered lipids involved since lipid metabolism is not isolated and the majority of the cellular lipids are interwoven [12, 13]. It would be life-saving if a software program could be developed to automatically process the lipidomics data to elucidate the corresponding lipid metabolism pathways to the altered lipid levels after perturbation. Although the strategies for analysis of a few isolated pathways have been described [14–17], unfortunately, such tool(s) are currently unavailable as outlined in Chapter 5. Therefore, interpretation of lipidomics data based on lipid metabolism pathways and networks has to be manually conducted at the current stage of the field development. To this end, it is important to get familiar with the pathways and networks of lipid metabolism for manual interpretation of the obtained lipidomics data.

At least three lipid metabolic networks and their relationships, which are common and essential to the majority of the lipids in mammals and plants, should be recognized. These networks include the one involving sphingolipid metabolism, the one related to GPL metabolism, and the one for glycerolipids.

16.3.1 Sphingolipid Metabolic Pathway Network

The sphingolipid metabolism network is illustrated in Figure 1.3. This network is centered at the class of Cer. Thus, accurately measuring the mass levels of Cer species is crucial for interpretation of sphingolipid metabolism. This network could be used to interpret the changes of the mass levels of these lipid classes as well as the subclasses containing different sphingoid backbones such as dihydrosphingolipids. As illustrated in Figures 1.3 and 1.6, differential selectivity of serine palmitoyltransferase (SPT) to fatty acyl CoA as well as amino acid substrates could yield different sphinganine analogs [18]; differential selectivity of Cer synthase to different acyl CoA species results in different dihydroceramide (DHCer) species and then Cer species; and different activities of DHCer desaturase could convert of DHCer to Cer in a different degree. The changed levels of different sphingolipid molecular species could be interpreted with differential activities of these enzymes as well as their different isoforms. For example, many Cer synthase isoforms exist in biological systems, which selectively synthesize Cer species containing different fatty acyl amide chains [19, 20]. Merrill and colleagues developed a powerful tool to comprehensively illustrate the interrelationships between these sphingolipid species [21, 22].

16.3.2 Network of Glycerophospholipid Biosynthesis Pathways

The networks of GPL biosynthesis pathways present in animals and plants are different from those of yeast or bacteria and thus they are separately illustrated (Figure 16.1). The gene(s) and protein(s) involved in each biosynthesis pathway can be found in review articles (e.g., [7]). It is important to recognize the cell type of a study prior to interpreting lipidomics data.

The main difference between animals/plants and yeast is the PS synthesis pathway. In the former, PS species are synthesized from PC and PE species through serine exchange with choline and ethanolamine by PS synthase 1 and 2, respectively, in

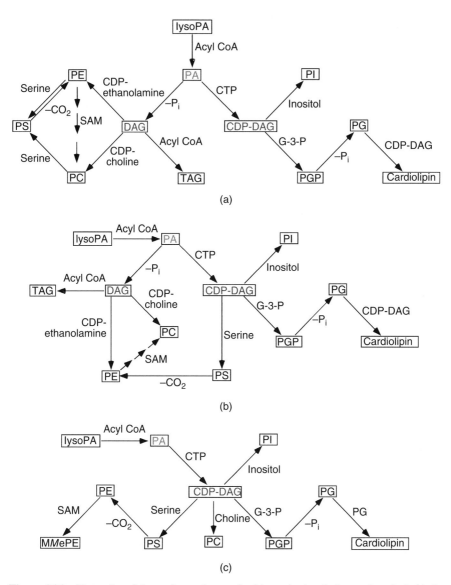

Figure 16.1 Networks of the major pathways for biosynthesis of glycerophospholipids in different species as indicated. SAM and PGP denote *S*-adenosylmethionine and phosphatidylglycerol phosphate, respectively. All other abbreviations can be found in the list of abbreviations. (a) Animals and plants; (b) yeast; and (c) bacteria.

mitochondria or mitochondria-associated membranes [23], whereas, in the latter, PS species are synthesized through condensation of serine with CDP-DAG catalyzed by PS synthase in endoplasmic reticulum (ER). It should be recognized that PS decarboxylation is an important pathway to synthesize PE species, particularly in mitochondria (see Ref. [24] for review).

In bacteria, all the GPL synthesis pathways are through CDP-DAG. For example, PC species are synthesized through condensation of choline with CDP-DAG [25]. PE species are obtained after decarboxylation from PS species, which are synthesized through condensation of serine with CDP-DAG. However, both PC and PE classes are synthesized through condensation of CDP-choline (-ethanolamine) with DAG, respectively, in both animals/plants and yeast. Another difference present in bacteria compared to animals/plants and yeast is the biosynthesis of CL. In bacteria, CL species are synthesized through condensation between PG species, whereas in animals/plants and yeast, this class of lipids is obtained from condensation of PG with CDP-DAG.

Unlike sphingolipid species, almost all of newly synthesized GPL species undergo fatty acyl chain remodeling either through a transacylase activity or through fatty acyl hydrolysis by phospholipases A_1 and A_2 (PLA_2) activities followed by acyltransferase activities. Thus, if changes only occur at the levels of individual molecular species of a class, both hydrolysis and remodeling of the class should be considered. If alterations are observed at the lipid class levels, the networks shown in Figure 16.1 should be used to interpret the results in combination with a possibly changed hydrolysis activity, which can usually be supported by the changes of the corresponding lysoGPL levels.

It is noteworthy that the biosynthesis of ether-containing GPL species, including plasmalogen species, is different from their diacyl counterparts. This synthesis is initiated from dihydroxyacetone phosphate (DHAP) in peroxisome (Figure 16.2) instead of glycerol-3-phosphate (G-3-P) in ER (Figure 16.1). After formation of 1-alkyl-*sn*-glycero-3-phosphate, other steps of the biosynthesis are essentially identical to those for the synthesis of their diacyl counterparts (Figure 16.2) and all occur in ER. It can be recognized from the pathways of ether-containing GPL biosynthesis, PC and PE species are as the major subclasses of ether-containing GPL species. Ether-containing PS species, which can be formed through exchange of serine with corresponding PC or PE head group, are present in low abundance. Alkylacylglycerophosphate (i.e., ether-containing PA) as an intermediate of the biosynthesis should be present in very low abundance. It should be emphasized that ether-containing GPL species are generally absent in plants. It should also be recognized that there exist a large number of ether-linked backbone in individual subclasses due to a certain degree of nonselective enzymatic activities involved in their biosynthesis (Figure 16.2) [26].

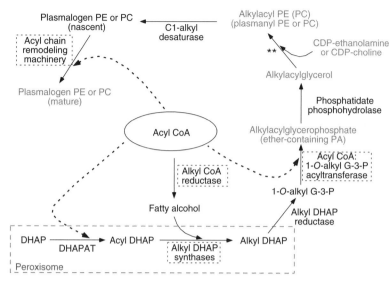

Figure 16.2 Pathways involved in the biosynthesis of ether-containing lipids. The enzymes that may be involved in nonselective utilization of acyl CoA pool are highlighted with broken-lined arrows. The formed ether-containing lipids are highlighted with light gray. The symbol of ** stands for CDP-ethanolamine:1-*O*-alkyl-2-acyl-*sn*-glycerol ethanolamine phosphotransferase or CDP-choline:1-*O*-alkyl-2-acyl-*sn*-glycerol choline phosphotransferase. DHAP denotes dihydroxyacetone phosphate. All other abbreviations can be found in the list of abbreviations.

16.3.3 Glycerolipid Metabolism

As illustrated in Figure 16.1, TAG species are reacylated from their corresponding DAG species through a DGAT activity. In addition to the major synthesis pathway as illustrated in Figure 16.1 (i.e., DAG is generated from PA through a dephosphatase activity), DAG species can also be generated from either reacylation of MAG species or hydrolysis of the head groups from a variety of GPL species through a PLC activity. Moreover, DAG can also be yielded from the TAG hydrolysis through a lipase activity. Thus, for interpretation of glycerolipid data, the following two aspects should always be kept in mind:

- DAG could be formed in different cellular compartments and thus the TAG species.
- The futile cycle of MAG, DAG, and TAG exists to a certain degree and should be balanced.

It should also be recognized that due to the presence of alkylacylglycerol (an analog of DAG) as an intermediate of plasmalogen biosynthesis (Figure 16.2) and alkenylacylglycerol (another analog of DAG) yielding from the hydrolysis of plasmalogen PC and PE, triglyceride species containing ether bonds at the *sn*-1 position occur naturally, which has been evidently demonstrated in biological samples [27, 28].

16.3.4   Interrelationship between Different Lipid Categories

In addition to the relationship between GPL and glycerolipids as illustrated in Figure 16.1 and discussed in the last section, the categories of GPL, glycerolipid, and sphingolipid are tightly connected through the following reaction in the biological systems:

$$PC + Cer \rightarrow SM + DAG \tag{16.1}$$

which is catalyzed by an SM synthase activity.

Moreover, acyl CoA is a key metabolite involving the majority of the synthesis, degradation, and remodeling activities in all of the categories of lipids. The availability of acyl CoA can directly influence the synthesis and remodeling of these lipids, thereby changing the lipid profiles. It is unfortunate that due to the differentially compartmental distribution of acyl CoA and very dynamic alteration in its content in individual compartments as well as in the entire cell, it is hard to interpret the lipidomics data by acyl CoA levels alone.

Taken together, interpretation of lipidomics data should not only extensively consider different pathways within a network of an individual lipid category but also expand the influence from other categories of lipids (i.e., consider the interrelationship between the networks). In order to do well on this, it is always important to conduct comprehensive analysis of cellular lipidomes, including individual molecular species and regioisomers.

16.4   RECOGNITION OF LIPID FUNCTIONS FOR DATA INTERPRETATION

In order to understand the significance of the determined lipid changes, recognizing the functions of individual lipid class, subclass, and molecular species is crucial. In general, lipids can be categorized into three major functions as cellular membrane components, energy storage depots (the majority of lipids in this category are involved in bioenergetics), and signaling molecules (Figure 16.3). A list of lipid classes in each category is tabulated (Table 16.1). In this section, a crosstalk between functions and lipids is discussed in some details.

16.4.1   Lipids Serve as Cellular Membrane Components

Basically, all of major polar lipid classes (e.g., PC, PE, PI, PS, PG, PA, SM, cerebroside including both GalCer and GluCer, ST, gangliosides) are the components of cellular membranes. Although some of their metabolism intermediates such as lysoGPL, DAG, Cer, NEFA are also present in membrane structures, these intermediates are

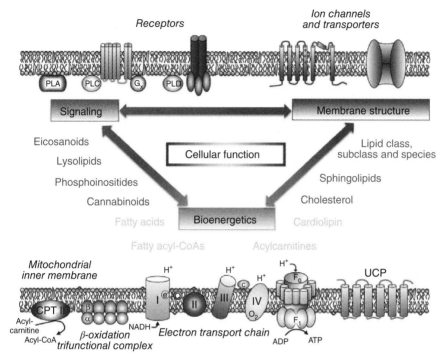

Figure 16.3 The pleiotropic roles of lipids in cellular functions. Lipids fulfill multiple roles in cellular function including cellular signaling (top left) through: (1) harboring latent second messengers of signal transduction that are released by phospholipases (PLA, PLC, and PLD enzymes); (2) covalent transformation of membrane lipids into biologically active ligands by kinases (e.g., PI 3,4,5-triphosphate); (3) providing molecular scaffolds for the assembly of protein complexes mediating receptor/effector coupling (e.g., G-protein-coupled receptors); and (4) coupling the vibrational, rotational, and translational energies and dynamics of membrane lipids to transmembrane proteins such as ion channels and transporters (top right), thereby facilitating dynamic cooperative lipid–protein interactions that collectively regulate transmembrane protein function. Moreover, lipids play essential roles in mitochondrial cellular bioenergetics (bottom) through the use of fatty acids as substrates for mitochondrial β-oxidation (bottom left) that result in the production of reducing equivalents (e.g., NADH). The chemical energy in NADH is harvested through oxidative phosphorylation whose flux is tightly regulated by mitochondrial membrane constituents including cardiolipins, which modulate electron transport chain (ETC) supercomplex formation. A second mechanism modulating mitochondrial energy production is the dissipation of the proton gradient by the transmembrane flip-flop of fatty acids in the mitochondrial inner membrane bilayer as well as the fatty acid-mediated regulation of uncoupling proteins (UCP). Source: Gross & Han 2011 [29]. Reproduced with permission of Elsevier.

TABLE 16.1 Summary of the Major Functions of Individual Lipid Class Played[a]

Cellular Functions	Lipid Classes
Membrane structural component	PC, PE, PI, PS, PG, PA, SM, CL, cholesterol, cerebroside (e.g., GalCer and GluCer), glycolipids, ST, gangliosides, etc.
Energy storage and metabolism	NEFA, TAG, DAG, MAG, acyl CoA, acylcarnitine, etc.
Signaling	All lysolipids, DAG, MAG, acyl CoA, acylcarnitine, NEFA, eicosanoids and other oxidized FA, ceramide, sphingosine, S1P, psychosine, steroids, *N*-acyl ethanolamine, etc.
Other special functions	Plasmalogen (*antioxidant*), acylcarnitine (*transport*), CL (*respiration*), PS (*cofactors, substrate of PE synthesis*), etc.

[a]All of abbreviations can be found in the list of abbreviations.

usually only present in very low levels so that they do not play significant roles in regulation of membrane structural function. However, accumulation of these intermediates could be crucial in alterations in membrane structural functions. Regardless of its nonpolar structure, cholesterol is also enriched in cellular membrane structure, particularly in plasma membranes.

Membrane lipids play many structural roles in cellular functions. Since membranes serve as compartment barriers at the cellular levels, changes of membrane lipid levels and compositions could alter membrane permeability to small neutral molecules such as water and even small ions and change in membrane fluidity. In general, increases in the amount of lipids containing polyunsaturated fatty acids as well as metabolic intermediates result in higher membrane permeability and fluidity. In contrast, higher amounts of lipids containing saturated fatty acids, cholesterol, sphingolipids, plasmalogen, etc. provide a barrier with lower membrane permeability and fluidity, but all of these components might favor the formation of lipid rafts (see Chapter 20).

Changes of lipid levels and compositions could alter the matrix in interactions with membrane proteins (e.g., ion channels), thereby influencing the protein configurations and functions. Moreover, alterations in membrane lipid components could also affect the microenvironments in cellular communication and cytosolic ion distribution. In this regard, glycolipids and gangliosides play important roles in the former while anionic lipids are the main players in the latter.

Different lipids possess different spatial shapes (e.g., cylinder, core, reverse core) and are favorable to form different phases such as bilayer, hexagonal, cubic, and micelle. Therefore, alterations in lipids possessing different spatial shapes could influence the processes of membrane fusion, vesicle transport, contact point formation, etc. For example, increased amounts of plasmalogen could accelerate membrane fusion; higher contents of cholesterol and CL in mitochondria favor the formation of contact points and in mitochondrial fusion and fission; higher levels of lysoGPL may lead to the formation of inverted micelles and breakage of cell integrity, etc.

Membrane lipids are the depots of lipid signaling molecules. Many lipid second messengers are released from membrane depots after hydrolysis or oxidation. For example, GPL, PI, and SM yield lysoGPL, DAG, and Cer, respectively, after enzymatically catalyzed hydrolysis. (Oxy)sterols and oxidized GPL species are some examples of oxidation. Therefore, alterations in the mass levels of these lipids impact the levels of signaling lipids.

Taken together, the majority of cellular lipids reside in the membranes and they play a variety of cellular functions. Altered mass levels and compositions of these membrane lipids directly impact on cellular functions. Interpreting the significance of the changed cellular lipids should consider all of these cellular functions. Moreover, it would be best to determine the predicted changes of cellular functions based on the lipidomics data, thereby validating the interpretation.

16.4.2 Lipids Serve as Cellular Energy Storage Depots

A few lipid classes are involved in energy metabolism, serving as energy storage when it excesses and as energy supplier when it demands. These lipid classes include TAG, DAG, NEFA, acylcarnitine, and acyl CoA. Altered levels of these lipid classes usually indicate an altered energy state in comparison to the control. Increased DAG and TAG contents indicate a state at which excessive energy is deposited in the storage depots. Increased levels of acylcarnitines suggest either mitochondrial dysfunction or excess fatty acid oxidation. Excess levels of NEFA and acyl CoA are generally related to either increased lipolysis or *de novo* biosynthesis. Changes of these lipid classes are usually interwoven. Accumulation of these lipid classes in animals and humans has been generally called "lipotoxicity" [30–33], which leads to obesity and insulin resistance.

The fatty acyl profiles of these lipid classes can provide wealthy information about fatty acid biosynthesis, metabolism, and uptake in mammalian systems. Here are a few examples. Increased mass levels of fatty acyls containing either n-3 or n-6 or both (e.g., 18:2, 20:4, and 22:6 FA as the most common n-6 (the first two) and n-3 fatty acids, respectively) in these lipid classes indicate either excess uptake of extracellular fatty acids or reduced utility of these fatty acids through oxidation. Accumulation of 16:0, 16:1, 18:0, and 18:1 fatty acyls largely suggests their increased *de novo* biosynthesis (see below) under patho(physio)logical conditions. The location of the double bond(s) gives rise to the information about the pathways of the *de novo* biosynthesis (Figure 16.4) and/or the uptake from the food carrying these isomers of a fatty acid. Similar information about FA biosynthesis and metabolism can also be provided from the FA profiles obtained from the analysis of these energy storage lipid classes.

The information about FA biosynthesis and/or uptake obtained from determining the location of double bonds can be clearly exemplified with the case of 18:1 FA. There are three major isomers of 18:1 FA in the majority of the biological samples, i.e., n-7, n-9, and n-12 18:1 FA isomers with minimal amount of others, particularly with a double bond in an even-numbered position [9]. The 18:1(n-9) FA isomer (commonly called "oleic acid") is by far the most abundant monoenoic FA in both plant and animal tissues. This isomer also serves the biosynthetic precursor of a family of fatty

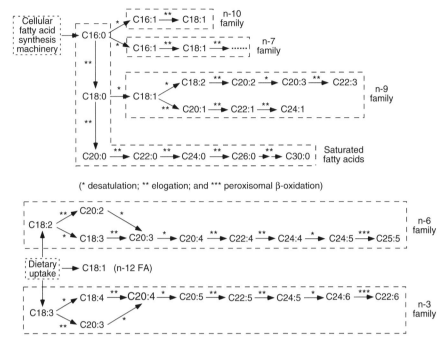

Figure 16.4 Long-chain and very long-chain fatty acid biosynthesis in mammals. The long-chain saturated fatty acids and unsaturated fatty acids of the n-10, n-7, and n-9 families (Top panel) can be synthesized from palmitic acid (C16:0) produced by the cellular fatty acid synthesis machinery. Long-chain fatty acids of the n-6 and n-3 families can only be synthesized from their respective precursors obtained from diets. The symbols of *, **, and *** stand for the involved activities of desaturation, elongation, and peroxisomal β-oxidation, respectively, in the steps. Many isoforms of the genes corresponding to these activities were identified (see the review [34] for details).

acids containing an n-9 terminal structure (Figure 16.4). The 18:1(n-7) FA isomer (trivially called *cis*-vaccenic acid) is a common monoenoic FA of bacterial lipids and is usually present as a minor component of most plant and animal tissues. Usually, the presence of this isomer in mammal tissues indicates the elongation of its 16:1(n-7) FA precursor. This isomer could further elongate to a family of fatty acids containing an n-7 terminal structure (Figure 16.4). The 18:1(n-12) FA isomer (i.e., petroselinic acid) occurs up to a level of 50% or more in seed oils of the Umbelliferae family, including carrot, parsley, and coriander [9]. Therefore, the presence of this 18:1 FA isomer clearly indicates the degree of the FA uptake of a mammal organ from the food. In summary, the wealthy information about FA biosynthesis and/or the source of the FA species can be obtained through analyzing the location of double bonds in fatty acyl chains of many energy storage lipid classes such as TAG, DAG, and NEFA.

The profile of TAG species in both plant and mammal tissues is usually very complicated, particularly in the latter. Thus, these profiles might not directly provide deep insights into the pathways and/or networks at the initial look. However, an approach through simulation of lipidomics data [17] based on the well-known pathways for TAG biosynthesis (Figure 5.2) could be employed to reveal the contribution of an individual pathway to the TAG pool. In addition, this method can serve as an auto-mated approach to identify the existence of TAG species in the analyzed samples as extensively validated [17]. It should be noted that this approach only consider the steady state of an organ although kinetic studies can be performed with this approach through analysis of the organ at different time points. Due to this consideration, this simulation model at its current version may not be fit to those systems that are very dynamic such as adipose tissues in which lipolysis and/or lipogenesis occur rapidly. However, this model can be readily modified to include the DAG content, which can be determined as previously described [35].

16.4.3 Lipids Serve as Signaling Molecules

A large number of lipid species serve as signaling molecules in any biological sig-naling event. Two types of signaling events involving lipids frequently occur in a biological system. In the first type of event, lipid species directly bind to and subse-quently activate a protein target (e.g., a receptor, kinase, or phosphatase), which in turn leads to a specific cellular function. In the second type of signaling event, changes of lipids could affect membrane structure and affect the interactions of membrane proteins with the membrane bilayers as described in Section 16.4.1. This type of sig-naling event is relatively slow and more complex than those seen in other signaling pathways [36]. Herein, only the first type of signaling is briefly discussed.

This type of signaling can be roughly categorized into four areas as follows:

The first category is those of sphingolipid-related second messengers, including Cer, ceramide-1-phosphate (C1P), sphingosine, sphingosine-1-phosphate (S1P), and more. Cer species can be either produced through *de novo* synthesis pathway or resulted from the hydrolysis of SM (Figure 1.3) [37]. The former is synthesized mainly with the catalysis of enzymes serine palmitoyl transferase (SPT) and Cer syn-thase in organelles such as ER and mitochondria-associated membranes. The latter converts SM to Cer by sphingomyelinase (SMase), which occurs at the plasma mem-brane interface and intracellular compartments (e.g., lysosome). Cer mediates many cell stress responses, including the regulation of programmed cell death (i.e., apop-tosis) [38, 39] and cell aging (i.e., senescence) [40]. Comprehensive reviews [36, 41, 42] on Cer-mediated signaling could be further consulted to the readers who are inter-ested in this area of work. C1P is produced by the action of Cer kinase on Cer (Figure 1.3). It was found that C1P can activate PLA_2 to release arachidonic acid through for-mation of ionophore [43]. C1P is also involved in many other cell processes such as vesicular trafficking, cell division and survival, and phagocytosis [44–46]. Sphingo-sine and its analogs are largely generated from Cer species (Figure 1.3) through a ceramidase activity in lysosome. Sphingosine can come out from lysosome through flip-flopping and diffusing to cytosol [36]. Sphingosine may play the roles in endo-cytosis, cell cycle, and apoptosis likely through interaction with protein kinases [47]. S1P is formed by phosphorylation of sphingosine (Figure 1.3) by sphingosine kinase.

S1P may be involved in cell survival, cell migration, and inflammation [48]. The involvement of S1P in these cellular processes is largely through interactions with the G-protein-coupled receptors, known as S1P receptors [48]. Inside the cell, S1P can induce calcium release, which is independent of the S1P receptors, the mechanism of which remains unknown. Other sphingolipids such as GluCer and gangliosides are also involved in cellular signaling processes [36, 49].

The second category of signaling lipids is that related to PI species. Specifically, PI can be phosphorylated to PIP_2, which itself directly agonizes inward rectifying potassium channels [50]. PIP_2 can be hydrolyzed into inositol triphosphate (IP_3) and DAG by a PLC activity, both of which serve as second messengers. The former interacts with an IP_3 receptor to release intracellular calcium ions stored in ER. The latter, which remains bound to the plasma membrane, activates the members of protein kinase C family [51, 52]. Moreover, PIP_2 can be further phosphorylated to produce PI (3,4,5)-triphosphate (PIP_3) by the action of phosphoinositide-3-kinase (PI3K). PIP_3 was shown to activate protein kinase B to increase binding to extracellular proteins and ultimately enhance cell survival [42].

The third category of signaling lipids are those serving as activators of G-protein-coupled receptors, including lysoGPLs (lysoPA in particular), S1P, platelet-activating factor (PAF), endocannabinoids, eicosanoids, fatty acid–hydroxy fatty acids, and retinol derivatives. Increased mass levels of these lipids may lead to the activation of the corresponding G-protein-coupled receptor(s).

Finally, steroid hormones, retinoic acid, eicosanoids, some NEFA, etc., can bind to nuclear receptors to activate transcription factors. For instance, retinoic acid can activate retinoic acid receptor to control differentiation and proliferation of many types of cells during development [53].

Overall, many lipids serve as signaling molecules. Any significant changes of these kinds of lipid species should consider their role in this direction and interpreted accordingly. Further cellular and molecular determination of their downstream changes such as kinase activation, transcription factor upregulation, and cell growth or apoptosis could confirm this type of cellular functions of the lipids.

16.4.4 Lipids Play Other Cellular Roles

In addition to those cellular lipid functions introduced earlier, some classes of these lipids possess some other unique functions in cellular processes, which should be recognized and are very useful for data interpretation. In this section, some examples of these lipid classes and their cellular functions (Table 16.1) are overviewed.

Acylcarnitines play an important role in fatty acyl transport in and out mitochondria. Two transport machineries are associated with this process, that is, carnitine palmitoyltransferase (CPT) I and II. In the context of inborn mitochondrial diseases, acylcarnitine production has been viewed as a detoxifying system that permits mitochondrial efflux of excess of acyl groups [54]. Unlike long-chain acylcarnitines, medium-chain species do not depend on the CPT system for transfer to the mitochondrial matrix [55].

The importance of CL in electron transport chain (ETC) for production of ATP in mitochondria is well recognized [56, 57]. It was called that CL species are the glue of the respiratory complexes [58]. This importance is evidenced in Barth syndrome, where both CL content is decreased and CL composition is changed in comparison to normal populations [59, 60]. This indicates that both content and composition of CL species are crucial for effective mitochondrial respiratory function. Many other studies particularly elucidate that the content of tetra18:2 CL and its affected CL composition are the key factor for the aforementioned importance [61]. Reduction of tetra18:2 CL content in the majority of mammalian mitochondria leads to mitochondrial malfunction. The underlying mechanism responsible for this type of malfunction is likely due to the overproduction of reactive oxygen species (ROS) resulted from the ineffective function of CL-associated complexes. Elimination of mitochondrial ROS levels with mitochondria-targeted antioxidant(s) could rescue the mitochondrial dysfunction to a great degree [62]. Accordingly, determination of a reduced CL content, particularly that of tetra18:2 CL, in comparison to controls generally suggests an overproduction of ROS in mitochondria and a malfunction of this organelle. A reduction of tetra18:2 CL content accompanied with an increase in lysoCL content suggests an increased activity of CL hydrolysis, probably due to the activation of calcium-independent $PLA_2\gamma$ [63] or an increased activity of monolysoCL acyltransferase [64, 65]. A reduction of tetra18:2 CL content along with increases in lysoCL content and the content of CL species containing relatively short FA chains such as 16:0 and 16:1 suggests the PLA_2 activation and a loss of CL remodeling activity (e.g., in the case of tafazzin mutation in Barth syndrome). A reduction of tetra18:2 CL content plus an increased content of CL species containing relatively short FA chains indicates the loss of CL remodeling activity such as the ALCAT activity [66, 67]. It should be particularly pointed out that the CL profiles in the matured organs are very different from those at their developmental stage, where the CL species are not fully remodeled and contain a large amount of species possessing relatively short FA chains. Similarly, the cell lines cultured from the neonatal or postneonatal organs show such CL profiles as the immature organs.

LysoPC species constitute an important class of bioactive molecules, which showed to be implicated in inflammatory disorders [68–71]. Some unsaturated lysoPCs especially those containing 20:4 and 22:6 [72, 73] may have anti-inflammatory properties compared to the saturated ones such as 14:0 and 16:0 lysoPC. LysoGPL species represent an interesting class of biochemical intermediates, which are involved in both biosynthesis and metabolism of diradyl GPL species that constitute the cellular lipid bilayer of all animal cells [74].

Plasmalogen species play many roles in cellular functions in addition to serving as a biomembrane component. Previous studies demonstrated that membranes comprised of plasmalogen species possess a compact membrane conformation and unique membrane dynamics in comparison to membranes comprised of its diacyl counterpart [75–78]. A central role of plasmalogen PE in facilitating membrane fusion and cell–cell communication [79, 80] was demonstrated and the propensity for membrane fusion is directly linked with the quantity of the double bonds present in the GPL acyl chains of the membrane [81]. The role of plasmalogen as an antioxidant in cellular membranes was well established (e.g., [82–87]). It was demonstrated that plasmalogen markedly delays the oxidative degradation of intrachain double bonds in polyunsaturated diacyl GPL species because the products of enol ether oxidation do not propagate the oxidation of polyunsaturated fatty acids. In addition to oxidative damage, other enzymatic or nonenzymatic factors that were demonstrated to induce membrane defects may also play an important role in causing plasmalogen deficiency (see [88, 89] for reviews). Therefore, determination of changed plasmalogen content in comparison to the controls may indicate the changes of one of the aforementioned cellular functions, which can be further examined through investigation of the corresponding cellular, molecular, biochemical, and/or physiological changes in the examined system. A reduction of plasmalogen content suggests either an increased oxidative stress or the activation of PLA_2 activity. In addition to examining the corresponding cellular responses including the measurement of oxidative stress, PLA_2 gene or protein overexpression, and/or PLA_2 phosphorylaiton or activity, the analysis of the lysoGPL profiles (particularly those of lysoPC and lysoPE) could provide definitive answers to the underlying mechanism(s) of the reduced plasmalogen content, at least an insight into the mechanism. Specifically, the increased levels of lysoplasmalogen species indicate the activation of PLA_2 activity, whereas the increased levels of acyl lysoGPL species (particularly those containing polyunsaturated FA) indicate the increased oxidative stress in the system.

In addition to its role as a membrane structural lipid, PS has many unique cellular functions. PS serves as an important precursor of mitochondrial PE in mammalian cells (see Section 16.3.2) [24]. PS exposure on the cell surface is an early event in apoptosis, which is believed to be one of the recognition signals by which apoptotic cells are removed by phagocytes [90, 91]. PS also serves as a cofactor in activation of several key signaling proteins, including protein kinase C, neutral SMase, Na^+/K^+ ATPase, and dynamin-1 (see review [23, 24] for original literature).

There are many other lipid classes in this category. For instance, acyl CoA species are involved in all the cellular processes related to lipid metabolism in addition to the involvement of energy metabolism; vitamins are the essential nutrients of mammals; wax serves as both chemical and physical barriers for plants; PA and DAG species are key intermediates for lipid biosynthesis in addition to their role in biomembrane, signal transduction, and energy metabolism as stated earlier. In summary, there is no doubt that recognition of the specific role(s) that an individual lipid class plays can clearly make the data interpretation better and insightful.

16.5 RECOGNIZING THE COMPLICATION OF SAMPLE INHOMOGENEITY AND CELLULAR COMPARTMENTS IN DATA INTERPRETATION

The existence of sample inhomogeneity and cellular compartments could lead to complications for data interpretation. In Chapter 13, we discussed the importance of careful sample preparation for accurate analysis of lipids and for developing the strategies to overcome the effects of sample inhomogeneity on acquisition of meaningful data. Here, we describe another level of sample complication, which may affect data interpretation. This complication comes from the sample itself, which contains mixed cell types, regardless of that the sample macroscopically appears homogeneity. Examples of this type of samples include numerous tissue samples from animals and plants such as brain cortex, hippocampus, the kidney, the pancreas, plant leaves, and plant stems.

The complication on data interpretation becomes obvious when these different types of cells differentially respond to a stimulus. Thus, the changed lipids could only represent the changes of lipidomes of the responded cell type, but not the entire organ or tissue section. This type of partially responded changes of cellular lipidomes could not only minimize the apparent changes but also complicate the results, particularly when different cell types contain different lipid profiles of individual class. For example, mouse hippocampus that is smaller enough has to be analyzed as the entire section, but it contains neurons, glial cells, and myelin sheath. Although all the neuronal cell bodies contain similar lipid profiles, which are not easy to be differentiated, they are different from those of myelin sheath. The neuronal cell bodies contain abundant polyunsaturated fatty acyls in PE species, whereas the presence of ST, cerebrosides, and plasmalogen PE species enriched with 18:1 fatty acyl is a unique feature of myelin. These features are helpful for data interpretation in recognizing the subtype of cells in response to a stimulus.

Similar to the effects of different cell types in a sample on interpreting lipidomics data, different lipid profiles present in different cellular compartments and microdomains are also a factor making data interpretation complicated. We extensively discuss the subcellular lipidomics in Chapter 20, which may aid in the understanding of this complication and interpreting the lipidomics data better.

In summary, the presence of sample inhomogeneity at different levels may complicate the interpretation of lipidomics data. Recognizing these complications as well as any unique features resulted from the complications should help us interpreting lipidomics data better.

16.6 INTEGRATION OF "OMICS" FOR DATA SUPPORTING

Data validation is important for lipidomics analysis. Utilization of an alternative approach to verify the obtained results is an ultimate strategy for the purpose. For example, the data obtained by shotgun lipidomics could be validated by LC-MS analysis and *vice versa*. Other alternative approaches, including GC-MS, NMR, TLC, or other chromatographic analysis, could also be employed under certain situations.

An alternative strategy for validation of lipidomics data is to determine the altered pathways at the gene and/or protein levels. From the altered lipidomics data, the changed pathways corresponding to the altered lipids could be speculated. If a prediction based on the lipidomics data is true, the levels of gene(s) and/or protein(s) should also be altered accordingly. Determination of such a change at the gene or protein should strongly support the lipidomics observation. It should be emphasized that acute activation of an enzymatic activity does not involve transcriptional and/or translational changes but only represent a change of posttranslational modification (e.g., phosphorylation) metabolic inhibitors/activators since changes of gene and/or protein levels need a relative long time (e.g., a day or longer). Accordingly, posttranslational modification should be examined to support the lipidomics data if a determined change of cellular lipidome occurs at a very short period of experimental time.

In fact, the majority of lipidomics experiments use the samples involving a relatively long period of time. Therefore, genomics and proteomics data should be very useful for supporting the lipidomics observations. While global proteomics analysis is still highly costly and takes time at its current development, powerful tools for genomic analysis have recently been developed amazingly [92] and genomic data can be obtained in a cost- and time-effective manner. Accordingly, in author's opinion, combination of genomics data with those obtained from lipidomics analysis should provide not only strong support for lipidomics findings but also deep insights into the biochemical mechanism(s) leading to the altered lipidomes. Moreover, the changes of gene expression derived from lipidomics data may alternatively inform genomic analysis as described in Section 5.5.2.

Thus, one of the logical future directions in the lipidomics field should be the development of tools capable of integrating lipidomics data with genes, transcripts, and enzyme data to perform biochemical pathway reconstruction and flux analyses. The reconstruction of lipid pathways will require strategies for pathway mapping of lipid data in a context-dependent manner. Given the structural diversity of lipid species, these tasks will be challenging and require a combination of novel and existing bioinformatics resources.

REFERENCES

1. Checa, A., Bedia, C., Jaumot, J. (2015) Lipidomic data analysis: Tutorial, practical guidelines and applications. Anal. Chim. Acta 885, 1–16.
2. Vaz, F.M., Pras-Raves, M., Bootsma, A.H., van Kampen, A.H. (2015) Principles and practice of lipidomics. J. Inherit. Metab. Dis. 38, 41–52.
3. Niemela, P.S., Castillo, S., Sysi-Aho, M., Oresic, M. (2009) Bioinformatics and computational methods for lipidomics. J. Chromatogr. B 877, 2855–2862.
4. Theodoridis, G., Gika, H.G., Wilson, I.D. (2011) Mass spectrometry-based holistic analytical approaches for metabolite profiling in systems biology studies. Mass Spectrom. Rev. 30, 884–906.
5. Sampaio, J.L., Gerl, M.J., Klose, C., Ejsing, C.S., Beug, H., Simons, K., Shevchenko, A. (2011) Membrane lipidome of an epithelial cell line. Proc. Natl. Acad. Sci. U. S. A. 108, 1903–1907.
6. Hu, C., Wang, Y., Fan, Y., Li, H., Wang, C., Zhang, J., Zhang, S., Han, X., Wen, C. (2015) Lipidomics revealed idiopathic pulmonary fibrosis-induced hepatic lipid disorders corrected with treatment of baicalin in a murine model. AAPS J. 17, 711–722.
7. Vance, D.E., Vance, J.E. (2008) Biochemistry of Lipids, Lipoproteins and Membranes. Elsevier Science B.V., City, Place pp 631.

8. Han, X., Yang, K., Gross, R.W. (2008) Microfluidics-based electrospray ionization enhances intrasource separation of lipid classes and extends identification of individual molecular species through multi-dimensional mass spectrometry: Development of an automated high throughput platform for shotgun lipidomics. Rapid Commun. Mass Spectrom. 22, 2115–2124.

9. Christie, W.W., Han, X. (2010) Lipid Analysis: Isolation, Separation, Identification and Lipidomic Analysis. The Oily Press, City. Place pp 448.

10. van Meer, G., Voelker, D.R., Feigenson, G.W. (2008) Membrane lipids: where they are and how they behave. Nat. Rev. Mol. Cell Biol. 9, 112–124.

11. Quehenberger, O., Armando, A.M., Brown, A.H., Milne, S.B., Myers, D.S., Merrill, A.H., Bandyopadhyay, S., Jones, K.N., Kelly, S., Shaner, R.L., Sullards, C.M., Wang, E., Murphy, R.C., Barkley, R.M., Leiker, T.J., Raetz, C.R., Guan, Z., Laird, G.M., Six, D.A., Russell, D.W., McDonald, J.G., Subramaniam, S., Fahy, E., Dennis, E.A. (2010) Lipidomics reveals a remarkable diversity of lipids in human plasma. J. Lipid Res. 51, 3299–3305.

12. Han, X., Gross, R.W. (2003) Global analyses of cellular lipidomes directly from crude extracts of biological samples by ESI mass spectrometry: a bridge to lipidomics. J. Lipid Res. 44, 1071–1079.

13. Han, X., Jiang, X. (2009) A review of lipidomic technologies applicable to sphingolipidomics and their relevant applications. Eur. J. Lipid Sci. Technol. 111, 39–52.

14. Henning, P.A., Merrill, A.H., Wang, M.D. (2004) Dynamic pathway modeling of sphingolipid metabolism. Conf. Proc. IEEE Eng. Med. Biol. Soc. 4, 2913–2916.

15. Kiebish, M.A., Bell, R., Yang, K., Phan, T., Zhao, Z., Ames, W., Seyfried, T.N., Gross, R.W., Chuang, J.H., Han, X. (2010) Dynamic simulation of cardiolipin remodeling: greasing the wheels for an interpretative approach to lipidomics. J. Lipid Res. 51, 2153–2170.

16. Zarringhalam, K., Zhang, L., Kiebish, M.A., Yang, K., Han, X., Gross, R.W., Chuang, J. (2012) Statistical analysis of the processes controlling choline and ethanolamine glycerophospholipid molecular species composition. PLoS One 7, e37293.

17. Han, R.H., Wang, M., Fang, X., Han, X. (2013) Simulation of triacylglycerol ion profiles: bioinformatics for interpretation of triacylglycerol biosynthesis. J. Lipid Res. 54, 1023–1032.

18. Pruett, S.T., Bushnev, A., Hagedorn, K., Adiga, M., Haynes, C.A., Sullards, M.C., Liotta, D.C., Merrill, A.H., Jr. (2008) Biodiversity of sphingoid bases ("sphingosines") and related amino alcohols. J. Lipid Res. 49, 1621–1639.

19. Levy, M., Futerman, A.H. (2010) Mammalian ceramide synthases. IUBMB Life 62, 347–356.

20. Stiban, J., Tidhar, R., Futerman, A.H. (2010) Ceramide synthases: roles in cell physiology and signaling. Adv. Exp. Med. Biol. 688, 60–71.

21. Kapoor, S., Quo, C.F., Merrill, A.H., Jr., Wang, M.D. (2008) An interactive visualization tool and data model for experimental design in systems biology. Conf. Proc. IEEE Eng. Med. Biol. Soc. 2008, 2423–2426.

22. Merrill, A.H., Jr., Stokes, T.H., Momin, A., Park, H., Portz, B.J., Kelly, S., Wang, E., Sullards, M.C., Wang, M.D. (2009) Sphingolipidomics: a valuable tool for understanding the roles of sphingolipids in biology and disease. J. Lipid Res. 50, S97–S102.

23. Vance, J.E. (2008) Phosphatidylserine and phosphatidylethanolamine in mammalian cells: two metabolically related aminophospholipids. J. Lipid Res. 49, 1377–1387.

24. Vance, J.E., Steenbergen, R. (2005) Metabolism and functions of phosphatidylserine. Prog. Lipid Res. 44, 207–234.

25. Sohlenkamp, C., Lopez-Lara, I.M., Geiger, O. (2003) Biosynthesis of phosphatidylcholine in bacteria. Prog. Lipid Res. 42, 115–162.

26. Yang, K., Zhao, Z., Gross, R.W., Han, X. (2007) Shotgun lipidomics identifies a paired rule for the presence of isomeric ether phospholipid molecular species. PLoS One 2, e1368.

27. Bartz, R., Li, W.H., Venables, B., Zehmer, J.K., Roth, M.R., Welti, R., Anderson, R.G., Liu, P., Chapman, K.D. (2007) Lipidomics reveals that adiposomes store ether lipids and mediate phospholipid traffic. J. Lipid Res. 48, 837–847.

28. Yang, K., Jenkins, C.M., Dilthey, B., Gross, R.W. (2015) Multidimensional mass spectrometry-based shotgun lipidomics analysis of vinyl ether diglycerides. Anal. Bioanal. Chem. 407, 5199–5210.

29. Gross, R.W., Han, X. (2011) Lipidomics at the interface of structure and function in systems biology. Chem. Biol. 18, 284–291.

30. Unger, R.H., Orci, L. (2000) Lipotoxic diseases of nonadipose tissues in obesity. Int. J. Obes. 24, S28-S32.

31. Lelliott, C., Vidal-Puig, A.J. (2004) Lipotoxicity, an imbalance between lipogenesis de novo and fatty acid oxidation. Int. J. Obes. Relat. Metab. Disord. 28 Suppl 4, S22-S28.

32. Schrauwen, P. (2007) High-fat diet, muscular lipotoxicity and insulin resistance. Proc. Nutr. Soc. 66, 33–41.

33. Drosatos, K., Schulze, P.C. (2013) Cardiac lipotoxicity: molecular pathways and therapeutic implications. Curr. Heart Fail. Rep. 10, 109–121.

34. Guillou, H., Zadravec, D., Martin, P.G., Jacobsson, A. (2010) The key roles of elongases and desaturases in mammalian fatty acid metabolism: Insights from transgenic mice. Prog. Lipid Res. 49, 186–199.

35. Wang, M., Hayakawa, J., Yang, K., Han, X. (2014) Characterization and quantification of diacylglycerol species in biological extracts after one-step derivatization: A shotgun lipidomics approach. Anal. Chem. 86, 2146–2155.

36. Hannun, Y.A., Obeid, L.M. (2008) Principles of bioactive lipid signalling: lessons from sphingolipids. Nat. Rev. Mol. Cell Biol. 9, 139–150.

37. Dbaibo, G.S., El-Assaad, W., Krikorian, A., Liu, B., Diab, K., Idriss, N.Z., El-Sabban, M., Driscoll, T.A., Perry, D.K., Hannun, Y.A. (2001) Ceramide generation by two distinct pathways in tumor necrosis factor alpha-induced cell death. FEBS Lett. 503, 7–12.

38. Obeid, L.M., Linardic, C.M., Karolak, L.A., Hannun, Y.A. (1993) Programmed cell death induced by ceramide. Science 259, 1769–1771.

39. Haimovitz-Friedman, A., Kolesnick, R.N., Fuks, Z. (1997) Ceramide signaling in apoptosis. Br. Med. Bull. 53, 539–553.

40. Venable, M.E., Lee, J.Y., Smyth, M.J., Bielawska, A., Obeid, L.M. (1995) Role of ceramide in cellular senescence. J. Biol. Chem. 270, 30701–30708.

41. Perry, D.K., Hannun, Y.A. (1998) The role of ceramide in cell signaling. Biochim. Biophys. Acta 1436, 233–243.

42. Prokazova, N.V., Samovilova, N.N., Golovanova, N.K., Gracheva, E.V., Korotaeva, A.A., Andreeva, E.R. (2007) Lipid second messengers and cell signaling in vascular wall. Biochemistry (Mosc.) 72, 797–808.

43. Pettus, B.J., Bielawska, A., Subramanian, P., Wijesinghe, D.S., Maceyka, M., Leslie, C.C., Evans, J.H., Freiberg, J., Roddy, P., Hannun, Y.A., Chalfant, C.E. (2004) Ceramide 1-phosphate is a direct activator of cytosolic phospholipase A2. J. Biol. Chem. 279, 11320–11326.

44. Gomez-Munoz, A., Kong, J.Y., Salh, B., Steinbrecher, U.P. (2004) Ceramide-1-phosphate blocks apoptosis through inhibition of acid sphingomyelinase in macrophages. J. Lipid Res. 45, 99–105.

45. Gomez-Munoz, A., Kong, J.Y., Parhar, K., Wang, S.W., Gangoiti, P., Gonzalez, M., Eivemark, S., Salh, B., Duronio, V., Steinbrecher, U.P. (2005) Ceramide-1-phosphate promotes cell survival through activation of the phosphatidylinositol 3-kinase/protein kinase B pathway. FEBS Lett. 579, 3744–3750.

46. Hinkovska-Galcheva, V., Boxer, L.A., Kindzelskii, A., Hiraoka, M., Abe, A., Goparju, S., Spiegel, S., Petty, H.R., Shayman, J.A. (2005) Ceramide 1-phosphate, a mediator of phagocytosis. J. Biol. Chem. 280, 26612–26621.

47. Smith, E.R., Merrill, A.H., Obeid, L.M., Hannun, Y.A. (2000) Effects of sphingosine and other sphingolipids on protein kinase C. Methods Enzymol. 312, 361–373.

48. Spiegel, S., Milstien, S. (2003) Sphingosine-1-phosphate: an enigmatic signalling lipid. Nat. Rev. Mol. Cell Biol. 4, 397–407.

49. Bektas, M., Spiegel, S. (2004) Glycosphingolipids and cell death. Glycoconj. J. 20, 39–47.

50. Hansen, S.B., Tao, X., MacKinnon, R. (2011) Structural basis of PIP2 activation of the classical inward rectifier K+ channel Kir2.2. Nature 477, 495–498.

51. Irvine, R.F. (1992) Inositol lipids in cell signalling. Curr. Opin. Cell Biol. 4, 212–219.

52. Nishizuka, Y. (1995) Protein kinase C and lipid signaling for sustained cellular responses. FASEB J. 9, 484–496.

53. Duester, G. (2008) Retinoic acid synthesis and signaling during early organogenesis. Cell 134, 921–931.

54. Ramsay, R.R. (2000) The carnitine acyltransferases: modulators of acyl-CoA-dependent reactions. Biochem. Soc. Trans. 28, 182–186.

55. Steiber, A., Kerner, J., Hoppel, C.L. (2004) Carnitine: a nutritional, biosynthetic, and functional perspective. Mol. Aspects Med. 25, 455–473.

56. Hoch, F.L. (1992) Cardiolipins and biomembrane function. Biochim. Biophys. Acta 1113, 71–133.

57. Chicco, A.J., Sparagna, G.C. (2007) Role of cardiolipin alterations in mitochondrial dysfunction and disease. Am. J. Physiol. Cell Physiol. 292, C33–C44.

58. Zhang, M., Mileykovskaya, E., Dowhan, W. (2002) Gluing the respiratory chain together. Cardiolipin is required for supercomplex formation in the inner mitochondrial membrane. J. Biol. Chem. 277, 43553–43556.

59. Schlame, M., Towbin, J.A., Heerdt, P.M., Jehle, R., DiMauro, S., Blanck, T.J. (2002) Deficiency of tetralinoleoyl-cardiolipin in Barth syndrome. Ann. Neurol. 51, 634–637.

60. Hauff, K.D., Hatch, G.M. (2006) Cardiolipin metabolism and Barth Syndrome. Prog. Lipid Res. 45, 91–101.

61. Schlame, M., Ren, M., Xu, Y., Greenberg, M.L., Haller, I. (2005) Molecular symmetry in mitochondrial cardiolipins. Chem. Phys. Lipids 138, 38–49.

62. He, Q., Harris, N., Ren, J., Han, X. (2014) Mitochondria-targeted antioxidant prevents cardiac dysfunction induced by tafazzin gene knockdown in cardiac myocytes. Oxid. Med. Cell. Longev. 2014, 654198.

63. Mancuso, D.J., Sims, H.F., Han, X., Jenkins, C.M., Guan, S.P., Yang, K., Moon, S.H., Pietka, T., Abumrad, N.A., Schlesinger, P.H., Gross, R.W. (2007) Genetic ablation of calcium-independent phospholipase A2gamma leads to alterations in mitochondrial lipid metabolism and function resulting in a deficient mitochondrial bioenergetic phenotype. J. Biol. Chem. 282, 34611–34622.

64. Ma, B.J., Taylor, W.A., Dolinsky, V.W., Hatch, G.M. (1999) Acylation of monolysocardiolipin in rat heart. J. Lipid Res. 40, 1837–1845.

65. Taylor, W.A., Hatch, G.M. (2003) Purification and characterization of monolysocardiolipin acyltransferase from pig liver mitochondria. J. Biol. Chem. 278, 12716–12721.

66. Cao, J., Liu, Y., Lockwood, J., Burn, P., Shi, Y. (2004) A novel cardiolipin-remodeling pathway revealed by a gene encoding an endoplasmic reticulum-associated acyl-CoA:lysocardiolipin acyltransferase (ALCAT1) in mouse. J. Biol. Chem. 279, 31727–31734.

67. Li, J., Romestaing, C., Han, X., Li, Y., Hao, X., Wu, Y., Sun, C., Liu, X., Jefferson, L.S., Xiong, J., Lanoue, K.F., Chang, Z., Lynch, C.J., Wang, H., Shi, Y. (2010) Cardiolipin remodeling by ALCAT1 links oxidative stress and mitochondrial dysfunction to obesity. Cell Metab. 12, 154–165.

68. Sevastou, I., Kaffe, E., Mouratis, M.A., Aidinis, V. (2013) Lysoglycerophospholipids in chronic inflammatory disorders: the PLA(2)/LPC and ATX/LPA axes. Biochim. Biophys. Acta 1831, 42–60.

69. Hung, N.D., Kim, M.R., Sok, D.E. (2011) 2-Polyunsaturated acyl lysophosphatidylethanolamine attenuates inflammatory response in zymosan A-induced peritonitis in mice. Lipids 46, 893–906.

70. Hung, N.D., Sok, D.E., Kim, M.R. (2012) Prevention of 1-palmitoyl lysophosphatidylcholine-induced inflammation by polyunsaturated acyl lysophosphatidylcholine. Inflamm. Res. 61, 473–483.

71. D'Arrigo, P., Servi, S. (2010) Synthesis of lysophospholipids. Molecules 15, 1354–1377.

72. Hung, N.D., Kim, M.R., Sok, D.E. (2009) Anti-inflammatory action of arachidonoyl lysophosphatidylcholine or 15-hydroperoxy derivative in zymosan A-induced peritonitis. Prostaglandins Other Lipid Mediat. 90, 105–111.

73. Huang, L.S., Hung, N.D., Sok, D.E., Kim, M.R. (2010) Lysophosphatidylcholine containing docosahexaenoic acid at the *sn*-1 position is anti-inflammatory. Lipids 45, 225–236.

74. Vance, J.E., Vance, D.E. (2004) Phospholipid biosynthesis in mammalian cells. Biochem. Cell Biol. 82, 113–128.

75. Pak, J.H., Bork, V.P., Norberg, R.E., Creer, M.H., Wolf, R.A., Gross, R.W. (1987) Disparate molecular dynamics of plasmenylcholine and phosphatidylcholine bilayers. Biochemistry 26, 4824–4830.

76. Han, X., Gross, R.W. (1990) Plasmenylcholine and phosphatidylcholine membrane bilayers possess distinct conformational motifs. Biochemistry 29, 4992–4996.

77. Han, X., Gross, R.W. (1991) Proton nuclear magnetic resonance studies on the molecular dynamics of plasmenylcholine/cholesterol and phosphatidylcholine/cholesterol bilayers. Biochim. Biophys. Acta 1063, 129–136.

78. Han, X., Chen, X., Gross, R.W. (1991) Chemical and magnetic inequivalence of glycerol protons in individual subclasses of choline glycerophospholipids: implications for subclass-specific changes in membrane conformational states. J. Am. Chem. Soc. 113, 7104–7109.

79. Glaser, P.E., Gross, R.W. (1995) Rapid plasmenylethanolamine-selective fusion of membrane bilayers catalyzed by an isoform of glyceraldehyde-3-phosphate dehydrogenase: discrimination between glycolytic and fusogenic roles of individual isoforms. Biochemistry 34, 12193–12203.

80. Glaser, P.E., Gross, R.W. (1994) Plasmenylethanolamine facilitates rapid membrane fusion: a stopped-flow kinetic investigation correlating the propensity of a major plasma membrane constituent to adopt an HII phase with its ability to promote membrane fusion. Biochemistry 33, 5805–5812.

81. Han, X., Gross, R.W. (1992) Nonmonotonic alterations in the fluorescence anisotropy of polar head group labeled fluorophores during the lamellar to hexagonal phase transition of phospholipids.[see comment]. Biophys. J. 63, 309–316.

82. Zoeller, R.A., Morand, O.H., Raetz, C.R. (1988) A possible role for plasmalogens in protecting animal cells against photosensitized killing. J. Biol. Chem. 263, 11590–11596.

83. Brosche, T., Platt, D. (1998) The biological significance of plasmalogens in defense against oxidative damage. Exp. Gerontol. 33, 363–369.

84. Hahnel, D., Beyer, K., Engelmann, B. (1999) Inhibition of peroxyl radical-mediated lipid oxidation by plasmalogen phospholipids and alpha-tocopherol. Free Radic. Biol. Med. 27, 1087–1094.

85. Sindelar, P.J., Guan, Z., Dallner, G., Ernster, L. (1999) The protective role of plasmalogens in iron-induced lipid peroxidation. Free Radic. Biol. Med. 26, 318–324.

86. Zoeller, R.A., Lake, A.C., Nagan, N., Gaposchkin, D.P., Legner, M.A., Lieberthal, W. (1999) Plasmalogens as endogenous antioxidants: somatic cell mutants reveal the importance of the vinyl ether. Biochem. J. 338, 769–776.

87. Murphy, R.C. (2001) Free-radical-induced oxidation of arachidonoyl plasmalogen phospholipids: Antioxidant mechanism and precursor pathway for bioactive eicosanoids. Chem. Res. Toxicol. 14, 463–472.

88. Farooqui, A.A., Horrocks, L.A., Farooqui, T. (2000) Glycerophospholipids in brain: their metabolism, incorporation into membranes, functions, and involvement in neurological disorders. Chem. Phys. Lipids 106, 1–29.

89. Klein, J. (2000) Membrane breakdown in acute and chronic neurodegeneration: Focus on choline-containing phospholipids. J. Neural Transm. 107, 1027–1063.

90. Fadok, V.A., de Cathelineau, A., Daleke, D.L., Henson, P.M., Bratton, D.L. (2001) Loss of phospholipid asymmetry and surface exposure of phosphatidylserine is required for phagocytosis of apoptotic cells by macrophages and fibroblasts. J. Biol. Chem. 276, 1071–1077.

91. Balasubramanian, K., Mirnikjoo, B., Schroit, A.J. (2007) Regulated externalization of phosphatidylserine at the cell surface: implications for apoptosis. J. Biol. Chem. 282, 18357–18364.

92. Metzker, M.L. (2010) Sequencing technologies - the next generation. Nat. Rev. Genet. 11, 31–46.

PART IV

APPLICATIONS OF LIPIDOMICS IN BIOMEDICAL AND BIOLOGICAL RESEARCH

17

LIPIDOMICS FOR HEALTH AND DISEASE

17.1 INTRODUCTION

As discussed, MS-based lipidomics has the power and flexibility of analyzing lipids at a large scale and in their intact forms from virtually any type of samples, provided appropriate procedures of extraction, cleanup, and/or derivatization are followed (see Chapter 13). On the other hand, since lipid metabolism is tightly regulated within a network and between networks (see Chapter 16), perturbation of any single synthesis or degradation pathway could lead to changes of lipid homeostasis of an entire system. Thus, studying the lipid homeostasis of a biological system before and after a stimulus utilizing lipidomics should provide deep insights into the mechanism(s) underpinning lipid changes in response to the stimulus.

Accordingly, it is not surprising that determining altered lipid states due to a disease and searching the pathogenic mechanism becomes a major task of lipidomics (and the study of lipids in general) (Figure 17.1). In addition to revealing the mechanism(s) responsible for the changes of lipid homeostasis between states, lipidomics can serve as a powerful tool to recapitulate these changes with model systems (e.g., animals, cells, plants) through gene manipulation to validate the revealed underlying mechanism(s) (Figure 17.1). Moreover, lipidomics may also be a very useful aid in the development of drugs for therapeutic intervention to the altered lipid homeostasis based on the revealed pathogenic mechanism(s) as well as testing the efficacy of any developed drugs (Figure 17.1). Finally, as a member of the "omics" family, lipidomics can always be exploited for biomarker discovery and development at any

Lipidomics: Comprehensive Mass Spectrometry of Lipids, First Edition. Xianlin Han.
© 2016 John Wiley & Sons, Inc. Published 2016 by John Wiley & Sons, Inc.

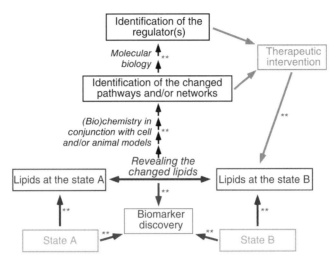

Figure 17.1 Schematic illustration of lipidomics applications for health and disease. The states A and B represent a pair of conditions for comparison. Ideally, there is only one variable such as a kind of disease, treatment with a nutrient/hormone/drug, and a genetic mutation between the two states. Lipidomics study could be involved in many processes as indicated with the symbols of **.

levels such as organelle, cells, organs, and entire body systems (Figure 17.1). For example, changes of CL species in content, composition, or combined in animals, which can sensitively reflect the efficiency of electron transport chains in mitochondria, represent a fine biomarker of mitochondrial function.

Taken together, applications of lipidomics have been spanned from observational studies to assessment of drug or nutritional supplementation, from gene phenotyping to biomarker discovery, from a specific organ to a whole body system, from homeostatic processes to diseases, etc. This chapter is not intended to comprehensively review these applications of lipidomics in health and disease since the topic is too broad to be covered in a single chapter. Any readers who are interested in a specific disease area are referred to the original reports given below and numerous valuable reviews in the topics [1–19]. Herein, only an overview of lipidomics studies on a few diseases and health conditions with selected examples is given.

17.2　DIABETES AND OBESITY

Lipidomics study has been broadly applied to explore the metabolic origin of diabetes and obesity complications. For example, lipidomic analysis of lipid extracts from white adipose tissue in young healthy monozygous twins with different body weights was performed [20]. The study led to a finding that obese co-twins compared to their lean counterparts had relatively higher content of ether-containing glycerophospholipids consisting of long-chain polyunsaturated fatty acids (PUFA), particularly arachidonic acid, despite lower dietary PUFA intake. The study revealed that the content of GPL species containing less number of double bonds and less number of carbon atoms tended to be lower in white adipose tissue of the obese co-twins, suggesting a vigorous futile cycle remodeling of these GPL species. Although the trigger of this remodeling remained unknown, available genomic, lipidomic, and clinical data suggest that the decreased PUFA intake may play a potential signaling role and suggest the importance of membrane lipids in initiating the metabolic comorbidities [20]. It was demonstrated in the study that the observed remodeling of membrane lipids in obesity involved three types of structural changes, which are as follows:

- The polar head group (e.g., the mass levels of PE *vs.* PC).
- The linkage of aliphatic chains at the *sn*-1 position (i.e., the changed content of plasmalogen *vs.* diacyl counterparts).
- The degree of unsaturation of the *sn*-2 fatty acyl chains.

These changes have an impact on the biophysical properties of cellular membranes, including membrane fluidity [20]. The investigators interpreted the observed remodeling as a form of allostatic adaptation, which was aimed at preserving the membrane properties in expanding adipocytes. It is unfortunate that allostatic adaptation carries a cost. In the case of lipid remodeling in white adipocytes, the allostatic load could lead to the increased vulnerability to inflammation because arachidonic acid present as a rich FA substituent in plasmalogens, which were enriched in obese co-twins, is the precursor of proinflammatory lipid mediators [21, 22]. Clearly, it is necessary to conduct further studies to elucidate the true underlying mechanism(s) driving to the lipid remodeling present in the white adipocyte tissue of obese individuals and to determine its association with metabolic disease.

In a similar study of monozygotic twins discordant for obesity to eliminate any genetic factors, the investigators demonstrated that obesity was related to distinct changes of the global serum lipid profile [23]. In comparison to nonobese co-twins, increased levels of lysoPC species and decreased levels of ether-containing GPL species (e.g., plasmalogens) were determined in serum of obese co-twins. Moreover, these lipid changes were well correlated with insulin resistance, a metabolic characteristic of acquired obesity in these healthy adult twins. As discussed in Chapter 16, lysoPC species are associated with proinflammatory and proatherogenic conditions and plasmalogens are known to exert antioxidative properties. Based on the findings, the researchers proposed that proper management of obesity, with a new generation of therapies targeted to the lipid metabolism pathways of these lipid classes and subclasses, could likely correct these abnormalities and favorably modify the risk course and outcome of diabetes and obesity. Other studies employing either targeted or shotgun lipidomics showed that intake of fatty fish could lead to lowering the levels of a variety of serum lipids in obese individuals [24, 25].

A lipidomics study also demonstrated that depletion of tetra18:2 CL species occurred at the very early stages of type I diabetic mouse model induced by streptozotocin injection [26, 27]. CL species are extensively remodeled at very early stages, leading to a diversified profile of CL molecular species [27]. Fatty acyl contents of 18:2 and 18:1 in CL, particularly the former, are dramatically decreased; in contrast, the content of 22:6 is substantially increased [28]. Therefore, the CL species containing longer FA chains than those of 18 carbon atoms are drastically increased [27]. Similar to the type I diabetic model, changes of CL species are also manifest in mouse heart of type II diabetes (e.g., insulin-resistant, leptin-deficient *ob/ob* mice) [27]. These kinds of CL profiles are ubiquitously present in mouse heart of type II diabetes mellitus (T2DM) such as the high-fat-induced, *db/db*, or protein kinase AKT2 knockout model (Han, unpublished data). It is believed that tetra18:2 CL plays a key role in cardiac mitochondrial function [29]. Accordingly, the extensive CL remodeling manifest in diabetic myocardium should have sequential impacts on mitochondrial and, thus, cardiac dysfunction.

It is important to recognize that mitochondrial and peroxisomal phospholipases are key players in the regulation of cellular bioenergetics and signaling [30, 31]. Mice deficient in calcium-independent $PLA_2\gamma$ (i.e., $iPLA_2\gamma^{-/-}$) are resistant to high fat diet-induced weight gain and hyperinsulinemia [32]. Specifically, shotgun lipidomics analysis of lipids present in white adipose tissue from wild-type mice demonstrated a twofold increase in TAG content after being fed with a high-fat diet as compared to $iPLA_2\gamma^{-/-}$ mice. It is known that tissue macrophage inflammatory pathways were shown to contribute to obesity-associated insulin resistance [33].

Type I diabetes is an autoimmune disease characterized by a relatively long symptom-free period that precedes the overt disease and caused by progressive loss of insulin-secreting capacity from pancreatic β-cells. Although genetic risk factors of this disease are well characterized, the biochemical mechanisms responsible for disease onset and progression remain elusive. Multiple lipidomics studies revealed that, compared to the controls, children who later progressed to type I diabetes had lower levels of PC species at birth as measured from cord blood and persistently diminished ether-containing GPL species throughout the follow-up [34–36]. It was found that these lipidomic changes are specific to progression of type I diabetes, but not to the progression of general autoimmunity [35]. Mechanistic study suggests that deficiency in choline might be a causal factor of this disease as choline plays many important roles in early development [36].

17.3 CARDIOVASCULAR DISEASES

Lipidomics research on cardiovascular disease is multiplexed, including profiling of serum/plasma or myocardial lipidomes, to reveal the underlying molecular mechanism(s) leading to the diseases, to assess the effectiveness of nutritional and drug treatments, and to discover and develop possible lipid biomarker for early diagnosis of the diseases as examples. Lipidomics was exploited to understand the biochemical basis of the effectiveness of n-3 PUFA in cardiovascular health [37–39].

In this aspect, novel anti-inflammatory mediators, including resolvins and protectins derived from docosahexaenoic acid (DHA, 22:6 FA), are discovered in association with the beneficial effects of dietary DHA on prevention of cardiovascular diseases [37, 38]. In contrast, lipidomics research further confirmed the association of a variety of eicosanoids, oxylipins, and endocannabinoids with initiation and progression of cardiovascular diseases [39]. These studies strongly suggest that the levels of n-3 and n-6 PUFA should be well balanced.

Lipidomics analysis of plasma samples demonstrated that perturbed lipid profiles were present in unstable coronary syndrome and endarterectomy patients [10, 40]. The studies suggest that the altered lipid profiles are antecedents to the onset of acute coronary syndrome. Therefore, it appears that many of those identified lipids might serve as useful biomarkers for differentiating the severity of unstable coronary artery disease. Lipidomics analysis of myocardial samples revealed that alterations in the content of mitochondria-specific lipids (i.e., CL species) could occur at the earliest stages of diabetes-induced cardiomyopathy [27], and other changed lipids occurred at the relatively late stages [26, 27, 41, 42]. These observations strongly suggest that mitochondrial malfunction should be among the earliest events of diabetic cardiomyopathy and maybe other cardiac failures.

Hypertension is a key risk factor in the progression of cardiovascular disease. A shotgun lipidomics approach was exploited to study the plasma lipidome from hypertension patients [43]. It was demonstrated that the levels of ether-containing GPL species and free cholesterol were decreased in the plasma of hypertensive patients and found that the reduced mass levels of ether-containing GPL species were specific to hypertension relative to obesity and insulin resistance as examined. Recently, lipidomics approaches by using both LC-MS and MALDI-MS were used for assessment of the efficacy of antihypertensive drug therapy [44, 45]. It was found that the levels of both PC and TAG species were altered in the plasma of hypertensive patients, suggesting the effects of hypertension pathogenesis on plasma lipid metabolism. Specifically, the contents of both PC and TAG classes were significantly increased in patients with hypertension, which tended to decrease after treatment with antihypertensive therapy [44]. In addition, it was also found that the total content of cholesteryl esters was significantly decreased in hypertensive patients after drug treatment [44].

17.4   NONALCOHOL FATTY LIVER DISEASE

Nonalcoholic fatty liver disease (NAFLD) is a common form of chronic liver disease affecting individuals at different ages [46–48]. NAFLD is associated with overaccumulation of lipids in the liver ranging from steatosis to nonalcoholic steatohepatitis or NASH characterized by the accumulation of fat in the liver along with evidence of liver cell damage, inflammation, and different degrees of scarring or fibrosis [48]. Although TAG accumulation is ubiquitous in NAFLD [49], many other classes of lipids, including NEFA, DAG, free cholesterol, cholesteryl ester, Cer, and GPL are also accumulated [50]. Therefore, it is not doubtful and was proved that lipidomics is the powerful tool to facilitate the determination of the changes of all these lipids and to reveal the possibly underlying mechanism(s) leading to the accumulation.

For example, altered lipids in liver tissue samples from NAFLD patients in comparison to normal controls were determined by GC [2]. It was found that the mean levels (nmol/g of tissue) of DAG (normal/NAFLD: 1922/4947) and TAG (13,609/128,585) were increased significantly in NAFLD, but the levels of NEFA remained unaltered (5533/5929). The study showed that there existed a stepwise increase in the mean TAG/DAG ratio from normal to NAFLD (7/26, $p < 0.001$). The total PC content decreased in NAFLD. The ratio of free cholesterol/PC increased progressively (0.34/0.69, $p < 0.008$). Taken together, NAFLD is associated with numerous changes in cellular lipidome of the liver.

In another study using flash-frozen liver biopsies obtained from 37 insulin-resistant, obese, nondiabetic individuals [51], it was found that hepatic DAG content in cytoplasmic lipid droplets was the best predictor of insulin resistance ($\gamma = 0.80$, $p < 0.001$), and it was responsible for 64% of the variability in insulin sensitivity. Mechanistic studies showed that hepatic DAG content was strongly correlated with activation of hepatic PKCε ($\gamma = 0.67$, $p < 0.001$), which impairs insulin signaling, and that there was no significant association between insulin resistance and other putative lipid metabolites or plasma or hepatic markers of inflammation. The investigators concluded that hepatic DAG content in lipid droplets was the best predictor of insulin resistance in humans, and the data supported the hypothesis that NAFLD-associated hepatic insulin resistance is caused by an increase in hepatic DAG content, which results in activation of PKCε.

Lipidomics was applied to analyze serum samples from a large number of well-characterized individuals in whom liver fat content was measured using ^{1}H NMR or liver biopsy [3]. It was found that two PUFA-containing PC species (i.e., 24:1-20:4 ether-containing and 18:1-22:6 PC) and a TAG species (i.e., 48:0 TAG) could serve as serum lipid signature to best estimate the liver fat composition, among which the PC species were negatively associated with the liver fat content and the TAG species was positively associated with the liver fat content. When this lipid signature was applied for the diagnosis of NAFLD, a sensitivity of >69% and a specificity of >75% were validated [3]. This outcome is well comparable with a previously established model comprising several clinical measures [52]. Moreover, the positive association of serum 48:0 TAG content with NAFLD may explain the association of this lipid species with increased risk of diabetes [41, 53] and insulin resistance [54], since NAFLD is a known risk factor for the development of diabetes and metabolic syndrome [55].

It appears that the availability of NEFA from circulation is the main determinant to the NAFLD hepatic lipid loading in patients [56]. A growing body of evidence mainly from experimental studies suggests that lipid compartmentalization in hepatocytes and in particular the type of lipids accumulating may play a crucial role in disease progress [57]. This novel concept has significant implications when considering potential alternative treatment strategies for patients with NAFLD.

17.5 ALZHEIMER'S DISEASE

Alzheimer's disease (AD), the most common cause of dementia worldwide, is a progressive neurodegenerative disorder characterized clinically by progressive cognitive impairment [58] and neuropathologically by the appearance of diffusible amyloid-beta (Aβ), neuritic plaques, and intraneuronal neurofibrillary tangles [59]. At present, no effective therapy has been shown to delay the onset or progression of AD. During the last decade, multiple large-scale AD clinical trials have failed, including multiple strategies mostly focused on targeting the amyloid precursor protein secretases, or its cleavage product Aβ, highlighting our incomplete knowledge of both cognitive impairment and the pathogenesis of AD [60–62]. Despite no therapy is currently available to prevent AD, early disease detection would still be of high importance in order to delay the onset of the disease through pharmacological treatment and/or lifestyle changes or for assessing the efficacy of potential AD therapeutic agents.

Numerous studies on AD-related lipidomics were performed (see reviews [1, 12, 63–66] for details). Different conclusions were drawn from the studies. These differences between the studies largely resulted from three possibilities: utilizing different analytical methods, employing samples at different AD stages, and with or without considering the inhomogeneities of brain samples. Different lipidomic platforms target the analysis of different lipid classes, subclasses, or individual molecular species with different coverage at the time when the research was conducted. It is obvious that different findings utilizing these different platforms can be uncovered. As AD progresses to the late stages, neurodegeneration becomes more and more severe and neuronal loss becomes more apparent; thus, broader lipid signals and membrane disruptions are involved in the development of pathology. Moreover, nutritional conditions are changed as AD becomes more severe. All of these factors lead to possible changes in brain lipid constituents and signaling. These significant lipid changes in brain samples of AD at the late stage were already revealed by conventional methods for lipid analysis [67–69]. The inhomogeneity of brain samples and its impact on lipidomics analysis are discussed in details in Section 13.2. It is suggested that the researchers and the readers should recognize the limitations of a particular study considering these aspects and make some justification accordingly.

Although no conclusion can be made from over a decade of lipidomics studies, an increasingly accurate picture of altered lipid metabolism in AD could be provided. Below a few aspects of findings in AD obtained from lipidomics are summarized.

The altered lipid levels of pure gray and white matter (i.e., considering the sample inhomogeneity) from different brain regions of subjects with very mild AD (i.e., mild cognitive impairment (MCI)) were determined by using the MDMS-SL platform at the emerging time of lipidomics field [70–72]. Three marked changes of lipid levels were revealed in the postmortem brain samples of subjects with MCI. These changes include the specific and substantial loss of sulfatides [70, 71], a significant increase and molecular species compositional change in Cer [70], and a significant loss of plasmalogen content [72]. The loss of plasmalogen content is well correlated with the severity of AD stages [72]. Intriguingly, the tendency of ST levels in CSF with AD severity is well correlated with that in the brain [73]. Further studies by using the MDMS-SL platform uncovered an apoE-mediated ST metabolism pathway in the CNS [74]. This discovery tightly links the alterations in apoE-mediated lipid trafficking and metabolism with AD pathogenesis [64].

Recently, to explore the hypothesis that alterations in ST, Cer, and PE plasmalogen are present at the stage of preclinical AD (i.e., cognitively normal at death, but with AD neuropathology), MDMS-SL was further exploited to analyze these lipid classes and individual molecular species in lipid extracts from postmortem brains of subjects with preclinical AD [75]. It was found that (1) ST levels were significantly lower in subjects with preclinical AD compared to those without AD neuropathology; (2) the levels of PE plasmalogens were marginally lower at this stage of AD; and (3) changes of the Cer levels were undetectable with the available samples. These results not only indicate that cellular membrane defects are present at the earliest stages of AD pathogenesis but also suggest that ST loss is among the earliest events of AD development, whereas alterations in the levels of PE plasmalogen and Cer occur relatively later in the disease.

The MDMS-SL approach was also applied to measure levels of over 800 species of GPL classes, SM, Cer, TAG, cholesterol, and cholesteryl esters in the plasma of AD individuals to explore potential lipid biomarkers for the diagnosis of AD [76]. Using plasma from 26 AD (17 with mild and 9 with moderate AD) and 26 controls, it was found that SM levels were significantly lower and Cer levels were higher in AD patients compared to the controls. Furthermore, it was also found that the rank of the changed mass levels of SM and Cer species was strongly correlated with the rank of AD severity ($p < 0.004$), which is consistent with other reports [77–79]. The higher plasma Cer levels [76, 80] are in line with other studies with different samples, including the middle frontal cortex [81], white matter [70], and CSF [77] from AD subjects compared to normal controls. Taken together, the findings across methods and sample varieties (brain tissue, CSF, and plasma) suggest that the sphingolipid pathway is perturbed in AD.

Deficiency in PE plasmalogen content in the late stages of AD is well documented by applying conventional approach for lipid analysis [82–84]. MDMS-SL showed its deficiency at the molecular species levels and present at the very early stages of AD as well as in animal models [72]. Further lipidomic studies showed the presence of plasmalogen deficiency in human plasma of AD patients at very early stages [85, 86]. These findings indicate that deficiency in plasma plasmalogen content could be used as a biomarker for diagnosis of AD. Further studies also indicate that peroxisomal dysfunction might be the underlying molecular mechanism, and improving peroxisomal plasmalogen biosynthesis may serve as a therapeutic drug target [87]. Other possible mechanisms leading to plasmalogen deficiency such as serving as endogenous antioxidants was also proposed [1]. This mechanism is strongly supported with the evidence that severe oxidative stress is manifest in AD [88]. Deficiency in plasmalogen that is very enriched in myelin sheath may represent a symbol of myelin loss. Research in this area remains warranted.

The loss of PC content in the late stages of AD was also well documented [69, 89]. The connection of PC loss with choline deficiency in AD may be interpreted based on the fact that choline is a rate-limiting factor for PC biosynthesis. Numerous lipidomics studies revealed the loss of PC content in the plasma of AD subjects, including very early stage of AD [76, 86, 90–92]. This finding indicates the role of choline in AD development and may suggest that PC deficiency in human plasma could serve as a biomarker for AD and supplying choline might be a nutritional approach to delay the progression of AD [92].

A 2D LC-MS approach was employed to examine plasma sterols between AD cases and controls [93]. In an initial study with plasma samples from 10 AD cases and 10 controls, the investigators uncovered that the levels of desmosterol, a precursor of cholesterol, were significantly lower in AD patients relative to those of controls ($p < 0.009$). This finding was then further confirmed with 26 MCI individuals, 41 AD subjects, and 42 age-matched controls. Moreover, the researchers found that the reduced levels of desmosterol were well correlated with the severity of AD and that the changes in desmosterol levels could most closely represent the pathological progression of AD.

Lipidomics studies suggested a potential link of decreased levels of DHA with AD [94] since many studies revealed that the metabolites resulted from n-3 PUFA, including DHA-derived neuroprotectins, possess neuroprotective properties in anti-inflammation [95–97]. It was recognized for a long time that gangliosides play an important role in AD pathogenesis [98]. A recent study [99] showed that reduction of the levels of ganglioside molecular species could be considered as a potential AD progression temporarily.

17.6 PSYCHOSIS

It has been long known that lipid abnormalities are associated with psychosis [5]. For example, abnormal GPL metabolism or altered brain membrane lipid composition is associated with the development of schizophrenia [100]. Importantly, the results from an intervention study with long-chain n-3 fatty acids clearly indicate the importance of lipids in the progression of psychosis since the study demonstrated that treatment with long-chain n-3 fatty acids could reduce the risk of progression to psychotic disorder [101].

In a cohort study by lipid profiling [102], it was found that schizophrenia is associated with the elevated serum levels of specific TAG species containing low number of double bonds and relatively shorter chains of fatty acyls. The study demonstrated how lipidomics could be a powerful tool for dissecting complex disease-related metabolic pathways as well as for identifying diagnostic and prognostic markers in psychiatric research. A lipidomic study of monozygous twin pairs discordant provided further evidence and possible mechanism in the direction. It was found that the patients with schizophrenia had higher TAG levels and were more insulin resistant than their co-twins [4]. Integrative analysis of magnetic resonance image with lipidomics revealed significant associations of the elevated TAG content in plasma with the decreased gray matter density. These results strongly suggest that blood-based molecular markers may be related and sensitive to the brain structural changes. In other studies on human blood of patients with schizophrenia and depression, reduced levels of endocannabinoids in disorders were also reported [103, 104], suggesting the involvement of lipid signaling in the disorders.

In lipidomics profilings of lipid extracts from postmortem brain tissues with psychoses, including schizophrenia, many lipid abnormalities, including alterations in the levels of NEFA and PC species in gray and white matter and an increase in Cer content in white matter, were found in the patients [105]. Along the same line of findings, elevated arachidonic acid-containing PC species were identified in brain tissue samples of a mouse model of depression [106].

Lipidomics was employed to determine the drug efficacy to psychosis [107, 108]. It was found that treatment of the patients with antipsychotic medication led to significant changes in serum lipidomic profiles in as little as 2 weeks of medication use. The investigators found different changes of serum lipids after treatment with different antipsychotic drugs. Intriguingly, these lipidomic findings were in good agreement with the studies of gene expression from which it was found that antipsychotic drugs strongly activated the genes involved in lipid metabolism [109, 110].

Owing to the inherent complications and challenges associated with the investigation of psychotic disorders, it is urgent to develop sensitive biomarkers for identification of individuals at very high risk of developing psychosis and for developing efficient preventive treatment for them, and for the early prediction of treatment response. To this end, it appears that the lipidomic studies might be able to identify lipid species, which have diagnostic potential in psychiatry, both as markers sensitive to disease progression and outcomes and as markers for predicting the treatment response [13].

17.7 CANCER

Although cancer in general is commonly described as a genetic disease, recent studies showed that tumorigenic phenotypes result from a host of mutational events that combine to alter multiple signaling pathways, both intrinsic and extrinsic, which converge to alter core cellular metabolism and provide support for the three basic needs of dividing cells: rapid ATP generation to maintain energy status; increased biosynthesis of macromolecules; and tightened maintenance of appropriate cellular redox status. Lipids play many key roles in all of these basic processes essential for tumor development [111] (also see Chapter 16). As examples, NEFA are the major building blocks for lipid biosynthesis and remodeling; cholesterol, GPL classes, and sphingolipids represent the major structural components of cellular membranes; TAG serves as the energy storage depot, which, along with acyl CoA and acylcarnitine, is involved in energy metabolism and ATP production; and CL species facilitate mitochondrial respiration chain efficiency. Bioactive lipids play important roles in signaling, functioning as second messengers and as hormones in cancer cells. For example, lysoGPL species are involved in cell proliferation, survival, and migration through the regulation of G-protein-coupled receptors [112]. Aberrant production of lysoPA can contribute to cancer initiation and progression [113]. The levels of NEFA arc associated with lipid hormone synthesis, affecting tumor-promoting signaling processes [114]. Hydrolysis products of PI and its phosphorylated derivatives (e.g., PIP, PIP_2, and PIP_3) or themselves are second messengers and cellular regulators to activate the PI_3K/AKT signaling pathway [115]. The significance of this pathway in chemotherapy and radiotherapy for human cancers is well recognized [115]. Changes in lipid metabolism affect numerous cellular processes, including cell growth, proliferation, differentiation, and motility important in tumorigenesis. Accordingly, it is not surprising that the cancer cells undergo profound changes in lipid metabolism, thus offering new diagnostic and therapeutic opportunities that could be unraveled by lipidomics.

The lipid profiles of body fluids reflect the general condition of the whole organism, which naturally serve as the resource of biomarkers predictive of the existence of different cancers. In addition to identification of novel biomarkers for early diagnosis of a certain type of cancers, qualitative and quantitative assessments of lipids in the blood and other body fluids may be also useful in monitoring the efficacy and toxicity of anticancer treatment.

A series of lipidomics studies related to cancers were conducted in recent years, such as prostate cancer [116, 117], breast cancer [118–120], hepatic carcinoma [121, 122], renal cell carcinoma [123], thyroid papillary cancer [124], and colorectal cancer [125]. Regarding the detail pictures of lipid molecular species, a few rather conclusive points could be made with respect to serving as potential biomarkers. LysoGPL species were linked to ovarian cancer [126]. Urinary GPL content may be useful for the diagnosis of prostate cancer [116]. GPL classes were also implicated in breast cancer, with marked differences in the expression of certain species as determined in both tissue and cell lines [118, 119]. Increased levels of eicosanoids were tightly linked to cancer progression and underlying inflammation [127–129]. To this end, the anti-inflammatory protective properties of n-3 PUFA have great interests in being explored as chemotherapeutic agents, formation of anti-inflammatory eicosanoids and docosanoids leading to reduced tumourigenesis appears promising in cases such as liver, neural, and colorectal cancers [130–132]. Available lipidomics data are very promising in determining novel biomarkers, but large cohort studies involving multiple centers are necessary to finalize the findings from these small-scale studies.

In a largest lipidomics study in cancer to date, 267 human breast tumor tissue samples covering different degrees of malignancy as well as normal breast tissues were examined [118]. It was found that the levels of GPL classes, predominantly PC species, containing those *de novo* synthesized fatty acids such as palmitate and myristate, were increased in tumors relative to normal controls and were also associated with cancer progression and patient survival. Further studies confirmed that the genes associated with FA biosynthesis such as stearoyl CoA desaturase and FA synthase were overexpressed in breast cancer, which could affect the cancer cell viability [118]. Another lipidomics study on breast tumor samples by MALDI-MS [133] determined an increase in the levels of PC(32:0), PC(34:1), and PC(36:2), which assume largely containing pamitate component. Thus, these two studies are well consistent.

In another study, lipidomics was employed to determine lipid profiles in fresh-frozen tissues and plasma samples from prostate cancer patients and controls [117]. A large number of lipid species were found changed under cancer conditions in comparison to the controls. Specifically, the levels of 78 plasma lipid species increased and 27 plasma species decreased; the levels of 56 lipid species from tissue samples were elevated, whereas the contents of 12 lipid species were reduced. The levels of lysoPC species increased in both plasma and cancer tissues of patients compared to the controls. Ceramide phosphoethanolamine decreased significantly in cancer tissues. Urinary GPLs from prostate cancer patients were also determined by lipidomics [116]. Significant differences of one PC, one PE, six PS, and two PI species between controls and patients were demonstrated. The results suggest that lipid alterations might serve as potential biomarkers for diagnosis of prostate cancer.

17.8   LIPIDOMICS IN NUTRITION

Because the lipidome is sensitive to many pathogenically relevant factors such as host genotype (i.e., metabolic pathways), gut microbiota, and diet, the accurate measurement of all lipids within an individual biofluid, tissue, or cell type positions lipidomics with the potential to revolutionize nutrition research. It is well known that diet is directly associated with "metabolic syndrome" [134, 135]. However, the connections of specific diets with health outcomes remain elusive. Metabolic regulation is different from person to person and at different ages. An optimal diet for one individual is not necessarily idea for the other. Accordingly, the primary goal of nutrition research is to understand the link between specific diets and health outcomes and to optimize dietary nutrition for human well-being, thereby delaying or preventing diet-related diseases. Lipidomics is therefore considered a powerful platform to study the contributions of genes, diet, nutrients, and human metabolism for health and disease.

Specifically, lipidomics is important for nutritional research in the following areas [136]:

- Monitoring individual nutritional status
- Follow-up of compliance, progress, and success of dietary guidance and intervention
- Identification of side effects, unexpected metabolic responses, or lack of response to specific dietary changes
- Recognition of metabolic shifts in individuals due to environmental changes or lifestyle modifications
- Normal progression of aging and maturation
- Applying for food research, such as development of food products and evaluation of food quality, functionality, bioactivity, and toxicity.

17.8.1 Lipidomics in Determination of the Effects of Specific Diets or Challenge Tests

Lipidomics was widely applied to determine the changes of lipidomes in response to caloric restriction. For example, it was demonstrated that modest caloric restriction results in accelerated GPL hydrolysis, membrane remodeling, and TAG accumulation in mouse heart [137]. Specifically, after a brief period of fasting (i.e., 4 and 12 h), substantial decreases in the choline and ethanolamine GPL pools in the heart were observed (a total of 39 nmol GPL/mg protein after 12-h fasting, representing ~25% of total GPL content). The same study also showed that shortening FA chain length was manifest in the major GPL classes after fasting, which further reduced endogenous myocardial energy stored in GPL FA chains, leading to changes in the physical properties of myocardial membranes. In contrast to reduction of GPL content, it was found that the TAG content was not decreased during fasting, but increased nearly three-fold after 12-h re-feeding. In contrast to the heart, no changes in GPL content were found, but a dramatic decrease in TAG mass occurred in the skeletal muscle after 12-h fasting. These results indicate that GPL classes could serve as a rapidly mobilizable energy source during modest caloric deprivation in mouse heart while the TAG pool is the major source of energy reserve in the skeletal muscle. Significantly larger differences of fasting-to-postprandial plasma acylcarnitine concentrations were detected in individuals with 6-month caloric restriction compared to the controls [138]. The study also revealed that the observed differences are related to improvements in insulin sensitivity [138]. In contrast to the peripheral system, alteration in ST homeostasis is the only change in mouse cerebral cortex after chronic caloric restriction as assessed by shotgun lipidomics [139].

Lipidomic profiles were determined for subjects with myocardial infarction or unstable ischemic attack after dietary intervention including either fatty fish or lean fish for 8 weeks [24, 140]. It was found that numerous lipid species, including Cer, lysoPC, lysoPE, DAG, and PC, were significantly decreased in the group fed with the fatty fish diet; whereas cholesteryl esters and specific long-chain TAG species are significantly increased in the group fed with the lean fish diet [24]. The decrease in lysoPC and Cer contents in the fatty fish group might be related to the anti-inflammatory effects of n-3 PUFA, since both lysoPC and Cer are the major bioactive lipid components involving inflammation [141].

Lipidomics profiling was used to determine the effects of plant sterols on lipid metabolism [142] since it is well known that plant sterol intervention could lead to reduction of both total cholesterol and LDL cholesterol in human [143]. In the study, two plant sterol-enriched yogurt drinks containing different fat content were given to healthy mildly hypercholesterolemic subjects for 4 weeks and then lipidomic analysis of serum samples were performed [142]. It was found that both drinks resulted in reduction of both total and LDL cholesterol levels [144]. In addition, the low-fat drink resulted in a reduction of the levels of several SM species. This reduction, which correlates well with the reduction of LDL cholesterol, could be due to the co-localization of SM and cholesterol on the surface layer of LDL lipoproteins.

Lipidomics was also used to determine the effects of probiotic intervention on global lipidomic profiles in humans [145]. Decreased mass levels of lysoPC, SM, and several PC species were detected after exposure of probiotic *Lactobacillus rhamnosus* GG bacteria for 3 weeks. These changes may be due to the metabolic events behind the beneficial effects of *L. rhamnosus* GG on gut barrier function [146].

17.8.2 Lipidomics to Control Food Quality

Lipidomics is also a powerful tool for controlling food quality and detecting fraud in food products. The specificity and sensitivity of MS-based methods is officially recognized by international quality system control agencies for detection of fraud and bad practices in food manipulation [147]. The term "foodomics" was coined to define studies in the food and nutrition domains through the application of advanced "omics" technologies to improve the health and well-being of consumers [148]. The MS-based strategies for foodomics have recently been reviewed [149].

Lipid species, particularly TAG species, present in oils and fats are important constituents of the human diet, largely depending on the degree of FA saturation. To this end, both MDMS-SL and multistage MS/MS approaches could be used to characterize the saturation and regiospecificity of those species in complex mixtures [150–153]. Examples in this area of research include the following: stereospecific analysis of FA in TAG found in olive oil and adzuki beans [154, 155]; analysis of GPL markers of milk quality [156]; imaging of rice varieties [157]; quantitation of GPLs and esterified fatty acids in avocados, eggs, various meats, and fish oils with or without nutritional supplementation of the animals producing these food stuffs [158–160]; and assessment of lysoGPL content in soy protein isolate for infant formulas [161]. Furthermore, analysis of SM species in human breast milk revealed the distribution of fatty acids in this lipid class, whereas increased levels of n-3 PUFA were found in human milk following fish oil nutritional supplementation [162].

Changes in GPL composition during storage are one of the most important changes affecting the freshness of fish, with oxidation and hydrolysis of GPL species being the main reasons for quality deterioration. Shotgun lipidomics was used for effective analyses of GPL species from fish muscle stored at room temperature [163]. For example, some PE species that are present in low concentrations in fresh samples are increased during storage. The investigators suggested that those species may come from microbiome breeding in the muscle, a phenomenon that has not been identified previously [163], implying its potential relevance as a marker of fish quality. MALDI-TOF/MS was also applied for the characterization of crude lipid extracts of several dietary products including cow milk, soymilk, and hen egg [164]. Finally, lipidomics can elucidate GPL FA composition in the food pathogen *Listeria monocytogenes* [165]. Such results could help to identify markers for monitoring the growth of this bacterium and lead to improved approaches to food preservation.

REFERENCES

1. Han, X. (2005) Lipid alterations in the earliest clinically recognizable stage of Alzheimer's disease: Implication of the role of lipids in the pathogenesis of Alzheimer's disease. Curr. Alzheimer Res. 2, 65–77.

2. Puri, P., Baillie, R.A., Wiest, M.M., Mirshahi, F., Choudhury, J., Cheung, O., Sargeant, C., Contos, M.J. and Sanyal, A.J. (2007) A lipidomic analysis of nonalcoholic fatty liver disease. Hepatology 46, 1081–1090.

3. Oresic, M., Hyotylainen, T., Kotronen, A., Gopalacharyulu, P., Nygren, H., Arola, J., Castillo, S., Mattila, I., Hakkarainen, A., Borra, R.J., Honka, M.J., Verrijken, A., Francque, S., Iozzo, P., Leivonen, M., Jaser, N., Juuti, A., Sorensen, T.I., Nuutila, P., Van Gaal, L. and Yki-Jarvinen, H. (2013) Prediction of non-alcoholic fatty-liver disease and liver fat content by serum molecular lipids. Diabetologia 56, 2266–2274.

4. Oresic, M., Seppanen-Laakso, T., Sun, D., Tang, J., Therman, S., Viehman, R., Mustonen, U., van Erp, T.G., Hyotylainen, T., Thompson, P., Toga, A.W., Huttunen, M.O., Suvisaari, J., Kaprio, J., Lonnqvist, J. and Cannon, T.D. (2012) Phospholipids and insulin resistance in psychosis: A lipidomics study of twin pairs discordant for schizophrenia. Genome Med. 4, 1.

5. Berger, G.E., Smesny, S. and Amminger, G.P. (2006) Bioactive lipids in schizophrenia. Int. Rev. Psychiatry 18, 85–98.

6. Fernandis, A.Z. and Wenk, M.R. (2009) Lipid based biomarkers for cancer. J. Chromatogr. B 877, 2830–2835.

7. Ekroos, K., Janis, M., Tarasov, K., Hurme, R. and Laaksonen, R. (2010) Lipidomics: A tool for studies of atherosclerosis. Curr. Atheroscler. Rep. 12, 273–281.

8. Smilowitz, J.T., Zivkovic, A.M., Wan, Y.J., Watkins, S.M., Nording, M.L., Hammock, B.D. and German, J.B. (2013) Nutritional lipidomics: Molecular metabolism, analytics, and diagnostics. Mol. Nutr. Food Res. 57, 1319–1335.

9. Stock, J. (2012) The emerging role of lipidomics. Atherosclerosis 221, 38–40.

10. Stegemann, C., Drozdov, I., Shalhoub, J., Humphries, J., Ladroue, C., Didangelos, A., Baumert, M., Allen, M., Davies, A.H., Monaco, C., Smith, A., Xu, Q. and Mayr, M. (2011) Comparative lipidomics profiling of human atherosclerotic plaques. Circ. Cardiovasc. Genet. 4, 232–242.

11. Watson, A.D. (2006) Thematic review series: Systems biology approaches to metabolic and cardiovascular disorders. Lipidomics: A global approach to lipid analysis in biological systems. J. Lipid Res. 47, 2101–2111.

12. Trushina, E. and Mielke, M.M. (2014) Recent advances in the application of metabolomics to Alzheimer's disease. Biochim. Biophys. Acta 1842, 1232–1239.

13. Hyotylainen, T. and Oresic, M. (2014) Systems biology strategies to study lipidomes in health and disease. Prog. Lipid Res. 55, 43–60.

14. Li, M., Yang, L., Bai, Y. and Liu, H. (2014) Analytical methods in lipidomics and their applications. Anal. Chem. 86, 161–175.

15. Murphy, S.A. and Nicolaou, A. (2013) Lipidomics applications in health, disease and nutrition research. Mol. Nutr. Food Res. 57, 1336–1346.

16. Han, X. and Zhou, Y. (2014) Application of lipidomics in nutritional research. In: Metabolomics as a tool in nutritional research (Sebedio, J.-L., Brennan, L. eds.). pp. 63–84, Elsevier Ltd, Cambridge, UK.

17. Kolovou, G., Kolovou, V. and Mavrogeni, S. (2015) Lipidomics in vascular health: Current perspectives. Vasc. Health Risk Manag. 11, 333–342.

18. Dehairs, J., Derua, R., Rueda-Rincon, N. and Swinnen, J.V. (2015) Lipidomics in drug development. Drug Discov. Today Technol. 13, 33–38.

19. Dawson, G. (2015) Measuring brain lipids. Biochim. Biophys. Acta 1851, 1026–1039.

20. Pietilainen, K.H., Rog, T., Seppanen-Laakso, T., Virtue, S., Gopalacharyulu, P., Tang, J., Rodriguez-Cuenca, S., Maciejewski, A., Naukkarinen, J., Ruskeepaa, A.L., Niemela, P.S., Yetukuri, L., Tan, C.Y., Velagapudi, V., Castillo, S., Nygren, H., Hyotylainen, T., Rissanen, A., Kaprio, J., Yki-Jarvinen, H., Vattulainen, I., Vidal-Puig, A. and Oresic, M. (2011) Association of lipidome remodeling in the adipocyte membrane with acquired obesity in humans. PLoS Biol. 9, e1000623.

21. Murphy, R.C. (2001) Free-radical-induced oxidation of arachidonoyl plasmalogen phospholipids: Antioxidant mechanism and precursor pathway for bioactive eicosanoids. Chem. Res. Toxicol. 14, 463–472.

22. Schmitz, G. and Ecker, J. (2008) The opposing effects of n-3 and n-6 fatty acids. Prog. Lipid Res. 47, 147–155.

23. Pietilainen, K.H., Sysi-Aho, M., Rissanen, A., Seppanen-Laakso, T., Yki-Jarvinen, H., Kaprio, J. and Oresic, M. (2007) Acquired obesity is associated with changes in the serum lipidomic profile independent of genetic effects - a monozygotic twin study. PLoS One 2, e218.

24. Lankinen, M., Schwab, U., Erkkila, A., Seppanen-Laakso, T., Hannila, M.L., Mussalo, H., Lehto, S., Uusitupa, M., Gylling, H. and Oresic, M. (2009) Fatty fish intake decreases lipids related to inflammation and insulin signaling--a lipidomics approach. PLoS One 4, e5258.

25. McCombie, G., Browning, L.M., Titman, C.M., Song, M., Shockcor, J., Jebb, S.A. and Griffin, J.L. (2009) omega-3 oil intake during weight loss in obese women results in remodelling of plasma triglyceride and fatty acids. Metabolomics 5, 363–374.

26. Han, X., Yang, J., Cheng, H., Yang, K., Abendschein, D.R. and Gross, R.W. (2005) Shotgun lipidomics identifies cardiolipin depletion in diabetic myocardium linking altered substrate utilization with mitochondrial dysfunction. Biochemistry 44, 16684–16694.

27. Han, X., Yang, J., Yang, K., Zhao, Z., Abendschein, D.R. and Gross, R.W. (2007) Alterations in myocardial cardiolipin content and composition occur at the very earliest stages of diabetes: A shotgun lipidomics study. Biochemistry 46, 6417–6428.

28. He, Q. and Han, X. (2014) Cardiolipin remodeling in diabetic heart. Chem. Phys. Lipids 179, 75–81.

29. Schlame, M., Ren, M., Xu, Y., Greenberg, M.L. and Haller, I. (2005) Molecular symmetry in mitochondrial cardiolipins. Chem. Phys. Lipids 138, 38–49.

30. Kinsey, G.R., McHowat, J., Beckett, C.S. and Schnellmann, R.G. (2007) Identification of calcium-independent phospholipase A2gamma in mitochondria and its role in mitochondrial oxidative stress. Am. J. Physiol. Renal Physiol. 292, F853–F860.

31. Gadd, M.E., Broekemeier, K.M., Crouser, E.D., Kumar, J., Graff, G. and Pfeiffer, D.R. (2006) Mitochondrial iPLA2 activity modulates the release of cytochrome c from mitochondria and influences the permeability transition. J. Biol. Chem. 281, 6931–6939.

32. Mancuso, D.J., Sims, H.F., Yang, K., Kiebish, M.A., Su, X., Jenkins, C.M., Guan, S., Moon, S.H., Pietka, T., Nassir, F., Schappe, T., Moore, K., Han, X., Abumrad, N.A. and Gross, R.W. (2010) Genetic ablation of calcium-independent phospholipase A2gamma prevents obesity and insulin resistance during high fat feeding by mitochondrial uncoupling and increased adipocyte fatty acid oxidation. J. Biol. Chem. 285, 36495–36510.

33. Xu, H., Barnes, G.T., Yang, Q., Tan, G., Yang, D., Chou, C.J., Sole, J., Nichols, A., Ross, J.S., Tartaglia, L.A. and Chen, H. (2003) Chronic inflammation in fat plays a crucial role in the development of obesity-related insulin resistance. J. Clin. Invest. 112, 1821–1830.

34. Oresic, M., Simell, S., Sysi-Aho, M., Nanto-Salonen, K., Seppanen-Laakso, T., Parikka, V., Katajamaa, M., Hekkala, A., Mattila, I., Keskinen, P., Yetukuri, L., Reinikainen, A., Lahde, J., Suortti, T., Hakalax, J., Simell, T., Hyoty, H., Veijola, R., Ilonen, J., Lahesmaa, R., Knip, M. and Simell, O. (2008) Dysregulation of lipid and amino acid metabolism precedes islet autoimmunity in children who later progress to type 1 diabetes. J. Exp. Med. 205, 2975–2984.

35. Oresic, M., Gopalacharyulu, P., Mykkanen, J., Lietzen, N., Makinen, M., Nygren, H., Simell, S., Simell, V., Hyoty, H., Veijola, R., Ilonen, J., Sysi-Aho, M., Knip, M., Hyotylainen, T. and Simell, O. (2013) Cord serum lipidome in prediction of islet autoimmunity and type 1 diabetes. Diabetes 62, 3268–3274.

36. La Torre, D., Seppanen-Laakso, T., Larsson, H.E., Hyotylainen, T., Ivarsson, S.A., Lernmark, A. and Oresic, M. (2013) Decreased cord-blood phospholipids in young age-at-onset type 1 diabetes. Diabetes 62, 3951–3956.

37. Serhan, C.N., Hong, S., Gronert, K., Colgan, S.P., Devchand, P.R., Mirick, G. and Moussignac, R.L. (2002) Resolvins: A family of bioactive products of omega-3 fatty acid transformation circuits initiated by aspirin treatment that counter proinflammation signals. J. Exp. Med. 196, 1025–1037.

38. Mas, E., Croft, K.D., Zahra, P., Barden, A. and Mori, T.A. (2012) Resolvins D1, D2, and other mediators of self-limited resolution of inflammation in human blood following n-3 fatty acid supplementation. Clin. Chem. 58, 1476–1484.

39. Balvers, M.G., Verhoeckx, K.C., Bijlsma, S., Rubingh, C.M., Meijerink, J., Wortelboer, H.M. and Witkamp, R.F. (2012) Fish oil and inflammatory status alter the n-3 to n-6 balance of the endocannabinoid and oxylipin metabolomes in mouse plasma and tissues. Metabolomics 8, 1130–1147.

40. Meikle, P.J., Wong, G., Tsorotes, D., Barlow, C.K., Weir, J.M., Christopher, M.J., MacIntosh, G.L., Goudey, B., Stern, L., Kowalczyk, A., Haviv, I., White, A.J., Dart, A.M., Duffy, S.J., Jennings, G.L. and Kingwell, B.A. (2011) Plasma lipidomic analysis of stable and unstable coronary artery disease. Arterioscler. Thromb. Vasc. Biol. 31, 2723–2732.

41. Han, X., Abendschein, D.R., Kelley, J.G. and Gross, R.W. (2000) Diabetes-induced changes in specific lipid molecular species in rat myocardium. Biochem. J. 352, 79–89.

42. Su, X., Han, X., Mancuso, D.J., Abendschein, D.R. and Gross, R.W. (2005) Accumulation of long-chain acylcarnitine and 3-hydroxy acylcarnitine molecular species in diabetic myocardium: Identification of alterations in mitochondrial fatty acid processing in diabetic myocardium by shotgun lipidomics. Biochemistry 44, 5234–5245.

43. Graessler, J., Schwudke, D., Schwarz, P.E., Herzog, R., Shevchenko, A. and Bornstein, S.R. (2009) Top-down lipidomics reveals ether lipid deficiency in blood plasma of hypertensive patients. PLoS One 4, e6261.

44. Hu, C., Kong, H., Qu, F., Li, Y., Yu, Z., Gao, P., Peng, S. and Xu, G. (2011) Application of plasma lipidomics in studying the response of patients with essential hypertension to antihypertensive drug therapy. Mol. Biosyst. 7, 3271–3279.

45. Stubiger, G., Aldover-Macasaet, E., Bicker, W., Sobal, G., Willfort-Ehringer, A., Pock, K., Bochkov, V., Widhalm, K. and Belgacem, O. (2012) Targeted profiling of atherogenic phospholipids in human plasma and lipoproteins of hyperlipidemic patients using MALDI-QIT-TOF-MS/MS. Atherosclerosis 224, 177–186.

46. Angulo, P. (2002) Nonalcoholic fatty liver disease. N. Engl. J. Med. 346, 1221–1231.

47. Wieckowska, A. and Feldstein, A.E. (2005) Nonalcoholic fatty liver disease in the pediatric population: A review. Curr. Opin. Pediatr. 17, 636–641.

48. Brunt, E.M. and Tiniakos, D.G. (2005) Pathological features of NASH. Front. Biosci. 10, 1475–1484.

49. Browning, J.D. and Horton, J.D. (2004) Molecular mediators of hepatic steatosis and liver injury. J. Clin. Invest. 114, 147–152.

50. Cheung, O. and Sanyal, A.J. (2008) Abnormalities of lipid metabolism in nonalcoholic fatty liver disease. Semin. Liver Dis. 28, 351–359.

51. Kumashiro, N., Erion, D.M., Zhang, D., Kahn, M., Beddow, S.A., Chu, X., Still, C.D., Gerhard, G.S., Han, X., Dziura, J., Petersen, K.F., Samuel, V.T. and Shulman, G.I. (2011) Cellular mechanism of insulin resistance in nonalcoholic fatty liver disease. Proc. Natl. Acad. Sci. U. S. A. 108, 16381–16385.

52. Kotronen, A., Peltonen, M., Hakkarainen, A., Sevastianova, K., Bergholm, R., Johansson, L.M., Lundbom, N., Rissanen, A., Ridderstrale, M., Groop, L., Orho-Melander, M. and Yki-Jarvinen, H. (2009) Prediction of non-alcoholic fatty liver disease and liver fat using metabolic and genetic factors. Gastroenterology 137, 865–872.

53. Rhee, E.P., Cheng, S., Larson, M.G., Walford, G.A., Lewis, G.D., McCabe, E., Yang, E., Farrell, L., Fox, C.S., O'Donnell, C.J., Carr, S.A., Vasan, R.S., Florez, J.C., Clish, C.B., Wang, T.J. and Gerszten, R.E. (2011) Lipid profiling identifies a triacylglycerol signature of insulin resistance and improves diabetes prediction in humans. J. Clin. Invest. 121, 1402–1411.

54. Kotronen, A., Velagapudi, V.R., Yetukuri, L., Westerbacka, J., Bergholm, R., Ekroos, K., Makkonen, J., Taskinen, M.R., Oresic, M. and Yki-Jarvinen, H. (2009) Serum saturated fatty acids containing triacylglycerols are better markers of insulin resistance than total serum triacylglycerol concentrations. Diabetologia 52, 684–690.

55. Adams, L.A., Waters, O.R., Knuiman, M.W., Elliott, R.R. and Olynyk, J.K. (2009) NAFLD as a risk factor for the development of diabetes and the metabolic syndrome: An eleven-year follow-up study. Am. J. Gastroenterol. 104, 861–867.

56. Donnelly, K.L., Smith, C.I., Schwarzenberg, S.J., Jessurun, J., Boldt, M.D. and Parks, E.J. (2005) Sources of fatty acids stored in liver and secreted via lipoproteins in patients with nonalcoholic fatty liver disease. J. Clin. Invest. 115, 1343–1351.

57. McClain, C.J., Barve, S. and Deaciuc, I. (2007) Good fat/bad fat. Hepatology 45, 1343–1346.

58. Waldemar, G., Dubois, B., Emre, M., Georges, J., McKeith, I.G., Rossor, M., Scheltens, P., Tariska, P. and Winblad, B. (2007) Recommendations for the diagnosis and management of Alzheimer's disease and other disorders associated with dementia: EFNS guideline. Eur. J. Neurol. 14, e1–e26.

59. Montine, T.J., Phelps, C.H., Beach, T.G., Bigio, E.H., Cairns, N.J., Dickson, D.W., Duyckaerts, C., Frosch, M.P., Masliah, E., Mirra, S.S., Nelson, P.T., Schneider, J.A., Thal, D.R., Trojanowski, J.Q., Vinters, H.V. and Hyman, B.T. (2012) National Institute on Aging-Alzheimer's Association guidelines for the neuropathologic assessment of Alzheimer's disease: A practical approach. Acta Neuropathol. 123, 1–11.

60. Galimberti, D., Scarpini, E. (2011) Disease-modifying treatments for Alzheimer's disease. Ther. Adv. Neurol. Disord. 4, 203–216.

61. Mangialasche, F., Solomon, A., Winblad, B., Mecocci, P. and Kivipelto, M. (2010) Alzheimer's disease: Clinical trials and drug development. Lancet Neurol. 9, 702–716.

62. Salomone, S., Caraci, F., Leggio, G.M., Fedotova, J. and Drago, F. (2012) New pharmacological strategies for treatment of Alzheimer's disease: Focus on disease modifying drugs. Br. J. Clin. Pharmacol. 73, 504–517.

63. Fonteh, A.N., Harrington, R.J., Huhmer, A.F., Biringer, R.G., Riggins, J.N. and Harrington, M.G. (2006) Identification of disease markers in human cerebrospinal fluid using lipidomic and proteomic methods. Dis. Markers 22, 39–64.

64. Han, X. (2010) Multi-dimensional mass spectrometry-based shotgun lipidomics and the altered lipids at the mild cognitive impairment stage of Alzheimer's disease. Biochim. Biophys. Acta 1801, 774–783.

65. Wood, P.L. (2012) Lipidomics of Alzheimer's disease: Current status. Alzheimer's Res. Ther. 4, 5.

66. Touboul, D. and Gaudin, M. (2014) Lipidomics of Alzheimer's disease. Bioanalysis 6, 541–561.

67. Svennerholm, L. and Gottfries, C.G. (1994) Membrane lipids, selectively diminished in Alzheimer brains, suggest synapse loss as a primary event in early-onset form (type I) and demyelination in late-onset form (type II). J. Neurochem. 62, 1039–1047.

68. Roth, G.S., Joseph, J.A., Mason and R.P. (1995) Membrane alterations as causes of impaired signal transduction in Alzheimer's disease and aging. Trends Neurosci. 18, 203–206.

69. Klein, J. (2000) Membrane breakdown in acute and chronic neurodegeneration: Focus on choline-containing phospholipids. J. Neural Transm. 107, 1027–1063.

70. Han, X., Holtzman, D.M., McKeel, D.W., Jr., Kelley, J. and Morris, J.C. (2002) Substantial sulfatide deficiency and ceramide elevation in very early Alzheimer's disease: Potential role in disease pathogenesis. J. Neurochem. 82, 809–818.

71. Cheng, H., Xu, J., McKeel, D.W., Jr. and Han, X. (2003) Specificity and potential mechanism of sulfatide deficiency in Alzheimer's disease: An electrospray ionization mass spectrometric study. Cell. Mol. Biol. 49, 809–818.

72. Han, X., Holtzman, D.M. and McKeel, D.W., Jr. (2001) Plasmalogen deficiency in early Alzheimer's disease subjects and in animal models: Molecular characterization using electrospray ionization mass spectrometry. J. Neurochem. 77, 1168–1180.

73. Han, X., Fagan, A.M., Cheng, H., Morris, J.C., Xiong, C. and Holtzman, D.M. (2003) Cerebrospinal fluid sulfatide is decreased in subjects with incipient dementia. Ann. Neurol. 54, 115–119.

74. Han, X., Cheng, H., Fryer, J.D., Fagan, A.M. and Holtzman, D.M. (2003) Novel role for apolipoprotein E in the central nervous system: Modulation of sulfatide content. J. Biol. Chem. 278, 8043–8051.

75. Cheng, H., Wang, M., Li, J.-L., Cairns, N.J. and Han, X. (2013) Specific changes of sulfatide levels in individuals with pre-clinical Alzheimer's disease: An early event in disease pathogenesis. J. Neurochem. 127, 733–738.

76. Han, X., Rozen, S., Boyle, S., Hellegers, C., Cheng, H., Burke, J.R., Welsh-Bohmer, K.A., Doraiswamy, P.M. and Kaddurah-Daouk, R. (2011) Metabolomics in early Alzheimer's disease: Identification of altered plasma sphingolipidome using shotgun lipidomics. PLoS One 6, e21643.

77. Satoi, H., Tomimoto, H., Ohtani, R., Kitano, T., Kondo, T., Watanabe, M., Oka, N., Akiguchi, I., Furuya, S., Hirabayashi, Y. and Okazaki, T. (2005) Astroglial expression of ceramide in Alzheimer's disease brains: A role during neuronal apoptosis. Neuroscience 130, 657–666.

78. Mielke, M.M., Haughey, N.J., Ratnam Bandaru, V.V., Schech, S., Carrick, R., Carlson, M.C., Mori, S., Miller, M.I., Ceritoglu, C., Brown, T., Albert, M. and Lyketsos, C.G. (2010) Plasma ceramides are altered in mild cognitive impairment and predict cognitive decline and hippocampal volume loss. Alzheimers Dement. 6, 378–385.

79. Mielke, M.M., Haughey, N.J., Bandaru, V.V., Weinberg, D.D., Darby, E., Zaidi, N., Pavlik, V., Doody, R.S. and Lyketsos, C.G. (2011) Plasma sphingomyelins are associated with cognitive progression in Alzheimer's disease. J. Alzheimers Dis. 27, 259–269.

80. Mielke, M.M., Bandaru, V.V., Haughey, N.J., Rabins, P.V., Lyketsos, C.G. and Carlson, M.C. (2010) Serum sphingomyelins and ceramides are early predictors of memory impairment. Neurobiol. Aging 31, 17–24.

81. Cutler, R.G., Kelly, J., Storie, K., Pedersen, W.A., Tammara, A., Hatanpaa, K., Troncoso, J.C. and Mattson, M.P. (2004) Involvement of oxidative stress-induced abnormalities in ceramide and cholesterol metabolism in brain aging and Alzheimer's disease. Proc. Natl. Acad. Sci. U. S. A. 101, 2070–2075.

82. Ginsberg, L., Rafique, S., Xuereb, J.H., Rapoport, S.I. and Gershfeld, N.L. (1995) Disease and anatomic specificity of ethanolamine plasmalogen deficiency in Alzheimer's disease brain. Brain Res. 698, 223–226.

83. Ginsberg, L., Xuereb, J.H. and Gershfeld, N.L. (1998) Membrane instability, plasmalogen content, and Alzheimer's disease. J. Neurochem. 70, 2533–2538.

84. Farooqui, A.A., Rapoport, S.I. and Horrocks, L.A. (1997) Membrane phospholipid alterations in Alzheimer's disease: Deficiency of ethanolamine plasmalogens. Neurochem. Res. 22, 523–527.

85. Goodenowe, D.B., Cook, L.L., Liu, J., Lu, Y., Jayasinghe, D.A., Ahiahonu, P.W., Heath, D., Yamazaki, Y., Flax, J., Krenitsky, K.F., Sparks, D.L., Lerner, A., Friedland, R.P., Kudo, T., Kamino, K., Morihara, T., Takeda, M. and Wood, P.L. (2007) Peripheral ethanolamine plasmalogen deficiency: A logical causative factor in Alzheimer's disease and dementia. J. Lipid Res. 48, 2485–2498.

86. Oresic, M., Hyotylainen, T., Herukka, S.K., Sysi-Aho, M., Mattila, I., Seppanan-Laakso, T., Julkunen, V., Gopalacharyulu, P.V., Hallikainen, M., Koikkalainen, J., Kivipelto, M., Helisalmi, S., Lotjonen, J. and Soininen, H. (2011) Metabolome in progression to Alzheimer's disease. Transl. Psychiatry 1, e57.

87. Wood, P.L., Smith, T., Lane, N., Khan, M.A., Ehrmantraut, G. and Goodenowe, D.B. (2011) Oral bioavailability of the ether lipid plasmalogen precursor, PPI-1011, in the rabbit: A new therapeutic strategy for Alzheimer's disease. Lipids Health Dis. 10, 227.

88. Markesbery, W.R. (1997) Oxidative stress hypothesis in Alzheimer's disease. Free Radic. Biol. Med. 23, 134–147.

89. Nitsch, R.M., Blusztajn, J.K., Pittas, A.G., Slack, B.E., Growdon, J.H. and Wurtman, R.J. (1992) Evidence for a membrane defect in Alzheimer disease brain. Proc. Natl. Acad. Sci. U. S. A. 89, 1671–1675.

90. Mapstone, M., Cheema, A.K., Fiandaca, M.S., Zhong, X., Mhyre, T.R., MacArthur, L.H., Hall, W.J., Fisher, S.G., Peterson, D.R., Haley, J.M., Nazar, M.D., Rich, S.A., Berlau, D.J., Peltz, C.B., Tan, M.T., Kawas, C.H. and Federoff, H.J. (2014) Plasma phospholipids identify antecedent memory impairment in older adults. Nat. Med. 20, 415–418.

91. Whiley, L., Sen, A., Heaton, J., Proitsi, P., Garcia-Gomez, D., Leung, R., Smith, N., Thambisetty, M., Kloszewska, I., Mecocci, P., Soininen, H., Tsolaki, M., Vellas, B., Lovestone, S. and Legido-Quigley, C. (2014) Evidence of altered phosphatidylcholine metabolism in Alzheimer's disease. Neurobiol. Aging 35, 271–278.

92. Hartmann, T., van Wijk, N., Wurtman, R.J., Olde Rikkert, M.G., Sijben, J.W., Soininen, H., Vellas, B. and Scheltens, P. (2014) A nutritional approach to ameliorate altered phospholipid metabolism in Alzheimer's disease. J. Alzheimers Dis. 41, 715–717.

93. Sato, Y., Suzuki, I., Nakamura, T., Bernier, F., Aoshima, K. and Oda, Y. (2012) Identification of a new plasma biomarker of Alzheimer's disease using metabolomics technology. J. Lipid Res. 53, 567–576.

94. Astarita, G. and Piomelli, D. (2011) Towards a whole-body systems [multi-organ] lipidomics in Alzheimer's disease. Prostaglandins Leukot. Essent. Fatty Acids 85, 197–203.

95. Stark, D.T. and Bazan, N.G. (2011) Synaptic and extrasynaptic NMDA receptors differentially modulate neuronal cyclooxygenase-2 function, lipid peroxidation, and neuroprotection. J. Neurosci. 31, 13710–13721.

96. Niemoller, T.D. and Bazan, N.G. (2010) Docosahexaenoic acid neurolipidomics. Prostaglandins Other Lipid Mediat. 91, 85–89.

97. Ji, R.R., Xu, Z.Z., Strichartz, G. and Serhan, C.N. (2011) Emerging roles of resolvins in the resolution of inflammation and pain. Trends Neurosci. 34, 599–609.

98. Yanagisawa, K. (2007) Role of gangliosides in Alzheimer's disease. Biochim. Biophys. Acta 1768, 1943–1951.

99. Valdes-Gonzalez, T., Goto-Inoue, N., Hirano, W., Ishiyama, H., Hayasaka, T., Setou, M. and Taki, T. (2011) New approach for glyco- and lipidomics--molecular scanning of human brain gangliosides by TLC-Blot and MALDI-QIT-TOF MS. J. Neurochem. 116, 678–683.

100. Horrobin, D.F. (1998) The membrane phospholipid hypothesis as a biochemical basis for the neurodevelopmental concept of schizophrenia. Schizophr. Res. 30, 193–208.

101. Amminger, G.P., Schafer, M.R., Papageorgiou, K., Klier, C.M., Cotton, S.M., Harrigan, S.M., Mackinnon, A., McGorry, P.D. and Berger, G.E. (2010) Long-chain omega-3 fatty acids for indicated prevention of psychotic disorders: A randomized, placebo-controlled trial. Arch. Gen. Psychiatry 67, 146–154.

102. Oresic, M., Tang, J., Seppanen-Laakso, T., Mattila, I., Saarni, S.E., Saarni, S.I., Lonnqvist, J., Sysi-Aho, M., Hyotylainen, T., Perala, J. and Suvisaari, J. (2011) Metabolome in schizophrenia and other psychotic disorders: A general population-based study. Genome Med. 3, 19.

103. De Marchi, N., De Petrocellis, L., Orlando, P., Daniele, F., Fezza, F. and Di Marzo, V. (2003) Endocannabinoid signalling in the blood of patients with schizophrenia. Lipids Health Dis. 2, 5.

104. Hill, M.N., Miller, G.E., Carrier, E.J., Gorzalka, B.B. and Hillard, C.J. (2009) Circulating endocannabinoids and N-acyl ethanolamines are differentially regulated in major depression and following exposure to social stress. Psychoneuroendocrinology 34, 1257–1262.

105. Schwarz, E., Prabakaran, S., Whitfield, P., Major, H., Leweke, F.M., Koethe, D., McKenna, P. and Bahn, S. (2008) High throughput lipidomic profiling of schizophrenia and bipolar disorder brain tissue reveals alterations of free fatty acids, phosphatidylcholines, and ceramides. J. Proteome Res. 7, 4266–4277.

106. Green, P., Anyakoha, N., Yadid, G., Gispan-Herman, I. and Nicolaou, A. (2009) Arachidonic acid-containing phosphatidylcholine species are increased in selected brain regions of a depressive animal model: Implications for pathophysiology. Prostaglandins Leukot. Essent. Fatty Acids 80, 213–220.

107. Kaddurah-Daouk, R., McEvoy, J., Baillie, R.A., Lee, D., Yao, J.K., Doraiswamy, P.M. and Krishnan, K.R. (2007) Metabolomic mapping of atypical antipsychotic effects in schizophrenia. Mol. Psychiatry 12, 934–945.

108. Kaddurah-Daouk, R., Bogdanov, M.B., Wikoff, W.R., Zhu, H., Boyle, S.H., Churchill, E., Wang, Z., Rush, A.J., Krishnan, R.R., Pickering, E., Delnomdedieu, M. and Fiehn, O. (2013) Pharmacometabolomic mapping of early biochemical changes induced by sertraline and placebo. Transl. Psychiatry 3, e223.

109. Ferno, J., Raeder, M.B., Vik-Mo, A.O., Skrede, S., Glambek, M., Tronstad, K.J., Breilid, H., Lovlie, R., Berge, R.K., Stansberg, C. and Steen, V.M. (2005) Antipsychotic drugs activate SREBP-regulated expression of lipid biosynthetic genes in cultured human glioma cells: A novel mechanism of action? Pharmacogenomics J. 5, 298–304.

110. Polymeropoulos, M.H., Licamele, L., Volpi, S., Mack, K., Mitkus, S.N., Carstea, E.D., Getoor, L., Thompson, A. and Lavedan, C. (2009) Common effect of antipsychotics on the biosynthesis and regulation of fatty acids and cholesterol supports a key role of lipid homeostasis in schizophrenia. Schizophr. Res. 108, 134–142.

111. Santos, C.R. and Schulze, A. (2012) Lipid metabolism in cancer. FEBS J. 279, 2610–2623.

112. Murph, M., Tanaka, T., Pang, J., Felix, E., Liu, S., Trost, R., Godwin, A.K., Newman, R. and Mills, G. (2007) Liquid chromatography mass spectrometry for quantifying plasma lysophospholipids: Potential biomarkers for cancer diagnosis. Methods Enzymol. 433, 1–25.

113. Mills, G.B. and Moolenaar, W.H. (2003) The emerging role of lysophosphatidic acid in cancer. Nat. Rev. Cancer 3, 582–591.

114. Nomura, D.K., Long, J.Z., Niessen, S., Hoover, H.S., Ng, S.W. and Cravatt, B.F. (2010) Monoacylglycerol lipase regulates a fatty acid network that promotes cancer pathogenesis. Cell 140, 49–61.

115. Fresno Vara, J.A., Casado, E., de Castro, J., Cejas, P., Belda-Iniesta, C. and Gonzalez-Baron, M. (2004) PI3K/Akt signalling pathway and cancer. Cancer Treat. Rev. 30, 193–204.

116. Min, H.K., Lim, S., Chung, B.C. and Moon, M.H. (2011) Shotgun lipidomics for candidate biomarkers of urinary phospholipids in prostate cancer. Anal. Bioanal. Chem. 399, 823–830.

117. Zhou, X., Mao, J., Ai, J., Deng, Y., Roth, M.R., Pound, C., Henegar, J., Welti, R. and Bigler, S.A. (2012) Identification of plasma lipid biomarkers for prostate cancer by lipidomics and bioinformatics. PLoS One 7, e48889.

118. Hilvo, M., Denkert, C., Lehtinen, L., Muller, B., Brockmoller, S., Seppanen-Laakso, T., Budczies, J., Bucher, E., Yetukuri, L., Castillo, S., Berg, E., Nygren, H., Sysi-Aho, M., Griffin, J.L., Fiehn, O., Loibl, S., Richter-Ehrenstein, C., Radke, C., Hyotylainen, T., Kallioniemi, O., Iljin, K. and Oresic, M. (2011) Novel theranostic opportunities offered by characterization of altered membrane lipid metabolism in breast cancer progression. Cancer Res. 71, 3236–3245.

119. Doria, M.L., Cotrim, Z., Macedo, B., Simoes, C., Domingues, P., Helguero, L. and Domingues, M.R. (2012) Lipidomic approach to identify patterns in phospholipid profiles and define class differences in mammary epithelial and breast cancer cells. Breast Cancer Res. Treat. 133, 635–648.

120. Denkert, C., Bucher, E., Hilvo, M., Salek, R., Oresic, M., Griffin, J., Brockmoller, S., Klauschen, F., Loibl, S., Barupal, D.K., Budczies, J., Iljin, K., Nekljudova, V. and Fiehn, O. (2012) Metabolomics of human breast cancer: New approaches for tumor typing and biomarker discovery. Genome Med. 4, 37.

121. Gorden, D.L., Ivanova, P.T., Myers, D.S., McIntyre, J.O., VanSaun, M.N., Wright, J.K., Matrisian, L.M. and Brown, H.A. (2011) Increased diacylglycerols characterize hepatic lipid changes in progression of human nonalcoholic fatty liver disease; comparison to a murine model. PLoS One 6, e22775.

122. Hou, W., Zhou, H., Bou Khalil, M., Seebun, D., Bennett, S.A. and Figeys, D. (2011) Lyso-form fragment ions facilitate the determination of stereospecificity of diacyl glycerophospholipids. Rapid Commun. Mass Spectrom. 25, 205–217.

123. Del Boccio, P., Raimondo, F., Pieragostino, D., Morosi, L., Cozzi, G., Sacchetta, P., Magni, F., Pitto, M. and Urbani, A. (2012) A hyphenated microLC-Q-TOF-MS platform for exosomal lipidomics investigations: Application to RCC urinary exosomes. Electrophoresis 33, 689–696.

124. Ishikawa, S., Tateya, I., Hayasaka, T., Masaki, N., Takizawa, Y., Ohno, S., Kojima, T., Kitani, Y., Kitamura, M., Hirano, S., Setou, M. and Ito, J. (2012) Increased expression of phosphatidylcholine (16:0/18:1) and (16:0/18:2) in thyroid papillary cancer. PLoS One 7, e48873.

125. Li, F., Qin, X., Chen, H., Qiu, L., Guo, Y., Liu, H., Chen, G., Song, G., Wang, X., Guo, S., Wang, B. and Li, Z. (2013) Lipid profiling for early diagnosis and progression of colorectal cancer using direct-infusion electrospray ionization Fourier transform ion cyclotron resonance mass spectrometry. Rapid Commun. Mass Spectrom. 27, 24–34.

126. Sutphen, R., Xu, Y., Wilbanks, G.D., Fiorica, J., Grendys, E.C., Jr., LaPolla, J.P., Arango, H., Hoffman, M.S., Martino, M., Wakeley, K., Griffin, D., Blanco, R.W., Cantor, A.B., Xiao, Y.J. and Krischer, J.P. (2004) Lysophospholipids are potential biomarkers of ovarian cancer. Cancer Epidemiol. Biomarkers Prev. 13, 1185–1191.

127. Freedman, R.S., Wang, E., Voiculescu, S., Patenia, R., Bassett, R.L., Jr., Deavers, M., Marincola, F.M., Yang, P. and Newman, R.A. (2007) Comparative analysis of peritoneum and tumor eicosanoids and pathways in advanced ovarian cancer. Clin. Cancer Res. 13, 5736–5744.

128. Chang, S.H., Liu, C.H., Conway, R., Han, D.K., Nithipatikom, K., Trifan, O.C., Lane, T.F. and Hla, T. (2004) Role of prostaglandin E2-dependent angiogenic switch in cyclooxygenase 2-induced breast cancer progression. Proc. Natl. Acad. Sci. U. S. A. 101, 591–596.

129. Mal, M., Koh, P.K., Cheah, P.Y. and Chan, E.C. (2011) Ultra-pressure liquid chromatography/tandem mass spectrometry targeted profiling of arachidonic acid and eicosanoids in human colorectal cancer. Rapid Commun. Mass Spectrom. 25, 755–764.

130. Gleissman, H., Yang, R., Martinod, K., Lindskog, M., Serhan, C.N., Johnsen, J.I. and Kogner, P. (2010) Docosahexaenoic acid metabolome in neural tumors: Identification of cytotoxic intermediates. FASEB J. 24, 906–915.

131. Hawcroft, G., Loadman, P.M., Belluzzi, A. and Hull, M.A. (2010) Effect of eicosapentaenoic acid on E-type prostaglandin synthesis and EP4 receptor signaling in human colorectal cancer cells. Neoplasia 12, 618–627.

132. Weylandt, K.H., Krause, L.F., Gomolka, B., Chiu, C.Y., Bilal, S., Nadolny, A., Waechter, S.F., Fischer, A., Rothe, M. and Kang, J.X. (2011) Suppressed liver tumorigenesis in fat-1 mice with elevated omega-3 fatty acids is associated with increased omega-3 derived lipid mediators and reduced TNF-alpha. Carcinogenesis 32, 897–903.

133. Kim, I.C., Lee, J.H., Bang, G., Choi, S.H., Kim, Y.H., Kim, K.P., Kim, H.K. and Ro, J. (2013) Lipid profiles for HER2-positive breast cancer. Anticancer Res 33, 2467–2472.

134. Moller, D.E. and Kaufman, K.D. (2005) Metabolic syndrome: A clinical and molecular perspective. Annu. Rev. Med. 56, 45–62.

135. Unger, R.H. and Scherer, P.E. (2010) Gluttony, sloth and the metabolic syndrome: A roadmap to lipotoxicity. Trends Endocrinol. Metab. 21, 345–352.

136. Stella, C., Beckwith-Hall, B., Cloarec, O., Holmes, E., Lindon, J.C., Powell, J., van der Ouderaa, F., Bingham, S., Cross, A.J. and Nicholson, J.K. (2006) Susceptibility of human metabolic phenotypes to dietary modulation. J. Proteome Res. 5, 2780–2788.

137. Han, X., Cheng, H., Mancuso, D.J. and Gross, R.W. (2004) Caloric restriction results in phospholipid depletion, membrane remodeling and triacylglycerol accumulation in murine myocardium. Biochemistry 43, 15584–15594.

138. Huffman, K.M., Redman, L.M., Landerman, L.R., Pieper, C.F., Stevens, R.D., Muehlbauer, M.J., Wenner, B.R., Bain, J.R., Kraus, V.B., Newgard, C.B., Ravussin, E. and Kraus, W.E. (2012) Caloric restriction alters the metabolic response to a mixed-meal: Results from a randomized, controlled trial. PLoS One 7, e28190.

139. Kiebish, M.A., Young, D.M., Lehman, J.J. and Han, X. (2012) Chronic caloric restriction attenuates a loss of sulfatide content in the PGC-1α-/- mouse cortex: A potential lipidomic role of PGC-1α in neurodegeneration. J. Lipid Res. 53, 273–281.

140. Schwab, U., Seppanen-Laakso, T., Yetukuri, L., Agren, J., Kolehmainen, M., Laaksonen, D.E., Ruskeepaa, A.L., Gylling, H., Uusitupa, M. and Oresic, M. (2008) Triacylglycerol fatty acid composition in diet-induced weight loss in subjects with abnormal glucose metabolism--the GENOBIN study. PLoS One 3, e2630.

141. Aiyar, N., Disa, J., Ao, Z., Ju, H., Nerurkar, S., Willette, R.N., Macphee, C.H., Johns, D.G. and Douglas, S.A. (2007) Lysophosphatidylcholine induces inflammatory activation of human coronary artery smooth muscle cells. Mol. Cell. Biochem. 295, 113–120.

142. Szymanska, E., van Dorsten, F.A., Troost, J., Paliukhovich, I., van Velzen, E.J., Hendriks, M.M., Trautwein, E.A., van Duynhoven, J.P., Vreeken, R.J. and Smilde, A.K. (2012) A lipidomic analysis approach to evaluate the response to cholesterol-lowering food intake. Metabolomics 8, 894–906.

143. De Smet, E., Mensink, R.P. and Plat, J. (2012) Effects of plant sterols and stanols on intestinal cholesterol metabolism: Suggested mechanisms from past to present. Mol. Nutr. Food Res. 56, 1058–1072.

144. Noakes, M., Clifton, P.M., Doornbos, A.M. and Trautwein, E.A. (2005) Plant sterol ester-enriched milk and yoghurt effectively reduce serum cholesterol in modestly hypercholesterolemic subjects. Eur. J. Nutr. 44, 214–222.

145. Kekkonen, R.A., Sysi-Aho, M., Seppanen-Laakso, T., Julkunen, I., Vapaatalo, H., Oresic, M. and Korpela, R. (2008) Effect of probiotic Lactobacillus rhamnosus GG intervention on global serum lipidomic profiles in healthy adults. World J. Gastroenterol. 14, 3188–3194.

146. Ng, S.C., Hart, A.L., Kamm, M.A., Stagg, A.J. and Knight, S.C. (2009) Mechanisms of action of probiotics: Recent advances. Inflamm. Bowel Dis. 15, 300–310.

147. Aiello, D., De Luca, D., Gionfriddo, E., Naccarato, A., Napoli, A., Romano, E., Russo, A., Sindona, G. and Tagarelli, A. (2011) Review: Multistage mass spectrometry in quality, safety and origin of foods. Eur. J. Mass Spectrom. 17, 1–31.

148. Cifuentes, A. (2009) Food analysis and foodomics. J. Chromatogr. A 1216, 7109.

149. Herrero, M., Simo, C., Garcia-Canas, V., Ibanez, E. and Cifuentes, A. (2012) Foodomics: MS-based strategies in modern food science and nutrition. Mass Spectrom. Rev. 31, 49–69.

150. Han, X., Yang, K. and Gross, R.W. (2012) Multi-dimensional mass spectrometry-based shotgun lipidomics and novel strategies for lipidomic analyses. Mass Spectrom. Rev. 31, 134–178.

151. Wang, M., Han, R.H. and Han, X. (2013) Fatty acidomics: Global analysis of lipid species containing a carboxyl group with a charge-remote fragmentation-assisted approach. Anal. Chem. 85, 9312–9320.

152. Hsu, F.-F. and Turk, J. (1999) Structural characterization of triacylglycerols as lithiated adducts by electrospray ionization mass spectrometry using low-energy collisionally activated dissociation on a triple stage quadrupole instrument. J. Am. Soc. Mass Spectrom. 10, 587–599.

153. Lin, J.T. and Arcinas, A. (2008) Analysis of regiospecific triacylglycerols by electrospray ionization-mass spectrometry(3) of lithiated adducts. J. Agric. Food Chem. 56, 4909–4915.

154. Vichi, S., Pizzale, L. and Conte, L.S. (2007) Stereospecific distribution of fatty acids in triacylglycerols of olive oils. Eur. J. Lipid Sci. Technol. 109, 72–78.

155. Yoshida, H., Tomiyama, Y., Yoshida, N., Shibata, K. and Mizushina, Y. (2010) Regiospecific profiles of fatty acids in triacylglycerols and phospholipids from Adzuki beans (Vigna angularis). Nutrients 2, 49–59.

156. Calvano, C.D., De Ceglie, C., Aresta, A., Facchini, L.A. and Zambonin, C.G. (2013) MALDI-TOF mass spectrometric determination of intact phospholipids as markers of illegal bovine milk adulteration of high-quality milk. Anal. Bioanal. Chem. 405, 1641–1649.

157. Zaima, N., Goto-Inoue, N., Hayasaka, T. and Setou, M. (2010) Application of imaging mass spectrometry for the analysis of Oryza sativa rice. Rapid Commun. Mass Spectrom. 24, 2723–2729.

158. Pacetti, D., Boselli, E., Lucci, P. and Frega, N.G. (2007) Simultaneous analysis of glycolipids and pholipids molecular species in avocado (Persea americana Mill) fruit. J. Chromatogr. A 1150, 241–251.

159. Zhou, L., Zhao, M., Ennahar, S., Bindler, F. and Marchioni, E. (2012) Liquid chromatography-tandem mass spectrometry for the determination of sphingomyelin species from calf brain, ox liver, egg yolk, and krill oil. J. Agric. Food Chem. 60, 293–298.

160. Pacetti, D., Boselli, E., Hulan, H.W. and Frega, N.G. (2005) High performance liquid chromatography-tandem mass spectrometry of phospholipid molecular species in eggs from hens fed diets enriched in seal blubber oil. J. Chromatogr. A 1097, 66–73.

161. Fang, N., Yu, S. and Badger, T.M. (2003) LC-MS/MS analysis of lysophospholipids associated with soy protein isolate. J. Agric. Food Chem. 51, 6676-6682.

162. Dunstan, J.A., Roper, J., Mitoulas, L., Hartmann, P.E., Simmer, K. and Prescott, S.L. (2004) The effect of supplementation with fish oil during pregnancy on breast milk immunoglobulin A, soluble CD14, cytokine levels and fatty acid composition. Clin. Exp. Allergy 34, 1237–1242.

163. Wang, Y. and Zhang, H. (2011) Tracking phospholipid profiling of muscle from Ctennopharyngodon idellus during storage by shotgun lipidomics. J. Agric. Food Chem. 59, 11635–11642.

164. Calvano, C.D., Carulli, S. and Palmisano, F. (2010) 1H-pteridine-2,4-dione (lumazine): A new MALDI matrix for complex (phospho)lipid mixtures analysis. Anal. Bioanal. Chem. 398, 499–507.

165. Mastronicolis, S.K., Arvanitis, N., Karaliota, A., Magiatis, P., Heropoulos, G., Litos, C., Moustaka, H., Tsakirakis, A., Paramera, E. and Papastavrou, P. (2008) Coordinated regulation of cold-induced changes in fatty acids with cardiolipin and phosphatidylglycerol composition among phospholipid species for the food pathogen *Listeria monocytogenes*. Appl. Environ. Microbiol. 74, 4543–4549.

18

PLANT LIPIDOMICS

18.1 INTRODUCTION

As the development of lipidomics field, mass spectrometric analysis of plant lipids has been advanced accordingly. The branch of lipidomics in analysis of plant cellular lipids utilizing mass spectrometry was specifically called "plant lipidomics" [1]. As described early in the book (Chapter 3), all kinds of MS-based lipidomics approaches, particularly those based on ESI-MS or ESI-MS/MS, could be used in plant lipidomics. The advantages of ESI-MS-based methods over "traditional" lipid analytical approaches for analysis of plant lipids were extensively discussed [1–3].

Both direct infusion-based shotgun lipidomics and LC-MS-based approaches are applicable in plant lipidomics. Product-ion analysis after CID is the essential approach for identification and characterization of plant lipids. Precursor-ion scan (PIS) and neutral-loss scan (NLS) are widely used to efficiently "filter" the individual molecular species of a lipid class or a category of lipid classes of interest, usually based on a common fragment such as the head group of the lipid class(es). Moreover, such kind of scanning could also serve as the preliminary scanning for further targeted analysis of the species by using an MRM approach after LC-MS. The majority of PIS and NLS for plant lipids are identical to those used for analysis of animal lipids [4, 5], but several additional scans are devised for plant-specific lipid classes [6–8].

From the analysis of PIS, NLS, and/or MRM, the intensities of the targeted lipid species are typically compared with those of one or more internal standards of the lipid class for quantification as previously described (Chapters 3 and 14).

Lipidomics: Comprehensive Mass Spectrometry of Lipids, First Edition. Xianlin Han.
© 2016 John Wiley & Sons, Inc. Published 2016 by John Wiley & Sons, Inc.

As described in Chapter 15, correction for different stable isotopologue distribution due to differences in the number of carbon atoms between the species of interest and the selected internal standard(s) and correction for baseline should be performed prior to the comparison of the intensities. It should be recognized that in the majority of the studies in plant lipidomics after direct infusion, MDMS-type analysis is generally not conducted; thus, the isobaric or isomeric species are not resolved as regrettable. A detailed protocol for profiling of polar lipids in plants after direct infusion can be found [9].

As development of plant lipidomics, high-throughput lipid profiling can be used to detect metabolic changes in response to developmental, environmental, and stress-induced physiological changes. Rapid analysis of individual lipid molecular species from normal and genetically altered cells or organisms can also provide the detailed information needed to elucidate the functions of genes affecting lipid metabolism, signaling, and homeostasis. Furthermore, lipidomic analyses can be exploited to serve as quality controls for the purpose of improving the quality of plant-based foods after genetic manipulations.

Different types of mass spectrometry-based lipidomics approaches (Chapters 2 and 3) are also exploited in plant lipidomics. For example, FTICR-MS has been employed in plant lipidomics to obtain chemical formulas of intact lipid ions [10] in taking the advantage of this instrument with high mass accuracy, good sensitivity, and simplicity in spectral analysis [11]. However, the disadvantage of this approach is the lack of structural identification; thus, additional data are required to identify specific compounds. Other types of mass spectrometers including QqQ [2] and QqTOF [12, 13] are commonly used in plant lipidomics with or without coupling with LC-MS. Direct infusion-based approaches were very successful for analyzing many of the plant lipid classes such as plant GPL classes and glycolipids [1, 9, 14].

In this chapter, a few lipid classes relatively specific to plant lipidomes in comparison to those of mammalian cellular lipidomes are first described. As described in Chapter 6, only the fragmentation patterns of individual lipid classes are summarized. Then, examples of the application of plant lipidomics for plant research are briefly given.

18.2 CHARACTERIZATION OF LIPIDS SPECIAL TO PLANT LIPIDOME

The majority of the plant lipids are similar to those present in the mammals. Thus, the fragmentation patterns used to identify and characterize the lipids described earlier (Part II), including GPLs, sphingolipids, glycerolipids, and bioactive lipids could also be applied for analysis of plant lipidomes. However, there exist a few subsets of lipid classes that are not or only minimally present in mammals such as some classes of galactolipids, sulfolipids, galactosyl group-containing sphingolipids, and plant sterols. In this section, the fragmentation characters of these special lipid classes are overviewed.

18.2.1 Galactolipids

Galactolipids generally name the lipids containing a hydrophobic core of MAG or DAG and a polar head group containing one, two, three, or more galactosyl moieties. They are the main component of plant membrane lipids, and monogalactosyl DAG (MGDG) and digalactosyl DAG (DGDG) are the two classes of most abundant galactolipids in plant membranes.

In the classification based on charge propensity, both MGDG and DGDG belong to the category of polar lipids carrying only with covalent bonds (i.e., no net charge(s) under physiological conditions) (Figure 2.3). Thus, these classes of lipids can be readily ionized as formation of adducts in both positive- and negative-ion modes.

In the positive-ion mode, these lipid species give rise to their alkaline or ammonium adducts (i.e., $[M+Alk]^+$ or $[M+NH_4]^+$, respectively). The adduct(s) that are formed is dependent on the concentrations of these adducted cations in the matrix. An adduct cation that is preferred to use for characterization of the samples should be introduced during lipid extraction or added prior to infusion of the solution in the case of shotgun lipidomics. In the case of LC-MS analysis, the preferred adduct cation (e.g., ammonium) should be present in the mobile phase.

These formed adducts of galactolipid species in the positive-ion mode can be selected, and fragmentation can be generated by CID on a variety of mass spectrometers in the product-ion analysis mode. The fragmentation pattern of MGDG species should essentially be identical to those of hexosyl DAG species as described in Chapter 9 since MS analysis in the positive-ion mode is generally unable to distinguish the different conformations of the sugar ring structures. The most abundant fragment yielded from these adducted ions is the hexose plus the adduct cation. The fragmentation pattern of DGDG species is also briefly described in Chapter 9. The ammoniated DGDG species result in two abundant fragment ions at $[M+NH_4-341]^+$ and $[M+NH_4-359]^+$, corresponding to the neutral loss of ammoniated digalactose minus a water molecule and ammoniated digalactose, respectively. These characteristic fragments can be used to profile DGDG species present in plant lipidomes after direct infusion [7]. In addition, the fatty acyl constituent(s) can also be determined from the one or two very abundant fragment ions, corresponding to the protonated FA substituent(s) esterified to glycerol minus H_2O, depending on whether the DGDG species contain an identical FA chain or two different FA chains, respectively [15]. Similarly, product-ion analysis of alkaline adducts of DGDG species can also be used to determine the fatty acyl constituents as previously described (Chapter 9 and Ref. [16]).

In the negative-ion mode, galactolipids can be ionized as formation of anion adducts (i.e., $[M+Y]^-$ ($Y^- = Cl^-$, $HCOO^-$, $^-OCOCH_3$, ...)). It should be recognized that as ionization conditions in the ion source become more harsh, the $[M-H]^-$ ion, generated from the source fragmentation of $[M+Y]^-$ after neutral loss of a HY molecule, becomes more intensive. Similar to their ionization in the positive-ion mode, the formation of an adduct ion depends on the concentration of the adducted anions in the matrix. An adduct anion that is preferred to use for characterization of the samples should also be introduced during lipid extraction or added prior to infusion of the solution in the case of shotgun lipidomics. In the case of LC-MS analysis, the preferred adduct anion (e.g., formate) should be present in the mobile phase. Thus, ammonium formate is frequently included as a modifier in the mobile phase in LC-MS analysis, whereas a small amount of (volatile) salt (e.g., NH_4HCO_2, NH_4Cl, HCO_2Li) could be added in the matrix in the case of shotgun lipidomics. It is worth noting that, similar to PC (Chapter 7), formation of galactolipid adducts with hydroxide is somehow not easy.

The anionic adducts of galactolipid species ionized in the negative-ion mode can be selected (usually by a quadrupole analyzer) and product-ion analysis can be performed through CID on a variety of mass spectrometers [6, 17]. The product-ion mass spectra of anion adducted galactolipid species (i.e., $[M+Y]^-$) generally display two kinds of fragment ions, one corresponding to the neutral loss of HY and the other corresponding to the fatty acyl carboxylate(s). The number of carboxylate ions depends on whether the galactolipid species contain identical or different fatty acyl substituents in each molecule. If the $[M-H]^-$ ion of galactolipid species is selected for product-ion analysis, only the type of FA carboxylate(s) is present in the product-ion mass spectra abundantly. These FA fragments should facilitate the identification of individual galactolipid species [13].

The fragments corresponding to the galactosyl ring(s) are usually displayed in low or very low abundance in the product-ion mass spectra, particularly yielded from the $[M-H]^-$ precursor ions. However, fragmentation of the anionic adduct may give rise to useful signature fragment ions, which could be used for characterization of these ring structures as previously demonstrated in the characterization of hexosyl Cer species in the negative-ion mode [18].

18.2.2   Sphingolipids

Plants contain many classes of sphingolipids, including Cer and hydroxy Cer, GluCer, inositol phosphorylceramide (IPC), and glycosyl inositol phosphorylceramide (GIPC) (i.e., phytoglycosphingolipid) [19]. However, it should be recognized that sphingomyelin and gangliosides do not occur in plants. Moreover, it has not been examined whether these plant sphingolipids should be particularly extracted in comparison to mammalian sphingolipids or to other plant lipids. However, an extraction procedure utilizing the lower phase of an isopropanol–hexane–water (11:4:5, v/v/v) mixture appears optimal for extracting sphingolipids [20]. Furthermore, to enrich sphingolipids present in lipid extracts and also to reduce the effects of "ion suppression" on ionization for MS analysis after direct infusion, the lipid extracts could be subject to hydrolysis under basic conditions prior to the lipid analysis, for example, by incubation with methylamine [19]. This procedure is equivalent to those for enriching animal sphingolipids through alkaline hydrolysis [21, 22].

Analysis of Cer, hydroxy Cer, and GluCer by ESI-MS and ESI-MS/MS is described in details in Chapter 7. But the analysis of IPC and GIPC, which is special to plant lipidomics relative to animal lipids, is discussed below.

IPC is an analog of phosphatidylinositol (PI) in glycerophospholipids. The molecular species of IPC contain a net negative charge under mild alkaline conditions. Thus, this class of lipid species is favorable to be ionized in the negative-ion mode than in the positive-ion mode (Figure 2.3). These species generally yield a molecular ion of $[M-H]^-$ in the negative-ion mode, although these can also be ionized as $[M+Alk]^+$ or $[M-H+2Alk]^+$ (Alk = Li, Na, K, ...) in the positive-ion mode under certain conditions.

Product-ion mass spectral analysis of IPC in both negative-ion and positive-ion modes has been extensively conducted by Hsu and colleagues [23]. Tandem MS analysis of $[M-H]^-$ (i.e., in the negative-ion mode) by CID displays multiple informative fragment ions in three categories:

- The fragments arising from the phosphoinositol moiety of the molecule, including the prominent ions at m/z 241, corresponding to an inositol-1,2-cyclic phosphate anion, and at m/z 259, corresponding to an inositol monophosphate anion, along with the ion at m/z 223 arising from m/z 241 after further loss of H_2O. (These ions are characteristic to IPC molecules and are similar to those observed for PI species (Chapter 7).)
- The fragment ions arising from the losses of a dehydrated inositol (162 amu) and inositol (180 amu) residues, respectively.
- The one resulting from the loss of an 18:0-fatty acyl substituent as a ketene. (The presence of this fragment ion is useful for the identification of the FA substituents in the IPC species and therefore unambiguous elucidating the structure of long-chain sphingoid base.)

Tandem MS mass spectra of $[M+Alk]^+$ (i.e., in the positive-ion mode) by CID also display numerous informative fragment ions as follows [23]:

- The major fragmentation process arising from the cleavage of inositol monophosphate residue, yielding a prominent ion corresponding to the elimination of inositol monophosphate residue and an ion at m/z $(260+Alk)^+$, corresponding to an alkaline-adducted ion of inositol monophosphate.
- Other ions equivalent to the alkaline-adducted ceramide ion and its further dehydrated ions.
- Two fragment ions arising from fatty acyl and long-chain sphingoid base (e.g., the ions at m/z 308 and 236 are generated from d16:1–18:0 species [23]).

The fragmentation patterns resulted from other IPC ions in the positive- or negative-ion mode are also very informative, all of which can be found in the study performed by Hsu and colleagues [23]. However, these fragmentation patterns may not be practically useful for profiling IPC species present in plant lipidomes.

As a lipid class carrying a net negative charge under physiological conditions, the GIPC species should be readily ionized in the negative-ion mode. However, analysis of these species in the negative-ion mode appears to be lacking. Analysis of GIPC species in the positive-ion mode was conducted [20]. Sodiated adducts of GIPC species are prominent in the survey scan although protonated and other alkaline-adducted ions coexist in the analysis [20]. Fragmentation of the sodiated adducts by low-energy CID yields numerous informative fragment ions, which are tabulated in the original report [20]. Ammoniated GIPC can be detected when ammonium salt is present in the matrix. Characterization of these GIPC ions demonstrates two abundant and useful fragments corresponding to the neutral loss of 179 and 615 amu, respectively, arising from the head group. These characteristic fragments can be used to profile GIPC species present in plant lipidomes after direct infusion.

18.2.3   Sterols and Derivatives

In contrast to animals and fungi, which contain primarily cholesterol and ergosterol, respectively, plants produce a complex mixture of sterols comprised of sitosterol, campesterol, stigmasterol, and additional minor sterols that include cholesterol [24]. Moreover, these sterols are also glycosylated with or without further acylation. Similar to the methods developed for the analysis of sterols present in animals, the methods through derivatization such as utilizing *N*-chlorobetainyl chloride to enhance ionization efficiency of these sterols are also developed [8, 25]. However, their glycosylated and acyl glycosylated derivatives (i.e., steryl glycosides (SG) and acyl steryl glycosides (ASG)) are special to plants relative to animals. Thus, analysis and characterization of these major plant sterol derivatives need to be discussed.

Recently, Schrick et al. [8] characterized the ammoniated SG and ASG species in the positive-ion mode by low-energy CID. Similar to the analysis of ammoniated cholesterol and cholesteryl esters, they found that both ammoniated SG and ASG ions give rise to a prominent fragment ion resulting from the loss of 197 amu (i.e., glucose plus ammonia) and acyl glucoside plus ammonia (the neutrally lost mass depending on the fatty acyl chains), respectively. The resulted fragment ion corresponds to the steryl cation. Therefore, the investigators demonstrated and concluded that NLSs of 197 and (197 + acyl ketene) can be used to efficiently profile all the SG and ASG species present in lipid extracts of plant samples, respectively, after direct infusion with the addition of ammonium salt in the matrix.

18.2.4   Sulfolipids

Sulfolipids are a category of sulfated glycolipids. The class of sulfated galactolipids is present in modest abundance in plants. The presence of the sulfate group makes this class of lipids carrying a net negative charge under physiological conditions and, thus, these lipid species can be readily ionized as deprotonated species in the negative-ion mode. These lipid species were extensively analyzed by both direct infusion ESI-MS and LC-MS as summarized previously [14].

Product-ion mass spectrometric analysis of the deprotonated sulfolipid ions in the negative-ion mode after low-energy CID demonstrates two types of abundant fragment ions:

- The fragment ion(s) corresponding to FA carboxylate anion(s), which are present in one or two ions depending on whether the two FA substituents in each molecule are identical [26, 27].

- A characteristic fragment ion at m/z 225 corresponding to a dehydrosulfoglycosyl anion. (Thus, PIS of this ion in the negative-ion mode can be used to profile all sulfolipid species present in lipid extracts of plant samples as demonstrated previously [7]).

18.2.5   Lipid A and Intermediates

Lipid A is a category of complex hexa-acylated disaccharide of glucosamine in bacteria. The existence of Lipid A metabolism pathway in plants was demonstrated by LC-MS [28]. The structure of the Lipid A-related compounds is confirmed by product-ion analysis utilizing QqTOF-MS [28].

18.3   LIPIDOMICS FOR PLANT BIOLOGY

18.3.1   Stress-Induced Changes of Plant Lipidomes

Plant membrane lipids undergo many changes when the plants are exposed to various stress conditions such as cold, heat, mechanical wounding, and phosphorus deficiency. Lipidomics allows researchers to reveal the lipid changes induced by the stress conditions. Furthermore, the great insights into the underlying mechanism(s) leading to these changes can also be facilitated in the most cases. Following are a few examples of the studies on stress-induced changes of plant lipidomes that are summarized.

18.3.1.1   Lipid Alterations in Plants Induced by Temperature Changes   Many plants experience very different temperature conditions in different seasons. It has been proposed that reorganization of plant cellular membranes is one of the prime targets in response to temperature stress [29]. Plant cells undergo upregulation of the content and/or composition of membrane lipids that contain different degree of fatty acyl unsaturation to modulate membrane fluidity to adapt the environment change of temperature. The regulation of lipid content and/or composition of lipids is usually achieved through cellular biosynthesis or remodeling processes or their combination. This type of cellular regulation to adapt environmental changes was well recognized [30, 31]. To adapt a low-temperature environment, one of the cellular responses is to increase membrane fluidity and reduce the propensity of cellular membranes to undergo freezing-induced nonbilayer phase formation, thus preserving the membrane integrity and cellular survival. To adapt a high-temperature environment, one of the cellular responses is to increase membrane rigidity to preserve cellular water content and optimize photosynthesis and other cellular processes [32–36]. Lipidomics naturally provides a powerful tool to monitor such kinds of cellular lipid changes in plants induced by temperature changes and to investigate the underlying molecular pathways leading to the changes.

For example, in a study conducted by Welti et al. [6], lipidomics approaches are used to not only detect changes in specific lipid molecular species of *Arabidopsis* as a result of low-temperature stress but also reveal that phospholipase Dα plays a critical role in reorganizing freezing-induced cellular lipid changes. The researchers found that the lipid species that contain two polyunsaturated acyl chains, such as 18:2–18:3 and di18:3 PC, 18:2–18:3 and di18:3 PE, 18:3–16:3 MGDG, and di18:3 DGDG are significantly increased in the plants during cold acclimation. They also found that the levels of more saturated species in those lipid classes such as 18:0–18:2, di18:1, 18:1–18:2, and 18:0–18:3 PC species are reduced. The results clearly indicate the upregulation of desaturation activity and concomitant increases in remodeling of both extraplastidic (PC and PE) and plastidic (MGDG and DGDG) lipid species through phospholipases A and D activities. In the study, the scientists also observed that the mass levels of specific molecular species of the lipid metabolites including lysoPC, lysoPE, and PA are significantly increased in response to cold exposure. These observations suggest that altered lipids in response to cold stress may play signaling functions since PA and lysoGPL are important bioactive lipids with potential regulatory functions, such as activation of target signaling proteins, regulation of cytoskeletal organization, and regulation of ion-channel function [37]. In conclusion, lipidomics analysis reveals that lipids might play both structural and regulatory roles in plant adaptation and survival during low-temperature exposure.

Further studies of a comparative analysis of the lipidomes of 15 *Arabidopsis* accessions throughout the Northern Hemisphere were conducted by using LC-FTMS, which allows the detection of 180 lipid species [38]. It was shown that after the plants exposed for 14 days of cold acclimation at 4 °C, the plants from most accessions accumulate massive amounts of storage lipids, with most of the changes occurring in the TAG species containing long-chain unsaturated fatty acyl chains, while the total amount of membrane lipids are only slightly changed. This is a sensible evidence to prove that the relative abundance of several lipid species is highly correlated with the freezing tolerance of the accessions, allowing the identification of possible marker lipids for plant freezing tolerance.

To improve the thermotolerance and investigate the underlying mechanisms leading to thermotolerance in plants, an exploited forward genetic approach is to identify a series of *Arabidopsis thaliana* thermosensitive mutants (*atts*) that fail to acquire thermotolerance after a brief exposure to heat stress (e.g., 38 °C) [39]. Such a heat stress condition could induce acquired thermotolerance in wild-type seedlings and enable the wild-type plants to survive the otherwise lethal high-temperature treatment. By using this approach, researchers have identified digalactosyl DAG synthase 1 (DGD1) as one of such *atts* factors [40]. DGD1 is the major enzyme that catalyzes the conversion of MGDG to DGDG in chloroplast. It was found that mutations in *DGD1* lead to plants both defective in acquired thermotolerance and susceptible to moderately elevated temperatures [40]. Further studies showed that these failures are not due to the changes in the transcript level of *DGD1* gene, but are caused by a single amino acid substitution in the gene, resulting in localized conformational changes in DGD1 protein and changes in DGD1 function in the mutants [40]. These results, therefore, suggest that thermotolerance in plants could be tightly associated with their content and/or composition of galactolipids produced by *DGD1* mutation.

To conform the genetic findings as described earlier, lipidomics analysis was employed to determine the changes of lipid content and fatty acid profiles induced with DGD1 mutations [6, 17]. It was found that the *DGD1* mutant plants grown at normal temperature retain 60–66% of the DGDG levels of the wild-type counterparts, depending on the mutants. Moreover, the reduction of DGDG species in the mutants is also manifest in both plastid membrane and ER. However, regardless of no significant change in the total levels of MGDG in the mutant plants, the content of plastid-derived MGDG is decreased while the ER-derived MGDG level is increased at normal temperature. As a result, the ratio of DGDG to MGDG decreases in the mutants. Furthermore, the substantially lower content of DGDG in the null mutant plants of *DGD1-1* in comparison to those present *DGD1* mutants indicates that the mis-sense mutations of *DGD1* altered rather than eliminated DGD1 function.

To understand the possible roles of DGDG and the ratio of DGDG to MGDG in acquired thermotolerance, changes of lipid composition in leaf tissue of wild-type plants were analyzed by ESI-MS/MS before and after a heat acclimation at 38 °C for 24 h [6, 17]. It was found that a significant increase in the relative amount of DGDG following heat acclimation treatment is present in the plants at 21 °C. Specifically, the relative content of DGDG in wild-type plants was increased by 23% after exposure to heat, whereas the relative content of MGDG was reduced from 47 to 38 mol%, which represents a significant increase in the ratio of DGDG to MGDG from 0.27 at 21 °C to 0.42 upon heat acclimation at 38 °C. These findings are consistent with those observations from other studies [41, 42]. For example, the ratio of DGDG to MGDG is increased in the bean plants (*Phaseolus vulgaris*) in response to elevated temperature treatments, and it is believed that the increased ratio plays an important role in acquisition of thermotolerance [41]. Furthermore, it is also suggested that the observed sharp increase in the ratio of DGDG to MGDG in wild-type plants plays a crucial role in the ability of plants to tolerate and adapt to high temperatures. From the biophysical chemistry point of view, DGDG molecules possess a higher polar head group than MGDG so that they are favorite of forming bilayers in aqueous environments and a high affinity to the water molecule, whereas MGDG molecules possess a smaller head group, which favors to form hexagonal phase structure and only retains a low amount of water [42, 43]. Thus, the increased ratio of DGDG to MGDG should be favorable to maintain chloroplast membrane integrity and normal membrane protein function at high temperatures.

As mentioned earlier and established previously [34, 35, 44], it is well known that the degree of fatty acyl saturation in lipid species is closely related to the heat tolerance. In the studies on heat acclimation, it was also found that an increased level of fatty acyl saturation as indicated by the double-bond index in galactolipids from wild-type plants after heat acclimation is also present [6, 17]. Specifically, the double-bond index of DGDG species is reduced from 2.58 to 2.34, and the double-bond index of MGDG decreased from 2.95 to 2.84 after a heat acclimation for 24 h at 38 °C. Increases in the levels of fatty acyl saturation among other major polar lipids in response to heat treatment are also observed in the studies [6, 17]. Collectively, these and other results [40] suggest that the acquired thermotolerance induced by the heat acclimation in wild-type plants is associated with an increase in the relative content of DGDG and the ratio of DGDG to MGDG as well as the moderate increase in fatty acyl saturation of the plant lipids.

It is well documented that the plants carrying some *DGD1* mutants (*DGD1-2* and *DGD1-3*) are unable to grow at moderately high temperatures of 30 °C or above and fail to acquire thermotolerance after an exposure at 38 °C for 90 min [40]. To determine whether the thermosensitive phenotype of these *DGD1* mutants is associated with the changes of galactolipid content and/or composition, lipidomics analysis was employed to profile lipid changes in response to heat acclimation in comparison to the wild-type plants [6, 17]. It was found that the composition of DGDG in these mutants did not increase after an exposure to heat acclimation, but a slight decrease in DGDG was observed in both mutants, which was in contrast to the changes of the wild-type plants. Moreover, the levels of DGDG in the leaves of these *DGD1* mutant plants were only 41% of those present in the wild-type plants. Furthermore, the studies showed that the levels of MGDG in both wild-type and *DGD1* mutant plants were no significant differences upon heat acclimation. These observations were linked to the decreases in the ratios of DGDG to MGDG in these *DGD1* mutant plants, but an increase in this ratio from 0.27 to 0.42 in wild-type plants. In addition, the studies also showed that the plants carrying these *DGD1* mutants did not increase the degree of fatty acyl saturation in their galactolipids after treatment. These results further confirmed that the levels of DGDG, the ratios of DGDG/MGDG, and the degree of fatty acyl saturation in these lipid classes play an essential role in both basal and acquired thermotolerance in *Arabidopsis*.

In another study, the changes of lipidomes in response to temperature were conducted as a temperature-dependent manner (i.e., at 4, 21, and 32 °C) by LC-MS [13]. In addition to the previous findings as described earlier, it was also found that an immediate decrease in unsaturated species of PG was present in plants under high-temperature conditions (i.e., 32 °C). This finding suggests that the reduction of unsaturation in PG species could be the first stages of adaptation to heat stress conditions.

The majority of the studies as described earlier identify that the major strategy by which plants adapt to temperature change is to decrease the degree of membrane lipid unsaturation under high-temperature conditions and to increase the extent of saturation under low-temperature conditions. This observation led researchers to hypothesize that this strategy cannot be adapted by plants in ecosystems and environments with frequent alterations between high and low temperatures, because changes in lipid unsaturation are complex and require a large amount of energy inputs. This hypothesis was tested by employing a lipidomics approach through profiling the changes in molecular species of membrane glycerolipids in two plant species sampled from an alpine environment and in other two plant species grown in a growth chamber, with the temperature cycling daily between heat and freezing [45]. It was found that six classes of GPL and two classes of galactolipids show significant changes, but the degree of unsaturation of total lipids and of three classes of lysoGPL remains unchanged. This pattern of changes in membrane lipids is distinct from that occurring during slow alterations in temperature as described earlier. These results led the scientists to propose two types of model for the adaptation of plants to temperature changes, which are as follows: (1) remodeling of membrane lipids, but maintenance of the degree of unsaturation is used to adapt to frequent temperature alterations; and (2) both remodeling and changes in the degree of unsaturation to adapt to infrequent temperature alterations are present [45].

Plants contain a large diversity of sphingolipid species. Typically, 85–90% of sphingolipid long-chain bases (LCB) in *Arabidopsis* leaves contain a Delta8 double bond produced by sphingoid LCB Delta8 desaturase (SLD). In addition to the lipidomics studies on GPL and glycolipids as described earlier, lipidomics analysis was also exploited to identify the role of altered sphingolipid content and composition in the plants in response to the cold and heat challenges. For example, comprehensive analysis of sphingolipids in rosettes of the SLD mutants by LC-MS [46] revealed a 50% reduction in the levels of GluCer species and a corresponding increase in those of GIPC species. Moreover, the mutants displayed significantly altered responses to the prolonged exposure to low temperatures. These results were consistent with a role for LCB Delta8 unsaturation in selective channeling of ceramides for the synthesis of complex sphingolipids and the physiological performance of *Arabidopsis*. Lipidomics profiling of sphingolipid species was also used to identify the function of several genes involved in sphingolipid metabolism [46–52].

18.3.1.2 Wounding-Induced Alterations in Plastidic Lipids

18.3.1.2 Wounding-Induced Alterations in Plastidic Lipids   Whether lipids are involved in plant wounding is an important topic in plant biology. If yes, then what kinds of lipids are associated with this process. Lipidomics analysis certainly facilitates the research in this area of plant biology [53]. The researchers employed ESI-QqToF-MS after CID to identify the structures of complex lipids including fatty acyl chains in combination with ESI-MS/MS in the precursor-ion mode. In the study, 17 species of PG, MGDG, and DGDG-containing oxylins, including (9*S*,13*S*)-12-oxo-phytodienoic acid (OPDA), dinor-oxophytodienoic acid ((7*S*,11*S*)-10-oxo-phytodienoic acid, dnOPDA), 18-carbon ketol acids, and 16-carbon ketol acids, were identified [53].

Quantitative analysis by ESI-MS/MS [53] showed that five OPDA- and/or dnOPDA-containing MGDG and two OPDA-containing DGDG species were significantly accumulated as a function of time in mechanically wounded leaves. In unwounded leaves, the levels of all oxylipin-containing lipid species were low. However, significant accumulation of numerous oxylipin-containing lipid species occurred within 15 min after mechanical wounding of the *Arabidopsis* leaves. Particular interest was the finding that the polar lipid species containing two oxylipins in the same molecule, such as diOPDA and OPDA-dnOPDA species, increased very quickly and to a greater extent than those species only carrying one oxylipin. These observations may suggest that lipid peroxidation occurs much faster in intramolecules than intermolecules, which is likely due to free radical propagation. However, studies to validate this possibility remain warranted.

18.3.1.3 Phosphorus Deficiency-Resulted Changes of Glycerophospholipids and Galactolipids   Phosphorus is an essential macronutrient that often affects the growth and development of plants. When this nutrient becomes deficient, lower levels of GPL classes, but higher levels of galactolipids (mostly those of DGDG species), are manifest in plants in comparison to those grown under normal phosphate conditions [54]. Lipidomics analysis was used to provide quantitative lipid profiling of the detailed changes in lipid molecular species resulted from phosphorus deficiency in plants [55].

It was found from lipidomics that the mass levels of DGDG increased 10-fold in roots and 72% in rosettes, whereas the contents of PC were reduced 51 and 17%, respectively, under phosphorus deficiency conditions. These observations not only confirmed the results from other studies, but also indicated that changes of lipid levels were more drastic in roots than in rosettes under phosphorus deficiency conditions. Moreover, it was also found that the increased mass levels in galactolipids quantitatively compensated the loss of GPL content in rosettes due to phosphorus deficiency. Specifically, *Arabidopsis* rosettes contained a total mass level of 86.7 nmol GPLs/mg dry weight, including PC, PE, PI, PG, PS, and PA, and a total content of 78.5 nmol galactolipids/mg dry weight, including MGDG and DGDG, under normal growth conditions. Thus, the total concentration of lipids, including GPLs and galactolipids, is 165 nmol/mg dry weight in usually grown rosettes. However, the total GPL and galactolipid levels were 67.0 and 97.7 nmol/mg dry weight, respectively. This made the total content of lipids in rosettes still at 165 nmol/mg dry weight under phosphorus deficiency conditions. These results indicated the importance of membrane lipid homeostasis in cellular function, at least in rosettes.

The molecular species profiling from lipidomics analysis revealed the changes of possible synthesis pathways of plant lipid classes in rosettes and roots under normal and phosphorus deficiency conditions [55]. In plants, DGDG species are clarified into three pools: the "plastidic pool" that is originally derived from the prokaryotic pathway and located in the plastid; the "ER–extraplastidic pool" that is derived from the eukaryotic pathway and located outside the plastid; and the "ER–plastidic pool" that is derived from the eukaryotic pathway but is located inside the plastid. Lipidomics analysis revealed that the major molecular species of GPL, including 34:2, 34:3, 36:4, 36:5, and 36:6, were generally lower in both rosettes and roots under phosphorus deficiency conditions. The compensated increase in DGDG species also contained these fatty acyl structures, but not the 34:6 and 34:5 DGDG species, which were originally derived from the prokaryotic, plastid pool. These observations not only suggest that the hydrolysis of GPL species supplies DAG moieties for DGDG biosynthesis under phosphorus deficiency conditions, but also indicate that the increased DGDG content during phosphorus starvation is essentially derived from the eukaryotic pathways in both rosettes and roots.

18.3.2   Changes of Plant Lipidomes during Development

18.3.2.1 Alterations in Lipids during Development of Cotton Fibers   Cotton fibers are single cells that initiate on the epidermal surface of ovules at anthesis

[56]. Their elongation occurs primarily during the first 15 days after anthesis, and the cells achieve a length of greater than 2 cm by the time. Then, the cells increase in thickness over the period of approximately 15–40 days after anthesis (i.e., fiber maturation) [35, 36]. Ultimately, the cell cytoplasm collapses and mature fibers desiccate when the fruit opens at ~45 days after anthesis. Obviously, this entire process of fiber development is tightly associated with rapid plasma membrane and vacuolar expansion so that synthesis of a large amount of membrane lipids occurs to accommodate this development. Therefore, determining the changes of the content and composition of lipid species during the development by lipidomics is an important step to better understand the lipid metabolism and function in fiber cell formation.

ESI-MS/MS analysis [57] revealed that the content of GPL (the major polar constituents of fibers) is generally similar between elongating and maturing periods. Specifically, PC, PE, and PI classes were the principal components in all stages of fiber cell development, together accounting for more than 70% of the total polar lipids, and PG, MGDG, and DGDG were only the relatively minor polar lipid constituents. However, the molecular composition of these major GPL classes was significantly different between the periods. Generally, the composition of GPL species containing saturated fatty acyl chains was higher in maturing fibers than that in elongating fibers. For example, a greater proportion of 34:2 PC and 34:2 PE was present in maturing fibers relative to elongating fibers. In contrast, relatively higher content of 36:6 and 36:4 PC, and 34:3 and 36:6 PE was present in elongating fibers compared to maturing fibers. These lipidomics data, in combination with those obtained from genomic studies, delineated the pathways likely to be important for membrane lipid synthesis in cotton fiber cells at the molecular level [57].

18.3.2.2 *Changes of Lipids during Potato Tuber Aging and Sprouting*   Potato tubers are model organs for aging studies because they can survive up to 3 years after harvest, if they are stored at low temperature. Analysis of GPLs and galactolipids by ESI-MS/MS showed a decrease in the levels of PC, PE, PI, DGDG, and MGDG during this sprouting phase period [58]. The decrease in the amount of linoleic acid in the complex lipids correlated well with the amount of its metabolites such as 9-hydroperoxy linoleic acid. These lipidomics observations indicate that the lipoxygenase pathway in combination with galactolipases and phospholipases plays an important role during this period.

18.3.3   Characterization of Gene Function by Lipidomics

18.3.3.1   *Role of Fatty Acid Desaturases and DHAP Reductase in Systemic Acquired Resistance*   Systemic acquired resistance (SAR) is an inducible defense mechanism in plants to confer resistance to a broad spectrum of pathogens [59, 60]. SAR is constitutively expressed in the *A. thaliana SSI2* mutant. Moreover, the *SSI2* gene encodes a stearoyl carrier protein desaturase that catalyzes the conversion of acyl carrier protein-conjugated stearic acid to oleic acid in the plastids [61]. The resulted oleoyl moiety of acyl carrier protein is subsequently transferred to the

synthesis of glycerolipids in the plastids or shunted to the cytosol as 18:1-CoA and subsequently incorporated into glycerolipids in the ER. In summary, the *SSI2* mutation-resulted lipid alterations should play an important role in SAR.

Thus, ESI-MS/MS analysis was employed to profile the lipid changes induced by *SSI2* mutation. It was shown that the *SSI2* mutation resulted in an overall reduction in the contents of glycerolipid species that were synthesized and localized in the plastids [62]. It was also found that the composition of some extraplastidic lipid species was changed to a certain extent. For example, the levels of PC, PE, and PI species containing combined 34:2 fatty acyl chains were lower in the leaves of the *SSI2* mutant than those in the wild-type control. In contrast, the levels of PC, PE, and PI species containing combined 36:2 and 36:3 fatty acyl chains were higher in the *SSI2* mutant plants than the wild-type plants. These observations indicate that regardless of the primary activity loss in the *SSI2* mutant is in the conversion of 18:0-carrier protein to 18:1-carrier protein, other compensatory changes occur. The impact of the *SSI2* mutation on SAR is likely due to the generation of a signal from the metabolites, which could modulate plant defense [61].

In contrast to the constitutive expression of SAR in the *SSI2* mutant, SAR is compromised in the *SFD1* mutant [63]. ESI-MS/MS analysis revealed that the composition of plastidic glycerolipids (particularly galactolipids) was altered in the *SFD1* mutant in comparison to the wild-type plants, suggesting that *SFD1* is required for lipid biosynthesis in the plastids. Specifically, the levels of a plastidic lipid species (i.e., 34:6 MGDG, which is the most abundant complex lipid species in *Arabidopsis* leaves) were lower in the *SFD1* mutant than the wild type. In contrast, the content of 36:6 MGDG (another plastidic lipid that is derived from DAG and shunted into the plastids from the ER) was significantly higher in the *SFD1* mutant than the wild-type leaves. This latter observation suggests that an increased flux of ER-synthesized lipid species into the plastid might compensate for the lipid composition deficiency in the *SFD1* mutant.

Lipidomics not only facilitates identification of gene function(s) but also provides some foundation for identifying mutated genes. For example, ESI-MS/MS profiling of polar lipid species in the *SSI2/SFD1* mutant plant suggested that the *SFD1* gene was involved in plastid lipid biosynthesis [62]. Alterations in plastid lipids induced *SFD1* mutant were confirmed by lipid profiles of the *SFD1* single mutant plant [63]. From these complementation experiments, therefore, the identity of *SFD1*, as a dihydroxy-acetone phosphate reductase, and its involvement in plant defense were determined. Similarly, lipidomics analysis aided in identification of the gene that contains the *SFD4* mutation [62]. The plants carrying this mutant contained high levels of plastidic galactolipid species that carried monounsaturated fatty acids such as palmitoleic (16:1) and oleic (18:1) acids, but contained lower levels of galactolipid species carrying polyunsaturated 16C and 18C fatty acyl chains. These findings suggest the *SFD4* gene may encode a ω6-desaturase that involves a desaturation activity to produce hexadecadienoic (16:2) and linoleic (18:2) acids, which are subsequently incorporated into plastid-synthesized galactolipids. Comparison of the lipid profile present in *SFD4* mutant to that of the *FAD6* mutant, which lacks a functional ω6-desaturase, further confirmed the finding from lipidomics analysis.

18.3.3.2   Roles of Phospholipases in Response to Freezing   Changes of lipid species are mediated by the activities of enzymes that involve in their catabolism and metabolism. Phospholipases, which are clarified mainly into PLA, PLC, and PLD, based on the site of cleavage, are the major groups of enzymes that involve GPL catabolism and metabolism [64]. Among these, PLD, which converts GPL to PA species, is the most prevalent family of phospholipase in plants [37]. There exist 12 *PLD* genes in *Arabidopsis*, which are clarified into six types: three isoforms of *PLDα*, two isoforms of *PLDβ*, three kinds of *PLDγ*, one *PLDδ*, one *PLDε*, and two isoforms of *PLDζ*. Each of the PLD isoforms possesses distinguishable biochemical properties, such as different requirements for Ca^{2+}, preferences for different GPL substrates, requirement of oleic acid for their activities, etc. They play different roles in response to different stresses.

To this end, lipidomics profiling of plant samples carrying *PLD* mutants in comparison to wild-type counterparts provided great insights into the cellular functions of different PLD isoforms [17, 55]. For example, lipidomics analysis greatly revealed the different roles of PLDs in freezing-induced lipid hydrolysis [6, 65].

It was found that the levels of GPL species resulted primarily from the PC, PE, and PG species carrying combined 36:4 and 36:5 fatty acyl chains are reduced the most at −8 °C (a sublethal-freezing temperature for cold-acclimated *Arabidopsis*), accompanying dramatic increases in their metabolite such as PA, lysoPC, and lysoPE [6]. However, the changes of the mass levels in PI molecular species were minimal and the contents of some PI species actually increased. Moreover, the levels of MGDG and DGDG were either minimally lost or no changes after exposure to this sublethal-freezing temperature. These lipidomics results clearly indicate that phospholipases are activated to a greater extent than galactolipases. With the low levels of the products of PLA (i.e., lysoGPLs) in comparison with the PLD product (i.e., PA), the studies indicate that PLD plays a major role in changes of lipids in frozen plant tissue. Finally, comparison of the profiles of PA species with other lipid classes aided to identify the substrates that give rise to PA species. Lipidomics analysis revealed that the majority of the increased PA species match with those of decreased PC species, indicating that the class of PC species was the major source of PLD substrates and GPL hydrolysis upon freezing.

So next question is which PLD isoform(s) play the essential role in plants in response to low temperatures, leading to the increase in the levels of PA species. Further studies by lipidomics revealed that suppression of *PLDα*1 led to a higher level of PC species, but a lower level of PA species [6]. These results indicated that PC was the major *in vivo* substrate for *PLDα*1 under freezing conditions. In fact, *PLDα*1 was responsible for ~50% of PC hydrolyzed under the freezing conditions tested [6]. The lack of difference between wild-type and *PLDα*1-deficient plants in the levels of PE and PG species suggests that other enzymes in addition to *PLDα*1 were responsible for the freezing-induced hydrolysis of PE and PG species.

18.3.3.3   Role of PLDζ in Phosphorus Deficiency-Induced Lipid Changes   As described earlier (Section 18.3.1.3), phosphorus deficiency leads to decreases of the GPL levels in plants in demanding phosphate for other cell functions and to yield

DAG species for galactolipid biosynthesis. Then, the question is what causes the loss of GPL content under phosphorus deficiency conditions. Further studies by lipidomics [65, 66] revealed that the loss of *PLDζ2*, but not *PLDζ1*, led to a significantly decreased accumulation of PA content in roots under phosphorus deficiency conditions and compromised the ability of plants to hydrolyze GPLs and to increase galactolipid levels. Moreover, plants null of *PLDζ1* and *PLDζ2* exhibited shorter primary roots than wild-type plants under the phosphorus deficiency conditions. These results indicate that *PLDζ1* and *PLDζ2* play an important role in regulating root development in response to nutrient limitation.

Multiple phosphorus conditions were examined [55, 67]. Under a normal growth condition (500 µM of phosphorus), disruption of *PLDζ1*, *PLDζ2*, or both did not result in the differences in the levels of PC, PA, DGDG, and root growth in comparison to the wild-type plants. However, under a phosphorus deficiency condition (e.g., 25 µM of phosphorus), disruption of both *PLDζ1* and *PLDζ2* resulted in a lower content of PA in roots, a retarded root growth, and unchanged levels of PC and DGDG in roots [67]. Under the phosphorus-free condition (0 µM of phosphorus), disruption of both *PLDζ1* and *PLDζ2* resulted in a lower level of PC with a correspondingly increased content of DGDG in roots. These results indicate that, under moderate phosphorus deficiency conditions (25 µM of phosphorus), *PLDζ1* and *PLDζ2* might function to modulate root growth for better nutritional absorption by increasing PA levels to stimulate root growth. But under severe phosphorus deficiency conditions (e.g., 0 µM of phosphorus), *PLDζ1* and *PLDζ2* might function to regulate lipid turnover between GPLs and galactolipids for efficient utilization of internal phosphorus availability. Collectively, *PLDζs* play both signaling and metabolic roles in plants under different phosphorus deficiency conditions [55].

18.3.4 Lipidomics Facilitates Improvement of Genetically Modified Food Quality

Lipidomics analysis could provide deep insights into the function of lipid metabolizing genes in transgenic plants, leading to producing high-quality foods such as with high content of ω-3 fatty acyl chains in plants [68]. This requires coordination of a number of heterologous enzymes to work together to produce the desired fatty acids and requires lipidomics analysis to provide comprehensive evaluations of the products (i.e., lipid profiles) after individual gene or a combination of multiple genes involving lipid metabolism. For example, ESI-MS/MS analysis was performed to evaluate the PC species from oilseed expressing heterologous desaturases and an elongase under the control of seed-specific promoters [68]. It was demonstrated that PC species acted as a substrate for the desaturases. Moreover, analysis of the lipid profiles indicated that the transgenic enzymes acted on all PC molecular species, rather than on a subset of PC species. Collectively, lipidomics provides the details of lipid species that are important for improving food quality.

REFERENCES

1. Welti, R., Shah, J., Li, W., Li, M., Chen, J., Burke, J.J., Fauconnier, M.L., Chapman, K., Chye, M.L. and Wang, X. (2007) Plant lipidomics: Discerning biological function by profiling plant complex lipids using mass spectrometry. Front. Biosci. 12, 2494–2506.

2. Welti, R. and Wang, X. (2004) Lipid species profiling: A high-throughput approach to identify lipid compositional changes and determine the function of genes involved in lipid metabolism and signaling. Curr. Opin. Plant Biol. 7, 337–344.

3. Isaac, G., Jeannotte, R., Esch, S.W. and Welti, R. (2007) New mass-spectrometry-based strategies for lipids. Genet. Eng. (N Y) 28, 129–157.

4. Brugger, B., Erben, G., Sandhoff, R., Wieland, F.T. and Lehmann, W.D. (1997) Quantitative analysis of biological membrane lipids at the low picomole level by nano-electrospray ionization tandem mass spectrometry. Proc. Natl. Acad. Sci. U. S. A. 94, 2339–2344.

5. Yang, K., Cheng, H., Gross, R.W. and Han, X. (2009) Automated lipid identification and quantification by multi-dimensional mass spectrometry-based shotgun lipidomics. Anal. Chem. 81, 4356–4368.

6. Welti, R., Li, W., Li, M., Sang, Y., Biesiada, H., Zhou, H.-E., Rajashekar, C.B., Williams, T.D. and Wang, X. (2002) Profiling membrane lipids in plant stress responses. Role of phospholipase Da in freezing-induced lipid changes in Arabidopsis. J. Biol. Chem. 277, 31994–32002.

7. Welti, R., Wang, X. and Williams, T.D. (2003) Electrospray ionization tandem mass spectrometry scan modes for plant chloroplast lipids. Anal. Biochem. 314, 149–152.

8. Schrick, K., Shiva, S., Arpin, J.C., Delimont, N., Isaac, G., Tamura, P. and Welti, R. (2012) Steryl glucoside and acyl steryl glucoside analysis of Arabidopsis seeds by electrospray ionization tandem mass spectrometry. Lipids 47, 185–193.

9. Shiva, S., Vu, H.S., Roth, M.R., Zhou, Z., Marepally, S.R., Nune, D.S., Lushington, G.H., Visvanathan, M. and Welti, R. (2013) Lipidomic analysis of plant membrane lipids by direct infusion tandem mass spectrometry. Methods Mol. Biol. 1009, 79–91.

10. Iijima, Y., Nakamura, Y., Ogata, Y., Tanaka, K., Sakurai, N., Suda, K., Suzuki, T., Suzuki, H., Okazaki, K., Kitayama, M., Kanaya, S., Aoki, K. and Shibata, D. (2008) Metabolite annotations based on the integration of mass spectral information. Plant J. 54, 949–962.

11. Southam, A.D., Payne, T.G., Cooper, H.J., Arvanitis, T.N. and Viant, M.R. (2007) Dynamic range and mass accuracy of wide-scan direct infusion nanoelectrospray fourier transform ion cyclotron resonance mass spectrometry-based metabolomics increased by the spectral stitching method. Anal. Chem. 79, 4595–4602.

12. Esch, S.W., Tamura, P., Sparks, A.A., Roth, M.R., Devaiah, S.P., Heinz, E., Wang, X., Williams, T.D. and Welti, R. (2007) Rapid characterization of the fatty acyl composition of complex lipids by collision-induced dissociation time-of-flight mass spectrometry. J. Lipid Res. 48, 235–241.

13. Burgos, A., Szymanski, J., Seiwert, B., Degenkolbe, T., Hannah, M.A., Giavalisco, P. and Willmitzer, L. (2011) Analysis of short-term changes in the *Arabidopsis thaliana* glycerolipidome in response to temperature and light. Plant J. 66, 656–668.

14. Samarakoon, T., Shiva, S., Lowe, K., Tamura, P., Roth, M.R. and Welti, R. (2012) *Arabidopsis thaliana* membrane lipid molecular species and their mass spectral analysis. Methods Mol. Biol. 918, 179–268.

15. Moreau, R.A., Doehlert, D.C., Welti, R., Isaac, G., Roth, M., Tamura, P. and Nunez, A. (2008) The identification of mono-, di-, tri-, and tetragalactosyl-diacylglycerols and their natural estolides in oat kernels. Lipids 43, 533–548.

16. Wang, W., Liu, Z., Ma, L., Hao, C., Liu, S., Voinov, V.G. and Kalinovskaya, N.I. (1999) Electrospray ionization multiple-stage tandem mass spectrometric analysis of diglycosyldiacylglycerol glycolipids from the bacteria Bacillus pumilus. Rapid Commun. Mass Spectrom. 13, 1189–1196.

17. Devaiah, S.P., Roth, M.R., Baughman, E., Li, M., Tamura, P., Jeannotte, R., Welti, R. and Wang, X. (2006) Quantitative profiling of polar glycerolipid species from organs of wild-type Arabidopsis and a phospholipase Dalpha1 knockout mutant. Phytochemistry 67, 1907–1924.

18. Han, X. and Cheng, H. (2005) Characterization and direct quantitation of cerebroside molecular species from lipid extracts by shotgun lipidomics. J. Lipid Res. 46, 163–175.

19. Markham, J.E. and Jaworski, J.G. (2007) Rapid measurement of sphingolipids from *Arabidopsis thaliana* by reversed-phase high-performance liquid chromatography coupled to electrospray ionization tandem mass spectrometry. Rapid Commun. Mass Spectrom. 21, 1304–1314.

20. Markham, J.E., Li, J., Cahoon, E.B. and Jaworski, J.G. (2006) Separation and identification of major plant sphingolipid classes from leaves. J. Biol. Chem. 281, 22684–22694.

21. Merrill, A.H., Jr., Sullards, M.C., Allegood, J.C., Kelly, S. and Wang, E. (2005) Sphingolipidomics: High-throughput, structure-specific, and quantitative analysis of sphingolipids by liquid chromatography tandem mass spectrometry. Methods 36, 207–224.

22. Jiang, X., Cheng, H., Yang, K., Gross, R.W. and Han, X. (2007) Alkaline methanolysis of lipid extracts extends shotgun lipidomics analyses to the low abundance regime of cellular sphingolipids. Anal. Biochem. 371, 135–145.

23. Hsu, F.F., Turk, J., Zhang, K. and Beverley, S.M. (2007) Characterization of inositol phosphorylceramides from Leishmania major by tandem mass spectrometry with electrospray ionization. J. Am. Soc. Mass Spectrom. 18, 1591–1604.

24. Benveniste, P. (2004) Biosynthesis and accumulation of sterols. Annu. Rev. Plant Biol. 55, 429–457.

25. Wewer, V., Dombrink, I., vom Dorp, K. and Dormann, P. (2011) Quantification of sterol lipids in plants by quadrupole time-of-flight mass spectrometry. J. Lipid Res. 52, 1039–1054.

26. Cedergren, R.A. and Hollingsworth, R.I. (1994) Occurrence of sulfoquinovosyl diacylglycerol in some members of the family Rhizobiaceae. J. Lipid Res. 35, 1452–1461.

27. Basconcillo, L.S., Zaheer, R., Finan, T.M. and McCarry, B.E. (2009) A shotgun lipidomics approach in Sinorhizobium meliloti as a tool in functional genomics. J. Lipid Res. 50, 1120–1132.

28. Li, C., Guan, Z., Liu, D. and Raetz, C.R. (2011) Pathway for lipid A biosynthesis in *Arabidopsis thaliana* resembling that of Escherichia coli. Proc. Natl. Acad. Sci. U. S. A. 108, 11387–11392.

29. Berry, J.A. and Bjorkman, O. (1980) Photosynthetic response and adaptation to temperature in higher plants. Annu. Rev. Plant Biol. 31, 491–543.

30. Uemura, M., Joseph, R.A. and Steponkus, P.L. (1995) Cold acclimation of *Arabidopsis thaliana* (Effect on plasma membrane lipid composition and freeze-induced lesions). Plant Physiol. 109, 15–30.

31. Thomashow, M.F. (1999) Plant cold acclimation: Freezing tolerance genes and regulatory mechanisms. Annu. Rev. Plant. Physiol. Plant. Mol. Biol. 50, 571–599.

32. Marcum, K.B. (1998) Cell membrane thermostability and whole plant heat tolerance of Kentucky bluegrass. Crop. Sci. 38, 1214–1218.

33. Gorver, A., Agarwal, M., Katiyar-Argarwal, S., Sahi, C. and Argarwal, S. (2000) Production of high temperature tolerance transgenic plants through manipulation of membrane lipids. Curr. Sci. 79, 557–559.

34. Falcone, D.L., Ogas, J.P. and Somerville, C.R. (2004) Regulation of membrane fatty acid composition by temperature in mutants of Arabidopsis with alterations in membrane lipid composition. BMC Plant Biol. 4, 17.

35. Larkindale, J. and Huang, B. (2004) Changes of lipid composition and saturation level in leaves and roots for heat-stressed and heat acclimated creeping bentgrass (Agrostis stolonifera). Environ. Exp. Bot. 51, 57–67.

36. Barkan, L., Vijayan, P., Carlsson, A.S., Mekhedov, S. and Browse, J. (2006) A suppressor of fab1 challenges hypotheses on the role of thylakoid unsaturation in photosynthetic function. Plant Physiol. 141, 1012–1020.

37. Wang, X., Devaiah, S.P., Zhang, W. and Welti, R. (2006) Signaling functions of phosphatidic acid. Prog. Lipid Res. 45, 250–278.

38. Degenkolbe, T., Giavalisco, P., Zuther, E., Seiwert, B., Hincha, D.K. and Willmitzer, L. (2012) Differential remodeling of the lipidome during cold acclimation in natural accessions of *Arabidopsis thaliana*. Plant J. 72, 972–982.

39. Burke, J.J., O'Mahony, P.J. and Oliver, M.J. (2000) Isolation of Arabidopsis mutants lacking components of acquired thermotolerance. Plant Physiol. 123, 575–588.

40. Chen, J., Burke, J.J., Xin, Z., Xu, C. and Velten, J. (2006) Characterization of the Arabidopsis thermosensitive mutant atts02 reveals an important role for galactolipids in thermotolerance. Plant Cell Environ. 29, 1437–1448.

41. Suss, K.H. and Yordanov, I.T. (1986) Biosynthetic cause of in vivo acquired thermotolerance of photosynthetic light reactions and metabolic responses of chloroplasts to heat stress. Plant Physiol. 81, 192–199.

42. Webb, M.S. and Green, B.R. (1991) Biochemical and biophysical properties of thylakoid acyl lipids. Biochim. Biophys. Acta 1060, 133–158.

43. Quinn, P.J. (1988) Effects of temperature on cell membranes. Symp. Soc. Exp. Biol. 42, 237–258.

44. Alfonso, M., Yruela, I., Almarcegui, S., Torrado, E., Perez, M.A. and Picorel, R. (2001) Unusual tolerance to high temperatures in a new herbicide-resistant D1 mutant from Glycine max (L.) Merr. cell cultures deficient in fatty acid desaturation. Planta 212, 573–582.

45. Zheng, G., Tian, B., Zhang, F., Tao, F. and Li, W. (2011) Plant adaptation to frequent alterations between high and low temperatures: Remodelling of membrane lipids and maintenance of unsaturation levels. Plant Cell Environ. 34, 1431–1442.

46. Chen, M., Markham, J.E. and Cahoon, E.B. (2012) Sphingolipid Delta8 unsaturation is important for glucosylceramide biosynthesis and low-temperature performance in Arabidopsis. Plant J. 69, 769–781.

47. Tsegaye, Y., Richardson, C.G., Bravo, J.E., Mulcahy, B.J., Lynch, D.V., Markham, J.E., Jaworski, J.G., Chen, M., Cahoon, E.B. and Dunn, T.M. (2007) Arabidopsis mutants lacking long chain base phosphate lyase are fumonisin-sensitive and accumulate trihydroxy-18:1 long chain base phosphate. J. Biol. Chem. 282, 28195–28206.

48. Chen, M., Markham, J.E., Dietrich, C.R., Jaworski, J.G. and Cahoon, E.B. (2008) Sphingolipid long-chain base hydroxylation is important for growth and regulation of sphingolipid content and composition in Arabidopsis. Plant Cell 20, 1862–1878.

49. Chao, D.Y., Gable, K., Chen, M., Baxter, I., Dietrich, C.R., Cahoon, E.B., Guerinot, M.L., Lahner, B., Lu, S., Markham, J.E., Morrissey, J., Han, G., Gupta, S.D., Harmon, J.M., Jaworski, J.G., Dunn, T.M. and Salt, D.E. (2011) Sphingolipids in the root play an important role in regulating the leaf ionome in *Arabidopsis thaliana*. Plant Cell 23, 1061–1081.

50. Roudier, F., Gissot, L., Beaudoin, F., Haslam, R., Michaelson, L., Marion, J., Molino, D., Lima, A., Bach, L., Morin, H., Tellier, F., Palauqui, J.C., Bellec, Y., Renne, C., Miquel, M., Dacosta, M., Vignard, J., Rochat, C., Markham, J.E., Moreau, P., Napier, J. and Faure, J.D. (2010) Very-long-chain fatty acids are involved in polar auxin transport and developmental patterning in Arabidopsis. Plant Cell 22, 364–375.

51. Markham, J.E., Molino, D., Gissot, L., Bellec, Y., Hematy, K., Marion, J., Belcram, K., Palauqui, J.C., Satiat-Jeunemaitre, B. and Faure, J.D. (2011) Sphingolipids containing very-long-chain fatty acids define a secretory pathway for specific polar plasma membrane protein targeting in Arabidopsis. Plant Cell 23, 2362–2378.

52. Saucedo-Garcia, M., Guevara-Garcia, A., Gonzalez-Solis, A., Cruz-Garcia, F., Vazquez-Santana, S., Markham, J.E., Lozano-Rosas, M.G., Dietrich, C.R., Ramos-Vega, M., Cahoon, E.B. and Gavilanes-Ruiz, M. (2011) MPK6, sphinganine and the LCB2a gene from serine palmitoyltransferase are required in the signaling pathway that mediates cell death induced by long chain bases in Arabidopsis. New Phytol. 191, 943–957.

53. Buseman, C.M., Tamura, P., Sparks, A.A., Baughman, E.J., Maatta, S., Zhao, J., Roth, M.R., Esch, S.W., Shah, J., Williams, T.D. and Welti, R. (2006) Wounding stimulates the accumulation of glycerolipids containing oxophytodienoic acid and dinor-oxophytodienoic acid in Arabidopsis leaves. Plant Physiol. 142, 28–39.

54. Hartel, H., Dormann, P. and Benning, C. (2000) DGD1-independent biosynthesis of extraplastidic galactolipids after phosphate deprivation in Arabidopsis. Proc. Natl. Acad. Sci. U. S. A. 97, 10649–10654.

55. Li, M., Welti, R. and Wang, X. (2006) Quantitative profiling of Arabidopsis polar glycerolipids in response to phosphorus starvation. Roles of phospholipases D zeta1 and D zeta2 in phosphatidylcholine hydrolysis and digalactosyldiacylglycerol accumulation in phosphorus-starved plants. Plant Physiol.. 142, 750–761.

56. Stewart, J.D. (1975) Fiber initiation on the cotton ovule (Gossypium hirsutum). Am. J. Bot. 62, 723–730.

57. Wanjie, S.W., Welti, R., Moreau, R.A. and Chapman, K.D. (2005) Identification and quantification of glycerolipids in cotton fibers: Reconciliation with metabolic pathway predictions from DNA databases. Lipids 40, 773–785.

58. Fauconnier, M.L., Welti, R., Blee, E. and Marlier, M. (2003) Lipid and oxylipin profiles during aging and sprout development in potato tubers (Solanum tuberosum L.). Biochim. Biophys. Acta 1633, 118–126.

59. Durrant, W.E. and Dong, X. (2004) Systemic acquired resistance. Annu. Rev. Phytopathol. 42, 185–209.

60. Shah, J. (2005) Lipids, lipases, and lipid-modifying enzymes in plant disease resistance. Annu. Rev. Phytopathol. 43, 229–260.

61. Kachroo, P., Shanklin, J., Shah, J., Whittle, E.J. and Klessig, D.F. (2001) A fatty acid desaturase modulates the activation of defense signaling pathways in plants. Proc. Natl. Acad. Sci. U. S. A. 98, 9448–9453.

62. Nandi, A., Krothapalli, K., Buseman, C.M., Li, M., Welti, R., Enyedi, A. and Shah, J. (2003) Arabidopsis sfd mutants affect plastidic lipid composition and suppress dwarfing, cell death, and the enhanced disease resistance phenotypes resulting from the deficiency of a fatty acid desaturase. Plant Cell 15, 2383–2398.

63. Nandi, A., Welti, R. and Shah, J. (2004) The *Arabidopsis thaliana* dihydroxyacetone phosphate reductase gene SUPPRESSSOR OF FATTY ACID DESATURASE DEFICIENCY1 is required for glycerolipid metabolism and for the activation of systemic acquired resistance. Plant Cell 16, 465–477.

64. Vance, D.E. and Vance, J.E. (2008) Biochemistry of Lipids, Lipoproteins and Membranes. Elsevier Science B.V., Amsterdam. pp 631.

65. Li, W., Li, M., Zhang, W., Welti, R. and Wang, X. (2004) The plasma membrane-bound phospholipase Ddelta enhances freezing tolerance in *Arabidopsis thaliana*. Nat. Biotechnol. 22, 427–433.

66. Misson, J., Raghothama, K.G., Jain, A., Jouhet, J., Block, M.A., Bligny, R., Ortet, P., Creff, A., Somerville, S., Rolland, N., Doumas, P., Nacry, P., Herrerra-Estrella, L., Nussaume, L. and Thibaud, M.C. (2005) A genome-wide transcriptional analysis using *Arabidopsis thaliana* Affymetrix gene chips determined plant responses to phosphate deprivation. Proc. Natl. Acad. Sci. U. S. A. 102, 11934–11939.

67. Li, M., Qin, C., Welti, R. and Wang, X. (2006) Double knockouts of phospholipases Dzeta1 and Dzeta2 in Arabidopsis affect root elongation during phosphate-limited growth but do not affect root hair patterning. Plant Physiol. 140, 761–770.

68. Abbadi, A., Domergue, F., Bauer, J., Napier, J.A., Welti, R., Zahringer, U., Cirpus, P. and Heinz, E. (2004) Biosynthesis of very-long-chain polyunsaturated fatty acids in transgenic oilseeds: Constraints on their accumulation. Plant Cell 16, 2734–2748.

19

LIPIDOMICS ON YEAST AND *MYCOBACTERIUM TUBERCULOSIS*

19.1 INTRODUCTION

Similar to the case of plant lipidomics (see Chapter 18), as development of lipidomics, MS-based lipidomics approaches for analysis of lipids present in yeast and bacteria have also been advanced accordingly. The branch of lipidomics in analysis of yeast membrane lipids was particularly named as "yeast lipidomics." Of course, all types of MS-based lipidomics approaches as described early in the book (Chapter 3), particularly those based on ESI-MS or ESI-MS/MS, could be used for the studies of yeast and mycobacterium lipidomes.

One of the yeasts, which were investigated the best, is the common bakers' yeast, *Saccharomyces cerevisiae*, for which both complete genome and detailed protein data sets are available. *S. cerevisiae* contains a relatively simple, but conserved network of lipid metabolism/catabolism pathways. This network controls the synthesis of hundreds of lipid molecular species [1], and therefore its full lipidome. As compared in Chapter 16, the biosynthesis and metabolism of GPLs in the yeasts are very similar to those of higher eukaryotes, with three main exceptions.

- PS species in the yeasts are mainly synthesized by the CDP-DAG pathway and not by PS synthase from PE species.
- Both Kennedy pathway through CDP-choline and successive methylation from PE to PC catalyzed by *N*-methyltransferases [2, 3] play equal roles in

Lipidomics: Comprehensive Mass Spectrometry of Lipids, First Edition. Xianlin Han.
© 2016 John Wiley & Sons, Inc. Published 2016 by John Wiley & Sons, Inc.

biosynthesis of yeast PC species, whereas the methylation pathway does not play crucial roles in mammals under normal physiological conditions [4, 5].

- The yeasts generally contain relatively low abundance of PUFAs, whereas PUFAs are completely absent in *S. cerevisiae*. This is due to the activities of fatty acid synthase [6], Δ9 desaturase [7], and fatty acid elongases [6, 8], which largely contribute to the diversified lipid species present in the yeasts and only produce saturated or monounsaturated fatty acids.

In addition to the relatively simple and conserved network for lipid synthesis, numerous genes involved in lipid metabolism can be mutated or deleted without apparent physiological consequences. Therefore, *S. cerevisiae* has been used as a prime model organism for studying the molecular organization and regulatory circuitry of eukaryotic lipidomes for long times [9–11], as well as lipidomics.

In this chapter, a general protocol for analysis of yeast lipidome by MS is described, followed by the examples for the analysis of yeast lipidomes. The applications of yeast lipidomics for determining the lipid phenotypes of yeast gene mutation and deletion are also discussed.

On the other hand, bacteria constitute a domain of prokaryotic microorganisms. There are approximately 5×10^{30} bacteria on Earth [12], which form a biomass exceeding that of all plants and animals. Most bacteria have not been characterized. However, many studies reveal that there exist some special lipids in bacteria, which play crucial roles in bacterial functions. For example, lipopolysaccharides (also known as lipoglycans and endotoxin) consist of a Lipid A core structure and a polysaccharide; they are located in the outer membrane of Gram-negative bacteria and elicit strong immune responses in animals. This family of lipids was well characterized by MS and was extensively reviewed [13]. Another enriched and special lipid family in bacteria is plasmalogen species. Plasmalogen possesses unique functions, but not yet fully understood, in organisms from bacteria to protozoans. Their biosynthetic pathways in many anaerobic bacterial species differ from those in aerobic and anaerobic organisms. Lipidomics analysis of this family of lipids was also extensively reviewed [14]. These contexts are not further discussed in this chapter. It is advised to read the referenced review articles if one is interested in those areas of lipidomics studies. However, one of the most studied bacteria utilizing lipidomics is the *Mycobacterium tuberculosis*. *M. tuberculosis* is the causative agent of tuberculosis, which synthesizes and secretes a wide array of biologically active polyketide lipids (see Lipid MAPS lipid classification for the structures of this family of lipids [15]), in addition to cellular membrane lipids. The lipidomics studies on *M. tuberculosis* are summarized in the chapter.

19.2   YEAST LIPIDOMICS

19.2.1   Protocol for Analysis of Yeast Lipidomes by Mass Spectrometry

An extensive protocol for analysis of yeast lipidomes by MS was described [16]. This protocol should be very helpful for anyone who would like to perform yeast lipidomics. Some outlines from the protocol in combination with other studies are given below.

The yeasts are generally cultured in synthetic defined media as previously described [17]. This type of media can be readily available commercially. The yeasts are grown in 500 ml shake-flask cultures containing 50 ml of media. Growth of yeasts was monitored by measurements of the optical density at a wavelength of 600 nm (OD_{600}). An OD_{600} value of 1.0 corresponds to a cell dry weight of about 0.17 g/L. Cells are commonly harvested at 10 OD_{600} units for lipid analysis and pelleted by centrifugation (e.g., 3000×g for 5 min). The yeast should be kept on ice to prevent any hydrolysis by the phospholipase activities hereafter.

A normalization parameter (e.g., cell dry weight or protein content) should be selected and determined [16] so that final results can be compared between the samples. It is advised that at least one internal standard of each lipid class and subclass of interest should be added to the cell lysates prior to lipid extraction for quantification, profiling, or relative comparison regardless of which approach is employed for lipid analysis. Lipids from the harvested yeasts are commonly extracted utilizing solvents. Different solvent systems were used for the purpose (e.g., Bligh–Dyer extraction [18], two-step extraction [19], and ethanol-based extraction [16]). It appears that all of the extraction methods described in Chapter 13 could be used.

With the lipid extracts from yeast cultures, either a shotgun lipidomics or an LC-MS (or LC-MS/MS) approach can be employed for lipid analysis. Generally, the shotgun lipidomics approach is more suitable for global lipidomics analysis [19], whereas LC-MS/MS in an MRM mode is usually more toward to the targeted analysis of lipid classes or individual molecular species [16]. However, high mass accuracy-based LC-MS utilizing FTICR-MS should be useful for global lipid analysis since the instrument provides an accurate mass measurement to the element composition as previously demonstrated [18]. It is worth noting that the MDMS-SL platform should be able to analyze the majority of the lipid classes, subclasses, and individual molecular species present in the yeasts although the platform was only used for analysis of their cardiolipin molecular species [20, 21].

From the acquired spectra, different approaches use different data processing tools to identify and quantify lipid ions or species. A few software packages are commercially available for this purpose (see Chapter 5). For example, LipidView available from Sciex could be used for the ion peak-based approaches (e.g., tandem MS-based shotgun lipidomics), top-down shotgun lipidomics, and targeted lipid species (molecular ion)-based (e.g., LC-MS/MS in an MRM mode); Lipid-Blast available from Dr. Oliver Fiehn laboratory could be employed for accurate mass-based (high mass accuracy MS-based or top-down approach) methods; and LipidSearch available from Thermo Fisher Scientific could be exploited to process any lipidomics data based on product-ion analysis (e.g., top-down shotgun lipidomics and LC-MS/MS in the data-dependent mode). However, many laboratories develop their home-made programs to process the lipidomics data from the acquired mass spectra. For example, the MDMS-SL platform uses their own AMDMS-SL program [22] and Welti's laboratory developed their own program to process their tandem MS-based shotgun lipidomics data [23].

19.2.2 Quantitative Analysis of Yeast Lipidome

Recently, a quantitative analysis of yeast lipidome was performed by high accuracy mass spectrometry-based shotgun lipidomics [19]. The researchers employed a two-step extraction procedure to maximize the recovery and relatively separate different lipid classes based on their polarities. This step greatly enhances the coverage of the analysis. Specifically, the scientists first extracted the yeast cell lysates spiked with lipid internal standards with chloroform/methanol (17:1, v/v) and then extracted the residual aqueous phase with chloroform/methanol (2:1, v/v). Through this two-step extraction procedure, 80–99% of lipid species among relatively apolar lipid classes, including ergosterol, TAG, DAG, PC, lysoPC, PE, PG, long-chain sphingoid base, and Cer, are recovered first, and 74–95% of species among relatively polar lipid classes, including PA, lysoPA, PS, lysoPS, PI, lysoPI, CL, long-chain sphingoid phosphate, IPC, mannosyl-inositolphosphoceramide (MIPC), and mannosyl-diinositolphosphoceramide (M(IP)$_2$C), were recovered in the second step. It was found that this two-step lipid extraction procedure increases the recovery of M(IP)$_2$C by fourfold. The scientists further found that with the internal standards, differential partition of individual lipid species of a class into the solvent phases is minimal and the effects of this differential partitioning can be ignored.

Lipid species recovered from the two-step extraction were then analyzed by six sequentially and automatically performed MS and MS/MS experiments followed by data processing using the dedicated software [24]. A broad dynamic quantification range covering three to 4 orders of magnitude and a detection limit in the low picomole range were obtained through addition of a small amount of modifier (0.2 mM ethylamine). The quantitative analysis covered 21 major lipid classes of the 30 yeast lipid classes and 162 molecular species present in wild-type *S. cerevisiae* BY4741, representing >95% of the yeast lipid content in estimation. In the analysis, several low-abundant biosynthetic intermediates, including phosphoinositides and ergosterol esters [9], were also detected, but not quantified due to the lack of applicable internal standards.

Among the analyzed lipid species, the most abundant one was ergosterol, accounting for 12.0 mol% of the lipidome (equivalent to 481 pmol/0.2 OD units), followed by 8.9 mol% M(IP)$_2$C 18:0;3/26:0;1. It was determined that the corresponding sphingolipid intermediates of this M(IP)$_2$C species, including Cer, IPC, and MIPC, are 128-, 14-, and sixfold, respectively, of less abundance. This result indicates that the IPC, MIPC, and M(IP)$_2$C synthetases efficiently shunt the corresponding substrates under steady-state conditions.

From the lipidomics analysis, the researchers found that the lipidome present in wild-type *S. cerevisiae* BY4741 contains 20.3, 14.9, 14.3, 8.3, and 1.8 mol% of PI, PE, PC, PA, and PS, respectively, relative to the all analyzed lipid content. In contrast to the previous belief that the content of PC dominates the yeast glycerophospholipidome [25], it was found that PI represents the major lipid class present in wild-type yeast strains of BY4742, BY4743, NY13, and CTY182, accounting for 17–30% of their lipidomes. It was also found that 16:1, 18:1, and 16:0 FA constituents represent the most-abundant FA moieties among GPL species, which is consistent with the previous findings [25, 26].

Lipidomics analysis also revealed that *S. cerevisiae* modulated its lipid composition in three major ways in response to an elevated growth temperature (e.g., 37 °C) in comparison to the growth at 24 °C. First, it produced higher contents of PI species, but less content of PE and TAG species. Second, it upregulated the levels of GPL species containing 16:0 and 18:1 FA constituents. Lastly, it synthesized higher contents of sphingolipids containing a longer chain sphingoid base (i.e., 20:0;3). For example, a 18-fold increase in 20:0;3/26:0;1 M(IP)$_2$C was detected in the study.

19.2.3   Comparative Lipidomics Studies on Different Yeast Strains

Recently, lipidomics analysis was performed to address if phylogenetically different yeast strains possess distinct GPL profiles and if genetically closely related strains show analogies in their composition of GPL species. The researchers employed an HPLC/LIT-FTICR-MS approach as previously described [27, 28] to profile GPL species of five different yeast strains, including *S. cerevisiae*, *Saccharomyces bayanus*, *Kluyveromyces thermotolerans*, *Pichia angusta*, and *Yarrowia lipolytica*, grown under identical environmental conditions to minimize the effects of external conditions on variations of GPL profiles [18]. Among these, four yeast strains with few genetic analogies and not closely related to each other were selected since *S. cerevisiae* and *S. bayanus* are the genetically related ones. In the study, over 100 molecular species among nine classes of GPL classes were quantified. It was found that a characteristic profile of GPL corresponding to each strain was present. However, genetically related yeast strains (e.g., *S. cerevisiae* and *S. bayanus*) showed their great similarities in their profiles.

In the comparative study, over 100 GPL molecular species of CL, PA, PE, *N*-monomethyl phosphatidylethanolamine (M*M*ePE), *N,N*-dimethylphosphatidylethanolamine (D*M*ePE), PC, PI, PS, and PG present in the aforementioned five yeast strains were profiled. Comparison between these profiles showed that significant differences in number, distribution, and relative levels of the identified GPL species existed among the phylogenetically different yeast strains. Generally, the number of determined species was less in *S. cerevisiae* and *S. bayanus*, whereas *K. thermotolerans*, *P. angusta*, and *Y. lipolytica* possessed a larger variety of GPL species. In addition to the difference of the numbers, the composition of the major GPL species in the classes was also significantly different between the genetically diverse yeasts, whereas the patterns of the related yeast strains showed great similarities. Specifically, the patterns of *S. cerevisiae* and *S. bayanus* were very similar, both of which contain four major species, possessing relatively short FA chains and less number of double bonds. For example, 32:2 and 34:2 PE, and 32:2 and 34:2 PC represented the major species in the yeast *S. bayanus*. The yeast *Y. lipolytica* possessed much longer FA chains and higher degree of unsaturation. In contrast, the lipid patterns present in the yeasts *K. thermotolerans* and *P. angusta* were rather complex, containing even longer FA chains and much higher number of double bonds compared to the other three strains.

19.2.4   Lipidomics of Yeast for Lipid Biosynthesis and Function

As a model system, yeast has many advantages for the study of lipid biology, which are as follows:

- Convenient experimental tractability, due to simple growth conditions and facile genetic manipulation compared to mammalian cells.
- Cultivated in completely defined medium, allowing a precise control of physical and biochemical parameters, while mammalian cells usually grow robustly using cultivation with lipid-rich serum-containing media, which is particularly relevant for studying lipid metabolism, since cells can take up molecules as nutrients and fatty acids from the media, making the interpretation of phenotypes less clear.
- A well-curated database and a large collection of plasmids and genomic libraries that are available.
- The high degree of conservation of many fundamental metabolic pathways between yeast and mammals.

The broad spectrum of available yeast mutants provides useful tools to discover and/or characterize mutants involving lipid metabolism, transport, and turnover. For example, Daum et al. [25] systematically analyzed yeast strains to screen potential defects in lipid metabolism; Proszynski et al. [29] determined the connections of sphingolipids and ergosterol with cell surface delivery; and Hancock et al. [30] identified a large number of mutants displaying an overproduction of inositol phenotype associated with misregulation of GPL biosynthetic genes, including *INO1* encoding inositol-3-phosphate synthase [31]. Moreover, the available yeast deletion collection enabled scientists to determine the effects of gene loss and gene interactions at a genome-wide scale under different growth conditions (see review [2] for details). Despite the presence of some limitations in probing the roles of specific lipid species in cell signaling with the yeast model system, the opportunity in this area is still great, particularly in conjunction with the newly developing technologies such as metabolomics and lipidomics.

From the studies with the yeast model system, it was demonstrated that the underlying molecular mechanisms as well as the individual components of these essential eukaryotic processes were conserved. Even the names of many genes involving these processes in mammalian cell biology now derived from the original discovery of their homologs in yeast. For example, the yeast *OLE1* gene encoded the sole Δ9 FA desaturase in yeast and could be functionally replaced by the rat stearoyl-CoA desaturase gene [32]. The HMG-CoA reductase genes present in hamster and human could be used to rescue the lethality of hmg1/hmg2 yeast cells [33].

It should be recognized that although much of the fundamental metabolism of lipids in yeast is similar to that in mammalian cells, there exist some key differences between the yeast and mammalian cells. For example,

- Yeast sphingolipids contain inositol rather than choline, as is the case for sphingomyelin in mammalian cells [34].

- In contrast to mammals, the yeasts produce phytosphingosine and phytoceramide [34].
- The yeasts contain a less complex mixture of fatty acids (lack of PUFA unless supplied with nutritional sources) than mammalian cells, which makes the yeast only a few hundred lipid species in its lipidome [19, 35], in comparison to thousands in mammalian cells [1, 36].
- The yeasts synthesize PS species through the reaction of CDP-DAG with free serine, but mammalian cells synthesize PS via an exchange reaction with PE. Therefore, in the yeasts, CDP-DAG serves as a precursor not only to PI and its derivatives, PG, PGP, and CL, as in mammalian cells, but also to PS, PE, and PC (through PE methylation).
- The yeasts synthesize ergosterol as the major sterol [37] as compared to cholesterol as the major one in mammalian cells.

These differences in fundamental lipid metabolism between the yeast and mammalian cells have implications for potential signaling mechanisms [38].

Overall, since lipids are indispensable molecules for physiology arose, studying the mechanisms that control lipid homeostasis, metabolism, trafficking, and function is a significant topic in cell biology. To this end, many detailed reviews on GPL, sphingolipid, and sterol metabolism have been published recently. Any readers who are interested in the topics should consult these articles for more details [11, 39, 40]. Below are a few examples studying altered lipid homeostasis, metabolism, trafficking, and function, and the potential molecular mechanism(s) leading to these changes by using lipidomics approaches.

Boumann et al. [41] employed ESI-MS/MS combined with stable isotope labeling to analyze different contribution of each of the two routes for PC biosynthesis to the FA composition of newly synthesized PC. The routes of PC biosynthesis include the PE methylation pathway and the CDP-choline pathway (also known as the Kennedy pathway) [42]. The researchers utilized mutants blocked in one of the two pathways and exclusively used the remaining route to allow the cells to synthesize PC species. It was found that, compared to PC species yielded from PE methylation route, the CDP-choline pathway-derived PC species were enriched for diunsaturated species because it was shown that the PE species containing two 16:1 FA chains was preferentially used as a precursor for methylation [41]. In the study, the investigators also revealed that PC species undergo remodeling at the sn-1 position of the glycerol backbone and that this remodeling is required to achieve the steady-state PC pattern.

Guan and Wenk [35] employed ESI-MS/MS to determine semiquantitative profiles of both GPL and sphingolipids from wild type and mutant $S.$ $cerevisiae$ strains after using a simple extraction procedure. The mutants studied include $slc1\Delta$, deletion of an acyl transferase, and $scs7\Delta$, lack of a lipid hydroxylase. It was shown that the $slc1\Delta$ strain yielded a GPL profile similar to that of the wild type except that the former did not produce short and saturated PI species when the cells were grown in media containing a long-chain sphingoid base. Differences in sphingolipids were also detected in the study, including an increase in short-chain sphingolipids, phytoceramides, and IPC species. In contrast, it was shown that the $scs7\Delta$ mutant contained a GPL profile similar to that of the wild type, but showed a drastic reduction of the levels of hydroxyl-containing and complex Cer species.

In another study, Boumann et al. [43] used ESI-MS/MS to analyze the levels of GPL species when the cells underwent depletion of cellular PC biosynthesis, which was achieved by using a mutant strain, *cho2Δ opi3Δ*, depleting the two methyl-transferases that catalyze the three-step methylation of PE to produce PC species. Thus, PC biosynthesis of this mutant strain completely depended on the CDP-choline pathway (i.e., the Kennedy pathway). This mutant strain stopped growing under the culture conditions without choline supplementation when PC levels dropped to <2% of total GPL [43]. It was found that as PC levels decreased, the levels of both PE and PI increased, and the remaining PC species underwent acyl chain remodeling, in which diunsaturated PC species were replaced by monounsaturated species. The researchers found that this remodeling process did not require the PLD encoded by the *PLD1/SPO14* gene. Moreover, as PC depletion progresses, cellular FA composition changed following a pattern with increases in short and unsaturated FA composition as occurred in PC species. Specifically, a 40% increase in FA chains containing 16 carbon atoms *vs.* 18 carbon FA chains was determined, accompanied by a 10% increase in the proportion of unsaturated FA species. The expression level of the FA desaturase encoded by the *OLE1* gene was increased as the FA composition changes [43]. Furthermore, as PC depleted, PE species showed the greatest change in FA chain composition, while neutral lipids were less affected. Finally, lipidomics results suggested that yeast cells possessed a regulatory mechanism to maintain the intrinsic membrane curvature since PE species containing the increased levels of shortening and saturated FA chains possessed less propensity to form nonbilayer structure [43].

Addressing the effects of nutritional deprivation or supplementation (e.g., inositol and/or choline) on lipid metabolism is always an interest topic in lipidology and the yeasts provide a powerful tool to study these effects [44, 45]. Lipidomics could provide deep insights into the biochemical mechanisms responsible for lipid changes under different nutritional conditions. For example, ESI-MS/MS analysis was performed to determine the effects of inositol deficiency on lipid metabolism and showed that the level of a PC species containing two acyl chains of myristic acid, i.e., dimyris-toyl PC (DMPC), was markedly increased under the conditions, accounting for a 40% of total PC [46]. In the study, it was demonstrated that similar low levels of DMPC were observed in cells grown in media containing inositol regardless of the presence of choline. Low levels of DMPC were also observed when cells were cultured in media lacking both inositol and choline. These observations suggest that the high DMPC levels that were present in cells grown under the condition of inositol deple-tion were uniquely associated with growth in the presence of choline. Furthermore, the investigators also observed a number of additional effects upon synthesis and turnover of both GPL and neutral lipids, suggesting that inositol addition exerted a profound effect upon the entire yeast lipidome.

Elo1, Elo2, and Elo3 are 3-ketoacyl-CoA synthases embedded in the ER [8] and involved in catalyzing the four-step cycle that successively elongates precursor acyl CoA by two methylene groups. Ejsing et al. [19] performed a comparative lipidome analysis of *elo1Δ*, *elo2Δ*, *elo3Δ*, and wild-type strains. In the analysis, the investigators quantified in a broad spectrum with a total of 250 molecular species among 21 lipid classes. They found that the contents of $M(IP)_2C$, MIPC, and IPC in *elo1Δ* cells were elevated by 1.4- to 3.8-fold, whereas the levels of PI, PC, and PE were reduced. Analysis of *elo1Δ* lipidome revealed that the deletion of Elo1 promoted elongation of shorter chain acyl CoA to C26:0 CoA mediated by Elo2 and Elo3 [19]. This promotion concomitantly enhanced the production of long-chain sphingoid bases for sphingolipid biosynthesis, and in turn reduced the concentration of substrate PI species. Furthermore, the study also showed that the *elo1Δ* mutant possessed more GPL species containing 14:0 and 14:1 fatty acid moieties, which corroborated the known substrate preference of Elo1 [47, 48]. Analyses of *elo2Δ* and *elo3Δ* lipidomes showed specific changes in the content and composition of sphingolipid and GPL species [19]. For example, the sphingolipidomes of *elo2Δ* and *elo3Δ* mutants contained less MIPC and $M(IP)_2C$ contents, but accumulated 2.4- to 9.0-fold more IPC compared with that of the wild type. Furthermore, the investigators revealed that despite the similar levels of PI and PC classes in the *elo2Δ* and *elo3Δ* cells, the composition of PI and PC molecular species were dramatically different. The *elo2Δ* mutant contained less PI and PC species with 16:0, 16:1, and shorter FA substituents. This reduction was compensated by increased content of the species containing 18:1 and 18:0 FA. In contrast, the *elo3Δ* lipidome showed differences primarily in the composition of PC species (and to a smaller extent within PE, PS, and PA) characterized by increased abundance of 16:1–16:1 and 14:1–16:1 PC, and reduced content of 16:0–16:1 and 16:0–18:1 PC compared with that of the wild type.

19.2.5   Determining the Effects of Growth Conditions on Yeast Lipidomes

Klose et al. performed a systematic lipidomics study to investigate the influence of a variety of commonly used growth conditions on the lipidome of wild-type strain BY4741 [49]. The growth conditions tested include temperature (i.e., 15, 24, 30, and 37 °C), carbon source (i.e., glucose, raffinose, and glycerol), different growth phases (i.e., early and middle logarithmic, as well as early and stationary phase), complete (YP) *vs.* synthetic complete (SC) medium, galactose-induced overexpression of a cytosolic, as well as a transmembrane protein, and the presence or absence of the selective drug G418.

The investigators found four major observations from the study as follows:

- The carbon source had a strong effect on the overall lipid composition. Specifically, the lipidomes of the cells grown in raffinose and glycerol clustered together, while the lipidomes of the glucose-grown cells formed a separate cluster.

- The progression through the growth phase (i.e., from early logarithmic to stationary phase) caused remarkable changes in the cellular lipidome. This was manifest with the very distant cluster formed by lipidomes of the mid-log (OD3.5), early stationary (OD6), and late stationary (ODstat) yeast cultures.

- There was a pronounced effect of decreasing the growth temperature to 15 °C on lipidomes. Specifically, the lipidomes of cells grown at 24, 30, and 37 °C showed great similarity and were within the "glucose cluster," whereas the lipidome of cells grown at 15 °C appeared to be significantly different.

- It appears that the galactose-induced overexpression of a cytosolic or a transmembrane protein (within an induction time of up to 3 h), the presence of the selective drug G418, and the different mating types did not cause pronounced changes in the lipid composition of the yeast cells. The lipidomes of galactose-induced cells were within the part of the "nonglucose cluster," while the lipidomes of the different mating types clustered tightly with the lipidomes of glucose-grown cells.

19.3 *MYCOBACTERIUM TUBERCULOSIS* LIPIDOMICS

Mycobacterium tuberculosis (*M. tuberculosis*) is a pathogenic bacterial species in the family Mycobacteriaceae and the causative agent of most cases of tuberculosis. Lipids play critical roles in the biology of *M. tuberculosis*. The majority of the lipid classes are essentially identical to those described for mammals, plants, and/or yeast, and, therefore, to the fragmentation patterns characterized for those lipids and analysis strategies designed for those species. However, *M. tuberculosis* contains some special lipid classes [50], which have not been discussed so far, such as mycolic acids and methylglucose lipopolysaccharides. Indeed, of the 58 lipid classes already identified in *M. tuberculosis*, 40 of them have no counterparts in eukaryotic cells or even Gram-negative bacteria [51].

One example of these special lipid classes is mycolic acids. Mycolic acids are long fatty acids (Figure 19.1) and found in the cell walls of the mycolata taxon, a group of bacteria that includes *M. tuberculosis*. They form the major component of the cell wall of mycolata species. *M. tuberculosis* produces three main types of mycolic acids: α-, methoxy-, and keto- (Figure 19.1). Alpha-mycolic acids comprise at least 70% of the mycolic acids present in the organism and contain several cyclopropane rings. Methoxy-mycolic acids, which contain several methoxy groups, comprise between 10% and 15% of the mycolic acids in the organism. The remaining 10–15% of the mycolic acids are keto-mycolic acids, which contain several ketone groups. As discussed in Chapter 3, carboxylate-containing mycolic acids can be readily ionized in the negative-ion mode, particularly under basic conditions. Hsu et al. extensively characterized the mycolic acids in the negative-ion mode after CID [52]. It was found that α-mycolic acid gave rise to an abundant fragment ion corresponding to the loss of a meroaldehyde from the meromycolate chain, leading to the formation of a carboxylate anion containing α-alkyl chain. The structural information from these fragment ions afforded structural assignment of the mycolic acids, including the lengths of the meromycolate chain and of the α-branch. Therefore, PIS of these fragment ions might provide a simple specific means for classification of mycolic species; whereas NLS of meroaldehyde residue provided a simple approach to

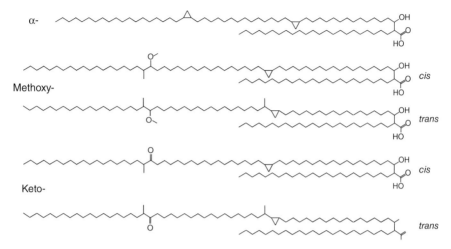

Figure 19.1   Representative molecular structures of mycolic acids. The representative structures of three subclasses of mycolic acids including α-, methoxy-, and keto-mycolic acids are depicted. Note that the number of carbon atoms in both meromycolate and α-branch chains can be varied.

reveal the mycolic acid molecules with specific meromycolate chain in mixtures as previously demonstrated [52].

Another example of the special lipid classes present in *M. tuberculosis* is methylglucose lipopolysaccharides. Characterization and analysis of methylglucose lipopolysaccharides, with bacterial lipopolysaccharides in general, were extensively reviewed [13], in which a wide variety of MS-based applications that were implemented to the structural elucidation of lipopolysaccharides were covered. Special description given in the review was the tandem mass spectrometric methods and protocols for the analyses of lipid A, the endotoxic principle of lipopolysaccharides. It is advised that advanced readers should look for the details from this invaluable review article [13].

Specific polyketide lipids present in *M. tuberculosis* interact with the host and are required for virulence. Mass spectrometric analysis was conducted to monitor this type of lipids [53]. It was discovered that the size and abundance of two lipid virulence factors, phthiocerol dimycocerosate and sulfolipid-1, were controlled by the availability of a common precursor, methyl malonyl CoA. These results suggested that growth of *M. tuberculosis* on fatty acids during infection led to increased flux of methyl malonyl CoA through lipid biosynthetic pathways, resulting in increased virulence lipid synthesis.

It is worthy noting that although Lipid MAPS database contained wealthy information about common lipids such as GPL and TAG species, many unique lipids found in Mycobacteria, such as sulfolipids and phenolic glycolipids, were not included in the database. To this end, Layre et al. [50] established a database similar to that of Lipid MAPS, termed MycoMass, which served to match ions identified by MS to known chemical structures. This database should serve as a powerful tool for the lipidomics study of *M. tuberculosis*. As more and more structures are identified from *M. tuberculosis,* as well as other lipid-rich bacteria, this tool should be of increasing benefit to the bacterial research community, since a comprehensive database certainly aids in elucidating their structures, identifying their biosynthetic pathways, and determining their role in *M. tuberculosis* biology.

REFERENCES

1. Yetukuri, L., Ekroos, K., Vidal-Puig, A. and Oresic, M. (2008) Informatics and computational strategies for the study of lipids. Mol. Biosyst. 4, 121–127.
2. Gaspar, M.L., Aregullin, M.A., Jesch, S.A., Nunez, L.R., Villa-Garcia, M. and Henry, S.A. (2007) The emergence of yeast lipidomics. Biochim. Biophys. Acta 1771, 241–254.
3. Carman, G.M. and Henry, S.A. (1999) Phospholipid biosynthesis in the yeast Saccharomyces cerevisiae and interrelationship with other metabolic processes. Prog. Lipid Res. 38, 361–399.
4. Vance, D.E., Li, Z. and Jacobs, R.L. (2007) Hepatic phosphatidylethanolamine N-methyltransferase, unexpected roles in animal biochemistry and physiology. J. Biol. Chem. 282, 33237–33241.
5. Wang, M., Kim, G.H., Wei, F., Chen, H., Altarejos, J. and Han, X. (2015) Improved method for quantitative analysis of methylated phosphatidylethanolamine species and its application for analysis of diabetic-mouse liver samples. Anal. Bioanal. Chem. 407, 5021–5032.
6. Tehlivets, O., Scheuringer, K. and Kohlwein, S.D. (2007) Fatty acid synthesis and elongation in yeast. Biochim. Biophys. Acta 1771, 255–270.
7. Martin, C.E., Oh, C.S. and Jiang, Y. (2007) Regulation of long chain unsaturated fatty acid synthesis in yeast. Biochim. Biophys. Acta 1771, 271–285.
8. Denic, V. and Weissman, J.S. (2007) A molecular caliper mechanism for determining very long-chain fatty acid length. Cell 130, 663–677.
9. Daum, G., Lees, N.D., Bard, M. and Dickson, R. (1998) Biochemistry, cell biology and molecular biology of lipids of Saccharomyces cerevisiae. Yeast 14, 1471–1510.
10. Carman, G.M. and Henry, S.A. (2007) Phosphatidic acid plays a central role in the transcriptional regulation of glycerophospholipid synthesis in Saccharomyces cerevisiae. J. Biol. Chem. 282, 37293–37297.
11. Dickson, R.C. (2008) Thematic review series: sphingolipids. New insights into sphingolipid metabolism and function in budding yeast. J. Lipid Res. 49, 909–921.
12. Whitman, W.B., Coleman, D.C. and Wiebe, W.J. (1998) Prokaryotes: the unseen majority. Proc. Natl. Acad. Sci. U. S. A. 95, 6578–6583.
13. Kilar, A., Dornyei, A. and Kocsis, B. (2013) Structural characterization of bacterial lipopolysaccharides with mass spectrometry and on- and off-line separation techniques. Mass Spectrom. Rev. 32, 90–117.
14. Rezanka, T., Kresinova, Z., Kolouchova, I. and Sigler, K. (2012) Lipidomic analysis of bacterial plasmalogens. Folia Microbiol. (Praha) 57, 463–472.
15. Fahy, E., Subramaniam, S., Brown, H.A., Glass, C.K., Merrill, A.H., Jr., Murphy, R.C., Raetz, C.R., Russell, D.W., Seyama, Y., Shaw, W., Shimizu, T., Spener, F., van Meer, G., VanNieuwenhze, M.S., White, S.H., Witztum, J.L. and Dennis, E.A. (2005) A comprehensive classification system for lipids. J. Lipid Res. 46, 839–861.

16. Guan, X.L., Riezman, I., Wenk, M.R. and Riezman, H. (2010) Yeast lipid analysis and quantification by mass spectrometry. Methods Enzymol. 470, 369–391.

17. Verduyn, C., Postma, E., Scheffers, W.A. and Van Dijken, J.P. (1992) Effect of benzoic acid on metabolic fluxes in yeasts: a continuous-culture study on the regulation of respiration and alcoholic fermentation. Yeast 8, 501–517.

18. Hein, E.M. and Hayen, H. (2012) Comparative lipidomic profiling of S. cerevisiae and four other hemiascomycetous yeasts. Metabolites 2, 254–267.

19. Ejsing, C.S., Sampaio, J.L., Surendranath, V., Duchoslav, E., Ekroos, K., Klemm, R.W., Simons, K. and Shevchenko, A. (2009) Global analysis of the yeast lipidome by quantitative shotgun mass spectrometry. Proc. Natl. Acad. Sci. U. S. A. 106, 2136–2141.

20. Claypool, S.M., Whited, K., Srijumnong, S., Han, X. and Koehler, C.M. (2011) Barth syndrome mutations that cause tafazzin complex lability. J. Cell Biol. 192, 447–462.

21. Baile, M.G., Sathappa, M., Lu, Y.W., Pryce, E., Whited, K., McCaffery, J.M., Han, X., Alder, N.N. and Claypool, S.M. (2014) Unremodeled and remodeled cardiolipin are functionally indistinguishable in yeast. J. Biol. Chem. 289, 1768–1778.

22. Yang, K., Cheng, H., Gross, R.W. and Han, X. (2009) Automated lipid identification and quantification by multi-dimensional mass spectrometry-based shotgun lipidomics. Anal. Chem. 81, 4356–4368.

23. Zhou, Z., Marepally, S.R., Nune, D.S., Pallakollu, P., Ragan, G., Roth, M.R., Wang, L., Lushington, G.H., Visvanathan, M. and Welti, R. (2011) LipidomeDB data calculation environment: online processing of direct-infusion mass spectral data for lipid profiles. Lipids 46, 879–884.

24. Ejsing, C.S., Duchoslav, E., Sampaio, J., Simons, K., Bonner, R., Thiele, C., Ekroos, K. and Shevchenko, A. (2006) Automated identification and quantification of glycerophospholipid molecular species by multiple precursor ion scanning. Anal. Chem. 78, 6202–6214.

25. Daum, G., Tuller, G., Nemec, T., Hrastnik, C., Balliano, G., Cattel, L., Milla, P., Rocco, F., Conzelmann, A., Vionnet, C., Kelly, D.E., Kelly, S., Schweizer, E., Schuller, H.J., Hojad, U., Greiner, E. and Finger, K. (1999) Systematic analysis of yeast strains with possible defects in lipid metabolism. Yeast 15, 601–614.

26. Schneiter, R., Brugger, B., Sandhoff, R., Zellnig, G., Leber, A., Lampl, M., Athenstaedt, K., Hrastnik, C., Eder, S., Daum, G., Paltauf, F., Wieland, F.T. and Kohlwein, S.D. (1999) Electrospray ionization tandem mass spectrometry (ESI-MS/MS) analysis of the lipid molecular species composition of yeast subcellular membranes reveals acyl chain-based sorting/remodeling of distinct molecular species en route to the plasma membrane. J. Cell Biol. 146, 741–754.

27. Hein, E.M., Blank, L.M., Heyland, J., Baumbach, J.I., Schmid, A. and Hayen, H. (2009) Glycerophospholipid profiling by high-performance liquid chromatography/mass spectrometry using exact mass measurements and multi-stage mass spectrometric fragmentation experiments in parallel. Rapid Commun. Mass Spectrom. 23, 1636–1646.

28. Hein, E.M., Bodeker, B., Nolte, J. and Hayen, H. (2010) Software tool for mining liquid chromatography/multi-stage mass spectrometry data for comprehensive glycerophospholipid profiling. Rapid Commun. Mass Spectrom. 24, 2083–2092.

29. Proszynski, T.J., Klemm, R.W., Gravert, M., Hsu, P.P., Gloor, Y., Wagner, J., Kozak, K., Grabner, H., Walzer, K., Bagnat, M., Simons, K. and Walch-Solimena, C. (2005) A genome-wide visual screen reveals a role for sphingolipids and ergosterol in cell surface delivery in yeast. Proc. Natl. Acad. Sci. U. S. A. 102, 17981–17986.

30. Hancock, L.C., Behta, R.P. and Lopes, J.M. (2006) Genomic analysis of the Opi- phenotype. Genetics 173, 621–634.

31. Greenberg, M.L. and Lopes, J.M. (1996) Genetic regulation of phospholipid biosynthesis in Saccharomyces cerevisiae. Microbiol. Rev. 60, 1–20.

32. Stukey, J.E., McDonough, V.M. and Martin, C.E. (1990) The OLE1 gene of Saccharomyces cerevisiae encodes the delta 9 fatty acid desaturase and can be functionally replaced by the rat stearoyl-CoA desaturase gene. J. Biol. Chem. 265, 20144–20149.

33. Basson, M.E., Thorsness, M., Finer-Moore, J., Stroud, R.M. and Rine, J. (1988) Structural and functional conservation between yeast and human 3-hydroxy-3-methylglutaryl coenzyme A reductases, the rate-limiting enzyme of sterol biosynthesis. Mol. Cell. Biol. 8, 3797–3808.

34. Perry, R.J. and Ridgway, N.D. (2005) Molecular mechanisms and regulation of ceramide transport. Biochim. Biophys. Acta 1734, 220–234.

35. Guan, X.L. and Wenk, M.R. (2006) Mass spectrometry-based profiling of phospholipids and sphingolipids in extracts from Saccharomyces cerevisiae. Yeast 23, 465–477.

36. Sampaio, J.L., Gerl, M.J., Klose, C., Ejsing, C.S., Beug, H., Simons, K. and Shevchenko, A. (2011) Membrane lipidome of an epithelial cell line. Proc. Natl. Acad. Sci. U. S. A. 108, 1903–1907.

37. Zinser, E., Paltauf, F. and Daum, G. (1993) Sterol composition of yeast organelle membranes and subcellular distribution of enzymes involved in sterol metabolism. J. Bacteriol. 175, 2853–2858.

38. Sturley, S.L. (2000) Conservation of eukaryotic sterol homeostasis: new insights from studies in budding yeast. Biochim. Biophys. Acta 1529, 155–163.

39. Hannich, J.T., Umebayashi, K. and Riezman, H. (2011) Distribution and functions of sterols and sphingolipids. Cold Spring Harb. Perspect. Biol. 3, a004762.

40. Carman, G.M. and Han, G.S. (2011) Regulation of phospholipid synthesis in the yeast Saccharomyces cerevisiae. Annu. Rev. Biochem. 80, 859–883.

41. Boumann, H.A., Damen, M.J., Versluis, C., Heck, A.J., de Kruijff, B. and de Kroon, A.I. (2003) The two biosynthetic routes leading to phosphatidylcholine in yeast produce different sets of molecular species. Evidence for lipid remodeling. Biochemistry 42, 3054–3059.

42. Kennedy, E.P. and Weiss, S.B. (1956) The function of cytidine coenzymes in the biosynthesis of phospholipides. J. Biol. Chem. 222, 193–214.

43. Boumann, H.A., Gubbens, J., Koorengevel, M.C., Oh, C.S., Martin, C.E., Heck, A.J., Patton-Vogt, J., Henry, S.A., de Kruijff, B. and de Kroon, A.I. (2006) Depletion of phosphatidylcholine in yeast induces shortening and increased saturation of the lipid acyl chains: evidence for regulation of intrinsic membrane curvature in a eukaryote. Mol. Biol. Cell 17, 1006–1017.

44. Kelley, M.J., Bailis, A.M., Henry, S.A. and Carman, G.M. (1988) Regulation of phospholipid biosynthesis in Saccharomyces cerevisiae by inositol. Inositol is an inhibitor of phosphatidylserine synthase activity. J. Biol. Chem. 263, 18078–18085.

45. Loewen, C.J., Gaspar, M.L., Jesch, S.A., Delon, C., Ktistakis, N.T., Henry, S.A. and Levine, T.P. (2004) Phospholipid metabolism regulated by a transcription factor sensing phosphatidic acid. Science 304, 1644–1647.

46. Gaspar, M.L., Aregullin, M.A., Jesch, S.A. and Henry, S.A. (2006) Inositol induces a profound alteration in the pattern and rate of synthesis and turnover of membrane lipids in Saccharomyces cerevisiae. J. Biol. Chem. 281, 22773–22785.

47. Toke, D.A. and Martin, C.E. (1996) Isolation and characterization of a gene affecting fatty acid elongation in Saccharomyces cerevisiae. J. Biol. Chem. 271, 18413–18422.

48. Schneiter, R., Tatzer, V., Gogg, G., Leitner, E. and Kohlwein, S.D. (2000) Elo1p-dependent carboxy-terminal elongation of C14:1Delta(9) to C16:1Delta(11) fatty acids in Saccharomyces cerevisiae. J. Bacteriol. 182, 3655–3660.

49. Klose, C., Surma, M.A., Gerl, M.J., Meyenhofer, F., Shevchenko, A. and Simons, K. (2012) Flexibility of a eukaryotic lipidome--insights from yeast lipidomics. PLoS One 7, e35063.

50. Layre, E., Sweet, L., Hong, S., Madigan, C.A., Desjardins, D., Young, D.C., Cheng, T.Y., Annand, J.W., Kim, K., Shamputa, I.C., McConnell, M.J., Debono, C.A., Behar, S.M., Minnaard, A.J., Murray, M., Barry, C.E., 3rd, Matsunaga, I. and Moody, D.B. (2011) A comparative lipidomics platform for chemotaxonomic analysis of Mycobacterium tuberculosis. Chem. Biol. 18, 1537–1549.

51. Chow, E.D. and Cox, J.S. (2011) TB lipidomics--the final frontier. Chem. Biol. 18, 1517–1518.

52. Hsu, F.F., Soehl, K., Turk, J. and Haas, A. (2011) Characterization of mycolic acids from the pathogen Rhodococcus equi by tandem mass spectrometry with electrospray ionization. Anal. Biochem. 409, 112–122.

53. Jain, M., Petzold, C.J., Schelle, M.W., Leavell, M.D., Mougous, J.D., Bertozzi, C.R., Leary, J.A. and Cox, J.S. (2007) Lipidomics reveals control of Mycobacterium tuberculosis virulence lipids via metabolic coupling. Proc. Natl. Acad. Sci. U. S. A. 104, 5133–5138.

20

LIPIDOMICS ON CELL ORGANELLE AND SUBCELLULAR MEMBRANES

20.1 INTRODUCTION

The function of biological cells is intimately associated with their internal organization, in which multiple subcellular structures have specialized roles. Such subcellular structures hold the secrets to normal cellular function, progression of disease, and the universe of interactions among biomolecules that define life. Initially observed in unicellular organisms, these subcellular structures were termed organelles, which literally mean "little organs," because of the parallel to the organ/body relationship in multicellular organisms [1]. In cell biology, an organelle is a specialized subunit within a cell that has a specific function. Individual organelles are usually separately enclosed within their own lipid bilayers. A eukaryotic cell contains many organelles, for example, the nucleus, ER, Golgi apparatus, mitochondrion, chloroplast (plastid). However, not all these organelles are found in only one cell or in an organism. The chloroplast, for instance, is abundant in plant cells but not in animal cells. The red blood cells (also known as erythrocytes) do not contain mitochondria. Prokaryotes, which were believed to have no organelles, have been recently described to possess "organelles." Examples are carboxysome (a protein-shell compartment for carbon fixation in some bacteria), chlorosome (a light harvesting complex in green sulfur bacteria), magnetosome (found in magnetotactic bacteria), and thylakoid (in some cyanobacteria) [2]. In order to understand the contribution of membrane lipid composition to the functionality of membrane-bound cellular processes, comprehensive and quantitative structural information on the organellar lipidome is essential [3].

Lipidomics: Comprehensive Mass Spectrometry of Lipids, First Edition. Xianlin Han.
© 2016 John Wiley & Sons, Inc. Published 2016 by John Wiley & Sons, Inc.

The subcellular membranes, on the other hand, refer to any membranes or membrane fractions present in cells. Studies on subcellular membranes pay more attention on the structures, compositions, and functions of the isolated membranes in comparison to the entire organelles. Unlike the organelles, a subcellular membrane may not carry a specific cellular function, although the majority of the subcellular membranes possess protein(s) that carry their properties. The subcellular membranes could be the membrane of an organelle or the parts of its membrane. For example, for the organelle of mitochondria, in addition to the mitochondrial membrane, outer and inner mitochondrial membranes as well as the membrane fraction of its contact points all belong to subcellular membranes. In addition to the membranes of the organelles mentioned earlier, plasma membrane and lipid rafts are examples frequently studied. Regarding the isolation of nonorganellar membranes, an approach that was widely utilized to characterize cellular membrane systems is the localization of enzymatic activities or marker proteins resident on the individual membranes. To this end, a few review articles for this purpose could be consulted [4–6].

In this chapter, an overview of the lipidomics research on organelles and subcellular membranes including those of cell organelles is provided, and characteristic features of the lipid composition of organelles and subcellular membranes are discussed.

20.2 GOLGI

The Golgi (also known as Golgi apparatus, Golgi complex, or Golgi body) is an organelle found in most eukaryotic cells, which was named after an Italian physician Camillo Golgi who identified the organelle in 1897 [7]. The Golgi apparatus packages proteins into membrane-bound vesicles inside the cell prior to sending to their destination. The Golgi resides at the intersection of the secretory, lysosomal, and endocytic pathways. This organelle plays multiple roles in cellular functions including lipid synthesis and transport, protein modification and sorting, lysosomal formation, etc. The unique function and membrane composition of the Golgi [8] make it great interest for lipidomics studies.

For instance, in a lipidomics study by shotgun lipidomics with a high mass accuracy mass spectrometer, lipid composition of Golgi-derived vesicles in the yeast strain KSY302 that overexpressed FusMidGFP through the fusion construct of FusMid and GFP was assessed [9]. In this study, Golgi-derived vesicles (i.e., post-Golgi secretory vesicles) and Golgi were separated by density-gradient centrifugation and immunoenriched with antibodies against FusMidGFP in order to study the raft-enriched lipid trafficking and the role of FusMid, which is a protein involved in vesicular transport from the Golgi to the plasma membrane. The immunoisolation procedure for specific recovery of the vesicles transporting a transmembrane raft protein (i.e., FusMid) from the Golgi to the cell surface in the yeast *Saccharomyces cerevisiae* was critical in the study. It was found that the vesicles contained 22.8 mol% of ergosterols and 11.1 mol% of $M(IP)_2C$, whereas the Golgi fraction consisted of 8.5 mol% of PI and 9.8 mol% of ergosterols. Relative to the Golgi fraction, the vesicles were enriched with 3.2-fold PA, 2.3-fold ergosterol, 2.6-fold M(IP)$_2$C, 2.2-fold IPC, and 2.4-fold MIPC. On the other hand, the Golgi fraction showed ~3-fold PS, ~2.5-fold PE, ~2-fold DAG, and ~1.5-fold PC enrichments relative to the vesicles. The observation that the immunoisolated vesicles exhibited a higher membrane order than the late Golgi membrane strongly suggested that lipid rafts played an important role in the Golgi-sorting machinery [9].

Sphingomyelin and cholesterol can assemble into domains and segregate from other lipids in the membranes, which can serve as platforms for protein transport and signaling [10]. Whether these types of domains exist in the Golgi membranes and whether they are required for protein secretion were the topics of another study involving lipidomics on Golgi membranes conducted by Duran et al. [11]. The researchers employed D-ceramide-C6 to manipulate lipid homeostasis of the Golgi membranes and utilized a lipidomics approach to profile the lipid species of the Golgi membranes isolated from D-ceramide-C6-treated HeLa cells. The study revealed the increased levels of C6-SM, C6-GluCer, and DAG. This was due to that D-ceramide-C6 treatment to HeLa cells led to inhibition of transport carrier formation at the Golgi membranes without affecting the fusion of incoming carriers. Moreover, the formed C6-SM prevented liquid-ordered domain formation in giant unilamellar vesicles and reduced the lipid order in the Golgi membranes of HeLa cells. These findings suggested the importance of a regulated production and organization of SM in the biogenesis of transport carriers at the Golgi membranes [11].

In higher eukaryotes, biosynthesis of GPL species and cholesterol mainly occurs in the ER [12]. However, (glyco)sphingolipids including SM and gangliosides are synthesized in the Golgi apparatus of the cells after Cer species are synthesized in the ER and are transported to the Golgi [13]. How these lipids are sorted and transported from the ER and the Golgi to plasma membrane through the secretory pathway is an interesting research topic in lipid biochemistry and cell biology. It was thought that transport and sorting along the secretory pathway of both proteins and most lipids were mediated by vesicular transport, with coat protein I (COPI) vesicles operating in the early secretory pathway. Lipidomics analysis was employed to determine the lipid content and composition of these vesicles since the protein constituents of these transport intermediates were well characterized [14]. Through quantitative analysis of lipid species present in COPI-coated vesicles and their parental Golgi membranes by using nanoESI-MS/MS, it was found that only low amounts of SM and cholesterol were present in COPI-coated vesicles compared with their donor Golgi membranes [15], indicating a significant segregation from COPI vesicles of these lipids and a sorting of individual SM molecular species. Moreover, the lipidomics data also provided the possible molecular mechanisms underlying this segregation, as well as implications on COPI function [15].

20.3   LIPID DROPLETS

Lipid droplets possess a neutral lipid core containing mainly TAG and cholesterol ester species, surrounded by a monolayer of mainly GPL species and proteins [16–18]. It has been more and more recognized that lipid droplets not only are simply storage vessels for hydrophobic compounds but also serve as active functional units for lipid metabolism, involving energy-linked metabolism, signaling, gene regulation, and autophagy [19–22]. Lipidomics analysis provides great insights into these structural and functional details of lipid droplets as well as the particular components in these organelles.

Hartler et al. [23, 24] performed lipidomics analysis of murine hepatocytic lipid droplets using UHPLC-MS in combination with MS/MS analysis, followed by automatic analysis of the MS data. The organelles were isolated after intervention studies involving nutritional stress (e.g., feeding with a high-fat diet for a prolonged time or fasting for a short term), genetic stress (i.e., knockout of adipocyte TAG lipase), or a combination of nutritional and genetic stresses (i.e., "super stress"). Lipidomics analysis at the levels of lipid classes and individual molecular species revealed not only very distinct patterns of TAG species but also those of DAG and PC species.

Specifically, the researchers found that murine hepatocytic lipid droplets contained a great amount of TAG content accounting for approximately 98 mol% of their total lipids, whereas remaining classes were, in quantitative order, DAG, PC, and other minor PL classes. Lipidomics analysis revealed that the TAG molecular species vary from C28:0 to C62:15 with distinct patterns under each condition. Subsequent principal component analysis (PCA) of TAG species contents readily separated the samples into nutritional, genetic, or superstresses. The researchers also found that calculation of an average number of carbons and double bonds in acyl constituents of total TAG species led to significantly distinct values under each physiological/metabolic condition of the animals. In the identical study, 26 DAG and 24 PC species were also determined with distinct patterns of these molecular species under different (patho)physiological conditions [23, 24].

Another study was conducted to investigate lipid composition of lipid droplets isolated from the embryos of *Gossypium hirsutum* (mature cotton) with or without genetically modified variety Bnfad2 among which the modifiers possessed a larger size of lipid droplets than the wild type [25]. Individual lipid droplets were selected under microscopy, and their lipid contents were analyzed by a novel technique called direct organelle mass spectrometry that couples direct visualization with detailed MS analysis of organelles. A multifaceted nanomanipulator, previously demonstrated to extract peptides from a single bead [26], was equipped with glass nanospray emitters prefilled with organic solvents (10 mM ammonium acetate in chloroform:methanol (1:1, v/v)) capable of extracting the lipid content out of lipid droplets. Different TAG molecular species were detected in both varieties of *G. hirsutum* but with different compositions. The lipid droplets from Bnfad2 variant animals obviously contained higher TAG content. In the same study, the lipidomes of lipid droplets from *A. thaliana* seeds and leaves were also determined. It was found that the seeds contained more eicosenoic (C20:1) acid-containing TAG species. Further study of lipid droplets from *A. thaliana* leaves demonstrated that these droplets had more 16:3 and 18:3 FA-containing TAG species that were minimal in the counterpart of the seeds.

20.4   LIPID RAFTS

Lipid rafts are currently defined as dynamic, nanometer-sized, and sterol- and sphingolipid-enriched protein assemblies [27]. The metastable resting state of rafts can be stimulated to coalesce into larger, more stable raft domains by specific lipid–lipid, protein–lipid, and protein–protein interactions [28]. When clustered, bilayer constituents are thought to be laterally stabilized according to their underlying affinity for pre-existing raft assemblies. In other words, clustering enhances the inclusion of proteins that partition into rafts and excludes those that segregate away. Furthermore, lipid rafts were not only found at the plasma membrane but also as part of the internal membrane of granules, Golgi complex, and even phagosomes [29, 30]. Lipidomics serves as a powerful tool to understand how diversified lipid classes and species are present in the rafts [31].

Early descriptions of lipid rafts noted their enrichment in cholesterol and glycosphingolipids and focused on their ability to resist extraction by nonionic detergents [32]. Later experiments showed that lipid rafts were a heterogeneous collection of domains that differ in protein and lipid composition as well as in temporal stability [33, 34]. The distinctive lipid composition of membrane rafts as demonstrated by lipidomics makes a clearer picture of membrane rafts.

Cholesterol levels in rafts were generally double from those found in their derived plasma membranes [35]. Similarly, the levels of SM species in the rafts were elevated by approximately 50% in comparison to that present in plasma membranes [35, 36]. Intriguingly, the elevated SM levels were offset with the decreased levels of PC [35, 36] so the total amount of choline-containing lipids was similar in rafts and plasma membranes. Such a composition of lipids present in the rafts led to a less fluid state than the surrounding membrane due to the tight packing of SM and cholesterol.

Lipidomics analyses of membrane rafts provided several other unexpected findings. *First*, the PS species levels were elevated twofold to threefold in rafts as compared with plasma membranes [34, 35]. This suggests that rafts may be a source for the rapid externalization of PS during apoptosis, platelet activation, or signaling transduction since PS species serve as the signaling source. *Second*, rafts were enriched in PE plasmalogens, particularly those containing arachidonic acid [34, 35]. Plasmalogens can function as antioxidants and the enrichment of this lipid subclass in rafts may serve to detoxify molecules that are internalized via lipid rafts or caveolae. Moreover, it is also possible that the rafts serve as an enriched source of arachidonic acid-containing GPL species for signaling purposes.

Lipidomic studies on lipid rafts were conducted with the raft fractions prepared by both detergent-free and detergent-containing protocols [34]. As anticipated, significant differences in lipid content and composition were present from the preparations with different protocols in a detergent-dependent manner [34]. For example, comparison of the content and composition of lipid species between protocols prepared with different detergents showed substantial different distributions of cholesterol and sphingolipids in the resultant fractions from individual detergent [33]. Thus, caution is warranted when assessing the results from lipidomics studies of raft fractions.

The majority of lipidomics studies on lipid rafts were conducted with samples collected from a certain range of fractions after centrifugation. Therefore, as noted earlier, the population of the isolated rafts was inhomogeneous. However, in a study performed by Brugger et al. [37], lipid rafts were isolated utilizing an immunoaffinity approach against the glycosylphosphatidylinositol (GPI)-anchored prion protein or the GPI-anchored Thy-1 protein. Lipidomics analyses of these isolated lipid rafts revealed that significant differences in the levels of cholesterol, PC, HexCer, and N-stearoyl Cer were present between Thy-1 protein- and prion protein-containing rafts. These findings confirmed the view that lipid rafts were heterogeneous in both protein and lipid contents. These observations also suggested that lipid rafts retained at least some of their biological differences after isolation.

It was frequently argued that the lipid rafts isolated by whatever method could contain artifacts resulted from the isolation procedures and did not reflect a real picture of lipid rafts in vivo regarding the content and composition of both lipids and proteins. This argument was challenged by the results from Brugger et al. [38] who studied the HIV lipidome by MS. The HIV virus is an enveloped retrovirus that buds from the membrane of infected cells. With the existence of raft marker proteins in the envelope of HIV, it was proposed that budding occurred from lipid rafts [39]. In the study, Brugger et al. [38] isolated budded HIV virus and determined that the budded HIV virus was enriched in cholesterol, sphingolipids, PS, and plasmalogen PE. The results indicated that the HIV membrane contains a lipid profile similar to that of lipid rafts isolated from the cells from which the virus buds and suggested that a membrane domain of this distinct composition must have existed in the cells at the location from which the virus buds. Accordingly, this study provided strong evidence for the existence of membrane rafts in intact cells.

Lipidomics studies of lipid rafts could provide insights into molecular mechanisms responsible for development of pathophysiological conditions, e.g., β-cell apoptosis and thus contributing to Type 2 diabetes induced with chronic saturated fatty acid exposure [40]. Through lipidomics studies, Boslem et al. [40] revealed that free cholesterol in the ER is reciprocally modulated by chronic palmitate and glucosylceramide synthase overexpression. This is consistent with the known coregulation and association of SM and free cholesterol enriched in lipid rafts. Further study through inhibition of SM hydrolysis led to partial protection against upregulation of ATF4/CHOP signaling pathway resulting from chronic palmitate exposure. These findings suggested that the loss of SM in the ER is a key event for initiating β-cell lipotoxicity, which leads to disruption of ER lipid rafts, perturbation of protein trafficking, and initiation of ER stress.

It is generally believed that the detergent-resistant membranes (DRMs) represent microdomains commonly occurred in plasma membrane. Whether these DRM microdomains are also present in the subcellular organelles has caught more and more interesting to biochemists and biologists. Lipidomics serves as a powerful tool to determine the content and composition of lipids present in these isolated DRMs in comparison to their parental membranes. For example, characterization of the microdomains from Golgi membranes was conducted [29]. It was found that these Golgi-derived detergent-insoluble microdomains had a low buoyant density compared to the cell-derived DRM and were highly enriched in lipids, containing 25% of total Golgi GPLs (including 67% of Golgi-derived SM) and 43% of Golgi-derived cholesterol. In contrast to the cell-derived DRM, these Golgi-derived microdomains only contained 10 major proteins, presented in nearly stoichiometric amounts, including the alpha- and beta-subunits of heterotrimeric G proteins, flotillin-1, caveolin, and subunits of the vacuolar ATPase. Intriguingly, the integrity of these Golgi-derived microdomains did not depend on their membrane environment, because these microdomains still could be isolated from a fused Golgi-ER organelle [29]. This finding indicated that the Golgi-derived microdomains are not in a dynamic equilibrium with neighboring membrane proteins and lipids.

20.5 MITOCHONDRION

The mitochondrion is a membrane-enclosed organelle present in most eukaryotic cells [41]. The number of mitochondria in a cell can vary widely by organism, tissue, and cell type. For instance, erythrocytes do not contain mitochondria, whereas hepatocytes could have thousands of mitochondria [42]. Mitochondria serve as "the powerhouse of the cell" since this organelle provides most of the cell's energy (i.e., ATP) supply. Moreover, mitochondria are also involved in other cellular functions such as signaling, cellular differentiation, cell death, as well as maintaining the control of the cell cycle and cell growth [43]. Therefore, mitochondrial dysfunction is associated with numerous human diseases, including mitochondrial disorders [44], cardiac dysfunction and heart failure [45], and neurodegeneration [46, 47].

Mitochondria constantly undergo fission and fusion [48]. Thus, they demand a constant and well-regulated supply of GPL species for membrane integrity [49]. Moreover, it is well known that mitochondria also play a role in interorganelle trafficking of the intermediates and products of GPL biosynthetic pathways, Cer and cholesterol metabolism, and glycosphingolipid anabolism [49, 50]. For example, decarboxylation of PS species to generate PE species in mitochondria plays a vital role in mammals [51]. Furthermore, mitochondria possess a unique class of GPL (i.e., cardiolipin).

Both the content and composition of CL are tightly associated with normal mitochondrial function [52–55]. Therefore, analysis of CL species can provide directly evaluation of mitochondrial bioenergetics. For example, in a recent study, mouse embryonic cells were treated with a modified fatty acid (i.e., triphenylphosphonium-linked octadecanoic acid) and studied under conditions leading to apoptosis [56]. CL was analyzed with LC-MS. It was found that the CL species contained an increased amount of octadecanoic acid when the cells were treated with the modified fatty acid. The study also revealed that treatment with this modified fatty acid conferred the cells with more resistance to apoptosis relative to untreated cells and cells treated with unmodified octadecanoic acid.

Lipidomics studies also revealed remarkable shifts of mouse myocardial mitochondrial PC and PE molecular species from 22:6(n−3) FA-containing species to 22:4 and 20:4(n−6) FA-containing species at the states of mitochondrial dysfunction, e.g., long-chain acyl CoA dehydrogenase deficiency (Han, unpublished data, Figure 20.1) or peroxisome proliferator-activated receptor γ coactivator-1 (PGC-1) deletion [57]. The association of decreased $n-3$ PUFA with GPL was reported previously [58]. It was suggested that the mitochondrion is a site for *de novo* biosynthesis of $n-3$ PUFA through a carnitine-dependent enzymatic pathway that was not fully delineated [59, 60], whereas $n-6$ PUFA is synthesized in the ER. Consistent with this notion, genetic defects in mitochondrial FA β-oxidation are associated with lower levels of 22:6 FA (Figure 20.1) [61, 62]. In the case of PGC-1 deletion, downregulation of the genes in the mitochondrial PUFA synthesis pathway

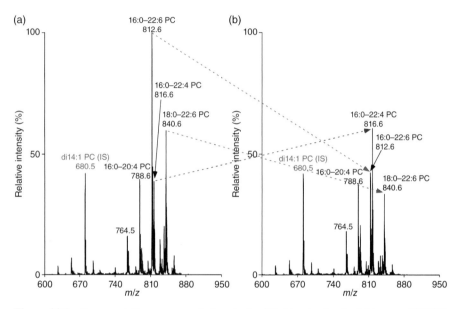

Figure 20.1 Representative mass spectral analysis of lipid extracts of wild-type and LCAD knockout mouse myocardium. Mass spectrometric analysis shows the presence of significantly different profiles of myocardial PC species between wild-type (a) and long-chain acyl CoA dehydrogenase (LCAD) knockout (b) mice, respectively. Both mass spectra are displayed after normalization to the internal standard (IS) ion peaks as indicated with broken lines. The arrows indicate the representative changes of PC species from wild-type mouse myocardium to LCAD knockout hearts. The mass levels of PC species containing 22:6 fatty acyl chains are specifically reduced in knockout myocardium in comparison to those in wild-type mouse heart. In contrast, the mass content of 16:0–22:4 PC is increased in knockout mice relative to wild-type mice. The significance of these findings is in fourfold. First, these results further verify the previous observations as described in the text. Second, these results suggest that LCAD not only involves in fatty acid oxidation but also affects membrane components. Third, the composition of the resultant membrane lipids has a large impact to both membrane fluidity and signaling, which might lead to membrane dysfunction or mitochondrial dysfunction since mitochondrial lipids contribute to a major part of membrane lipid mass levels, and ultimately contribute to the development of pathological conditions in LCAD knockout mice. Finally, these specific lipid changes could serve as biomarker(s) for early detection of LCAD deficiency-induced pathological conditions.

resulted from the PGC-1 deletion [63] might contribute to the observed decrease in $n-3$ PUFA-containing PE and PC species in mouse hearts.

Seyfried's group in collaboration with the author's laboratory conducted a few studies comparing the lipidomes of nonsynaptic and synaptic mitochondria [64–67]. In the first study, the researchers achieved a strategy to separate the two mitochondria populations free of myelin contamination [66]. Specifically, separation of nonsynaptic mitochondria from synaptosomes was first conducted with Ficoll density-gradient centrifugation. The synaptosomes were subsequently disrupted, and the released mitochondria were further purified by sucrose-gradient centrifugation. In contrast, the nonsynaptic mitochondria were directly purified by sucrose-gradient centrifugation after the Ficoll gradient. Lipids in each mitochondria type were then extracted and analyzed by MDMS-SL [68, 69]. It was found that increased levels of PS and Cer species are present in synaptic mitochondria, whereas the CL levels in nonsynaptic mitochondria were higher than those of synaptic mitochondria.

In the second study, lipids extracted from nonsynaptic mitochondria isolated from VM and B6 mice were analyzed [64]. It should be mentioned that the VM mice has a 210-fold increase in incidence of brain tumors. Lipidomics analysis revealed that regardless of similar total CL content present in nonsynaptic mitochondria between the VM and the B6 mice, the distribution of CL species differed drastically between the strains. B6 nonsynaptic mitochondria contained 95 CL species that were symmetrically distributed over seven major groups based on the mass-to-charge ratios. In contrast, VM nonsynaptic mitochondria contained only 42 CL species that were distributed asymmetrically. Furthermore, relative to B6 mice, the VM mouse model contained higher levels of both diacyl and plasmalogen PE, PI, PS, and Cer. However, a lower level of PC was present in the VM mouse model. Biochemical analysis showed that the activities of complex I, I/III, and II/III enzymes were lower, whereas the activity of complex IV was higher in the mitochondria of VM mice than in B6 mice. These results suggested a tight association of mitochondria lipid composition with the altered electron transport chain activities and brain cancer incidence manifest in VM mice compared to B6 mice.

In the third study, the lipidomes of mouse mitochondria isolated from cancerous and noncancerous brain cortices, and nonsynaptic mouse mitochondria isolated from cultured astrocytes and mouse tumor cells were determined and compared [67]. Mitochondria were isolated and purified as described earlier. Lipidomics analyses revealed surprising findings that the *in vitro* cell culture environments led to marked differences of the mitochondrial lipid profiles in comparison to those prepared from solid tumor grown *in vivo*. It was found that the *in vitro* growth environment produced abnormalities of lipids and electron transport chains in cultured nontumorigenic astrocytes that were similar to those associated with tumorigenicity. It appears that the culture environment obscured the boundaries of the Crabtree and the Warburg effects. These results indicated that *in vitro* growth environments could produce abnormalities in mitochondrial lipids and electron transport chain activities, thus contributing to a dependency on glycolysis for ATP production.

Lipidomics studies were also applied for determining the changes of lipids present in brain nonsynaptic mitochondria of rats after treatment with tacrine and its analogs [70], which are cholinesterase inhibitors used for treatment of Alzheimer's disease. In the study, lipid classes in lipid extraction of mitochondria were first separated by thin-layer chromatography (TLC) and quantified with a phosphorus assay prior to characterizing the lipid species with ESI-MS. Lipid classes of SM, PC, PI, PS, PS, CL, and ceramide-1-phosphate were determined in the study. It was found that tacrine could induce significant perturbations in the mitochondrial GPL profiles, including PC, PE, PI, and CL, and led to the presence of oxidized PS species. It was also found that the perturbation in lipid content and molecular composition of brain mitochondrial GPL species induced by tacrine analogues was in a lesser extent compared to that after tacrine treatment. Moreover, the results clearly indicated that abnormalities in CL content and the amount of oxidized PS species were associated with significant impairment of mitochondrial bioenergetics, mainly complex I. Collectively, the study suggested that tacrine and its analogs impaired mitochondrial function and bioenergetics, thus compromising the activity of brain cells. The study suggested a promising avenue to elucidate the potency and therapeutic window of tacrine and its analogs.

It is noteworthy to mention that a streamlined method was developed for detecting mitochondrial lipids without the need of a lipid extraction procedure [71]. In the method, the mitochondrial fraction was directly mixed with the MALDI matrix 9-aminoacridine, and lipids present in the fraction were analyzed by MALDI-TOF/MS. The investigators employed this method for analysis of mitochondrial lipids of various sources, including bovine heart and *S. cerevisiae*. The method enabled the investigators to detect CL, PA, PE, PG, PS, and PI classes. Clearly, the method could be used for quick screen of lipids from enriched mitochondrial fractions.

In another development on measuring mitochondrial lipids, Bird et al. [72] employed LC-MS with a Q-Exactive mass spectrometer and operated at high-energy CID for extensive MS/MS analysis. They detected the species of prenolipid, sterol, sphingolipid, fatty acyl, CL, PC, lysoPC, PE, and PS classes in the study.

20.6   NUCLEUS

The nucleus is a membrane-bound organelle present in eukaryotic cells. It contains most of the cell's genetic material [42]. The function of the nucleus is to maintain the integrity of the genes and to control the activities of the cell by regulating gene expression. It is usually termed that the nucleus is the control center of the cell. The main structures of the nucleus include the nuclear envelope (a double membrane that encloses the entire organelle and isolates its contents from the cellular cytoplasm) and the nucleoskeleton (a network within the nucleus that adds mechanical support) [42]. The nuclear membrane is impermeable to large molecules. Nuclear pores are required to regulate nuclear transport of molecules across the envelope. Thus, nuclear membrane integrity plays an important role in normal nuclear function and the nuclear lipidome likely has a key role in the molecular regulation of gene expression, although recent studies have also suggested that specific GPL classes bind and regulate specific transcription factors [73].

The nucleus contains a pool of lipids known as endonuclear lipids. An important lipid class in this environment is DAG (produced from PI and PC hydrolysis), which has been suggested to play a key role in the nuclear biosynthetic pathway of nuclear PC species [74]. The research group along with their collaborators developed a method to monitor the synthesis of endonuclear PC by using ESI-MS/MS after stable isotope labeling [75]. Specifically, mouse embryonic fibroblasts were treated with choline-d_9 leading to accumulation of intracellular pools of deuterium-labeled lipids. Cells were then homogenized in the presence of Triton X-100, which removed the nuclear envelope and contaminating PC upon fractionation by differential centrifugation. After isolation of the nuclear pellet, lipids were extracted and analyzed with ESI-MS/MS. The levels of endonuclear PC (deuterium labeled and wild type), SM, PE, and PI (labeled and wild type) were all analyzed. It was found that the PC, PE, and PS classes contain more saturated species in the nuclei-enriched fractions than in those of whole cell homogenates.

Lipidomics studies on myocardial nuclei were extensively described from the preparation of myocardial nuclei to the analyses of nuclear lipidome from rats and rabbits previously [76]. It was shown that rabbit myocardial nuclear lipidome contained relatively higher ratio of plasmalogen PC *vs.* diacyl PC in comparison to that observed in the rat myocardial nuclear lipidome. Moreover, the composition of the rat myocardial nuclear PC pool was relatively enriched with molecular species containing 20:4 and 22:6 FA constituents relative to that in the rabbit myocardial nuclear PC pool. While PE species of rabbit myocardial nuclei were enriched with 20:4 FA substituents and plasmalogens, the PE profile from rat myocardial nuclei showed less plasmalogen and more species containing 22:6 FA. Finally, it was found that significant differences of PE species between the nuclei and the mitochondria were present in rabbit myocardium. These studies clearly provided rich information on the molecular profiles of nuclear lipidomes and the potential role of the myocardial nuclear lipidomes on long-term cardiac cell function.

20.7 CONCLUSION

Lipids play a key role in organelle functions. Different organelles contain very different content and composition of different lipid classes. Moreover, any physiological and/or patho(physio)logical perturbation could lead to significant changes of organelle lipidomes. Therefore, lipidomics could provide detail dissents of these organelles, thereby revealing the possible origin, structure, and function of individual organelle. Fingerprinting the changes of lipid species of an organelle after a perturbation could connect the organelle function and involvement with the perturbation, thereby providing insights into molecular mechanism(s) of lipid changes and altered functions to a certain degree. Dynamic labeling could determine the turnover rate of individual lipid species under a specific condition, thereby providing a motion picture of individual organelle in an event. It should be recognized that lipidomics studies on organelles still remain a challenge for the majority of the laboratories largely due to the difficulty to isolate a large quantity of relatively pure organelles. However, this obstacle will be overcome as the development of more and more sensitive mass spectrometers and the advances in isolation techniques such as the utilization of magnetic beads and immunoaffinity reagents. On the other hand, methodologies of lipidomics analysis are getting more and more advanced and matured. These technologies and platforms could be readily applied for the lipidomics studies of organelles. Accordingly, it can be foreseen that lipidomics at the levels of cell organelles and subcellular membranes will become routine laboratory work and could serve as a key component for cellular characterization in the near future.

REFERENCES

1. Satori, C.P., Henderson, M.M., Krautkramer, E.A., Kostal, V., Distefano, M.D. and Arriaga, E.A. (2013) Bioanalysis of eukaryotic organelles. Chem. Rev. 113, 2733–2811.

2. Murat, D., Byrne, M. and Komeili, A. (2010) Cell biology of prokaryotic organelles. Cold Spring Harb. Perspect. Biol. 2, a000422.

3. Klose, C., Surma, M.A. and Simons, K. (2013) Organellar lipidomics--background and perspectives. Curr. Opin. Cell Biol. 25, 406–413.

4. Pasquali, C., Fialka, I. and Huber, L.A. (1999) Subcellular fractionation, electromigration analysis and mapping of organelles. J. Chromatogr. B 722, 89–102.

5. Pertoft, H. (2000) Fractionation of cells and subcellular particles with Percoll. J. Biochem. Biophys. Methods 44, 1–30.

6. Lee, Y.H., Tan, H.T. and Chung, M.C. (2010) Subcellular fractionation methods and strategies for proteomics. Proteomics 10, 3935–3956.

7. Fabene, P.F. and Bentivoglio, M. (1998) 1898-1998: Camillo Golgi and "the Golgi": one hundred years of terminological clones. Brain Res. Bull. 47, 195–198.

8. van Meer, G. (1989) Lipid traffic in animal cells. Annu. Rev. Cell Biol. 5, 247–275.

9. Klemm, R.W., Ejsing, C.S., Surma, M.A., Kaiser, H.J., Gerl, M.J., Sampaio, J.L., de Robillard, Q., Ferguson, C., Proszynski, T.J., Shevchenko, A. and Simons, K. (2009) Segregation of sphingolipids and sterols during formation of secretory vesicles at the trans-Golgi network. J. Cell Biol. 185, 601–612.

10. Vance, D.E. and Vance, J.E. (2008) Biochemistry of Lipids, Lipoproteins and Membranes. Elsevier Science B.V., Amsterdam. pp 631.

11. Duran, J.M., Campelo, F., van Galen, J., Sachsenheimer, T., Sot, J., Egorov, M.V., Rentero, C., Enrich, C., Polishchuk, R.S., Goni, F.M., Brugger, B., Wieland, F. and Malhotra, V. (2012) Sphingomyelin organization is required for vesicle biogenesis at the Golgi complex. EMBO J. 31, 4535–4546.

12. Fagone, P. and Jackowski, S. (2009) Membrane phospholipid synthesis and endoplasmic reticulum function. J. Lipid Res. 50 Suppl, S311–S316.

13. Gault, C.R., Obeid, L.M. and Hannun, Y.A. (2010) An overview of sphingolipid metabolism: from synthesis to breakdown. Adv. Exp. Med. Biol. 688, 1–23.

14. Walsby, A.E. (1994) Gas vesicles. Microbiol. Rev. 58, 94–144.

15. Brugger, B., Sandhoff, R., Wegehingel, S., Gorgas, K., Malsam, J., Helms, J.B., Lehmann, W.D., Nickel, W. and Wieland, F.T. (2000) Evidence for segregation of sphingomyelin and cholesterol during formation of COPI-coated vesicles. J. Cell Biol. 151, 507–518.

16. Tauchi-Sato, K., Ozeki, S., Houjou, T., Taguchi, R. and Fujimoto, T. (2002) The surface of lipid droplets is a phospholipid monolayer with a unique fatty acid composition. J. Biol. Chem. 277, 44507–44512.

17. Brasaemle, D.L. and Wolins, N.E. (2012) Packaging of fat: An evolving model of lipid droplet assembly and expansion. J. Biol. Chem. 287, 2273–2279.

18. Thiam, A.R., Farese, R.V., Jr. and Walther, T.C. (2013) The biophysics and cell biology of lipid droplets. Nat. Rev. Mol. Cell Biol. 14, 775–786.

19. Thiele, C. and Spandl, J. (2008) Cell biology of lipid droplets. Curr. Opin. Cell Biol. 20, 378–385.

20. Walther, T.C. and Farese, R.V., Jr. (2012) Lipid droplets and cellular lipid metabolism. Annu. Rev. Biochem. 81, 687–714.

21. Liu, K. and Czaja, M.J. (2013) Regulation of lipid stores and metabolism by lipophagy. Cell Death Differ. 20, 3–11.

22. Sahini, N. and Borlak, J. (2014) Recent insights into the molecular pathophysiology of lipid droplet formation in hepatocytes. Prog. Lipid Res. 54, 86–112.

23. Chitraju, C., Trotzmuller, M., Hartler, J., Wolinski, H., Thallinger, G.G., Lass, A., Zechner, R., Zimmermann, R., Kofeler, H.C. and Spener, F. (2012) Lipidomic analysis of lipid droplets from murine hepatocytes reveals distinct signatures for nutritional stress. J. Lipid Res. 53, 2141–2152.

24. Hartler, J., Köfeler, H.C., Trötzmüller, M., Thallinger, G.G. and Spener, F. (2014) Assessment of lipidomic species in hepatocyte lipid droplets from stressed mouse models. Scientific Data 1, 140051.

25. Horn, P.J., Ledbetter, N.R., James, C.N., Hoffman, W.D., Case, C.R., Verbeck, G.F. and Chapman, K.D. (2011) Visualization of lipid droplet composition by direct organelle mass spectrometry. J. Biol. Chem. 286, 3298–3306.

26. Brown, J.M., Hoffmann, W.D., Alvey, C.M., Wood, A.R., Verbeck, G.F. and Petros, R.A. (2010) One-bead, one-compound peptide library sequencing via high-pressure ammonia cleavage coupled to nanomanipulation/nanoelectrospray ionization mass spectrometry. Anal. Biochem. 398, 7–14.

27. Lingwood, D. and Simons, K. (2010) Lipid rafts as a membrane-organizing principle. Science 327, 46–50.

28. Lingwood, D., Ries, J., Schwille, P. and Simons, K. (2008) Plasma membranes are poised for activation of raft phase coalescence at physiological temperature. Proc. Natl. Acad. Sci. U. S. A. 105, 10005–10010.

29. Gkantiragas, I., Brugger, B., Stuven, E., Kaloyanova, D., Li, X.Y., Lohr, K., Lottspeich, F., Wieland, F.T. and Helms, J.B. (2001) Sphingomyelin-enriched microdomains at the Golgi complex. Mol. Biol. Cell 12, 1819–1833.

30. Dermine, J.F., Duclos, S., Garin, J., St-Louis, F., Rea, S., Parton, R.G. and Desjardins, M. (2001) Flotillin-1-enriched lipid raft domains accumulate on maturing phagosomes. J. Biol. Chem. 276, 18507–18512.

31. Shevchenko, A. and Simons, K. (2010) Lipidomics: Coming to grips with lipid diversity. Nat. Rev. Mol. Cell Biol. 11, 593–598.

32. Brown, D.A. and Rose, J.K. (1992) Sorting of GPI-anchored proteins to glycolipid-enriched membrane subdomains during transport to the apical cell surface. Cell 68, 533–544.

33. Schuck, S., Honsho, M., Ekroos, K., Shevchenko, A. and Simons, K. (2003) Resistance of cell membranes to different detergents. Proc. Natl. Acad. Sci. U. S. A. 100, 5795–5800.

34. Pike, L.J., Han, X. and Gross, R.W. (2005) Epidermal growth factor receptors are localized to lipid rafts that contain a balance of inner and outer leaflet lipids: a shotgun lipidomics study. J. Biol. Chem. 280, 26796–26804.

35. Pike, L.J., Han, X., Chung, K.N. and Gross, R.W. (2002) Lipid rafts are enriched in arachidonic acid and plasmenylethanolamine and their composition is independent of caveolin-1 expression: a quantitative electrospray ionization/mass spectrometric analysis. Biochemistry 41, 2075–2088.

36. Fridriksson, E.K., Shipkova, P.A., Sheets, E.D., Holowka, D., Baird, B. and McLafferty, F.W. (1999) Quantitative analysis of phospholipids in functionally important membrane domains from RBL-2H3 mast cells using tandem high-resolution mass spectrometry. Biochemistry 38, 8056–8063.

37. Brugger, B., Graham, C., Leibrecht, I., Mombelli, E., Jen, A., Wieland, F. and Morris, R. (2004) The membrane domains occupied by glycosylphosphatidylinositol-anchored prion protein and Thy-1 differ in lipid composition. J. Biol. Chem. 279, 7530–7536.

38. Brugger, B., Glass, B., Haberkant, P., Leibrecht, I., Wieland, F.T. and Krausslich, H.G. (2006) The HIV lipidome: A raft with an unusual composition. Proc. Natl. Acad. Sci. U. S. A. 103, 2641–2646.

39. Ono, A. and Freed, E.O. (2005) Role of lipid rafts in virus replication. Adv. Virus Res. 64, 311–358.

40. Boslem, E., Weir, J.M., MacIntosh, G., Sue, N., Cantley, J., Meikle, P.J. and Biden, T.J. (2013) Alteration of endoplasmic reticulum lipid rafts contributes to lipotoxicity in pancreatic beta-cells. J. Biol. Chem. 288, 26569–26582.

41. Henze, K. and Martin, W. (2003) Evolutionary biology: essence of mitochondria. Nature 426, 127–128.

42. Alberts, B., Johnson, A., Lewis, J., Margan, D., Raff, M., Roberts, K. and Walter, P. (2014) Molecular Biology of the Cell. Garland Science, New York. pp 1464.

43. McBride, H.M., Neuspiel, M. and Wasiak, S. (2006) Mitochondria: more than just a powerhouse. Curr. Biol. 16, R551–R560.

44. Gardner, A. and Boles, R.G. (2005) Is a 'mitochondrial psychiatry' in the future? A review Curr. Psychiatry Rev. 1, 255–271.

45. Lesnefsky, E.J., Moghaddas, S., Tandler, B., Kerner, J. and Hoppel, C.L. (2001) Mitochondrial dysfunction in cardiac disease: Ischemia--reperfusion, aging, and heart failure. J. Mol. Cell. Cardiol. 33, 1065–1089.

46. Sherer, T.B., Betarbet, R. and Greenamyre, J.T. (2002) Environment, mitochondria, and Parkinson's disease. Neuroscientist 8, 192–197.

47. Schapira, A.H. (2006) Mitochondrial disease. Lancet 368, 70–82.

48. Twig, G., Elorza, A., Molina, A.J., Mohamed, H., Wikstrom, J.D., Walzer, G., Stiles, L., Haigh, S.E., Katz, S., Las, G., Alroy, J., Wu, M., Py, B.F., Yuan, J., Deeney, J.T., Corkey, B.E. and Shirihai, O.S. (2008) Fission and selective fusion govern mitochondrial segregation and elimination by autophagy. EMBO J. 27, 433–446.

49. Osman, C., Voelker, D.R. and Langer, T. (2011) Making heads or tails of phospholipids in mitochondria. J. Cell Biol. 192, 7–16.

50. Lebiedzinska, M., Szabadkai, G., Jones, A.W., Duszynski, J. and Wieckowski, M.R. (2009) Interactions between the endoplasmic reticulum, mitochondria, plasma membrane and other subcellular organelles. Int. J. Biochem. Cell Biol. 41, 1805–1816.

51. Steenbergen, R., Nanowski, T.S., Beigneux, A., Kulinski, A., Young, S.G. and Vance, J.E. (2005) Disruption of the phosphatidylserine decarboxylase gene in mice causes embryonic lethality and mitochondrial defects. J. Biol. Chem. 280, 40032–40040.

52. Dowhan, W. (1997) Molecular basis for membrane phospholipid diversity: why are there so many lipids? Annu. Rev. Biochem. 66, 199–232.

53. Zhang, M., Mileykovskaya, E. and Dowhan, W. (2002) Gluing the respiratory chain together. Cardiolipin is required for supercomplex formation in the inner mitochondrial membrane. J. Biol. Chem. 277, 43553–43556.

54. Chicco, A.J. and Sparagna, G.C. (2007) Role of cardiolipin alterations in mitochondrial dysfunction and disease. Am. J. Physiol. Cell Physiol. 292, C33–C44.

55. He, Q. and Han, X. (2014) Cardiolipin remodeling in diabetic heart. Chem. Phys. Lipids 179, 75–81.

56. Tyurina, Y.Y., Tungekar, M.A., Jung, M.Y., Tyurin, V.A., Greenberger, J.S., Stoyanovsky, D.A. and Kagan, V.E. (2012) Mitochondria targeting of non-peroxidizable triphenylphosphonium conjugated oleic acid protects mouse embryonic cells against apoptosis: role of cardiolipin remodeling. FEBS Lett. 586, 235–241.

57. Lai, L., Wang, M., Martin, O.J., Leone, T.C., Vega, R.B., Han, X. and Kelly, D.P. (2014) A role for peroxisome proliferator-activated receptor gamma coactivator 1 (PGC-1) in the regulation of cardiac mitochondrial phospholipid biosynthesis. J. Biol. Chem. 289, 2250–2259.

58. Ovide-Bordeaux, S., Bescond-Jacquet, A. and Grynberg, A. (2005) Cardiac mitochondrial alterations induced by insulin deficiency and hyperinsulinaemia in rats: targeting membrane homeostasis with trimetazidine. Clin. Exp. Pharmacol. Physiol. 32, 1061–1070.

59. Infante, J.P. and Huszagh, V.A. (1997) On the molecular etiology of decreased arachidonic (20:4n-6), docosapentaenoic (22:5n-6) and docosahexaenoic (22:6n-3) acids in Zellweger syndrome and other peroxisomal disorders. Mol. Cell. Biochem. 168, 101–115.

60. Infante, J.P. and Huszagh, V.A. (2001) Impaired arachidonic (20:4n-6) and docosahexaenoic (22:6n-3) acid synthesis by phenylalanine metabolites as etiological factors in the neuropathology of phenylketonuria. Mol. Genet. Metab. 72, 185–198.

61. Harding, C.O., Gillingham, M.B., van Calcar, S.C., Wolff, J.A., Verhoeve, J.N. and Mills, M.D. (1999) Docosahexaenoic acid and retinal function in children with long-chain 3-hydroxyacyl-CoA dehydrogenase deficiency. J. Inherit. Metab. Dis. 22, 276–280.

62. Gillingham, M., Van Calcar, S., Ney, D., Wolff, J. and Harding, C. (1999) Dietary management of long-chain 3-hydroxyacyl-CoA dehydrogenase deficiency (LCHADD). A case report and survey. J. Inherit. Metab. Dis. 22, 123–131.

63. Lai, L., Leone, T.C., Zechner, C., Schaeffer, P.J., Kelly, S.M., Flanagan, D.P., Medeiros, D.M., Kovacs, A. and Kelly, D.P. (2008) Transcriptional coactivators PGC-1alpha and PGC-1beta control overlapping programs required for perinatal maturation of the heart. Genes Dev. 22, 1948–1961.

64. Kiebish, M.A., Han, X., Cheng, H., Chuang, J.H. and Seyfried, T.N. (2008) Brain mitochondrial lipid abnormalities in mice susceptible to spontaneous gliomas. Lipids 43, 951–959.

65. Kiebish, M.A., Han, X., Cheng, H., Chuang, J.H. and Seyfried, T.N. (2008) Cardiolipin and electron transport chain abnormalities in mouse brain tumor mitochondria: Lipidomic evidence supporting the Warburg theory of cancer. J. Lipid Res. 49, 2545–2556.

66. Kiebish, M.A., Han, X., Cheng, H., Lunceford, A., Clarke, C.F., Moon, H., Chuang, J.H. and Seyfried, T.N. (2008) Lipidomic analysis and electron transport chain activities in C57BL/6 J mouse brain mitochondria. J. Neurochem. 106, 299–312.

67. Kiebish, M.A., Han, X., Cheng, H. and Seyfried, T.N. (2009) In vitro growth environment produces lipidomic and electron transport chain abnormalities in mitochondria from non-tumorigenic astrocytes and brain tumours. ASN Neuro 1, e00011.

68. Yang, K., Cheng, H., Gross, R.W. and Han, X. (2009) Automated lipid identification and quantification by multi-dimensional mass spectrometry-based shotgun lipidomics. Anal. Chem. 81, 4356–4368.

69. Han, X., Yang, K. and Gross, R.W. (2012) Multi-dimensional mass spectrometry-based shotgun lipidomics and novel strategies for lipidomic analyses. Mass Spectrom. Rev. 31, 134–178.

70. Melo, T., Videira, R.A., Andre, S., Maciel, E., Francisco, C.S., Oliveira-Campos, A.M., Rodrigues, L.M., Domingues, M.R., Peixoto, F. and Manuel Oliveira, M. (2012) Tacrine and its analogues impair mitochondrial function and bioenergetics: A lipidomic analysis in rat brain. J. Neurochem. 120, 998–1013.

71. Angelini, R., Vitale, R., Patil, V.A., Cocco, T., Ludwig, B., Greenberg, M.L. and Corcelli, A. (2012) Lipidomics of intact mitochondria by MALDI-TOF/MS. J. Lipid Res. 53, 1417–1425.

72. Bird, S.S., Marur, V.R., Sniatynski, M.J., Greenberg, H.K. and Kristal, B.S. (2011) Lipidomics profiling by high-resolution LC-MS and high-energy collisional dissociation fragmentation: focus on characterization of mitochondrial cardiolipins and monolysocardiolipins. Anal. Chem. 83, 940–949.

73. Chakravarthy, M.V., Lodhi, I.J., Yin, L., Malapaka, R.R., Xu, H.E., Turk, J. and Semenkovich, C.F. (2009) Identification of a physiologically relevant endogenous ligand for PPARalpha in liver. Cell 138, 476–488.

74. Hunt, A.N., Clark, G.T., Attard, G.S. and Postle, A.D. (2001) Highly saturated endonuclear phosphatidylcholine is synthesized in situ and colocated with CDP-choline pathway enzymes. J. Biol. Chem. 276, 8492–8499.

75. Hunt, A.N. and Postle, A.D. (2006) Mass spectrometry determination of endonuclear phospholipid composition and dynamics. Methods 39, 104–111.

76. Albert, C.J., Anbukumar, D.S., Monda, J.K., Eckelkamp, J.T. and Ford, D.A. (2007) Myocardial lipidomics. Developments in myocardial nuclear lipidomics. Front. Biosci. 12, 2750–2760.

INDEX

4-hydroxyalkenal, 48

acyl CoA
  fragmentation patterns, 245
  functions, 368
  lipid synthesis and remodeling, 360
  solubility, 4
acyl phosphatidylglycerol, 194
acylcarnitine, 244
  lipid transport, 366
aggregates, 14, 92, 97, 310–311, 336–337
  concentration, 336
  homodimer, 324, 336
alkenyl lipids *see* plasmalogens and ether lipids
alkyl-and alkenylglycerols, 358–359
Alzheimer's disease, 385
amphiphilic, 4
anionic lipids, 58
anionic lysoglycerophospholipids, 193
annotation, lipid species, 133–136
artifacts
  contaminants, 299
  extraction, 286, 299
autoxidation, minimizing, 287

bacterial lipids, 436–438
baseline
  correction, 137
  noise, 338, 343
BHT *see* butylated hydroxytoluene
bile acids, 8–9, 25, 239
bioactive lipids, 5, 243–253
  signaling, 365
bioinformatics, 121
  modeling, 143
  pathway analysis, 140–143
  simulation, 140
  statistics, 134
  tools for LC-MS, 132
  visualization, 134
bis(monoacylglycero)phosphate, 194
Bligh–Dyer extraction, 296
  modified, 296
building block
  concept, 7, 10
  database, 123
  monitoring, 102
BUME extraction, 298
butylated hydroxytoluene, 287

Lipidomics: Comprehensive Mass Spectrometry of Lipids, First Edition. Xianlin Han.
© 2016 John Wiley & Sons, Inc. Published 2016 by John Wiley & Sons, Inc.

WILEY SERIES ON MASS SPECTROMETRY

Series Editors

Dominic M. Desiderio

Departments of Neurology and Biochemistry University of Tennessee Health Science Center

Joseph A. Loo

Department of Chemistry and Biochemistry UCLA

Founding Editors

Nico M. M. Nibbering (1938–2014)

Dominic Desiderio

John R. de Laeter · *Applications of Inorganic Mass Spectrometry*

Michael Kinter and Nicholas E. Sherman · *Protein Sequencing and Identification Using Tandem Mass Spectrometry*

Chhabil Dass · *Principles and Practice of Biological Mass Spectrometry*

Mike S. Lee · *LC/MS Applications in Drug Development*

Jerzy Silberring and Rolf Eckman · *Mass Spectrometry and Hyphenated Techniques in Neuropeptide Research*

J. Wayne Rabalais · *Principles and Applications of Ion Scattering Spectrometry: Surface Chemical and Structural Analysis*

Mahmoud Hamdan and Pier Giorgio Righetti · *Proteomics Today: Protein Assessment and Biomarkers Using Mass Spectrometry, 2D Electrophoresis, and Microarray Technology*

Igor A. Kaltashov and Stephen J. Eyles · *Mass Spectrometry in Structural Biology and Biophysics: Architecture, Dynamics, and Interaction of Biomolecules, Second Edition*

Isabella Dalle-Donne, Andrea Scaloni, and D. Allan Butterfield · *Redox Proteomics: From Protein Modifications to Cellular Dysfunction and Diseases*

Silas G. Villas-Boas, Ute Roessner, Michael A.E. Hansen, Jorn Smedsgaard, and Jens Nielsen · *Metabolome Analysis: An Introduction*

Mahmoud H. Hamdan · *Cancer Biomarkers: Analytical Techniques for Discovery*

Chabbil Dass · *Fundamentals of Contemporary Mass Spectrometry*

Kevin M. Downard (Editor) · *Mass Spectrometry of Protein Interactions*

Nobuhiro Takahashi and Toshiaki Isobe · *Proteomic Biology Using LC-MS: Large Scale Analysis of Cellular Dynamics and Function*

Agnieszka Kraj and Jerzy Silberring (Editors) · *Proteomics: Introduction to Methods and Applications*

Ganesh Kumar Agrawal and Randeep Rakwal (Editors) · *Plant Proteomics: Technologies, Strategies, and Applications*

Rolf Ekman, Jerzy Silberring, Ann M. Westman-Brinkmalm, and Agnieszka Kraj (Editors) · *Mass Spectrometry: Instrumentation, Interpretation, and Applications*

Christoph A. Schalley and Andreas Springer · *Mass Spectrometry and Gas-Phase Chemistry of Non-Covalent Complexes*

Riccardo Flamini and Pietro Traldi · *Mass Spectrometry in Grape and Wine Chemistry*

Mario Thevis · *Mass Spectrometry in Sports Drug Testing: Characterization of Prohibited Substances and Doping Control Analytical Assays*

Sara Castiglioni, Ettore Zuccato, and Roberto Fanelli · *Illicit Drugs in the Environment: Occurrence, Analysis, and Fate Using Mass Spectrometry*

Ángel Garciá and Yotis A. Senis (Editors) · *Platelet Proteomics: Principles, Analysis, and Applications*

Luigi Mondello · *Comprehensive Chromatography in Combination with Mass Spectrometry*

Jian Wang, James MacNeil, and Jack F. Kay · *Chemical Analysis of Antibiotic Residues in Food*

Walter A. Korfmacher (Editor) · *Mass Spectrometry for Drug Discovery and Drug Development*

Alejandro Cifuentes (Editor) · *Foodomics: Advanced Mass Spectrometry in Modern Food Science and Nutrition*

Christine M. Mahoney (Editor) · *Cluster Secondary Ion Mass Spectrometry: Principles and Applications*

Despina Tsipi, Helen Botitsi, and Anastasios Economou · *Mass Spectrometry for the Analysis of Pesticide Residues and their Metabolites*

Xianlin Han · *Lipidomics: Comprehensive Mass Spectrometry of Lipids*